Longwall Mining

Longwall Mining

Third Edition

Syd S. Peng

Professor emeritus
Department of Mining Engineering
College of Engineering and Mineral Resources
West Virginia University
USA

Distinguished professor
School of Energy Science and Engineering
Henan Polytechnic University
China

Distinguished professor
China University of Mining and Technology
China

CRC Press
Taylor & Francis Group
Boca Raton London New York Leiden

CRC Press is an imprint of the
Taylor & Francis Group, an **informa** business

A BALKEMA BOOK

CRC Press/Balkema is an imprint of the Taylor & Francis Group, an informa business

First issued in paperback 2021

© 2020 Taylor & Francis Group, London, UK

Typeset by Apex CoVantage, LLC

Library of Congress Cataloging-in-Publication Data
Applied for

Published by: CRC Press/Balkema
           Schipholweg 107c, 2316 XC Leiden, The Netherlands
           e-mail: Pub.NL@taylorandfrancis.com
           www.crcpress.com – www.taylorandfrancis.com

ISBN: 978-0-367-20192-0 (hbk)
ISBN: 978-1-03-208600-2 (pbk)
ISBN: 978-0-429-26004-9 (eBook)
DOI: https://doi.org/10.1201/9780429260049

Contents

Preface

The first edition of this book was published in 1984 by Wiley, New York. At that time long-wall mining in the United States was still in the initial development stage. In fact, the first-ever longwall mine by design went into production in 1984. Therefore, the book was mainly a review of longwall mining technology then available worldwide. The second edition was published in 2006 by me. It summarized the longwall mining technology as practiced in the US. It was most appropriate for me to undertake such a book-writing endeavor because my research work had been concentrating on it since its massive introduction from Europe in the mid-1970s and grew with its development to become the worldwide standard today. Probably because of that, the second edition has been very popular for all types of professionals involved with longwall mining. It was not unusual, whenever I visited a coal mine, that mine management pulled out my book to enhance discussion or ask for my signature.

Although there have been no major changes in longwall mining technology and operations in the past 13 years since the publication of *Longwall Mining, 2nd edition* in 2006, there have been many incremental developments in the whole system as well as various subsystems of the existing longwall mining operational technologies as detailed in the second edition; these changes have been added to this edition.

Major developments are automation, and health and safety technology, as well as equipment reliability, thereby greatly increasing productivity and cutting cost. In particular, the longwall system can now run automatically cut by cut forever without operators' intervention provided that the geology allows it. Other health and safety features such as LASC, personal proximity detection, color lighting, automatic shield water sprays, and remote shearer control are fully operational. There are more than 7000 sensors installed in current longwall mining systems. The big data obtained and fast communication technology have been fully utilized to improve and solve operational problems in real time. Those features are fully documented in the new edition.

In the pursuit of high productivity and cutting cost, life cycle management that increases equipment reliability has been implemented by the original equipment manufacturers (OEMs). Automation improvement such as tail end automatic chain tensioner greatly extends the armored face conveyor (AFC) chain's service life.

Other incremental improvements including dust and methane controls, entry development, panel design, and face move are addressed. Additional operational issues such as extension of panel width, life cycle management, and compatibility tests are also discussed.

Since the last plow longwall mine was closed in 2018, the chapter on plow longwalling has been dropped, and in its place "Automation of Longwall Components and System" has been added. Also a new chapter, "Longwall Top Coal Caving Mining (LTCC)," for thick seams has been added due to its successful application in Australia since 2005 and in China since the mid 1980s.

Acknowledgments

In preparation for the 3rd edition, I visited the OEMs and coal mines to discuss the latest technologies and their application. I am indebted to their generous assistance, including:

- **Al Hefferan and his group (Caterpillar)** for providing voluminous original materials, quick responses to my requests/questions any time and several face-to-face meetings:
- **Nigel Goff and Chuck Ficter (Komatsu Mining)** for providing a suite of original photos with captions on Komatsu mining equipment (longwall and R&P), all of which were specially made for the book. Also Komatsu shearer manufacturing group for meetings and plant tour.
- **Ryan Murray and his staff (Murray Energy Corporation); Ryan** for discussion on FCT trials for entry development; **David Bartsch** for updating Section 11.2, "Longwall Ventilation"; **Tim Eddy** for review and updating Chapter 13, "Power Distribution".
- **Wang, J.C. and Yang Shengli (CUMT Beijing)** for drafting Chapter 15, "Longwall Top Coal Caving (LTCC)" and bearing with me for many re-drafts.
- **Chris Mark (MSHA)** for drafting Section 4.5.1(1), "coal pillar stability".
- Australian friends. **Mark Dunn (CSIRO-Mining)** for drafting a short paragraph on CSIRO's LASC system on Section 9.2.2; **Ismet Cabulat (UNSW)** for searching information; and **Dan Payne (BMA)** for discussions about and got permission for me to use LTCC figures pertaining to Broadmeadow Mine, BMA.
- **John Steffenino (Tensar International)** for making a special trip to Morgantown, WV to discus Tensar Minex materials.
- **Tom Hutchinson (retired)** for coming out of retirement to have a meeting with me discussing the latest on "Monorail".
- **Uli Paschedag** (TUAS Georg Agricola, Bochum, Germany) for supplying German longwall equipment historical and recent development materials.
- My former students: **Brian Cappellini** (CB Mining CAT) for discussion meetings and materials on AFC; **Cheng Jingyi** (CUMT) for drafting Section 5.5, "Shield Performance"; **Jay Colinet** (NIOSH) for reviewing Section 11.4, "Dust Control"; **Du Feng** for drafting Fig. 9.3.1, "Development Chronology of Longwall Automation"; **Guo Wenbing** (HPU) for review Chapter 14, "Surface Subsidence"; **R.J. Matetic** (NIOSH) for info on Section 11.5, "Noise Control" and Section 14.5.2, "Subsidence Impact on Water"; **Gamal Rashed** (NIOSH) for updating new data for Section 3.3.2, "Field Measurements"; **Jack Trackemas** (NIOSH) for many discussion meetings and supplying and searching for materials and contacts; **Xu Feiya** (HPU) for keeping the Coal Age's US longwall Census statistics up to date, organizing the initial draft of "References" and

updating Section 14.5.2, "Subsidence Impact on Water"; **Li Yang** (CUMT Beijing) for recheck Section 4.9.1, "Development Ratio"; and **Peter Zhang** (NIOSH) for drafting Section 4.7.3, "Pumpable Cribs" and additional comments on Section 10.11, "Gas Well" failure factors, **Yang Jian** for proof reading.

- **Brijes Mishra** (WVU) for allowing his graduate students, **Xue Yuting** and **Shi Qingwen,** to help with redrafting figures and tables and formatting the whole manuscript.

Syd S. Peng
Morgantown, WV, USA
June, 2019

Chapter 1

US longwall mining

1.1 Introduction

The concept of longwall mining in the United States dates back to the late 19th century when wide faces were used for coal extraction in the Pittsburgh seam. But modern, fully mechanized longwall mining employing plows and frame supports was first introduced from West Germany in the early 1950s in southern West Virginia. It was used mainly for thin seam mining. During the 1950s and 1960s, the practice did not expand. In fact, due to its poor performance, only a few mines in southern West Virginia employed this technique. But in the mid-1970s when the shield supports were introduced, again from West Germany, to a northern West Virginian mine, resulting in increasing production and safety records, longwall mining was then recognized by the coal industry as a viable mining technique.

Thereafter the number of longwall mines grew steadily and rapidly (Fig. 1.1.1). A maximum of 118 longwall faces existed in 1982, but the number decreased steadily during the 1980s and 1990s and stabilized to around 50 in the first decade of the 21st century. It was gradually reduced further to the high 40s in the late 2000s and stabilized in recent years in the low 40s. It must, however, be emphasized that although the number of longwalls was decreasing, production increased tremendously during this period due to improvements in face equipment automation and reliability, use of heavy duty face equipment, better panel layout, skilled labor forces, and aggressive management.

The growth of US longwall mining production in the past four decades has been phenomenal. In 1975, there were 76 longwall faces, but their production was a mere 4% of underground coal mining production. In 2017, there were only 43 operating longwall faces, but they accounted for 62% of underground coal production. Proportion of longwall production has been steady even though the number of longwalls has decreased slightly in the past decade.

Throughout the past 40 years the basic elements and fundamental principles of longwall mining have remained the same, except that the panel has become much larger and the equipment has become larger, heavier, more powerful, and more reliable, and that automation has gotten more sophisticated and reliable.

US longwalls are highly productive and have established numerous world records since 1990. Productivity increased steadily. It reached the highest in 2000 at around 5.2 tons per man-hour (mine wide) or around 11,600 tons per man-year (Fig. 1.1.2). It began to dip in 2004 due to the hiring of more new young unskilled miners to replace the experienced retiring miners. It started to increase again in 2011 due to industry-wide layoffs and closure of

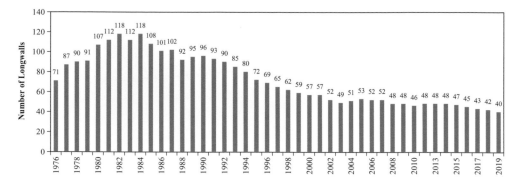

Figure 1.1.1 Yearly trends of number of longwalls in the United States since 1976

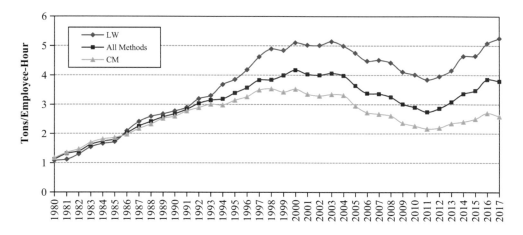

Figure 1.1.2 Productivity trend of US longwall mining

uneconomic mines resulting from a combination of abundant cheap shale gas and Obama Administration's policy "War on Coal." It reached the previous high in 2016 and continued to increase.

According to the 2019 US longwall census, there were 37 active longwall faces producing from slightly more than one million to 7.23 million clean tons of coal in 2018. It averaged about 4.56 million clean tons or approximately 7.01 million raw tons per longwall, assuming an average of 65% clean coal recovery.

From a production point of view, today's longwall mining has radically changed the conventional concept of coal mining in terms of reserve, production, safety, and mine planning.

Today, longwall mining has firmly established its role as the most productive and safe underground coal mining method and, consequently, in all cases there is no dispute that longwall mining is the preferred method whenever geological conditions and reserves are suitable.

1.2 Requirements and constraints for high production longwalls

Most US longwall mines operate one longwall unit in a mine and produce three to seven million clean tons annually, although several mines routinely operate two longwalls units occasionally; a few mines even operated three units in order to either keep up with or increase production. The advantages of one single longwall in a mine include a simpler mine infrastructure and high productivity, which is one of the major reasons why modern longwall is the most productive and cost-effective method.

Several conditions are necessary to ensure productivity of modern longwalls. (1) The equipment must be heavy duty with high installed power, have fast machine speed, be highly reliable, be quick and easy to maintain, and have a minimum life of at least one longwall panel without a major lost production breakdown. (2) In order to utilize the equipment mentioned above, longwall panels must be huge, i.e., wider and longer. (3) Favorable geological conditions must exist including reasonable methane gas content, easy to cave roof strata in the gob, and a firm floor and, in some cases, adequate seam thickness.

The major constraints for modern longwall mining are the following: (1) It requires a very large capital investment as compared to other methods, such as room-and-pillar mining. The industry norm is to fully depreciate the longwall equipment in five to seven years. (2) It requires a fairly large-sized reserve, say a minimum of 30–40 million tons, to recover the capital investment. The reserve is preferably a continuous block, but isolated blocks have been proven feasible. (3) Longwall mining is highly rigid in mine plan. Panels are laid out at least two years in advance and normally are extremely impractical, if not impossible, to change later on.

The reasons that US longwalls have become the most productive in the world within the 10- to 15-year period since its introduction from Europe, and acceptance by US coal industry, are as follows:

1   Favorable geological conditions – Coal seams are flat, fairly uniform and shallower, mostly less than 2100 ft (640 m), and the roof strata are generally easy to cave.
2   Excellent mining plan and equipment layout – US longwalls employ multi-entry gateroad development using continuous miners, which allows rapid gateroad development, and the layout creates minimum abutment pressure interaction between adjacent panels. Specific improvements to equipment and layout unique to US longwalls are addressed in Section 7.2, "Layout of Longwall Face Equipment Using the Shearer."
3   State-of-the-art equipment – All equipment are heavy duty and highly reliable. Individual subsystem reliability is greater than 90% without, increasing to close to 100%, with industry/OEM partnership services programs. The whole system reliability is more than 85% without a partnership program.
4   Highly efficient method of face move – For a 1200 ft (366 m) wide panel, a complete face move can be done in five to seven days (Note, normally the shearer and AFC and stage loader are pre-installed in a new face.). A complete face move is up to two weeks.
5   Excellent management and skilled work force in a highly competitive world market.
6   Production and safety incentive programs.

Among the six reasons mentioned here, items two through four will be addressed in detail in this book.

1.3 Panel layout

During the 1970s–1980s, US longwall mining has evolved from almost a wholly borrowed system, mainly that of Europe, to a unique system – one that is conducive to high-speed mining and subjected to strict safety and health standards. During that period, demonstration projects on advancing longwall mining, multi-slicing longwall mining, thick-seam longwall mining, and inclined seam longwall mining methods had all been tried in Colorado and Wyoming, while thin-seam longwall mining (using the shearer) was experimented with in Kentucky, Virginia, and West Virginia, mostly under federal government funding. The results were not favorable due to various reasons, and their applications were discontinued, except a thick-seam (14.5 ft or 4.4 m) longwall operation in Wyoming that lasted until 1988, and an inclined-seam longwall mine (up to 25°), also in Wyoming, which was discontinued in 2000 when its reserve became too expensive to mine. The face equipment setup and panel layout (either parallel to the strike or the dip direction) in the aforementioned thick and inclined seams were similar to those in the flat seams.

A longwall mine with a gently dipping seam, up to 7–9°, continues to operate in Utah.

After continuing improvements in the 1970s–1990s, the panel layout and face equipment and their setup were more or less standardized in the mid-1990s and remain so up until now. The current US longwall mining method has the following unique features (for special features in equipment and its layout, see Section 7.2):

1   The process uses a retreating longwall mining method with natural roof caving in the gob.
2   All longwalls are single slice or single seam and operate in flat or nearly flat coal seams.
3   Panel development uses the room-and-pillar method in which multiple entries (two to four entries, mostly three entries) are developed simultaneously in-seam in a rectangular shape by continuous miners. Those multiple entries are parallel to each other and interconnected by crosscuts at fixed intervals. Chain pillars between those entries are left unmined. All entries are roof-bolted in cycle (primary support) and/or supplemented with additional supports (supplementary support) later. These features allow fast mine, and especially panel, development.
4   Generally, seam height is the mining height and/or entry height. However, since the proven longwall face equipment is most convenient to operate with the minimum height greater than 65–70 in. (1.65–1.78 m) in range, longwall mining in seams thinner than this usually involves cutting into roof or floor or both (mostly roof) in order to make the height. Therefore, it is not unusual that clean coal recovery is as low as 30% of run-of-mine raw coal.
5   All mined-out gobs are ventilated with a bleeder system that must be inspected regularly, except the bleederless system is employed in four longwalls in New Mexico and Utah that are liable to spontaneous combustion.

Figure 1.3.1 shows a typical example of US longwall mine development. A vertical shaft, partitioned into two compartments, is sunk from the surface to the seam level and is used for both air intake and transportation of man and supplies, respectively. At the shaft bottom, entries and crosscuts are driven around and away from it. Mains that consist of 7 to 10 entries are developed first. Longwall panels are then developed perpendicular (occasionally oblique) to, and either on one or both sides of, the mains. The raw coal is transported all the

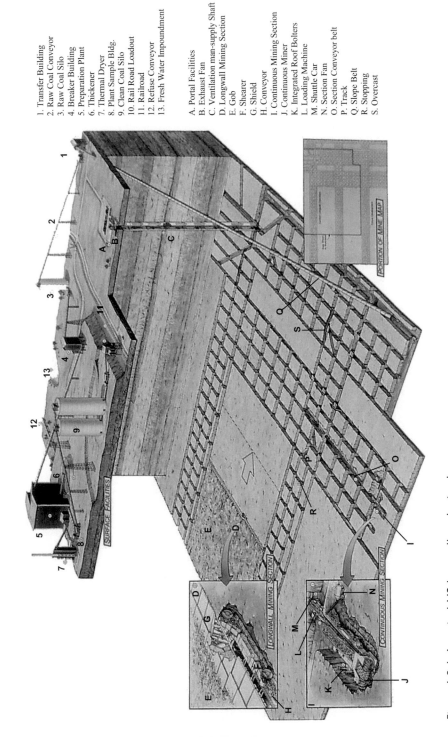

1. Transfer Building
2. Raw Coal Conveyor
3. Raw Coal Silo
4. Breaker Building
5. Preparation Plant
6. Thickener
7. Thermal Dryer
8. Plant Sample Bldg.
9. Clean Coal Silo
10. Rail Road Loadout
11. Railroad
12. Refuse Conveyor
13. Fresh Water Impoundment

A. Portal Facilities
B. Exhaust Fan
C. Ventilation man-supply Shaft
D. Longwall Mining Section
E. Gob
F. Shearer
G. Shield
H. Conveyor
I. Continuous Mining Section
J. Continuous Miner
K. Integrated Roof Bolters
L. Loading Machine
M. Shuttle Car
N. Section Fan
O. Section Conveyor belt
P. Track
Q. Slope Belt
R. Stopping
S. Overcast

Figure 1.3.1 A typical US longwall coal mine layout

Source: Courtesy of CONSOL Energy, Inc.

way by belt conveyor from the longwall face via the headgate and mains to the slope bottom where it is moved up through the slope to the surface and then to the preparation plant by belt conveyor. The slope is generally less than or equal to 16% (9.1°) in grade and consists of two compartments, one on top of the other. The upper compartment is for the belt conveyor for coal transportation out of the mine, while the lower compartment holds the track for heavy equipment transportation into and out of the mine. If the overburden is strong and massive strata such as sandstone, the two compartments may be arranged side by side resulting in a wider slope, e.g., 28 ft (8.5 m).

Figure 1.3.2 shows a typical panel layout of a US longwall section. The panels are blocked out by developing panel entries or gateroads (called the development section) perpendicular (occasionally oblique) to the main entries on one or both sides of the main entries. The panel width or face length, W, varies from 500 to 1580 ft (152–482 m) with the most common ones larger than 1200 ft (366 m) (Fig. 1.3.3).

When the panel entries on both sides have been developed to the designed length or the reserve boundaries, they are connected by the bleeder entries that are parallel to the main entries. The mains consist of 7 to 10 parallel entries, while the bleeder entries normally consist of three entries. The panel length, L, varies from 3200 to 22,500 ft (976–6,860 m) with more than half ranging from 8000 to 14,000 ft (2439–4268 m) (Fig. 1.3.4). The shortest panel length is generally dictated by property boundaries or odd reserve configuration. The trends are toward wider and longer panels so that there is a maximum of one longwall move per fiscal year.

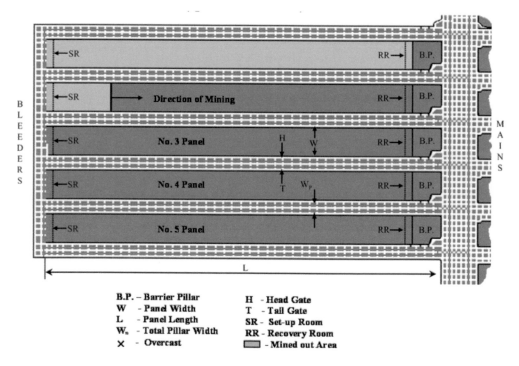

B.P. – Barrier Pillar   H - Head Gate
W   - Panel Width   T - Tail Gate
L   - Panel Length   SR - Set-up Room
W$_n$ - Total Pillar Width RR - Recovery Room
×   - Overcast   ☐ - Mined out Area

Figure 1.3.2 A typical panel layout of a US longwall mining section

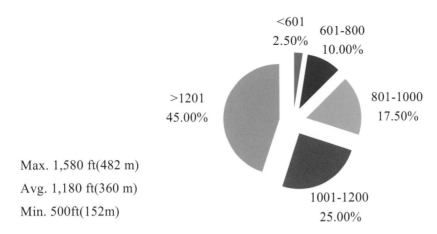

Max. 1,580 ft(482 m)

Avg. 1,180 ft(360 m)

Min. 500ft(152m)

Figure 1.3.3 Distribution of panel widths of US longwalls in 2019

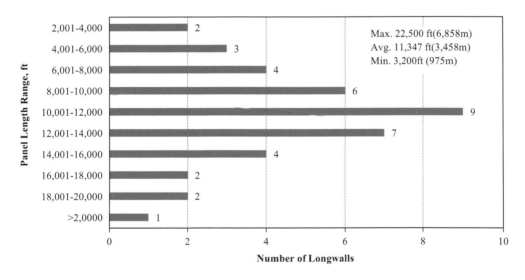

Figure 1.3.4 Distribution of panel length of US longwalls in 2019

The total width of the panel entries (or gateroads) depends on the number of entries and varies from 131 to 442 ft (40–135 m). The panel entries or gateroads consist of two, three, or four entries. The most common one is the three-entry system, followed by the four-entry and two-entry systems. All of the current US longwalls are of the retreating type. In the retreating method, longwall mining starts from the bleeder entries side (i.e., far end of the longwall panel) and proceeds toward the main entries on the opposite end of the panel. Longwall mining begins at the setup room or entry, SR (Fig. 1.3.2), where all of the face equipment is set up. In the early years using the bleeder system, there was a barrier pillar between the bleeder entries and the setup room or entry, which ranged from 200 to 500 ft (61–152 m), depending on the seam depth. In recent years, however, most designs have eliminated the barrier pillar.

In such cases, the setup room is located immediately adjacent to the outby side chain pillars of the bleeder entries.

When the retreating longwall face has reached the designated termination point on the main entry side, a recovery room or entry (RR in Fig. 1.3.2) is established from which all face equipment is recovered and moved to the setup room for the second panel. This is the longwall move. A barrier pillar from 200 to 500 ft (61–152 m) wide is left between the recovery room and the main entries. In order to save longwall move time, some designs include a pre-driven recovery room on the main entries side. A pre-driven recovery room eliminates the need to mesh the roof but requires sufficient roof support to cope with the incoming front abutment pressure. Both the setup and recovery rooms are located at the crosscut intersections at the respective ends of the panel to facilitate equipment setup and removal.

As longwall mining begins, the immediate entry on each side of the panel serves a special function. One is used for the transportation of coal, men, and supplies into and out of the face and is called the headgate. The headgate is always located on the solid coal block side of the reserve where the next panel is located so that it is not subject to the side abutment pressure generated by the previous mined out panel. The entry on the other side of the panel is used for the passage of return air and escapeway and is called the tailgate. Therefore, a development section or a gateroad system consists of, on the opposite side, one headgate (or headentry) for the panel being mined and one tailgate for the next (or future) panel.

A longwall is called a left-hand longwall if the headgate is on the left side of the longwall panel when one stands in the longwall face facing outby (or mining direction). The opposite is called a right-hand longwall. Some equipment, such as conveyor head drives and tail drives, are designed specifically either for the left-hand or right-hand longwalls, but can be rehanded by changing some components during a rebuild. However, all pans are bi-directional now.

1.4 Mining technique

Figure 1.4.1 is a cutaway view showing the panel layout and equipment setup in a longwall face. Coal at the face is cut by a shearer loader or simply shearer. The shearer generally rides on the pan of an armored face conveyor (AFC), which is laid on the floor parallel to the faceline which is the straight line of the exposed coal face of the panel coal block from headgate-corner to tailgate-corner. The face or longwall face refers to the space enclosed by the faceline in the front side and shield's caving shield (or hydraulic leg props) in the rear side and from headgate to tailgate T-junctions in which coal production operation is performed.

The shearer is hauled by two self-contained electric motors on each end of the machine turning a series of sprocket wheels that run on a specialized track laid on the gob side of the armored face chain conveyor (AFC). Thus the AFC serves as a track for the shearer to move on back and forth. Two cutting drums, one at each end of the shearer (see Fig. 7.4.1), similar in some respect to those used in a continuous miner but larger in diameter and narrower in width, are mounted on the face side of the shearer. The cutting position or height of a drum can be hydraulically adjusted by the ranging arms. The cutting force is provided by the rotating torque available at the axis of the drum. The width of the cut or web made by the shearer is about 30 to 42 in. (0.76–1.07 m) wide.

Coal cut by the shearer is loaded onto the AFC and transported to the headgate T-junction. A T-junction is the intersection of the longwall face and the headgate or tailgate. The AFC consists of a series of pans of more often 6.28 ft (2 m) or 5.75 ft (1.75 m) or 6.56 ft (2 m) long, depending on the width of the shield supports. A special connector is used to connect

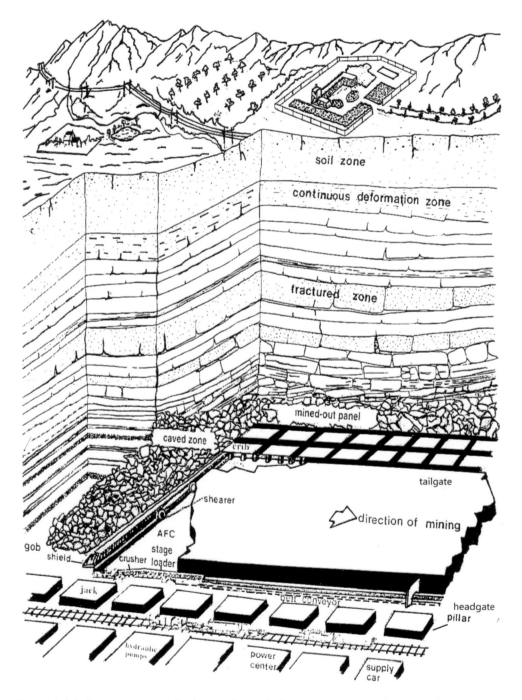

Figure 1.4.1 Cut-away view of a longwall panel showing equipment layout and overburden movement

the adjacent pans such that the whole panline can be bent vertically and horizontally to some extent without losing its stability and integrity.

At the headgate T-junction, coal is dumped, via a cross frame, onto a stage loader, which in turn empties to the entry belt conveyor some distance outby the T-junction. A stage loader or beam stage loader (BSL) is a shorter chain conveyor like an armored face conveyor, but it is mobile and capable of moving along with the face. Since the AFC is laid on the mine floor and the entry belt conveyor is supported by the belt structure high above the mine floor, the unloading end of a stage loader where it dumps coal into the entry belt conveyor is raised like a goose neck such that it can dump coal onto the belt and be pushed to overlap the entry belt conveyor for 12–15 ft (3.7–4.6 m). Stage loaders are 75–230 ft (22.8–70 m) long. At the inby end of the stage loader a crusher is installed to reduce oversized coal/rock materials.

The shield supports are used to support the roof at the face. They are set in one-web back locations and can be advanced one-web cut distance immediately after the shearer cuts and passes by in order to support the newly exposed roof as soon as possible (Fig. 1.4.2).

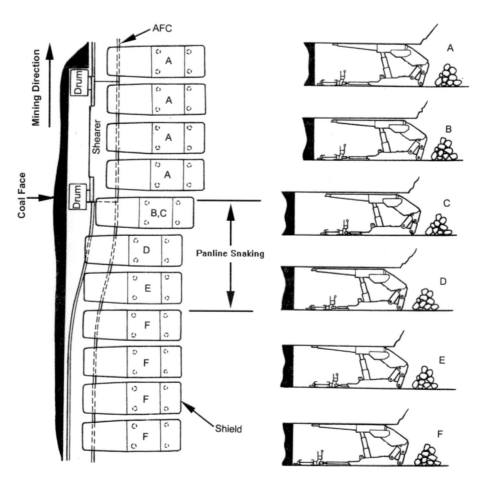

Figure 1.4.2 Operations of longwall face equipment including sequential steps of shield advance

In normal supporting condition (Fig. 1.4.2 A), the canopy of the shield is set tightly against the immediate roof strata by the supporting resistances of the hydraulic legs. When the leading drum of the shearer cuts and passes several shield units beyond the shield in question, the shield is advanced. A shield advance consists of the following four steps: in step 1, the shield legs are lowered 2 to 4 in. (51–102 mm) (Figure 1.4.2 B), and then in step 2, the shield is pulled forward for a distance equal to the width of the cut (web) by extending the double-acting hydraulic advancing ram (or DA ram) (Fig. 1.4.2 C and D). During shield advance the DA ram acts against the AFC panline, the position of which is held unchanged by the advancing rams' forces of the shields that are set on both sides of the shield to be advanced. As soon as the shield has been advanced to the designed position (Fig. 1.4.2 D), it is immediately reset against the roof in step 3 (Fig. 1.4.2 E). Finally, the advancing ram of the shield in question is extended to push the conveyor forward and becomes ready for the next cut in step 4 (Fig. 1.4.2 F).

Notice that the cross-sections B, C, D, and E are for illustration purpose because they are transient steps in the shield advance operation. In normal operation those steps are extremely short and usually not shown. The sequential steps described above are for the advance of an individual shield support in the direction perpendicular to the faceline. But along the faceline, the conveyor cannot be advanced in a sharp step immediately after the shearer has passed it due to the limited flexibility of the panline. Thus, the fourth step (Fig. 1.4.2 F) is usually sometime after the shearer's pass, resulting in a curved or snaked section of the panline. This is called panline snaking (Figure 1.4.2). The length of snaking required depends on pan width, pan length, and web width and ranges from five to seven pan lengths.

During panel development, the panel entries or gateroads are roof bolted as usual. But in the tailgate, one or two rows of cribs or other type of standing supports are erected when the previous panel is being mined. Cribs erection is kept 500 to 1000 ft (152–304 m) ahead of the retreating longwall face. Cribs are supplementary supports designed to cope with the abutment pressures created by longwall mining.

They are designed to cope with the side abutment pressure from the previous mine-out panel plus the front abutment pressure from the current panel. During retreat mining, the roof at the headgate T-junction area (up to 200 ft or 61 m outby the face) is generally reinforced with supports of some type (e.g., wood posts or hydraulic jacks in a 4 × 4 ft pattern). Other types of standing supports have also been used including cans, cribs, concrete blocks, pumpable foam blocks, etc., to increase support density to withstand the moving front and side abutment pressures.

1.5 Features and trends of US longwall mining

The following statistics were compiled from the annual US longwall census performed by the U.S. Department of Energy (1976–1980), *Coal Age* (1976–1981), *Coal Mining and Processing* (1982–1983), *Coal Mining* (1984–1987), *Coal* (1988–1996), *Coal Age* (1997–2019), and Reid (1991).

It must be noted that the contents of those statistics vary somewhat from year to year, but the trends are similar. Slightly more than two thirds of the longwall panels are located in the Appalachian coalfield (eastern Kentucky and Ohio, southwestern Pennsylvania and Virginia, and West Virginia). The remainder are located in Alabama, Illinois, Colorado, and Utah. In recent years, longwalls in Kentucky and Wyoming have been closed.

Mining height is generally the seam height, but in many cases the roof is also cut to make room for equipment and for ease of miners' travel. Mining height for the shearer faces varied from 4 to 13 ft (1524–3962 mm) with the majority at 5 to 8 ft (1524–2438 mm) (Fig. 1.5.1). Note that in Figure 1.5.1 the minimum mining height represents those for plowing longwalls, the last one of which was closed in 2018, and that is why the minimum mining height jumped to 5.5 ft (1674.4 mm) in 2019.

Mining or seam depth ranges from 200 to 3000 ft (61–915 m) with the majority in the 331–1312 ft (101–400 m) range (Fig. 1.5.2). Panel width increased gradually from an average of 460 ft (140 m) in 1976 to 1180 ft (360 m) in 2019 (Fig. 1.5.3). In recent years, panel width is usually larger than 700 ft (213 m) unless some geological, property, or other

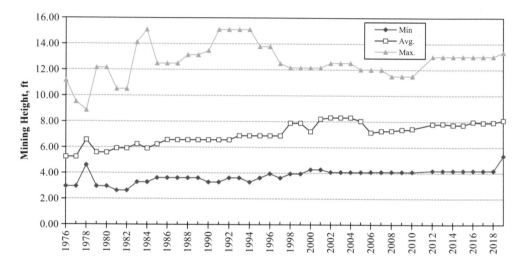

Figure 1.5.1 Yearly trends of mining height for US longwalls since 1976

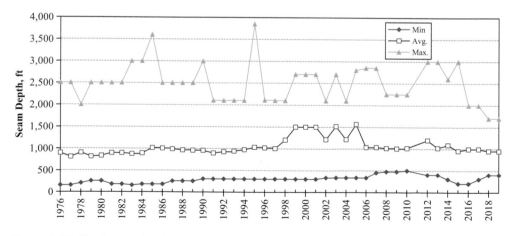

Figure 1.5.2 Yearly trends of mining depth for US longwalls since 1976

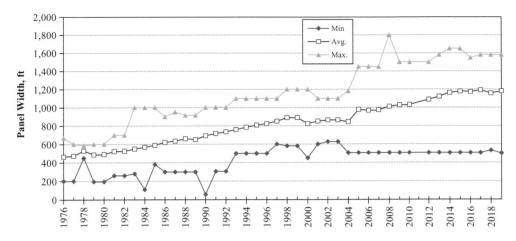

Figure 1.5.3 Yearly trends of panel width (or face length) for US longwalls since 1976

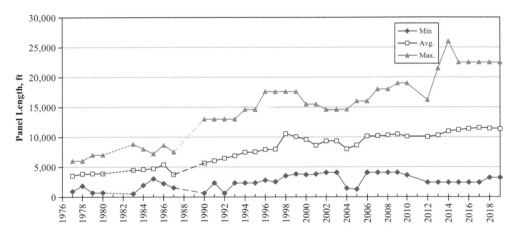

Figure 1.5.4 Yearly trends of panel length for US longwalls since 1976

constraints exist. The historical maximum panel width was 1650 ft (503 m) in 2013–2015, but the current maximum in 2019 is 1580 ft (482 m) wide.

Panel length grows from an average of 3700 ft (1128 m) in 1976 to 12,000 ft (3659 m) in 2019 (Fig. 1.5.4). Panel length is rarely less than 3000 ft (915 m) long unless special circumstances (e.g., end of reserves, important surface structures that must be protected, etc.) call for its setup. The current maximum length is 22,500 ft (6860 m) in 2019.

There is a consistent trend that panel width and length become wider and longer year by year even though in recent years the rate of increase seems to have stabilized somewhat. This trend will continue as long as the original equipment manufacturers can provide ever-increasing heavier, more powerful and reliable face equipment. The general concept is that face equipment should be able to work continuously without the need for major repair or rebuild until after a panel has been completed. A wider and longer panel reduces the number of face moves for a fixed area of reserves thereby cutting down or eliminating the non-productive days dedicated for face move.

The coal cutting machine includes both the shearer and plow. In the 1970s, the plow was more popular, but in recent years more than 98% of longwall panels employ the shearer. The only plow face operating after 2001 was closed in 2018 (Fig. 1.5.5). Therefore, longwall mining by plowing operation is not covered in this third edition. And among the shearers, the double-ended ranging drum shearer (DERDS) is used exclusively. The total installed power in a shearer increased from an average of 350 hp (261 kW) in 1976 to an average of 1950 hp (1413 kW) in 2019 (Fig. 1.5.6). In 2019 the maximum power is 2805 hp (2093 kW) with most having more than 1341 hp (1000 kW). A shearer employs the rack-type electric haulage and is equipped with multiple motors. Its haulage and cutting speeds can reach up to 150 ft (45.7 m) per minute. Drum cutting (web) width ranges from 30 to 42 in. (0.76 to 1.07 m)

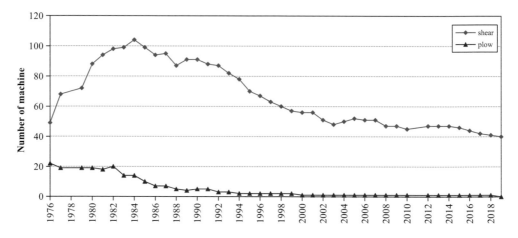

Figure 1.5.5 Yearly trends of coal cutting machines for US longwall since 1976

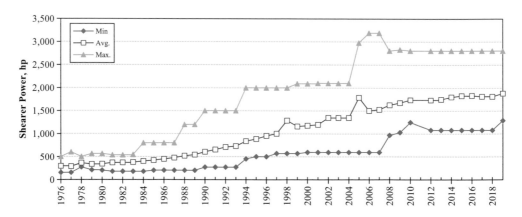

Figure 1.5.6 Yearly trends of installed shearer power for US longwalls since 1976

with more than 70% at 42 in. (1.07 m) (Fig. 1.5.7). As the face grows wider, it requires heavier equipment, and electric voltage drop across the face is dramatic when using the traditional 950–1000 volt electric power. Therefore, in order to reduce problems associated with heavy-duty equipment and handling of a large-sized cable and a large voltage drop across a wider face, all longwalls have switched to high voltages of 4180 volts except four faces at 2300 volts in 2019 (Fig. 1.5.8).

For the powered roof supports at the face, two-legged shields are exclusively used (Fig. 1.5.9) due to their ability to control caving between canopy tip and face and prevent caved coal/rock in the gob from entering the face area, and their simplicity in maintenance. The average yield load capacity was 490 tons in 1976 and increased to 1050 tons in 2019

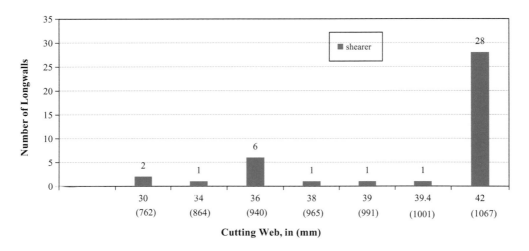

Figure 1.5.7 Distribution of cutting webs for US longwalls in 2019

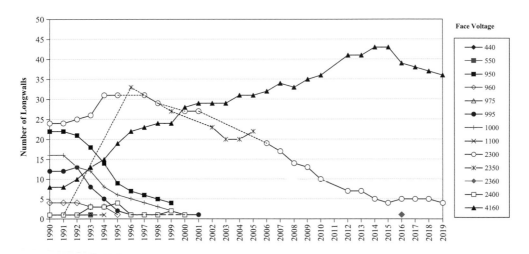

Figure 1.5.8 Yearly trends of face voltage for US longwalls since 1990

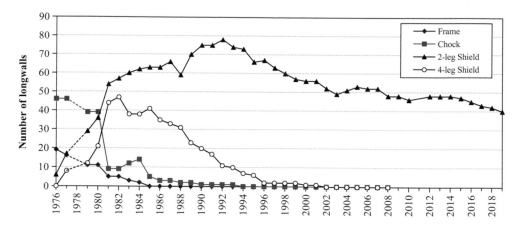

Figure 1.5.9 Yearly trends of types of power supports for US longwalls since 1976

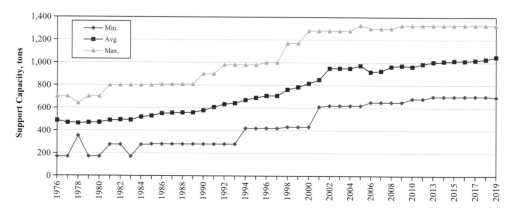

Figure 1.5.10 Yearly trends of power support capacity for US longwalls since 1976

with a maximum of 1328 tons (Fig. 1.5.10). It must be pointed out that the increase in yield load capacity over the years was not completely due purely to increase in yield load because increase of shield width from the original 1.5 m to 1.75 m and to now 2 m was not considered in the statistics, and furthermore not all mines changed at the same time. All shields are equipped with electrohydraulic control and shearer initiation options. The use of electrohydraulic control provides the first step for face automation. Shields are now remotely controlled by at most three push-button control. The shield cycle time has been reduced to six to eight seconds thus enabling fast shearer cutting. All shields are either 5.74 ft (1.75 m) or 6.56 ft (2 m) wide. A few old longwalls still use the 5 ft (1.52 m) shields. Since 1990 all new shields have been the two-leg type.

For the armored face conveyor (AFC), the chain size ranged from 0.71 to 1.18 in. (18–30 mm) in diameter in 1976. But in 2019 the chain size ranged from 1.34 to 2.05 in. (34–52 mm) in diameter with all except three faces being larger than 1.65 in. (42 mm) in diameter (Fig. 1.5.11). Currently the twin-center chain strands (TIB) are exclusively used (Fig. 1.5.12).

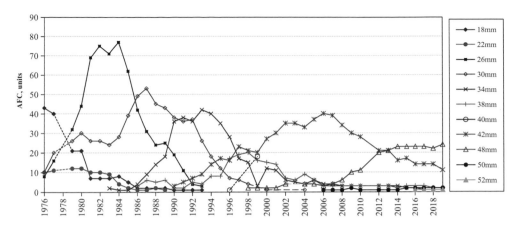

Figure 1.5.11 Yearly trends of types of AFC chain size for US longwalls since 1976

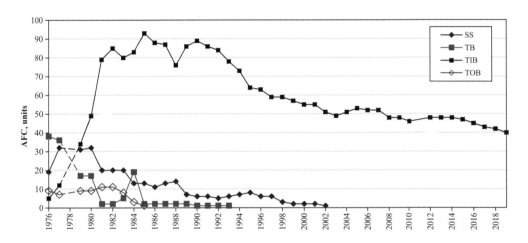

Figure 1.5.12 Yearly trends of AFC chain strand for US longwalls since 1976

The average total drive power of AFC in 1976 was 280 hp (155 kW), but in 2019 it is 4400 hp (3188 kW) (Fig. 1.5.13) with a maximum of 6600 hp (4925 kW) and a minimum of 1400 hp (1045 kW).

The pan width also increased, especially after 1986. In 2019 the pan width ranged from 34 in. (864 mm) to 53 in. (1346 mm) with the most common one at 39.4 in. (1000 mm) (Fig. 1.5.14). It must be noted that different OEMs specifies their pan width slightly different.

Chain speed has also increased (Fig. 1.5.15). It ranged from 210 to 310 ft per minute (1.07 to 1.58 m sec) with most running at 250 ft per minute (1.27 m sec) in 1976 to 230 to 451 ft per minute (1.17 to 2.29 m sec), with more than half greater than 381 ft per minute (1.94 m sec) in 2019. The designed carrying capacity of AFC was 400–600 tons per hour in 1976, but that is up to 7000 tons per hour in 2019.

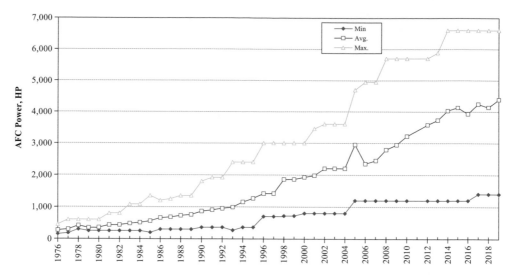

Figure 1.5.13 Yearly trends of AFC installed power for US longwalls since 1976

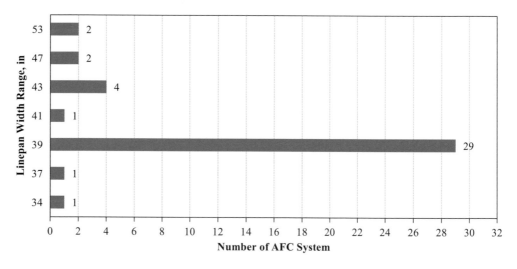

Figure 1.5.14 Distribution of AFC pan widths for US longwalls in 2019

Cross frame is exclusively used at the headgate T-junction to transfer coal from AFC to the stage loader and then onto the entry belt conveyor.

The armored chain conveyor is also used in the stage loader or BSL, but in order not to hinder material flow in AFC, the pan width and chain speed in the stage loader are wider and running faster than those in AFC. The pan width and chain speed in the stage loader stay essentially the same in the last seven years and expected to remain so in the near future

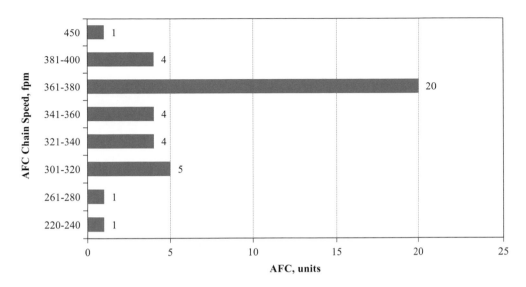

Figure 1.5.15 Distribution of AFC chain speeds for US longwalls in 2019

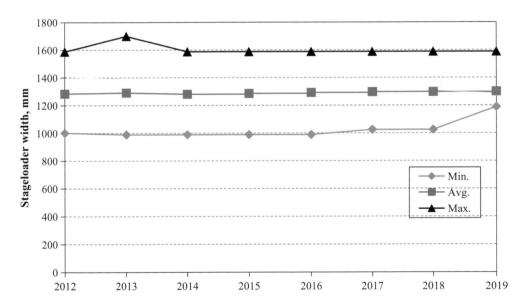

Figure 1.5.16 Yearly trends of stage loader width for US longwalls since 2012

(Figs 1.5.16 and 1.5.17). Crushers on the BSL were not common in the 1970s but came into wide use later in the 1980s and proved to be a good complement to the cross frame drive. Crusher is a very critical component in stage loader. In recent years it is getting much more powerful due to more large pieces of coal/rock dislodged or sloughed off the face.

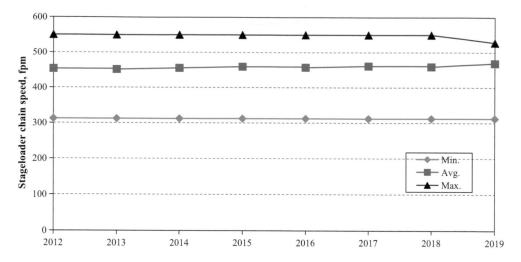

Figure 1.5.17 Yearly trends of stage loader chain speed for US longwalls since 2012

1.6 Summary of US longwall development trends

A longwall mining system is a coal manufacturing processing plant operating under various mining and geological conditions. It processes coal from an in-situ coal block onto a conveyor belt with the following steps and processes:

1   Cutting
2   Conveying
3   Roof control
4   Crushing
5   Ventilation
6   Dust and gas controls
7   Computer control

A set of longwall mining system costs from US$ 60 to more than US$ 100 million depending on the mining height and panel width. Therefore, only a large corporation that owns a large coal reserve (typically 5 × 10 miles or 8 × 16 km dividing into 15–20 panels in size) can afford to employ longwall mining. A longwall mine employs 350–600 people including surface coal preparation plant.

Currently with a total average price US$ 100 million, the typical longwall system in a typical longwall mine (an average panel width of 1150 ft or 351 m) has the following sets of equipment:

1   185–210 two-leg shields depending on shield width.
2   Two AFCs (each 1150 ft or 351 m long).
3   Two stage loaders (each 100 ft or 30 m long or more).
4   Two shearers.

5 Two crushers.
6 Two crawler mounted belt tailpieces.
7 Two monorail systems.
8 Two hydraulic power packs for shields.
9 Two sets of electrical control starting switchgears.

Normally only one of the two systems is used at one time. The second system is used to set up in the next panel to save face equipment move time. Only one set of shield supports is purchased due to its high cost, while two sets of all other equipment are purchased.

Table 1.6.1 summarizes the longwall development trends in the United States in the past four decades. It is obvious everything is getting bigger, heavier, and running faster in the pursuit of high efficiency production, thereby the service demand has more than tripled.

Table 1.6.1 US longwall trend comparison

Feature	Early 1980s longwall	Early 2010s longwall
Number of LW faces	112–118	43–48
Average panel width (ft)	530	1030
Average panel length (ft)	4000	10,030
Typical production (clean st d^{-1})	3500	20,000
Typical panel belt width (in.)	42	72
Typical panel belt speed (ft min^{-1})	500	1000
Typical depth of cut (in)	28	42
Power on face (v)	995	4160
Average powered support capacity (st)	500	1000
Powered support width (m)	1.5	2.0
AFC Power (hp)	600	5000
AFC Chain (# x mm)	2 × 26	2 × 48
AFC Drive	Fluid coupling	Soft start versions
AFC Deck thickness (mm)	20	75
AFC Bottom plate in use	None	All
AFC width (mm)	700	1100
Shearer design	SERD, DERD	DERD
Shearer power (hp)	500	2500
Shearer speed (ft min^{-1})	15	+100
Shearer haulage	Haulage chain being replaced	Rack-bar type
Emulsion pump (# x gpm)	2 × 52	4 × 100
Pump controls	Hydraulic unloader	Electric unloader, some VFDs
Monorails in use	Few	All
Monorail length (ft)	–	1700
Hydraulic pressure hoses (# x in)	1 × 1	3 × 2
Return hoses (# x in)	1 × 1–1/4	3 × 2–1/2
Water hoses (# x in)	1 × 1	3 × 2–1/2

Source: modified from Hutchinson, 2013

Table 1.6.2 Comparison of face section between 2005 and 2019

Feature	2005	2019
Prop free front from face to leg connection at canopy level – before shearer cutting, mm	4093	4111
after shearer cutting, mm	4933	5178
Distance between face and gob side sigma center, mm	2610	2772
Drum diameter, mm	1524	1651
Shearer height, mm	1145	1295
Underframe clearance, mm	462	455
Line pan sigma height, mm	326	370
Line pan width, mm	890	988
Haulage rack height, mm	924	1060

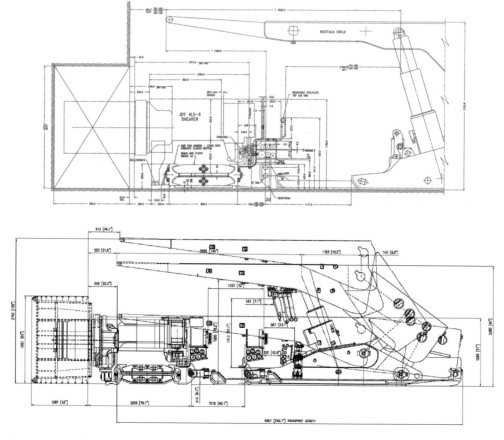

Figure 1.6.1 Comparison of face cross-section in the Pittsburgh seam between 1995 and 2016

This trend continues with no end in sight. The rate of increase was most obvious in the 1980s and 1990s, coinciding with high rate of production increase. The rate of increase decreases significantly and stabilizes in early 2000s and thereafter. Among the increases, the most interesting part is that the powered support capacity was 500 tons in the early 1980s and it was 1000 tons average in the early 2010s, twice as much. What made or required the change in terms of ground control theory, coal production, miners' safety or others?

The results of increased use of heavier equipment are most dramatic when examining the cross-section geometry of a modern longwall face. Niederriter (2005) compared the height increase due to increase in shearer installed power between 1990 and 2005. The total additional height to maintain the same height clearance was 8.5 in. (216 mm) while the width from coal face to the gob side furniture or cable handling channel increased 25.5 in. (648 mm) from 105.6 in. (2682 mm) to 131.1 in. (3330 mm). If the most common pan width of 39.4 in. (1000 mm) was used, the width increase would be 29.8 in. (757 mm).

Table 1.6.2 shows continuing changes of face and equipment dimensions from 2005 to 2019. All dimensions have increased except the underframe clearance.

Figure 1.6.1 compares the longwall face cross-section of 1992 and 2016 in the Pittsburgh seam. The face cross-sectional areas in front of the hydraulic legs are 81 ft^2 (7.5 m^2) and 122.88 ft^2 (11.42m^2), machine cross-section (shearer + AFC) are 17.4 ft^2 (1.62 m^2) and 28 ft^2 (2.6 m^2), walkways are 575 mm and 587 mm, and shield canopy lengths are 13 ft (3986 mm) and 15.8 ft (4820 mm) for the 1992 and 2016 faces, respectively.

1.7 Management and human factors in high production longwalls

Finally there is no doubt that advanced new equipment goes hand in hand with high productivity. But there were cases (Forrelli, 1999; Caudill, 2005) in which used equipment was employed and yet their productivity stayed consistently among the best. This trend is even more noteworthy in recent years in that rebuilt equipment is being used in new startup longwalls with record production. Obviously new and advanced equipment is critical to high productivity, but equipment is only part of the success story. Dedicated work forces and harmonious labor and management relations account for the remainder of the formula for a high production longwall. This issue is getting more attention in the industry in recent years due to the retirement of baby boomers.

Chapter 2

Longwall mine design

2.1 Introduction

Longwall mining design in the United States typically consists of two parts in sequential order: (1) panel layout by mine management and (2) selection of face equipment by mine management in consultation with OEMs or used equipment manufacturers. Panel layout includes panel distribution in the reserve area, panel width, entry development system (two-, three- and four-entry system), chain pillar size, entry width and roof supports, while equipment selection includes cutting machine DERD (shearer model), two-leg shield, and coal transportation, which includes in turn the armored face conveyor (AFC), stage loader, and panel and outby belt conveyors. For preliminary panel layout, mine management usually follows the longwall mining practices and experience gained in the same coal seam in the same coalfield, if available, and refined/confirmed with ground control design. If it is a virgin coal coalfield/coal seams for starting longwall mining, the first step is an in-depth ground control design.

Face equipment selection is discussed in Chapters 5 through 8, each of which is dedicated to a specific piece of equipment or a specific aspect of a particular piece of equipment. This chapter discusses the design of panel layout from mostly ground control point of view.

2.2 General consideration of panel layout

In laying out longwall panels, in addition to coal property boundaries, there are several commanding factors that will dictate the overall panel layout. The major ones are discussed here.

2.2.1 Reserve

Modern longwall mining has radically changed the concept of coal reserves. Prior to the mid-1980s, a 20-million-ton reserve may have been designed with a mine life of 15 to 20 years in mind. Today that same reserve only lasts two to four years at the most.

Modern longwall takes US$ 60–100 million or more (for an average panel width of 1100 ft or 335 m) to equip each panel, which, according to US Tax Code, usually requires at least seven years to amortize. During this period, a longwall may produce 18 to 50 million tons of clean coal, and, therefore, a large coal reserve is needed in order to make the investment feasible. This reserve, however, does not have to be a single, contiguous block. Experience has shown that they can be isolated blocks, for which, everything being equal, the critical factor is panel development rate.

2.2.2 Panel dimension

The general concept is to design the panels to be as large as possible, mainly to reduce the development ratio (the total linear footage cut by continuous miner in the gateroad per foot of longwall face advance) and the number of non-production days per fiscal year. Ideally, the panel dimensions shall be designed such that there will be a maximum of only one longwall face move or none at all. Currently the average panel width in 2019 is about 1180 ft (360 m), and the maximum width is 1580 ft (482 m) (Figs 1.3.3 and 1.5.3). The average panel length is about 11,347 ft (3459 m), and the maximum length is 22,500 ft (6860 m) (Figs 1.3.4 and 1.5.4). Experience has shown that modern longwall equipment can handle a panel with four to five million clean tons of coal without the need for major overhaul. So depending on seam thickness and reserve boundary, these numbers may be a useful guideline for laying out the panel dimensions.

2.2.3 Geology

Geology is the single most important factor in high production, high efficiency longwall mining. But geology has not received the attention it deserves, even though in recent years this traditional attitude seems to have changed from an investment point of view but unfortunately not from an engineering point of view. One case in point is that many coal companies are publicly owned now. They often report that a dip in their coal production is due to poor geology in their longwall mine(s). In this respect, most longwall mine have one or more geologists on their staff now who would map the gateroad geology including roof, coal, and floor as the gateroads are being developed. Some even include drilling short roof boreholes, 10–20 ft (3–6 m) deep at fixed intervals, and construct the gateroad roof cross-section showing the stratigraphic changes if any along the panel gateroads.

The geological factors affecting panel layout include seam inclination, roof and floor strata, and geological anomalies (Peng, 2008).

2.2.3.1 Seam inclination

As stated in Section 1.2 (p. 3), all current longwalls in the United States are operating in flat seams with local seam inclination less than 5–10% (3–6°) or gently inclined seams up to 7–9°. Experience gained in the 1970s–1980s had shown that under then-prevailing coal market conditions, seams with dips larger than 15° to 20° cannot be economically longwall-mined. Inclined seam longwalls were operating in Utah and Wyoming areas. But they were terminated one after another in the 1980s, and the last one was closed in early 2000s due to market and reserve conditions.

In a so-called flat seam, seam dip of less than 10% is common. When laying out panels under this condition, the mining direction should go updip such that water from both natural and man-made sources will flow back into the gob of previous panels. This way, water will not accumulate on the face and in the gob of the panel being mined. In the latter case, if water flows back into the gob being left behind, it may flood the bleeder system and pumping will be required. Since modern longwalls employ a considerable amount of water for dust suppression at the face, keeping the face dry, especially when the floor rock is soft and susceptible to water, is extremely important in order not to impede the fast-moving longwalls.

2.2.3.2 Roof and floor strata

Good roof and floor conditions are the prerequisite for a fast-moving longwall. A dip in monthly or quarterly production can easily be attributable to the occurrence of local poor roof or floor or both. In this respect, determination of rock mechanics properties of near-seam strata in conjunction with exploration boreholes cannot be over-emphasized. The rock properties determined should not be confined to common characteristics such as UCS (uniaxial compressive strength) and tensile strength. If the rock is shale, weathering susceptibility, such as lamination strength and water effect, should be determined. Furthermore, the conventional way of siting the exploration borehole grids in 1- to 5-mile spacing is not adequate and carries a high risk in virgin coal fields and some risk even in coal fields known for their uniformity, such as the Pittsburgh seam. Conversely, if the roof is hard and unlikely to cave following shield advance, greater-than-normal shield capacity may be needed, and in severe cases, artificially induced gob caving or narrower panels or both may be required. For best operating conditions, the floor should be hard and firm. If it is soft, the floor should be kept as dry as possible, and shields with a floor lifting device must be used. The combination of hard roof and soft floor is the worst to deal with.

Very often, the immediate roof is weaker than the coal. In this case, leaving a protective top coal layer under the immediate roof is the most common practice, provided the seam height is sufficiently thick. Conversely, if the immediate roof is so weak that it falls off right after or soon after undercutting, it would be beneficial to cut it out by the shearer drum as part of the mining height.

2.2.3.3 Geological anomalies

Other geological conditions that affect production, and thus panel layout, include faults, folds, sandstone channels, clay veins, and fractures (mountain seams or hill seams). Their locations should be mapped far in advance by geophysical methods for panel layout consideration or at the least be determined by geologists panel by panel during gateroad development and extrapolation into the panel interior made through deposition theories or simple linear extrapolation method.

Faults are fractures both sides of which have experienced relative movement, while joints are those without relative movement. Fractures may be filled, usually with weak materials, or remain open. If the offset of the fault is smaller than the mining height, a longwall can cut through, although with considerably reduced rate of advance. But if the fault is larger than the mining height, it is advisable to exclude the fault in the panel layout by using it as a panel boundary. If this cannot be done, special measures must be planned in advance and implemented to deal with the fault (see Section 10.10).

Fractures, either open or filled, are weak. If a longwall face is oriented parallel to a fracture, the roof may collapse in large blocks when the face reaches the area directly underneath it. Therefore, the face should be oriented perpendicular or sub-perpendicular to the fracture, at least 15°.

Folding results in anticlines and synclines, meaning a coal seam may range from flat to steeply inclined or declined. When seam inclination varies gradually, longwall panels may be laid out on the dip direction and retreat downdip. This way movement of the shield and panline is easier to control, especially at the steeper area. But if water is abundant, it must be handled aggressively, otherwise it may flood the face. On the other hand, if the seam dips uniformly, it is common to lay out the panels on strike and mining in retreat panel by panel in the downdip direction.

Sandstone channels that dip into the coal seam are normally very hard and reduce the minable seam height. Since the regular cutter bits are designed for coal and cannot cut hard sandstone effectively, production will slow down or even stop, even though the modern shearer is very powerful. If the intrusion cuts too deeply into the coal, the face equipment will not be able to advance under it, and the face has to be relocated. Sandstone channels should be identified prior to panel layout. However, many sandstone channels are localized in size, ranging from a few feet to more than one panel width. If the channel is small and less than the panel width, then it will be difficult to identify accurately.

Clay veins are intrusions filled with brecciated rock fragments. They are weak and tend to fall off as soon as undermined. Clay veins occur randomly and therefore are difficult to identify in advance. In severe cases, polyurethane grouting ahead of mining may be required.

In western coalfields, there are igneous intrusions that are much harder than the host coals. Therefore they should also be mapped in advance of mining.

Occasionally, the roof strata contain pockets of loose gravel or weak materials that are broken into minor cubes once exposed due to undermining (Peng and Finfinger, 2001). This type of roof is very difficult to control.

2.2.4 In-situ (horizontal) stress

Ground control theory dictates that entries or gateroads that are perpendicular to the maximum principal horizontal stress will be much less stable than those parallel to it. Therefore, it is advisable to orient the gateroads parallel or intersect at small angles to the direction of maximum principal horizontal stress (Peng, 2008).

In northern Appalachia, where modern high production longwall technology was developed, there were reports in the early 1970s (Dahl and Parson, 1972) that entries in the north-south direction had many roof falls, while those in the east-west direction remained stable due to the existence of the presumed "high east-west tectonic horizontal in-situ stress." Subsequently, all longwall panels and gateroads have been laid out in or near the east-west direction, and crosscuts either at 60 to 75° toward the inby direction or 90° from the east-west direction. With this type of panel layout, roof falls have been noticeably reduced in all phases of longwall mining in the Pittsburgh seam. Ground control research shows that change in panel orientation might have helped, but that two additional measures could be the major contributing factors: reduction of entry width from 20 to 16 or 15 ft (6.1 to 4.9 or 4.6 m) and enhancement of roof bolting systems.

Can longwall panels be successfully mined if they are not oriented in the east-west direction? The answer is positively "yes." There are numerous examples where panels not in the east-west direction have been mined successfully, even in the Pittsburgh seam. It is common in recent years to attribute roof control problems to the existence of "high horizontal stress." High horizontal stress cannot and should not be blamed for every roof control problem. In fact, high horizontal stress is the norm everywhere deep underground except in above-drainage mines, e.g., drift mines high in mountain ridges. Too often ground control problems have been treated two-dimensionally based on which solutions are developed. In fact, all ground control problems are three-dimensional, the solutions of which are different from if the problem is considered in two-dimensions.

Section 10.15 discusses the stress distribution around multiple longwall panels subject to high horizontal stress at various directions and the effects of the stress distribution on entry stability.

2.2.5 Multiple seam mining

If multiple seams of mineable coal exist in a reserve property, potential interaction of extracting one seam with the others must be considered in order to maximize safety and coal recovery. Adverse effects of seam interaction are mainly due to stress concentration created by improper location of remnant pillars and coal block boundaries in the previously mined-out seams. The severity of adverse interaction effects depends on seam depth, size, shape, and location of remnant pillar, rock type, and thickness and number of laminations/fractures of the interburden strata (Hsiung and Peng, 1987; Peng, 2008; Westman et al., 1997).

Seams should be extracted in descending order. This way, panel gateroads in the lower seam can be located directly underneath the gob of the immediate super-adjacent seam so that they are located within the destressed zone of the super-adjacent panels. If the mined-out panels in the upper seam are flooded, water must be drained or pumped out in order not to flood the lower seams when the seam interval is less than 30 to 60 times the lower seam mining height.

If different mining methods in adjacent seams are planned (i.e., room and pillar in the upper seam and longwall in the immediate lower seam or vice versa), then factors such as pillar size and location, rock strata stratigraphy, and thickness of interburden between the seams must be considered to determine the influence zone of high abutment pressure and its effects. It follows that there are so many scenarios that may exist, more definite conclusions can only be obtained by three-dimensional numerical computer analysis on a case-by-case basis.

The delayed time for mining directly over or under a coal seam that has been previously longwall-mined should be at least six months and preferably one year. This delayed time is required to allow the broken rock strata to settle and heal, especially if clay materials are abundant in the strata involved.

2.2.6 Rivers/streams or lineaments

Bad roof is known to occur directly underneath surface rivers and streams, although there are many exceptions. Whether a surface river or stream represents a bad roof or not probably depends on the depositional history of the overburden strata. It cannot be determined positively without detailed analysis of local and/or regional rock/coal deposition history. But when the roofs of gateroads are within 100 ft (30.5 m) of a river/stream, it may require special attention in roof support. Experiences also have shown that seams less than 150 ft (45.7 m) deep may not be suitable for longwall mining because the overburden strata may break in blocks from the seam level all the way to surface. Note that the previous general statements of experience are derived from longwall mining height of less than 10 ft (3 m).

Lineaments are linear surface features evidence in high-altitude photos, such as those taken by a LANDSAT satellite. They are mostly associated with rivers/streams, mountain ridges/valleys, faults, and fractures. Whether or not a lineament represents a bad roof depends on whether or not these features cut through the overburden and reach the coal seam roof. There have been reports that the intersection of multiple lineaments represents a very bad roof or unsupportable roof. Therefore lineaments, if confirmed, could serve as a rough guide for panel layout.

2.2.7 Oil or gas wells

In the Appalachian and central coalfields, there are numerous gas and oil wells, nearly all of which pass through the coal seams of interest because coal seams are much shallower than gas and oil reservoirs. Some of those old wells may be uncharted, or if charted, their locations may not be accurate. Since the early 2000s, due to the successful implementation of hydraulic fracturing and horizontal drilling, many shale gas wells have been drilled through the coal reserves designated for longwall mining in the northern Appalachian coalfield, resulting in greatly complicating the design of panel layout (Peng, 2008).

Several state laws require that when an oil or gas well is encountered, a square coal pillar of 200 ft (61 m) in size with the well at the center must be left for protection. But oil and gas wells are usually plugged off as prescribed by MSHA (Mine Safety and Health Administration of US Department of Labor) so that panels can be laid out as if no oil/gas wells exist. Thus, if many oil/gas wells exist within a panel or panels, this could be a very expensive operation.

If an oil/gas well is not plugged off, longwall panels can be laid out so that the wells are located in gateroad development sections, or on either end of the panel beyond the influence zone. If this happens, panel size and shape may vary from panel to panel. However, if there are many gas/oil wells, panels may be designed so that some oil/gas wells are plugged off and some are left in the gateroads. In the latter case, pillar design must follow state regulations that were established for room and pillar mining with pillar extraction in the 1950s–1960s, and that may not be universally applicable to modern longwall mining. Research is being conducted to correct this problem, and recent findings are discussed in Section 10.11 (p. 360).

2.2.8 Surface subsidence

Subsidence is a very critical public-relation issue in longwall panel design and, for that reason, in longwall mining itself. Almost every longwall mining region has one or more citizen groups advocating in conjunction with national environmental organizations, against longwall mining. The key problem is that every subsidence-related complaint, be it technically right or wrong, requires a lengthy legal process to resolve. If the longwall has to stop production, it is a very expensive process. The key to avoiding this problem is to take proactive measures to open dialogue with the surface property owners and start collecting pre-mining data far in advance, at least two years ahead. Techniques have been developed and proven successful case after case for all kinds of surface structures (see Chapter 14 for details).

When a longwall panel of sufficient size is excavated it creates four distinct zones in the overburden, i.e., caved, fractured, continuous deformation, and soil zones in ascending order (see Section 3.2, p. 45). On the surface, subsidence creates a trough formed by mains, a gateroad development section on both sides of the panel and the bleeder system on the rear end of the panel. When adjacent panels are mined, the surface above the gateroad development sections also subsides to some degree but usually not sufficient to cause structural damages.

If an aquifer or a water well is located in the caved zone or the lower portion of the fractured zone, the aquifer or well will be drained off after mining and may never recover or may take a long time to recover. If the source water is located in the upper portion of the fractured zone, the water level will fluctuate or be lost temporarily as the longwall face is within some distance before, during and passing it; if it is located in the continuous deformation zone, the water level may fluctuate temporarily, but mostly likely will not be affected permanently; if it is located within the 14° zones of influence from the panel edge, the water level may decrease temporarily. In the soils zone, along the panel edges where permanent cracks exist, surface water may disappear. Otherwise, there should be no subsidence effect.

If a house is located on the surface above the panel center, cracks may open and close at strategic points during mining, but there will be little or no permanent damage. If the house is located near but inside the panel edges, cracks may remain open and damages may be permanent. Houses located above the gateroad development sections will in general suffer little or no permanent damage. If a rectangular house within the panel has its longer dimension oriented parallel to the faceline, damage may be minimal.

All high-pressure large-diameter gas/oil transmission pipelines, if extended across the panel, should preferably be dug out and monitored for zones of high stress concentration, especially in steep terrains; high-voltage power transmission towers, if located above the panel development section, will not be affected to the extent that repairs/adjustments are necessary.

Surface streams, dams, ponds, and lakes have been successfully undermined without harm by longwall mining if they are located in the continuous deformation or soil zone or even in the upper portion of the fractured zone where clay partings or clayey shales are abundant.

2.2.9 Exact panel width

Panel width or face length either represents the actual panel coal block width from one side of the rib to the other side of the rib or from the center of headgate to the center of tailgate. The former definition is one entry (gateroad) width less than the latter one. In design the panel width, it is a common practice to use an even number, for example 1000 ft (304.9 m) or 1050 ft (320.1 m). The selected even number of convenient width most frequently leave a fraction unit of shield and pan to fill which tends to complicate the stop point location with respect to panel edge when the shearer approaches the headgate and tailgate, respectively. Furthermore, the last shield at tail end may stick out partial width into the tailgate colliding with standing supports during production operations and delaying face advance. Even with this obvious disadvantage, some longwalls still prefer this type of equipment layout.

The alternative is to layout the shield and pans first and develop the panel width to match the exact width of selected shields and pans. Figure 2.2.1 shows mine management's original designed panel width was 1479 ft (450.91 m) rib-to-rib or 1495 ft (455.68 m) center-to-center. After laying out the following sets of AFC and shields of 223 units: 6 HG shields @1.75m width, 213 face shields @2m width, and 4 TG shields @1.75m width, it was found that the tail end of panel block edge of 33.5 in. (850 mm) was left unsupported. So the actual panel width during development was adjusted to 1479–33/12 = 1476 ft (450 m) rib-to-rib or 1492 ft (455 m) center-to-center.

Maintaining an exact panel width for a long panel requires high discipline in entry development.

2.2.10 Spontaneous combustion

US coal mine ventilation covers face and bleeder ventilation systems requiring that both the face and bleeder in the gob must be ventilated.

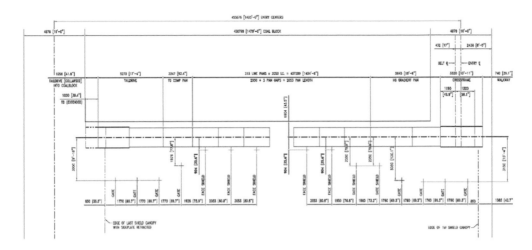

Figure 2.2.1 Design of exact panel width to match equipment layout

In the past decade spontaneous combustion of gob occurred in western longwall coal mines, New Mexico and Utah in particular. In order to inhibit coal's spontaneous combustion in the gob, MSHA allows the bleederless ventilation system on a mine by mine basis. In addition, CO is injected into the gob to suppress the potential for self-combustion. Section 11.2.2 (p. 392) described briefly the typical bleederless system.

2.3 Ground control consideration

2.3.1 General

Since the US coal industry was a newcomer in modern fully mechanized longwall mining in the 1970s and early 1980s, there had been quite a few failures in the startup of longwall mining, some of which resulted in closure of the mines permanently. The major reasons could be attributed to the neglect and/or ignorance of ground control issues. First and foremost was what kind of roof and floor strata were economically suitable for the face equipment then available. Those cases of failures included, for example, the following: (1) the powered supports were not properly designed or selected both in capacity and strength of structural components as well as joints welds that caused numerous production delays; (2) the roof was so weak that it fell off immediately after undercutting before the powered support could be advanced to support it at the face; (3) the longwall face was oriented in such a way that it collapsed when the face reached a large joint set parallel to the face; (4) the immediate roof did not cave following the face advance from the set up room and overhanged for more than 300 ft (91 m) and finally crushed all or parts of the powered supports at the face; (5) the chain pillars in the tailgate side were not properly sized such that a few panels into the district, the tailgate was badly deformed such that the AFC tail drive could not be freely advanced; (6) the headgate and tailgate T-junction areas were not properly supported such that either massive roof falls or excessive roof convergence or floor heave occurred hindering the movement of head and tail drives; (7) sandstone channels were found cutting into the coal seam inside the coal panel block, causing large production loss; and (8) in the late 1980s, as the industry consider production increase by extending longwall panel due to poor market condition the routine question was this: was the capacity of existing powered supports sufficient if the panel width was extended 100 ft (30.5 m), 200 ft (61 m), 300 ft (91.5 m) – or more?

The general experience had been that if the longwall mining project was going to succeed and continue, the first panel must be successful, i.e., no or little ground control problems that caused excessive production and safety delays.

The initiation and continuation of the annual *International Conference on Ground Control in Mining* (ICGCM) starting in 1981 provided an excellent forum whereby all professionals interested in ground control, in particular relating to longwall mining, gather to exchange information relating to ground control techniques in mining. They included university professors, government researchers and regulators, equipment manufacturers, mining consultants and services personnel, and above all, mine operators, not only from all parts of the United States but also from all major coal producing countries. Any new techniques and products presented at the conferences were quickly adopted by the mine operators, and if they were not successful, then they were abandoned immediately. If the trials were successful, they would be accepted and spread industry wide quickly. It was mainly through this process, the series of ground control problems associated with the introduction of European longwall technology were resolved one by one in a short period of time. The result was that ground control became the first part of mine design project whenever a new longwall mine or a new panel in an existing mine was planned, thereby setting up ground control as the leading factor in the success or failure of a longwall mining project.

Major ground control factors in longwall mining design include the following (Peng, 2008):

1   Optimum panel width (or face length) for the geological condition.
2   Type and yield capacity of hydraulic powered supports.
3   Rows and size of chain pillars in development section and number of entries.
4   Entry (gateroad) supports (primary and supplementary) including T-junctions.
5   Surface subsidence prediction and protection of structures and water bodies, both surface and underground.

After ground control design, the selection of face equipment follows. Since the OEMs of face equipment have standard models developed, it is really more or less selecting the closest models plus customized items for the designed panel layout. After nearly two decades of longwalling in the late 1990s, the ground control parameters for successful US longwalling have been proven and known industry wide, which reduces equipment selection and cost.

The structural design of longwall panel layout includes the determination of panel size (i.e., width and length), entry width, and arrangement of chain pillars (i.e., number of rows and size). The principle of structural design is that the structures must remain stable, either by nature or by artificial controls, throughout the active mining period. In determining the stability of a structure, one must first obtain the stress distribution in each element of the structure under the conditions in which the structure is expected to serve. Second, the magnitude of stress distribution in each element is compared with the strength of that element. In order that the structure be stable, the stress induced in and the strength of the elements must satisfy certain relationships defined by the selected failure criteria. In this respect, the structural design of a longwall panel layout involves the determination of abutment pressure around the panel edges and their effects on the determination of the number of rows and size of chain pillars. Since geological conditions in underground coal mines vary from place to place, it is not feasible, nor desirable, to define a single trend. In fact, stress distributions around the panel vary considerably with changes in stratigraphic sequence in both vertical and horizontal directions. Therefore, the results presented in any numerical computer modeling is only applicable under the specified mining and geological conditions, but may be considered at best as approximate general trend for other similar conditions. Therefore, panel layout design and the exact trend can only be obtained on a case-by-case basis.

In practice, among the structural elements of panel layout, only chain pillars are truly subject to ground control design, because entry width is dictated more by equipment selected for its development. For instance, entries developed by in-place miners are 15–16 ft (4.6–4.9 m) wide and those by the conventional continuous miner (CM) are 16–20 ft (4.9–6.1 m) but mostly 18–20 ft (5.5–6.1 m) in consideration of CM maneuverability. Pillar design is addressed in Section 4.6 (p. 90). So only the effect of panel width will be addressed here.

The need to know the effect of panel width comes from the concern that when mine management planned to extend their panel width for better productivity in the 1980s–1990s, their first question was always this: would it increase shield loading? If it did, how much more load would add to the existing shields and would the existing shields have enough capacity? Finite element modeling by Peng and Tsang (1994) shows that (1) the shield support load distribution in a longwall face is mainly affected by roof condition, degree of gob compaction, and interaction between shield and roof strata. In general the maximum shield load occurs toward the tailgate side due to previous panel gob unless the width of gateroad is sufficiently large such that the side abutment pressure from the previous mined-out panel is greatly reduced or eliminated, and (2) that increase in panel width will increase shield load, the amount of which depends on panel width, roof condition, and gob compaction. Overall the percentage of load increment ranges from 3.5 to 8.5% with an average of 6% per 100 ft (30 m) increment of panel width. When the panel width is larger than 700 ft (213.4 m), the load increment becomes smaller; and the effect of panel width on gateroad chain pillars is very small. After 2005, the width of most US longwall panels are either "critical" or "super-critical" which means further expansion in panel width would not induce additional roof load on the shields at the panel center. Industry practices have demonstrated that the total pillar width used on the development section is sufficient to isolate the panel or reduce sufficient amount of side abutment pressure from the immediate previous mined-out panel in terms of abutment pressure.

2.3.2 A case example of longwall panel design

As stated earlier, the established longwall panel design method consists of two sequential steps: (1) ground control panel layout design and (2) face equipment selection. This example case illustrates the first step. The second step is addressed in various other chapters in this book and therefore it will not be covered here.

The coal seam considered in this example is a virgin coal seam located across the state line between West Virginia and Virginia (Peng, 2013). The seam has never been longwall-mined before. So ground control design has to start with development and collection of basic information, including the following:

1   Geology: Development of stratigraphic distribution of overburden and floor within the reserve area including the seam of interest through corehole drilling and loggings.
2   Determination of rock properties of all major rock strata, including UCS (uniaxial compressive strength), tensile strength, and slake durability.

2.3.2.1 Geology

Figure 2.3.1 shows the preliminary panel layout based on the surface borehole drilling data. There are 32 boreholes, the stratigraphy of all of which was logged in detail and cores of all rock types and layers were obtained for rock property tests. The coal seam dipped less than 3.3 degrees.

It is well-known that there are many geological discontinuities (anomalies) in the overburden, including sandstone channels, faults, folds, joints, and clay veins. The occurrence of those geological anomalies may either delay the mining operations or cause safety problems if proper longwall panel layouts and ground control strategies are not implemented. Therefore, a good understanding of geology in the proposed longwall panels is very important in modern longwall operations, in which any production delay incurs a large economic loss.

Several techniques can be used to predict the existence of the geological anomalies. Some of those techniques may not be applicable to this longwall project because it is a virgin area with no underground development. Therefore, the fence diagram technique is performed to identify and locate some of the potential geological anomalies or discontinuities.

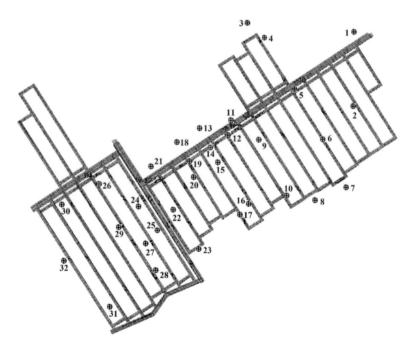

Figure 2.3.1 Preliminary longwall panel layout and surface borehole locations

For the proposed longwall panels, four overburden stratigraphic cross-sections are selected to construct the fence diagram. More cross-sections should be selected for more detailed geological analysis:

1   Cross-section 1: Boreholes 21 > 22 > 23.
2   Cross-section 2: Boreholes 18 > 19 > 20 > 17.
3   Cross-section 3: Boreholes 11 > 12 > 10.
4   Cross-section 4: Boreholes 31 > 28 > 23 > 17 > 10.

These four cross-sectional fence diagrams show the following:

1   Geological discontinuities in the roof may exist in cross-sections 1, 2, and 3.
2   Coal seam in deep cover area is thicker than in the shallow cover area. The thinnest part is only 2.38 ft (0.73 m) (borehole #18), while the thickest part is 4.47 ft (1.36 m) (borehole #23).
3   With all factors considered, the potential cutting height ranges from 5.74 ft (1.75 m) (borehole #11) to 7.73 ft (2.36 m) (borehole #23), while the maximum and minimum coal recovery rate (CRR) for each borehole area are 36.2% (borehole #19) and 62.8% (borehole #17), respectively.
4   With all factors considered, for the most efficient working conditions of underground miners, the mining height should be 6.0–6.5 ft (1.83–1.98 m) or more, if feasible.

2.3.2.2 Rock property determination

Laboratory tests on rock core specimens from boreholes 10, 12, 15, 18, 19, 22, 23, 28, and 30 with a total of 1267 tests (UCS 771, Brazilian tensile 250, slake durability 10, and point load 236) were selected and performed.

The purpose of the testing program was to evaluate the strength and physical properties of rock strata to provide input data for shield design, computer modeling for longwall panel layout, pillar design, and entry support, and determination of cutting horizons regarding the planned longwall mining.

The test results showed that strength of the same rock type at different elevations in a borehole varies considerably (Table 2.3.1). However, the strength of the same rock type at about the same horizon above the coal seam is relatively close to each other. This important feature is used to identify the rock layer whenever there is suspicion about the driller's logs.

The results of slake durability tests show that weathering will not affect the gateroads in any significant way. Floor digging of shields during shield advance is unlikely.

2.3.2.3 Design of shield support

The thickness of coal seam varies from 0 ft to 4.49 ft (0–1.37 m) with the overburden varying from 79 to 1326 ft (24.1–404.3 m) in the whole reserve areas. However, considering the minimum working height required for modern longwall system and suitability of miners' working height, three mining heights, 6, 6.5, and 7 ft (1.83, 1.98, and 2.13 m) that were much larger than the actual seam thickness, were used for shield capacity calculation.

For US longwall mining, the modified detached roof block method (Section 6.4, p. 184) has been proven to be most applicable, so it is used here to calculate the shield capacity for three mining heights. Shield load capacity has been determined for 32 boreholes (Fig. 2.3.1) which are located within the proposed longwall panels.

A. DETERMINATION OF SHIELD SUPPORT LOAD CAPACITY

The two-leg shields have been used exclusively in US longwall mining since the early 1990s (Peng, 2006), and adopted for use in modern high-production, high-efficiency longwalls all over the world. Therefore, the 1.75-m-wide two-leg shields are recommended for the proposed longwall mining.

There are many models available for shield load capacity design (Peng, 2006, 2008). Most of them can be approximated by the modified detached roof block model as shown in Figure 6.4.2 (p. 187). In this model, each individual stratum in the roof is considered

Table 2.3.1 UCS of all rock strata in the overburden

SJ-12-31C

Strata	Depth, ft		Thickness, ft	UCS, psi	Average UCS, psi	
	From	To				
Sandstone	568.31	580.68	12.37	10866	10,866	
Clay shale	580.68	582.19	1.51	1500	1500	
Sandy shale	582.19	585.63	3.44	5716	5602	
				5489		
Shale w/ sandstone streaks	585.63	587.83	2.20	7745	7471	
				6481		
				8188		
Shale	587.83	592.33	4.50	10,085	11,313	
				12,541		
Sandy shale w/ sandstone streaks	592.33	597.42	5.09	7483	10,031	Four times difference at different level
				12,578		
Sandstone w/ shale streaks	597.42	599.83	2.41	23,338	21,162	
				18,985		
shale	599.83	600.60	0.77	5018	5018.08	
Sandstone w/ shale streaks	600.60	609.64	9.04	25,479	22,173	
				20,423		
				26,507		
				20,733		
				17,722		
Shale w/ coal streaks	609.64	610.78	1.14	3464	3464.36	
Clay shale	610.78	617.69	6.91	6122	6655.35	
				8665		
				5178		
Shale	617.69	619.39	1.70	4030	5921	
				7811		
Sandstone w/ shale streaks	619.39	621.70	2.31	11,430	11,430	
Shale w/ siderite layer	625.44	627.97	2.53	14,326	14,326	
Coal seam	632.88	635.59	2.71			
Sandy shale w/ sandstone streaks	652.64	653.94	1.30	8188	9871	
				11,554		

Source: Note the strength of clay shale varies as much as four times at different elevations

a cantilever beam with fixed-end point at the coal face. In the shield load calculation, the self-supporting length of each beam is determined first. They are then summarized to obtain the total load on the shield (Peng, 2006). The model and its computational procedures are illustrated in Chapter 6, p. 186.

Shield capacities of three mining heights, 6, 6.5, and 7 ft (1.83, 1.98, and 2.13 m) were calculated, respectively. These three mining heights were selected considering the overall seam thickness distribution in the whole coal reserve area. The height (or thickness) of the modified detached roof block is controlled by the bulking factor of the roof strata. A conservative average bulking factor of 1.125 or 8 times mining height (48 ft [14.6m] for mining height 6 ft [1.83 m], 52 ft [15.8 m] for mining height 6.5 ft [1.98 m], and 56 ft [17.1 m] for mining height 7 ft [2.13 m], respectively) was used.

In the calculation process, the roof load or the required shield support load capacity is determined borehole by borehole. The results showed that the required support load capacity ranges from 512 tons (465.5 mt) (borehole 13) to 1384 tons (1258.2 mt) (borehole #5), from 555 tons (borehole 4) to 1384 tons (borehole 5), and from 597 tons (borehole 15) to 1.384 tons (borehole 5), respectively, for the three mining heights.

For all three mining heights, the required shield yield load for borehole 5 is the largest. The extra-large roof loading can be attributed to the fact that borehole 5 is located in the shallow cover area, and at the same time, there is a very thick (24.8 ft or 7.56 m) sandstone stratum within the caving height of concern. After mining, this sandstone stratum will overhang more than 37 ft (11.3 m) before caving, producing a very large additional weight on the shield. Similar reasoning applies to borehole 13 when mining height is 6.5 (1.98 m) and 7 ft (2.13 m).

Figures 2.3.2 and 2.3.3 are the isopach maps of the required yield load distribution for the two mining heights 6 and 7 ft (1.8 and 2.1 m), respectively. It can be seen that boreholes 5 and 13 are located outside the projected longwall mining area. So it will have little effect on longwall retreating mining.

Considering the overall distribution of the shield yield load requirement and panel mining sequence, it is recommended that the two-legged shield should have a yield load of 900 tons. The 900-ton shield will cover all but borehole 5.

B. CALCULATION OF FLOOR PRESSURE UNDER THE BASE PLATE

For the two-leg shields, when the immediate floor stratum is soft, the shield's base plates may dig into the floor during shield advance. This will cause difficulties in shield advance. Therefore, the bearing capacity of the immediate floor stratum must be known so that the shield dimensional configurations can be designed such that it produces a peak toe pressure less than the floor bearing capacity. If this is not feasible, floor lifting device must be properly designed and used.

In the determination of floor pressure in this section, the shield's base plates were assumed to be elastic. Based on the finite element computer analysis, Peng (2008) developed a regression equation to determine the maximum floor pressure as follows:

$$[p_{max} = 1233 + 213x_1 + 2.72x_2 - 38x_3]$$

Where x_1 is the Young's modulus of the floor material in 10^6 psi, x_2 is the vertical resultant load in tons, and x_3 is the contact width between the plate and the floor along the faceline direction in inches.

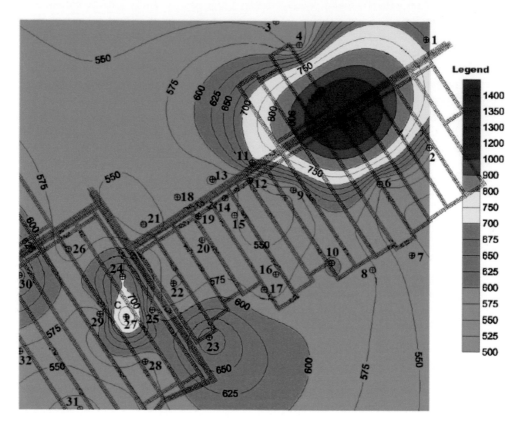

Figure 2.3.2 Shield capacity isopach map (tons) when mining height is 6 ft (1.83 m)

Three factors (i.e., Young's modulus of the floor rock, vertical resultant force, the contact width between the base plate and the floor) are used to analyze the floor pressure. Young's modulus of the floor rock is from the rock property tests for this project, and the rock type of the immediate floor is based on the results of the determination of cutting horizon in geology analysis. The vertical resultant force from the modified detached roof block method is used, and the contact width between the base plate and the floor is assumed to be 5.25 ft (1.6 m).

The great majority of the immediate floor strata are sandy shale, sandy shale with sandstone streaks, shale, shale with sandstone streak(s), and shale with coal streak(s). Since the *in-situ* bearing capacity of the floor rocks is not available, the UCSs of the floor rocks are used for estimating their floor bearing capacities by linear reduction method. The results demonstrate that the selected floor strata of mining horizon within the proposed longwall mining area are strong enough to prevent shields from digging into the floor. But it is important to note that fireclay and sandy fireclay are usually hard and firm when dry, but become soft and muddy when wet. So, it is very important to keep the floor as dry as possible, especially when the rock type is fireclay. On the other hand, slake durability tests of those floor rocks indicate that those floor rocks are resistant to

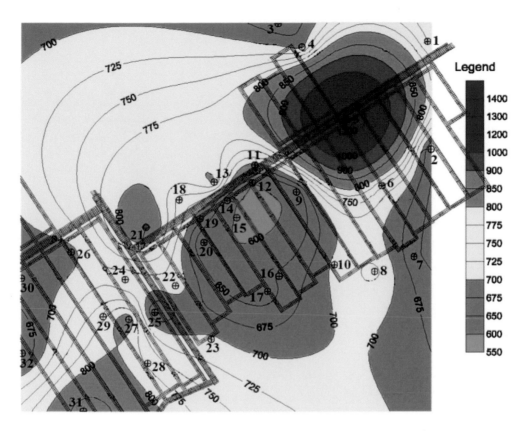

Figure 2.3.3 Shield capacity isopach map (tons) when mining height is 7 ft (2.13 m)

water. Perhaps, although so named, those floor rocks contain little clay materials, thereby becoming resistant to water weathering.

2.3.2.4 Surface subsidence

According to the proposed longwall layout map, there is one large and three smaller power transmission lines and a stream located within the proposed longwall panels. The structures that require subsidence control are 2 towers (towers 107 and 110) for the large transmission line and 25 towers or electric posts for the three small transmission lines which are located within the first seven panels of the proposed longwall panels. All of them will be affected by the surface subsidence process associated with the mining of the longwall panels. In this section, the preliminary assessments of subsidence influence of the two big towers (towers 107 and 110), one of the 25 small towers or electric posts and the stream are made. Panel layout and locations of towers and stream are shown in Figure 2.3.4. The method of subsidence prediction, assessment of subsidence effect and development of mitigation measures (control) follow those illustrated in Chapter 14 "Surface Subsidence." So separate detailed procedures are not discussed here in this section.

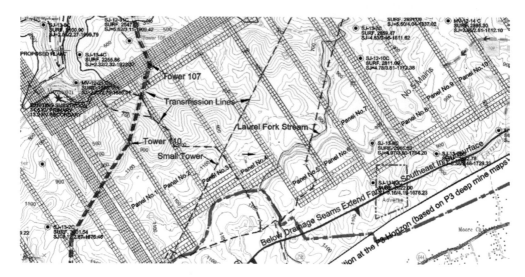

Figure 2.3.4 Overview of panel layout and locations of the power transmission line towers and stream

CISPM (comprehensive and integrated subsidence prediction model) is used for subsidence prediction (Peng and Luo, 1992).

A. SUBSIDENCE PREDICTION OF TOWER 110

Tower 110 is located on the chain pillars between the proposed panel 1 and 2. It is oriented at an angle of about 27° from the panel longitudinal direction. The power transmission tower is a steel lattice structure. Its height is about 100 ft (30.5 m), and its base is about 25 × 25 ft (7.6 × 7.6 m).

(1) Differential surface movements at the tower legs   The vertical and horizontal components of the surface movement at the four tower legs for the 10 prediction stages are predicted. For the high tower, only the differential movements among the legs could disturb the tower. The differential movement is the largest when the face in longwall panel 2 passes the tower a distance of 200 ft (61 m) at the assumed face advance rate of 45 ft d^{-1} (13.7m d^{-1}). Therefore, the time period when the face in panel 2 passes by the tower within this distance is the most critical one for the tower. Close attention should be exercised during that time period.

(2) Assessment of subsidence influence on tower 110   The effects of subsidence on power transmission tower 110 due to mining of longwall panels 1 and 2 are assessed based on the predicted final and dynamic surface movement and deformation indices at the four legs of the tower 110.

The high-altitude bird's eye view show that the height of the tower is much larger than the dimension of its base (Fig. 2.3.5). So the possible effect of subsidence-induced slope on

Figure 2.3.5 Satellite view of surface topography around Towers 110 and 107

tower 110 is the reduction of its safety factor. The maximum subsidence-induced slope at the bases of the tower is about 0.6%, reducing the stability safety factor by about 1%. Therefore, the subsidence process will not affect the stability of tower 110 to any significant level.

The predicted maximum differential subsidence of 0.238 ft (72 mm) and maximum differential resultant horizontal displacement of 0.372 ft (113 mm) also indicate the possibility of disturbance to the tower structure.

(3) Subsidence prediction of Tower 107   Same assessment procedures as illustrated previously for Tower 110 is applied to Tower 107.

The predicted maximum differential subsidence of 0.135 ft (41 mm) and maximum differential resultant horizontal displacement of 0.1 ft (30 mm) indicate that the tower structure might be disturbed too.

B. SUBSIDENCE PREDICTION OF STREAM

One stream is subjected to ground movements in the proposed mining area. The stream enters panel 5 from the west side, flows from southwest to northeast, and reaches the recovery room of panel 9. It finally exits the longwall panel over the recovery line of panel 9 (Fig. 2.3.6). The depth of mining below the stream is about 700 ft (213 m). Surface subsidence changes surface topography, thereby affecting surface water flow.

The final surface subsidence profile along the stream line is predicted when all 10 panels have been mined out. The maximum final subsidence along the stream is about 2.34 ft (0.71 m). The length of the stream where its final surface subsidence is more than 2 ft (0.61 m) is 4350 ft (1326 m) or 41% of the total length of the stream. There are five locations of large surface subsidence: (1) between points B and C, (2) around point F, (3) around point I, (4) between points M and N, and (5) near the middle point between points P and Q. Most of those areas are located in the central parts of the panels.

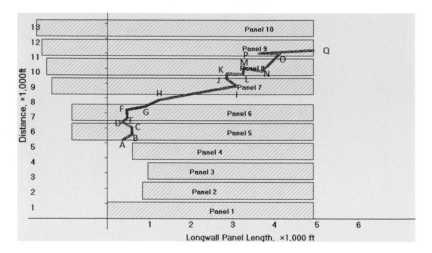

Figure 2.3.6 Longwall panels map showing the stream enters panel 5, flows through panels 6–9 and exits from panel 9

The maximum horizontal displacement, 0.777 ft (237 mm), is located near the headentry of panel 5 between points A and B.

The distribution of the slope profile is very similar to those of the surface horizontal displacement. The maximum surface slope, about 1%, is located near the headentry of panel 5.

For the predicted final surface strain profile, there are two peak tension and two peak compression zones located in the panel 5. The maximum tensile strain, 4.9×10^{-3} ft/ft, is located near point A. The second highest tensile strain, about 4.47×10^{-3} ft/ft, is located between G and H. The maximum compressive strain, 5.05×10^{-3} ft/ft, is located between points A and B.

The distribution of the curvature profile is similar to those of final strain. The maximum convex and concave curvatures located between points A and B are about 6.33×10^{-5} 1/ft and 6.5×10^{-5} 1/ft, respectively.

(1) Assessment of subsidence influence   Generally speaking, the height of the mining-induced water-flowing fractured zone is no more than 50–60 times the mining height. In this case, the overburden depth over the proposed longwall panels ranges from 650 to 700 ft (198–213 m) along the stream and the mining height is 6.5 ft (1.98 m). So, the overburden thickness is 100 to 108 times the mining height and the river water will not leak into the mining panels.

Cracks will occur on the surface soil layers, if the strain, especially the tensile strain, reaches 0.012 ft (Peng, 2008). In this case, the maximum tensile strain along the profile is 0.0049 ft ft. So, the cracks will not occur along the stream after mining, and the water will not leak into underground strata.

It should be noted that there are five sections of the stream (from points B to D, from points F to H, middle part of section between points N and O, from points I to J and second half of section between points P and Q (Fig. 2.3.6) where surface subsidence at the upstream of the sections are larger than that at the downstream of the sections. The differential surface subsidence for the five sections are 1.844, 2.081, 1.881, 1.86 and 2.11 ft (0.56, 0.63, 0.57,

0.57, and 0.64 m), respectively. If subsidence-induced slope is larger than the natural slope of the stream, lowering of the land surface upstream might cause the water body to pond, reducing the effective water flow.

C. SUBSIDENCE PREDICTION FOR ELECTRIC POSTS

A transmission line with 17 electric posts is located in the first three panels. Most of the 17 electric posts are located, by panel layout design, on or near the pillars, or on the edges of longwall panels, where ground subsidence process will not affect the stability of those posts significantly. The surface subsidence of 6 of the 17 electric posts located in the central part of panel 3 is assessed (Fig. 2.3.7). Post 1 is about 585 ft (178 m) from the setup room of panel 3. The base of the posts is about 1000 ft (305 m) above the coal seam.

The predicted surface movement and deformation indices after mining longwall panels 3 and 4 show that all types of subsidence indices are minor within the tolerance of the posts due mainly to thick overburden or large mining depth.

D. SUMMARY OF PREDICTED SURFACE SUBSIDENCE ON SURFACE STRUCTURES

(1) Surface subsidence prediction for towers   Comparison of the predicted dynamic and final surface movement and deformation indices at the base of towers caused by mining of proposed longwall panels indicates that the stability of the towers will not be affected by the ground subsidence process to any significant level.

The surface strain and differential movements at the bases of Tower 110 and small towers when the faces are passing by them might cause some problems to the tower structure if no precautionary mitigation measures are taken.

Figure 2.3.7 Surface topography around transmission line posts

After longwall panels 2 (for tower 110) and 4 (for small towers) have been mined, there should be no problems to the towers.

(2) Surface subsidence prediction for the stream   Generally speaking, the height of the mining-induced water-flowing fractured zone is no more than 50–60 times the mining height. The overburden depth along the stream over the proposed longwall panels far exceeds the height of the mining-induced water-flowing fractured zone. So the river will not leak into the mining panels.

Cracks will not occur along the stream after mining. The water will not leak into underground strata.

There are some sections where the surface subsidence at the upstream sections is larger than that at the downstream ones. Lowering of the land surface might cause the river water to ponding.

2.3.2.5 Design of panel width, gateroad chain pillars, and roof bolting

Several methods are available for chain pillar design. Among them, empirical and computer modeling are the most popular. The procedures and techniques of computer modeling for design panel width and chain pillar size have been addressed in detail elsewhere (Peng, 2008) and they are beyond the scope of this book. So they will not be discussed here.

The design of a roof-bolting system involves the determination of bolt type, bolt diameter, bolt length, and installation pattern. For the primary support, the scope of the installation pattern is restricted, because roof bolts installed at 4 × 4 ft (1.2 × 1.2 m), or 4 × 5 ft (1.2 × 1.5 m), or occasionally 5 × 5 ft (1.5 × 1.5 m) has been the rules in the US underground coal mines for the past few decades. Therefore, the design of a roof-bolting system is reduced to the selection of bolt type, bolt diameter, and bolt length. Operational experience in the past two decades has demonstrated that roof bolting design using only the aforementioned three parameters can be very effective.

Several techniques can be used to design the roof bolting parameters, including numerical modeling, theoretical calculation, and empirical method (Peng, 2008). The empirical method derived from experience may not be applicable for this case because there is no underground development yet. In addition, the theoretical calculation is not sufficient to analyze the complex interaction between the roof strata and the bolts. Therefore, the numerical modeling is recommended (Peng, 2008, Chapter 12).

Chapter 3

Strata mechanics

3.1 Introduction

As the coal in a longwall panel is being extracted by the shearer slice by slice, the surrounding strata are forced to move toward and attempt to fill the voids left by the extracted coal. This process induces a series of activities which include (1) movements of the rock strata between the roofline and the surface, resulting in surface subsidence; (2) abutment pressures on both sides of the panel, in front of the faceline, and in the bleeder end of the panel; and (3) roof-to-floor convergence in the gateroads and face area. All of these activities are discussed separately in this chapter except surface subsidence, which is discussed in Chapter 14.

3.2 Overburden movement

When a longwall panel of sufficient width and length is excavated, the overburden roof strata are disturbed in order of severity from the immediate roof toward the surface. Figure 3.2.1 shows the four zones of disturbance in the overburden strata in response to the longwall mining. The caved zone, which is the immediate roof before it caves, ranges in thickness from two to eight times the height of extraction (or mining height). In this zone, the strata fall on the mine floor and, in the process, are broken into irregular but platy shapes of various sizes. They are crowded in a random manner. Thus, the rock volume in its broken state is considerably larger than that of the original intact strata. The volume ratio of broken strata to its original intact strata is called the expansion ratio or bulking factor. Expansion ratio is a very important factor because it determines the height of the caved zone. This is illustrated in Section 3.2.1.2. There are various estimates of expansion ratios for various rock types.

Above the caved zone is the fractured zone. In this zone, the strata are broken into blocks by vertical and/or subvertical fractures and horizontal cracks due to bed separation. The adjacent blocks in each broken stratum still maintain contact either fully or partially across the vertical or subvertical fractures. Thus, there is a horizontal force that is transmitted through and remains in these strata. With this horizontal force, the individual blocks in these broken strata cannot move freely without affecting the movements of the adjacent blocks. These broken strata are called the force-transmitting beams. The thickness of the fractured zone ranges from 28 to 52 times the mining height (or its upper limit is 30 to 60 times the mining height above the roofline). The combined thickness of the caved and fractured zones ranges from 30 to 60 times the mining height (Dahl and Von Schonfeldt, 1976).

Between the fractured zone and the surface is the continuous deformation zone. In this zone, the strata deform gently without causing any major cracks that extend long enough to

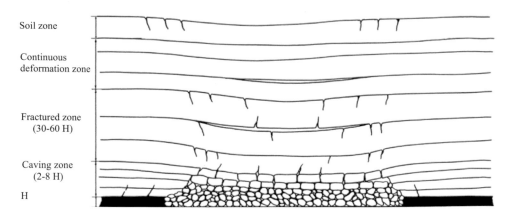

Figure 3.2.1 Overburden movement resulting from longwall mining

cut through the thickness of the strata, as in the fractured zone. Therefore, the strata behave essentially like a continuous or intact medium.

On the surface, there is a soil zone of varying depth depending on the location. In this zone, cracks are opened and closed as the longwall face comes and goes. In general, cracks on and near the panel edges tend to remain open permanently, whereas those in and around the center of the panel will close back up when the longwall face has passed by. Cracks vary from less than one to 3–4 ft (0.3 to 0.91–1.22 m) wide and from less than 1 ft (0.3 m) deep to as deep as the soil zone.

The formation of separate zones in the overburden has been reproduced in laboratory experiments (Kuzyniazouv, 1954). When the fall height of the falling strata is more than 2.0–2.5 times the thickness of the falling strata, the strata will fall rapidly and rotate during falling, resulting in irregular rock fragments on the floor (Fig. 3.2.2 C). When the fall height is between 1 and 2.0–2.5 times the thickness of the falling strata, the falling broken strata will lose contact with the surrounding strata, thereby losing the horizontal force (Fig. 3.2.2 B). If the thickness of immediate roof is less than the falling height, the falling strata will break at both ends and fall without losing horizontal contacts between the adjacent blocks, forming the force-transmitting beams in the lower portion of the fracture zone height (Fig. 3.2.2 A).

Thus, if the immediate roof is thinly laminated or bedded and it separates and falls along those laminations or bedding planes, satisfying the condition illustrated in Figure 3.2.2 C, the caved zone is likely to grow thicker. As the caved zone grows upward, the gap between the top of the rock piles on the floor and the lowest unbroken stratum in the roof becomes smaller, making it similar to the condition shown in Figure 3.2.2 B which simulates the lower portion of the fractured zone. Similarly, above this zone, the conditions for Figure 3.2.2 A are satisfied and result in the upper portion of the fractured zone.

Movement of the four zones discussed earlier has different degrees of effect on roof control at the longwall face. The effect decreases as the strata are located farther upward from the roof line. Those strata, the movement of which will affect roof control at the longwall faces, can be classified into two types, immediate and main roofs. The *immediate roof* is that portion of the overburden strata lying immediately above the coal seam top, approximately two to eight times that of the mining height (Fig. 3.2.3). Above the immediate roof, the strata in the

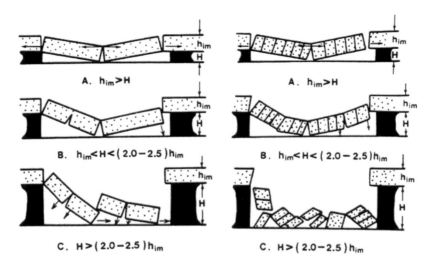

Figure 3.2.2 Characteristics of roof caving relative to falling height
Source: Modified from Kuzyniazouv, 1954

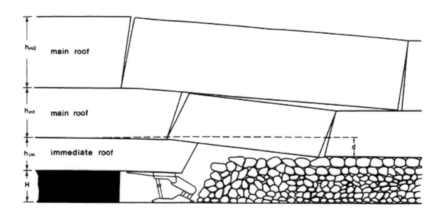

Figure 3.2.3 Immediate and main roofs

lower portion of the fractured zone are called the *main roof*. The definition of immediate and main roofs in this case refers to their location with respect to the coal seam top or roof line, as generally perceived. In other words, the immediate roof caves and breaks apart as soon as the shield advances, whereas the main roof overhangs and breaks periodically at certain interval. This definition also complies with the general concept that immediate roof is weak and caves immediately or soon after the shield advances, while main roof is hard rock and tend to overhang for some distance in the gob. Everything being equal, their severity of impact on roof control at the longwall face is proportional to their distance from the coal seam top or roof line.

3.2.1 Immediate roof

If the strata in the immediate roof will cave and fall in the gob following the advance of shield support, and after having fallen, they are broken up and cannot transmit the horizontal force along the direction of mining, then the shield support needs only to support that portion of the weight of the roof strata covered by the shield canopy area as soon as they are undermined. Conversely, if the strata immediately above the coal seam roof will not cave and fall in the gob following the advance of shield support, then the rock weight that needs to be supported by the shield support covers not only the canopy area but also the overhung portion of the strata. Therefore, the strata immediately above the coal seam roof is the key to roof control. Its rock type and thickness or density of lamination and its bonding strength in this part of the roof are the major factors governing the selection of shield capacity and various roof control techniques.

3.2.1.1 Classification and cavability of the immediate roof

There are many ways of classifying the roof for purpose of powered support evaluation for longwall mining (Kidybinski, 1977). From the production point of view, the strata immediately above the coal seam top can be divided into three groups, that is, unstable, medium stable, and stable (Peng and Chiang, 1982). When modern longwall mining was first introduced to US coal industry in the 1970s and up until late 1980s, all three types of roofs were encountered. The stable roof caused the most problems then as the powered support capacity was low and it was avoided as much as possible. Nowadays, most roofs are in between unstable and medium stable ones.

A   Unstable immediate roof (Fig. 3.2.4) – an unstable roof has the following features:

1   It consists of clay, clay shale, soft carbonaceous shale, slickensided shale, and well-jointed or fractured sandy shale.
2   Right after the shearer's cutting, if the support is not advanced immediately, the unsupported roof between the faceline and the canopy tip will fall in a short period of time (less than 5–10 minutes). Head coal may be required to stabilize the roof.
3   The roof in the gob caves immediately after support advance, i.e., there is no overhang. The caved fragments are small and the gob is well compacted.
4   Since the canopy length is generally four to six times the drum cutting width (or web), the roof, before its passage over the rear edge of the shield supports and onto the floor of the gob, is subjected to cyclic loading and unloading by the shield supports four to six times. This loading and unloading tends to break the roof, induce cavities, and move the caving line forward, resulting in the loading at the front half of the canopy being larger than at the latter half (Fig. 3.2.4).

B   Medium stable immediate roof (Fig. 3.2.5) – a medium stable roof has the following features:

1   It consists of hard or firm shale, sandy shale, and weak sandstone (e.g., fossilized or laminated sandstone). Joint and fractures are not well defined.
2   Under normal conditions, the exposed roof area and time duration without roof fall allow coal cutting in a single direction, that is, shields are not advanced during the shearer's cutting trip. Instead they are advanced during the return trip.

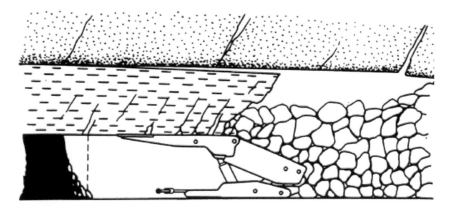

Figure 3.2.4 Unstable immediate roof

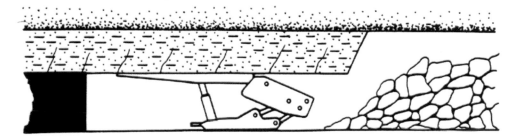

Figure 3.2.5 Medium stable immediate roof

3 Generally the roof caves in either immediately or shortly after shields have been advanced. The caved fragments are larger in size. The caving line is located near the rear edge of the canopy or sometimes extends a little bit into the gob (Fig. 3.2.5).

4 Canopy loading is distributed more uniformly between the front and rear parts of the shield support.

5 Cyclic loading and unloading by the shield support will also affect the roof stability if the shield is not advanced rapidly and continuously. If the face sloughs severely and the type of shield selected is not suitable, an inherently medium stable roof may become an unstable one.

C Stable immediate roof – there are three cases of stable immediate roof:

1 The immediate roof is sandy shale or sandstone which is thick and strong. It can be left unsupported for more than five to eight hours and still remain intact; cyclic loading and unloading by the shield support does not affect it; the resultant force acts toward the rear edge of the canopy; the roof cantilevers into the gob; and the roof in the gob falls periodically in large pieces. But if they fall near the rear end of

the shield support, the impact force might cause damages to the components of the shield support. Massive (thick) firm shale also belongs to this category.

2   The immediate roof is hard sandstone or conglomerate, which is very strong and massive. The roof will not only overhang in a large area, sometimes reaching 70,000–80,000 ft² (6503–7432 m²), but also will remain overhung for a long period of time (Fig. 3.2.6). During the non-weighting period, heavy roof pressure may not be obvious. But, once the periodic weighting period arrives, the roof over the gob acts vigorously en masse and produces an extremely large impact loading whenever caving occurs. Such impact loading occurs in a very short time and requires a very large capacity yield valve to relieve the pressure. In addition, the sudden caving of a very large area generates a stormy wind (or wind blast) in the face area damaging equipment and causing injuries/fatalities. The induced strong shock waves may be felt miles away. For this type of roof, it is necessary to perform properly induced caving, in terms of both time and operation.

3   The immediate roof is hard limestone or sandstone or firm sandy shale and thicker than the mining height. Joints and fractures are well developed and become the weak planes along which the rock breaks. As the face advances, it gradually sags and breaks in blocks. Further sagging causes the blocks to form a semi-arch, with the gob end eventually rested on the gob floor. Thus, during the non-weighting period the roof pressure may not be obvious. But, as the width of the semi-arch grows, it will break at sufficient length, creating periodic weighting, if certain conditions are satisfied. But in general, this type of roof strata does not cause roof stability problems.

Both operational experience and research results have demonstrated that roof stability is relative. For an unstable roof, certain techniques are required to control and change the factors contributing to the unstable conditions and to upgrade its stability. For a medium stable roof, a powered support is required, properly selected and applied, to maintain roof stability. Very often, a medium stable roof becomes an unstable one due to periodic weighting, poor roof conditions, or improperly selected powered supports. For the stable roof, with the exception of the gradually sagging one, its stability must be destroyed periodically and systematically by artificial means to avoid a sudden large area caving.

3.2.1.2 Caving height or thickness of the immediate roof

The thickness of the immediate roof that caves immediately or soon after shield advance is the basis for designing roof control techniques. The immediate roof is not necessarily the

Figure 3.2.6 Stable immediate roof

same throughout for a coal seam. Rather, it varies with stratigraphic sequence and method of mining.

Normally, caving initiates from the lowest strata in the roof immediately above the coal seam top and propagates upward into the fractured zone. The process of caving in each stratum is that the stratum sags downward as soon as it is undermined. When the downward sagging of the stratum exceeds the maximum allowable limit, it breaks and falls. As it falls, its volume increases; therefore, the gap between the top of the rock piles on the floor of the gob and the sagged but lowest un-caved stratum in the roof continues to decrease as the caving propagates upward. When the gap vanishes (i.e., the lowest un-caved roof strata have received the support of the caved rock piles), the caving stops. Thus the height of caving must satisfy the following condition (Fig. 3.2.3):

$$H - d = h_{im}(K - 1) \tag{3.2.1}$$

$$\text{and } d \le d_0 \tag{3.2.2}$$

Where H = mining height, d = sagging of the lowest un-caved strata, d_0 = maximum allowable sagging (without breaking and falling down) of the lowest un-caved strata, h_{im} = thickness of the roof strata that cave or caving height, and K = bulking factor of the roof strata that cave.

From Equation (3.2.1), the caving height can be determined by:

$$h_{im} = (H - d) / (K - 1) \tag{3.2.3}$$

It must be noted that d and K must be determined at the same location. The ideal place is within the area beginning from the point when the lowest sagging but un-caved stratum first makes contact with the rock piles to the point when the rock piles have been compressed to the final stage. Obviously, for safety reason, this is extremely difficult to measure.

From Equation (3.2.3), it can be seen that there are several factors that affect the caving height:

1   Mining height, H. Based on Equation (3.2.3), the caving height is proportional to the mining height. This relation can be used to control the caving height. But it must be noted that the caving height will not change immediately or continuously as soon as the mining height is changed. Rather, the caving height probably changes in leap steps in response to a sudden change in mining height.

2   Maximum allowable sagging of the lowest un-caved strata, d_0. Based on Equation (3.2.3), $d \le d_0$ has a profound effect on the height of caving. For instance, if $d = d_0 = H$, then $h_{im} = 0$, which means no caving; that is, the roof sags gradually until it touches the floor. On the other hand, if $d = d_0 = 0$, Equation (3.2.3) becomes:

$$h_{im} = H / (K - 1) \tag{3.2.4}$$

Equation (3.2.4) assumes that the strata break and fall without any sagging. In this case, it predicts the largest caving height.

Based on field studies (Song and Deng, 1982), the actual sagging, d, is related to the mining height, H, by:

$$d = cH \tag{3.2.5}$$

Where c is the ratio of the actual sagging of the strata before caving to the mining height. For very strong sandstone, $c = 0.1$–0.15; for medium and fine sandstone, $c = 0.15$–0.25; for sandy shale, $c = 0.35$–0.40; for shale and marl, $c = 0.40$–0.50; for well-jointed limestone, $c^3 1$.

3   Bulking factor, K. Bulking factor is defined as the ratio of the volume of the broken rock strata to the original volume of the same strata before they are broken and cave. Since the volume of the broken strata is always larger than that of the original intact strata, bulking factor is normally larger than one.

It must be noted, however, that the bulking factor varies with rock type, shape, and size of the caved rock fragments, the ways in which the caved rock fragments are piled up, and the pressure imposed on the rock fragments.

The size of the caved fragments and the uniformity of arrangement of the fragments are closely related to the caving height. When the caving height is large and the caved fragments are small, their arrangement is very disorderly, resulting in a larger bulking factor. Conversely, if the caving height is small and the caved fragments are large, their arrangement is very orderly and uniform, resulting in a smaller bulking factor. Therefore, different strata in the overburden result in different caving characteristics.

If the caved fragments have an orderly arrangement, the bulking factor is smaller; conversely, if the fragments have a disorderly arrangement, the bulking factor will be larger. For caved fragments with an orderly arrangement, the bulking factor of the stronger and harder rock will be smaller, because it will result in larger fragments; conversely, the weaker and softer rock will result in smaller fragments and, consequently, a larger bulking factor. For caved fragments with a disorderly arrangement, the bulking factor of the stronger and harder rocks will be larger, because they will result in larger fragments; conversely, the weaker and softer rock will result in smaller fragments and, consequently, a smaller bulking factor.

The caved fragments directly behind the supports usually have a higher bulking factor, because of lower height of the rock piles and less or no weight from the overlying strata. This is the original bulking factor, K_o. But as they are left further behind into the gob, the caved fragments continue to pile on top of them. As a result, the volume continues to decrease. At the place when the rock piles begin to make contact with the lowest un-caved stratum that sags down, the bulking factor is called semi-residual bulking factor, K_{ro}. As the rock piles move further deep into the gob, the weight of the overburden eventually acts on them. At this time the volume reduces to the minimum, that is, they are very well compacted. The bulking factor from this location onward is the residual bulking factor, K_r. Table 3.2.1 lists the original and residual bulking factors for various coal measure rocks.

Table 3.2.1 Bulking factors

Rock Type	Bulking Factor	
	Original, K_o	Residual, K_r
Sand	1.06–1.15	1.01–1.03
Clay	<1.20	1.03–1.07
Broken coal	<1.30	1.05
Clay shale	1.40	1.10
Sandy shale	1.60–1.80	1.25–1.35
Sandstone	1.50–1.80	1.30–1.35

Fayol (1913) made several laboratory studies and showed that, regardless of rock types, the bulking factor varies with particle size and applied pressure. In general, the bulking factor increases with particle size, while it decreases with increase in applied pressure.

Peng (1980) measured the volumes of underground roof fall cavities in the entries developed in the Pittsburgh seam and compared them with the volumes of the rock fragments piled up on the floor. He found that the bulking factor for the roof shale ranges from 1.25 to 1.30 with an average of 1.28.

4   Methods of roof control in the gob area. Different techniques for controlling roof caving in the gob area can change the height of caving. For instance, if the gob is backfilled with waste rocks or tailing sands, the same effect is achieved as reducing the mining height. Thus Equation (3.2.3) becomes:

$$h_{im} = \frac{(H - h_f) - d}{K - 1} \tag{3.2.6}$$

where h_f is the height of backfilling material and K is the bulking factor of the immediate roof. Obviously, when the waste materials are backfilled to full mining height, i.e., when $H = h_f$, $h_{im} = 0$, and there will be no caving. Due to various reasons (economic, safety, and environmental), backfilling of the gob of any types is not employed in US coal industry. Artificially induced caving is not used either.

3.2.2 Main roof

Above the immediate roof, the strong and stiff strata in the lower portion of the fractured zone are called the *main roof*. The main roof generally refers to the strata located in the lower portion of the fractured zone that are broken but un-caved and can still transmit horizontal forces, although the rear end of the strata is generally lower than the front end. Note the rear end of the strata is in the gob area, while the front end is located above the shield support and the immediate roof. The main roof may either be one or more layers depending on the stratigraphic characteristics and the gap of strata separation in that area. Its movements will greatly affect the stability of the immediate roof and thus the supports in the face area. Above the main roof, the strata movements are too far away to have a significant impact on the face area unless the face stops and parks there for some length of time. The main roof is generally believed to break periodically along the direction of face advance and imposes periodic roof weightings on the face area.

The location and thickness of the main roof can be determined by examining the stratigraphic columns above the coal seam. The method involves the determination of number, location, and thickness of the force-transmitting strata beams in the fractured zone. The guidelines are as follows:

1   Strata separation will occur along the bedding planes and laminations, because these are the weakest planes other than open fractures.
2   Strata separation and downward sagging occur first at the lowest stratum in the fractured zone and propagate upward.
3   The delay time for separation and downward sagging between adjacent strata depends primarily on the thickness and strength of the strata. If the upper strata are strong and thick, their movement is delayed far behind that of the lower strata. Conversely, if the upper strata are weak and thin, they will move simultaneously with the lower strata.

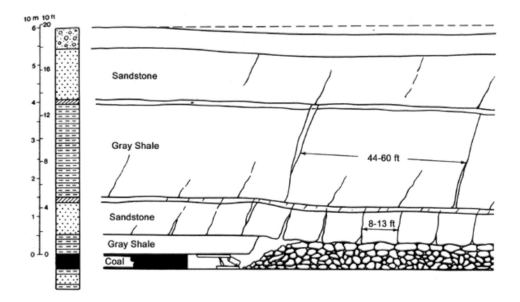

Figure 3.2.7 Immediate and main roof behavior in the Lower Kittanning seam

Figure 3.2.7 shows an actual case study in the Lower Kittanning seam. The immediate roof was gray shale, 2 ft (0.61 m) thick. The main roof consisted of two force-transmitting strata beams. The lower one was the 18 ft (5 m) thick sandstone plus the coal partings immediately above it, while the upper one was the 80 ft (24 m) thick gray shale plus the coal partings above it. The sandstone stratum above the 80 ft (24 m) thick gray shale was broken but retained to form a force-transmitting strata beam. However, it was so far from the immediate roof that its movement would not cause any significant effect on the movement of the immediate roof.

Sometimes, the lower portion of the main roof is easily mistaken for the immediate roof, especially when the immediate roof is rather thick. The best criterion to differentiate one from the other is whether or not it has the features of the periodic breakage and produces the periodic weighting pressure in the face area.

A thick hard rock stratum or combination of strata will overhang when it loses support underneath. Whether its failure will affect the powered support loading at and stability of the face depends on its thickness and more importantly the distance above the coal seam top. Its distance may be too far from the seam top to have effect on the panel being mined, it will affect subsequent panels cyclically in a district (see Section 10.12, p. 361). It must also be pointed out that a very thick firm but weak stratum could induce the same effect. In other words, stiffness, not only the hardness, of the rock strata beam counts. Stiffness considers hardness and thickness of the beam.

3.2.3 Sequences of overburden movements in a longwall panel

As longwall mining moves along the direction of mining in a longwall panel, there are two distinctive phases of overburden movements. The first phase of movement includes the distance from the setup entry to the point when the main roof first begins to break or, if the immediate roof does not cave right after support advance, the final stage of an interval that begins with a large-area caving of the immediate roof and lasts until the complete breakage of the upper strata in the main roof. During such interval, the maximum roof pressure measured at the face area is called the *first weighting* (Fig. 3.2.8). However, if additional pressures due to the breakage of the main roof strata are not significant, the first weighting refers to the roof pressure immediately before and during the large-area caving of the immediate roof. The distance from the setup entry to the first weighting is defined as the *first weighting interval, L_0*.

In unstable immediate roof, there is no first weighting, because the roof caves immediately following support advance.

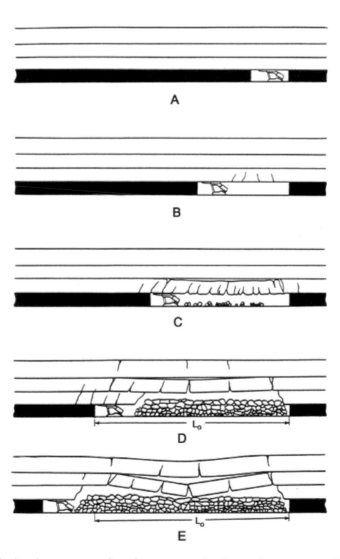

Figure 3.2.8 Idealized sequences of roof movements leading to first caving and first weighting

The second phase begins right after the first weighting and extends to the completion of the panel mining. During this period the roof pressure at the face area increases and decreases cyclically due to the cyclical breakage of the immediate roof or the main roof or both. This phenomenon is called the *periodic roof weighting* (Fig. 3.2.9). The maximum pressure occurring in each period is the *periodic roof weighting pressure*, and the distance between two consecutive roof weightings is called the *periodic roof weighting interval, L_p*. The severity of first and periodic weighting depends very much on mining and geological conditions.

When the immediate roof is weak and thick, the first and second phases may not show significant difference. But, in general, the first weighting pressure and interval are larger than the periodic roof weighting pressure and interval, respectively, unless the roof stratigraphic sequence changes to thicker and harder roof well inside the panel.

The first and periodic weighting were visibly identifiable when manual setting of powered supports were used in the 1970s and 1980s. However, when automatic setting by

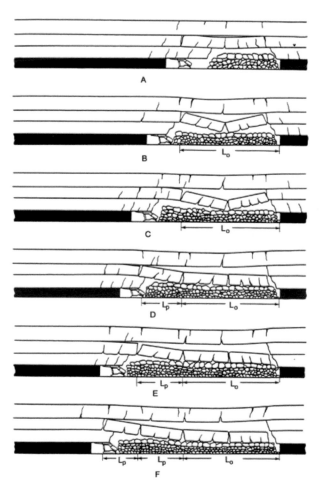

Figure 3.2.9 Idealized sequence of roof caving illustrating first caving and periodic caving

electrohydraulic control gained popularity and support capacity increased in the early 1990s, weighting became unidentifiable, although shield leg pressure monitoring still showed cyclic loading phenomena, though much weaker.

3.2.3.1 First weighting

Before longwall mining begins in a panel, a setup room of approximately 24–32 ft (7.3–9.8 m) wide is driven across the face at the starting end (or bleeders side) of the panel. The face equipment, including shields, shearer, and chain conveyor, is installed in the setup room (Fig. 3.2.8 A). In the initial period, as the mining proceeds, the roof remains intact and the gob is left open (Fig. 3.2.8 B). As the width of the gob widens, the immediate roof begins to bend and sag. Eventually, it separates from the main roof along its bedding plane of contact when the gob span increases further. At this time, the immediate roof behaves like a fixed-ended beam (or more exactly a thin plate anchored firmly at four edges); cracks begin to form on the upper surface at both ends of the beam (Fig. 3.2.8 C). The cracks rapidly propagate through the thickness of the beam. As this happens, the beam becomes a simple-ended one. The location of the maximum axial tensile stress immediately switches to the lower surface at the mid-span where the crack initiates and propagates through the thickness. After this, the immediate roof splits into two sections, both of which lose supports and fall. In the falling process, they rotate and break into different sizes and shapes upon landing on the floor (Fig. 3.2.8 D). The duration of this period depends on the stability of the roof immediately over the coal seam top, being from 0–20 ft (0–6.1 m) for the unstable roof to more than 300 ft (91 m) for the stable roof.

As the face moves farther toward the first weighting interval, L_0, the lower stratum in the main roof also separates from the upper one. Cracks form at both ends and the beam sags further until the mid-span touches the top of the rock piles resulting from the caving of the immediate roof (Fig. 3.2.8 E). The roof pressure reaches the maximum right before this happens. This is the *first weighting*. As the face moves beyond L_0, the cracks in the front section of the main roof beam cut through the thickness causing the beam to break away. As this happens, the roof pressure drops suddenly and the overburden movements enter into the second phase.

The first weighting interval can be estimated by the following (Fig. 3.2.8):

For the immediate roof:

$$L_{oim} = \sqrt{\frac{2h_{im}(T_{im})}{\gamma_{im}}} \tag{3.2.7}$$

For the main roof:

$$L_{om} = \sqrt{\frac{2h_m(T_m)}{\gamma_m}} \tag{3.2.8}$$

Where L_{oim} and L_{om} = first weighting interval for the immediate roof and main roof, respectively.

h_{im} and h_m = thickness of the immediate roof and main roof, respectively.

T_{im} and T_m = tensile strength of the immediate roof and main roof, respectively.

γ_{im} and γ_m = weight per unit volume of the immediate roof and main roof, respectively.

The larger of the two values from Equation (3.2.7) and Equation (3.2.8) should be used. But, in general, $L_{om} > L_{oim}$, as stated earlier.

3.2.3.2 Periodic weighting

After the first roof weighting (D and E in Fig. 3.2.8 and A and B in Fig. 3.2.9), the longwall face enters into the second phase of the overburden movement (Fig. 3.2.9 C–F). In this phase, the immediate roof, depending on rock types, caves immediately or with a little delay behind the shields. The lower stratum of the main roof breaks periodically causing periodic high roof pressure at the face area. The breaking length or interval such as L_{P1}, L_{P2}, and so on, varies with the strength, thickness, and joint conditions of the strata and the gap between the caved rock piles and the un-caved immediate or main roof, while the intensity of the periodic roof weighting depends on the breaking interval. On top of this, the upper stratum also breaks periodically, if sufficient gaps between broken and immediately overlying unbroken strata exist.

The sequences of overburden movements illustrated in Figures 3.2.8 and 3.2.9 are idealized based on certain stratigraphic sequence, for example, IIa and IIb in Figure 3.2.12. Actual movement sequence varies depending on the stratigraphic sequence.

As stated earlier, underground operational experience indicates that periodic weighting seems to become less intense or disappear as the faces advance faster and shield capacity becomes larger in high-production faces.

A. METHODS OF IDENTIFYING OR PREDICTING PERIODIC ROOF WEIGHTING.

There are three events in the variational characteristics of the resistances or pressure of the shield supports that can be used to identify/predict the periodic roof weighting (see Section 5.5 for a detailed description of the terminologies):

1   Significant increase of time-weighted average resistance. Both the final resistance and time-weighted average resistance (TWAR) of a support cycle rise greatly during periodic roof weighting. Since the final support resistance or pressure occurs almost instantaneously only when the neighboring support is being advanced, it is less desirable for identifying or predicting the onset of periodic roof weighting. On the other hand, the variation pattern for the time-weighted average resistance, which reflects resistance changes in a mining cycle, is more representative. Figure 3.2.10 shows the variations of TWAR as measured for the Pittsburgh seam (Peng et al., 1982) for a distance of 78 ft (24 m) face advance. The periodic weighting interval $L_{P1} = 14.7$ ft (4.5 m) consists of two parts, that is:

$$L_{P1} = a_1 + b_1 \tag{3.2.9}$$

Where $a_1 = 4.2$ ft (1.3 m) is the distance duration in which periodic roof weighting occurs, and $b_1 = 10.5$ ft (3.2 m) is the duration of relative equilibrium during which the roof pressures are relatively low.

The maximum TWAR occurring during periodic roof weighting is called maximum periodic weighting resistance, P_{tm}, which is related to the *periodic weighting intensity factor, I,* by

$$I = \frac{P_{tm}}{P_t} \tag{3.2.10}$$

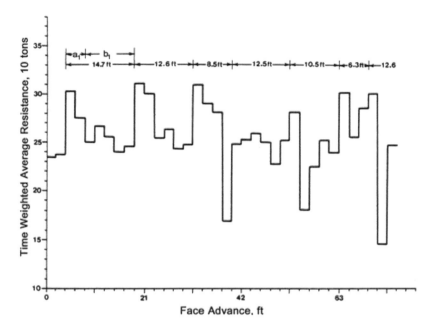

Figure 3.2.10 Variations of time-weighted-average-resistance (TWAR) measured in the Pittsburgh seam

Where P_t is the average TWAR during the nonperiodic roof weighting period immediately preceding the periodic roof weighting where P_{tm} is measured.

2   High setting resistance or rapidly rising rate of resistance increase. The rate of increase in support resistance after setting is one of the important markers in roof behavior. During the nonperiodic weighting period, the rate of resistance increase generally increases gradually after setting but becomes much larger immediately preceding support advance. Conversely, during periodic roof weighting, the rate of resistance increase immediately before support advance is relatively low but increases significantly faster right after setting (Fig. 3.2.11). This feature indicates that the roof moves vigorously during periodic weighting and that, immediately after setting, the supports rapidly receive the continuously increasing roof pressure. Therefore, the rate of resistance increase after support setting can be used to identify/predict the periodic roof weighting.

3   Features of periodic roof weighting. Two terms are most useful in describing the rules of changes in periodic roof weighting. One is the periodic weighting interval, L_p, which is the distance of face advance between two consecutive periodic weightings, and the other one is the time interval between them, the periodic weighting period, T_p.

The common feature concerning the main roof caving is that for every two to four smaller periodic roof weightings there is a larger periodic roof weighting. The smaller periodic roof weightings can be attributed to the deformation and caving of the lower portion of the main roof, and last for one or two mining cycles. On the other hand, the severe movements from the upper portion of the main roof caused the larger periodic roof

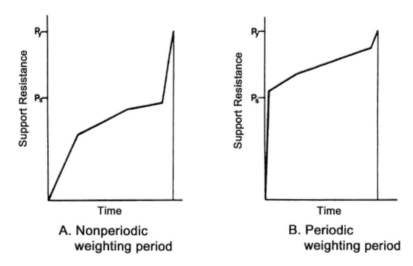

A. Nonperiodic
weighting period

B. Periodic
weighting period

Figure 3.2.11 Example patterns of the rate of resistance increase during non-periodic and periodic roof weighting

weightings, which last for three to seven mining cycles. Since there are differences in strata composition, and consequently strata behavior, the magnitude and characteristics of the roof weighting induced by the main roof are different from seam to seam or sometimes even from panel to panel in a coal seam.

The maximum periodic weighting resistance, P_{tm} is affected by the periodic weighting interval, L_p:

$$P_{im} = a + bL_p \qquad (3.2.11)$$

Where $a = 2.15$–2.41 is a constant depending on the properties of the main roof strata and rock pile conditions in the gob, and $b = 4.5$–17 is a constant related to the periodic weighting interval.

Equation (3.2.11) indicates that P_{im} is linearly increased with an increase in L_p. This is only valid within a certain limit of L_p. The relation will change once L_p exceeds this limit, due to the interaction between the fallen rock fragments and the broken main roof blocks, the conditions in which gob piles exist, and so on.

On the other hand, the relation between the periodic weighting intensity factor, I, and L_p is nonlinear:

$$I = \frac{L_p}{pL_p - q} \qquad (3.2.12)$$

Where $p = 0.80$–0.87 and $q = 0.43$–0.64 are constants related to overburden strata, and rock piles in the gob, and the average TWAR during non-weighting period. Equation (3.2.12) shows that, within a certain limit, I decreases as L_p increases, although the change is rather small.

3.2.4 Effects of stratigraphic sequences

It is well known that the stratigraphic sequence above a coal seam varies from seam to seam, mine to mine, panel to panel, and frequently even between different parts of the same panel. Differences in stratigraphic sequence will induce different degrees of overburden movement.

Review of numerous field studies in various coalfields in different parts of the world led to the conclusion that there are approximately five types of stratigraphic sequences (Fig. 3.2.12), each of which induces different types, and different degrees of severity, of strata movement, and, thus, of roof pressure at the longwall face area. Different intensity of roof pressure and roof activity requires different methods of roof control. Therefore, analysis of the stratigraphic sequence and its changes within the reserves of interest is the prerequisite for shield support design.

3.2.4.1 Type I stratigraphic sequence

In this type, the immediate roof is weak and very thick, usually more than four to five times the mining height. Within this range, it may be a homogeneous shale or soft sandy shale with

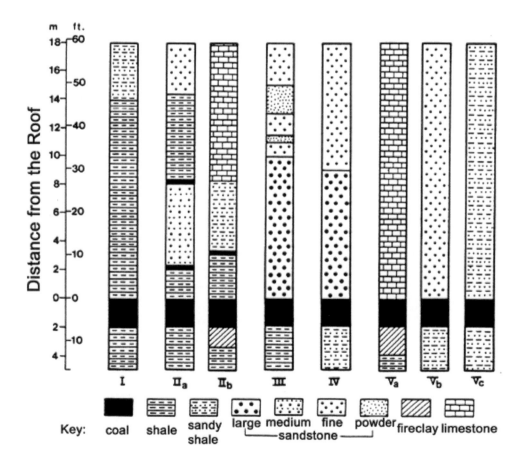

Figure 3.2.12 Five types of generalized stratigraphic sequence

or without laminations, or it may be thin-layered shale interbedded with thin streaks of coals, clay bands, and sandstones, or it may be thinly laminated coarse-grained sandstone due to preferred orientation of mica sheets or fossils.

This type of immediate roof caves immediately behind shield advance and fills up the whole gob space. Therefore, the main roof receives immediate support from the caved rock piles and maintains relative equilibrium all times. As a result, there will either be no periodic roof weighting, or, if any, a minor one that will not be identifiable without precise instrumentation.

Sometimes the immediate roof is not sufficiently thick, and the main roof, although hard and strong, is well jointed. Its sagging and resting on the gob piles will have little effect on the roof pressure at the face area.

3.2.4.2 Type II stratigraphic sequence

In this type, there is a weak immediate roof, but its thickness is less than four to five times the mining height. Consequently, the caved rock piles from the immediate roof cannot fill up the gob space, leaving a large gap between the top of the caved rock piles and the lowest un-caved stratum or strata in the main roof. Under such conditions, the main roof, lacking support, will move actively, the severity of which depends on the stiffness of the strata, and breaks periodically. This type of stratigraphic sequence will induce a clear but mild periodic weighting.

3.2.4.3 Type III stratigraphic sequence

The immediate roof in this type is hard and strong, i.e., a strong roof stratum of medium thickness rests directly on the coal top. It breaks only when it overhangs for a certain distance in the gob. Right before it breaks, the roof pressure at the face area is very high. Thus this type of stratigraphic sequence induces a clear and strong periodic weighting.

3.2.4.4 Type IV stratigraphic sequence

In this type, the roof immediately above the coal top is not only very hard and strong but also massive (i.e., homogenous and very thick). Conglomerates and thick fine-grained sandstones are typical examples of this type (Mills *et al.*, 2000). The roof tends to overhang in the gob for a very large area, sometime up to more than 107,600 ft^2 (10,000 m^2) before it caves. When it caves, it is accompanied by strong and stormy winds that may further cause roof stability problems and safety hazards to face crew. The shock waves may be felt as earthquake tremors miles away!

3.2.4.5 Type V stratigraphic sequence

In this type, the immediate roof is strong and thicker than the mining height. But it is either well jointed or not sufficiently stiff to overhang for a large distance. Thus it can sag gradually until it touches on the floor in the gob and form a semi-arch. The semi-arch will also break periodically. But it is usually not strong enough to be detected. This type of slow sagging roof is more likely to occur in thinner coal seams.

In classifying the aforementioned five types of stratigraphic sequence, only the roof strata are considered. However, it must be emphasized that floor rock property is also a critical factor for successful longwall mining. In high production faces, a strong, firm floor stratum is also required. In a soft floor, the shield base tends to sink into the floor. This causes difficulties for a shield to advance and slows down the speed of face advance. The degree of difficulty for shield advance depends on the portion of the support base that sinks and its depth of sinking. The worst roof and floor strata combination is a soft floor in Type IV, III, and II roof stratigraphic sequences in descending order.

It must also be noted that the aforementioned classification is based on strata strength and stiffness, not the commonly referred individual rock strength and stiffness. Shale as an individual rock is weaker than sandstone. But a very thick shale stratum is stiffer than a thin sandstone stratum and, thus, will overhang longer. Consequently, both strength and thickness (or spacing of laminations) of a rock stratum must be considered in the aforementioned classification.

3.2.5 Effects of time and longwall retreating rate

It is well-known that laboratory experiments show that rocks exhibit time-dependent behavior, and that the stronger the rock is the longer it lasts. This time-dependent behavior is more obvious in underground coal mines when in-situ strata are involved.

Extensive underground measurements of shield leg pressures in various longwall mines show that if a longwall stands idle over the weekend, the shield leg pressure will increase quickly within the first 24 hours and become stabilized thereafter (Peng *et al.*, 2019)

Shield leg pressure data also show that in the 1980s, when powered support capacity was relatively low, periodic weighting was easily identifiable because it exhibited roof deterioration. But in the 1990s, electrohydraulic control of shield ensured that the designed setting pressure is set, thereby ensuring a more uniform loading across the face than manual operations. Furthermore, when shield capacity became larger and the faces moved much faster, the intensity of periodic weighting was either reduced considerably or disappeared, because it was not clearly observable anymore.

3.3 Abutment pressures, gob caving, and gateroad convergence

3.3.1 Computer modeling

Conceptually, a softer coal seam is sandwiched between the relatively stronger roof and floor rocks and loaded by the weight of overburden. Stress is uniformly distributed in the coal seam under such conditions. When the panel gateroads are developed, the equilibrium conditions are destroyed due to the presence of openings. Stress distribution in the area has to be readjusted in order for a new state of equilibrium to be achieved. As a result, a de-stressed zone occurs in the roof of entries, and the load is transferred into the neighboring solid coal both in the panel and the pillars. Zones where the vertical pressure exceeds the average overburden pressure before mining are created in or near the edges of the panel and pillars. These zones are called the abutments, and the above-average pressures are the abutment pressures.

When longwall mining proceeds, abutment pressures will form around the edges of the gob and superimpose on those created during gateroad development. Figure 3.3.1 A shows the vertical stress distribution in the coal seam from a three-dimensional finite element modeling of multiple longwall panels using three-entry gateroad system.

The abutment pressure in front of the faceline is called the front abutment pressure: those along both sides of the panel in the gob area are the side abutment pressures. In the gob area,

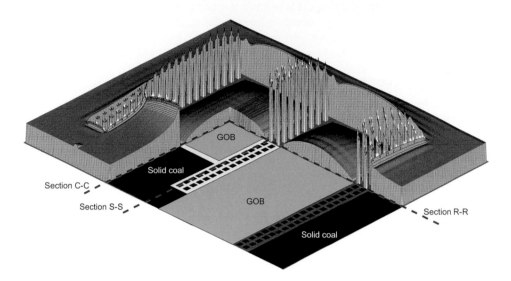

Figure 3.3.1A Three-dimensional view of vertical stress distribution in the coal seam

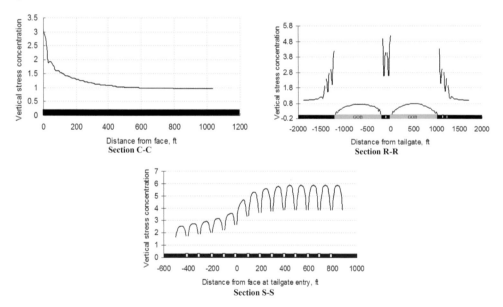

Figure 3.3.1B Vertical stress concentration at Sections C-C, R-R, and S-S

the maximum pressure realized is the overburden pressure, and thus, the existence of a rear abutment in the gob moving with the longwall face is unlikely except at the set up entry on the bleeder chain pillar side where a stationary abutment exists.

The front and side abutments intersect at the corners of the panel and superimpose on each other. Both the front and side abutment pressures decrease exponentially away from the edges of the panel and return to the overburden pressure some distance away (CC and RR sections in Fig. 3.3.1 B). The side abutment pressure near the ribs of the headgate and

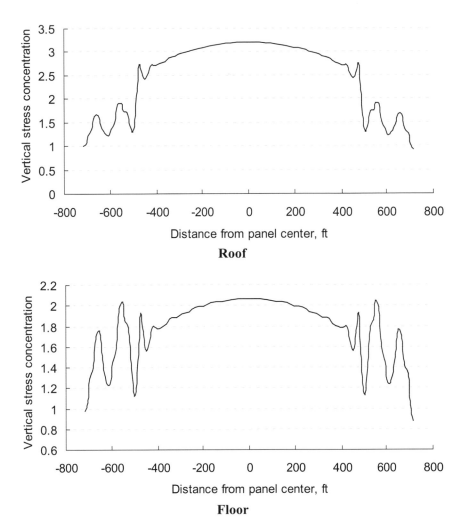

Figure 3.3.1C Vertical stress concentration in the immediate roof and floor

tailgate begins to increase when the face is some distance inby. It increases continuously and reaches the maximum value after the face has passed it. Thereafter it stabilizes, although in some cases yielding of pillars occurs (SS section in Fig 3.3.1 B).

The vertical stress distributions in the roof and floor are quite similar (Fig. 3.3.1 C) except the stress magnitude in the floor may be either smaller or larger, depending on the Young's modulus of the immediate roof.

3.3.2 Field measurements

Tables 3.3.1 and 3.3.2 summarize all underground ground control instrumentation conducted so far. Specifically, Table 3.3.1 covers those performed prior to 1982 (old data) and

Table 3.3.1 List of panels where measured ground control data are available (before 1982)

Panel No.	Mine	Location	Seam Name	Panel Thickness (ft)	Panel Name	Panel Width (ft)	Panel Length (ft)	Instrumentation Location[a] (ft)	Depth (ft)	Entry Width (ft)	Pillar Size (ft × ft)	Caving	Pillar Stress	Panel Stress	Convergence	Remark
1	VaBey Camp No3	West Virginia	Pittsburgh	5.25	2 left	150	2655	655–755	875	17	5[b]: 40 × 70			✓	✓	Shortwall (Peng and Park, 1977)
2A	Hendrix No 22	Kentucky	Elkhom #3	5	34	150	2000	990–1035	450	20	2: 30 × 55	✓	✓	✓		Shortwall (Wright et al.,1979)
2B	Hendrix No 22	Kentucky	Elkhorn #3	5	1.2	150	2000	1947–1988	450	20	2: 30 × 55		✓	✓	✓	Shortwall (Wright et al., 1979)
3A	Old Ben No.24	Illinois	Herrin #6	7.5	1	460	1735	770–780	620	20	2: 80 × 70	✓	✓	✓	✓	Longwall (Conroy, 1977)
3B	Old Ben No.24	Illinois	Herrin #6	7.5	1	460	1735	200	620	20	2: 80 × 70		✓	✓		Longwall (Conroy, 1977)
4	Old Ben No.24	Illinois	Herrin #6	7.5	2	462	1713	80	620	20	2: 80 × 70			✓	✓	Longwall (Conroy, 1979)
5A	York Canyon	New Mexico	York Canyon	7	4	560	1860		400					✓	✓	Longwall (Gentry, 1979)
5B	York Canyon	New Mexico	York Canyon	7	5	400	2100	965	400					✓	✓	Longwall (Gentry, 1979)
5C	York Canyon	New Mexico	York Canyon	7	7 left				460	18		✓		✓		Longwall (Lu, 1982)
6	York Canyon	West Virginia	Pittsburgh	7	2	400		80	920	18	2 42 × 100	✓	✓	✓		Longwall (Lu, 1982)
7	Olga No. 1	West Virginia	Pocahontas #4	4	9 left	360	4200	750	630	18	2: 40 × 75; 1: 125 × 75		✓		✓	Longwall (Peng, 1976)
8A	Quarto No. 4	Ohio	Pittsburgh	6.5	4 left			3270	520	15	1: 95 × 95; 1: 36 × 95		✓			Longwall (Acjarua, 1982)
8B	Quarto No. 4	Ohio	Pittsburgh	6.5	5 left	490	4100	300–400	610	15	1: 95 × 95; 1: 36 × 95		✓			Longwall (Acharya, 1982)
8C	Quarto No. 4	Ohio	Pittsburgh	6.5	5 left	490	4100	3000	760	15	1: 95 × 95; 1: 36 × 95		✓	✓		Longwall (Archarya, 1982)
9	Capco	Ohio	Pittsburgh	6	1	484	3100	480	570	17	2: 60 × 70		✓			Longwall (Roscoe and Hartshom, 1980)
10	Blacksville No.1	West Virginia	Pittsburgh	6					600			✓				Longwall (Davis and Krickovic, 1973)
11	Shoemaker	West Virginia	Pittsburgh	3.5					600			✓				Longwall (Dahl and Von Schonfeldt, 1976)

a Distance from setup entry
b Number of rows of chain pillars

Table 3.3.2 List of panels where measured ground control data are available (1985–2018)

Case #	Panel #	Location	Seam Name	Thickness, ft	panel Name	Width, ft	Location, ft	Instrumentation	Depth, ft	Entry width, ft	Pillar size rib-to-rib, ft	Caving stress	Pillar stress	Panel stress	Conver-gence	Remarks
U.S. mines																
1	12	WV	Pocahontas No. 3	7.5	12 right	1170	5000		700	16	99 × 121 and 84 × 259	✓				Gearhart et al. (2018)
2	3	WV	Lower Kittanning	8	2C	1200	1700		600	18	82 × 130	✓				Gearhart et al. (2017)
3	2	CO	DU	10	2	800	5200		2250	20	170 × 180	✓		✓		Larson and Whyatt (2012); NIOSH (2016)
4	1	CO	DU	11	1	800	6000		2250	20	73 × 180	✓		✓		Larson and Whyatt (2012); NIOSH (2016)
5	24	PA	Pittsburgh	7	E24	1500	x-cut 24		600	18	125 × 275 and 60 × 137	✓			✓	Su (2016)
6	1	OH	Pittsburgh	6.5	4 left	416	1460		450	16	110 × 110 and 40 × 110	✓			✓	Mark (1987)
7	2	OH	Pittsburgh	6.5	4 left	416	3875		520	16	110 × 110 and 40 × 110	✓			✓	Mark (1987)
8	3	OH	Pittsburgh	6.5	5 left	416	1350		610	16	110 × 110 and 40 × 110	✓			✓	Mark (1987)
9	3	OH	Pittsburgh	6.5	5 left	416	3300		760	16	110 × 110 and 40 × 110	✓			✓	Mark (1987)
10	–	KY	Harlan	11	–	500	500		630	18	90 × 90 and 50 × 90	✓			✓	Schuerger (1985)
11	–	KY	Harlan	11	–	500	1050		1560	18	90 × 90 and 50 × 90	✓			✓	Schuerger (1985)
12	–	KY	Harlan	11	–	500	500		480	18	90 × 90 and 50 × 90	✓			✓	Schuerger (1985)
13	–	KY	Harlan	11	–	500	1050		1410	18	90 × 90 and 50 × 90	✓			✓	Schuerger (1985)
14	4	PA	Pittsburgh	7.5	B4 LW	1450	Recovery		600	16	84 × 168	✓			✓	Barczak et al. (2007)
15	1	PA	Pittsburgh	7	1A	1500	–		1100	16	70 × 137 and 150 × 275	✓			✓	Lu and Hasenfus (2018)

(Continued)

Table 3.3.2 (Continued)

Case #	Panel #	Location	Seam	panel			Instrumentation		Entry width, ft	Pillar size rib-to-rib, ft	Caving	Pillar stress	Panel stress	Panel Convergence	Remarks
			Name	Thickness, ft	Name	Width, ft	Location, ft	Depth, ft							
16	9	VA	Pocahontas No. 3	5.5	S–9	600	3700	2100	20	20 × 80 and 120 × 180		✓	✓	✓	Campoli et al. (1993)
17	11	UT	Hiawatha	10	IIW	730	3500	1800	20	30 and 80–105		✓	✓		Maleki et al. (2009)
18	1	CO	wolf creek		site–1	640	–	1100	20	30 × 70 and 70 × 70	✓	✓		✓	Haramy and Fejes (1992)
19	1	CO	Wadge	10	site–2	640	–	1100	20	30 × 70 and 70 × 70	✓	✓		✓	Haramy and Fejes (1992)
20	–	Ut	Hiawatha	9	–	–	Bleeder	620	20	30 × 80		✓		✓	Maleki and Agapito (1988)
21	–	UT	Upper Blind Canyon	9	–	–	Bleeder	2000	20	30 × 80		✓		✓	Maleki and Agapito (1988)
22	6	UT	Wattis	10	–	–	Bleeder	1400	20	30 × 80		✓		✓	Maleki and Agapito (1988)
23	6	VA	Pocahontas No. 3	5.5	S–6	600	4700	1950	20	30 × 80 and 80 × 80		✓		✓	Campoli et al. (1990)
24	7	VA	Pocahontas No. 3	5.5	S–7	600	4700	1950	20	30 × 80 and 80 × 80		✓		✓	Campoli et al. (1990)
25	8	VA	Pocahontas No. 3	5.5	S–8	600	4600	2050	20	20 × 80 and 120 × 180		✓		✓	Campoli et al. (1990)
26	3	IL	Herrin No. 6	8.5	–	960	5980	650	15.5	45 × 105	✓	✓		✓	Yu et al. (1993)
27	–	wv	Pittsburgh	5.3	–	–	–	640		22 × 130		✓	✓		Mark et al. (1988)
28	–	PA	Pittsburgh	6.5	–	–	–	525		8 × 60		✓	✓		Mark et al. (1988)
29	1	PA	Pittsburgh	6.5	3A	600	–	573	20	45 × 80		✓			Listak and Zelanko (1987)
30	2	PA	Pittsburgh	6.5	3B	600	–	600	20	20 × 85 and 85 × 95		✓		✓	Listak and Zelanko (1987)

										Reference
31	3	PA	Pittsburgh	6.5	3C	600 –	455 20	20 × 85 and 85 × 95	✓	Listak and Zelanko (1987)
32	–	VA	Pocahontas No. 3	7.5		702 4000	2000 20	30 × 132 and 155 × 432	✓ ✓	Esterhuizen et al. (2018)
Australian mines										
33	26	NSW	Lithgow	8	26	840 2700	1066 15	116 × 313	✓ ✓	Colwell (2006)
34	8	Queenland	Lilyvale	11	8	902 2000	502 15.7	99 × 411	✓ ✓	Colwell (2006)
35	205	Queenland	German Creek	8.8	205	836 5000	722 16.4	Mines 115 × 308	✓ ✓	Colwell (2006)
36	21	Queenland	German Creek	8.5	21	1000 6000	688 17	82 × 310	✓ ✓	Colwell (2006)
37	–	Queenland	German Creek	9.5	SLW1	853 6500	640 17	98 × 311	✓ ✓	Colwell (2006)
38	5	NSW	Buli	9.5	5A4	820 4200	1574 15.7	122 × 394	✓ ✓	Colwell (2006)
39	27	NSW	West borehole	10	27	557 2600	558 16.4	131 × 360	✓	Colwell (2006)

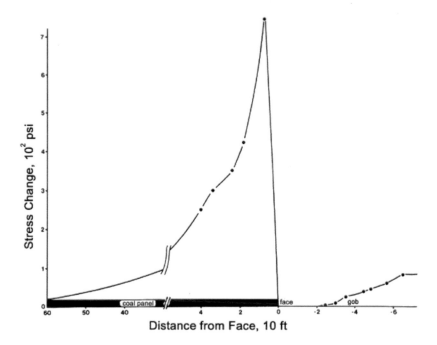

Figure 3.3.2 Measured front abutment pressure in front of the longwall face

Source: Peng and Park (1977)

Table 3.3.2 covers from 1985 to 2019 (New data). The old data had 17 panels 150–560 ft (46–171 m) wide and 400–875 ft (122–267 m) deep vs. 32 panels 416–1500 ft, mostly less than 1000 ft (305 m) wide and 455–2250 ft (139–688 m), mostly less than 800 ft (244 m) deep in the new data. The panel width to depth ratio was 0.17–1.0 and 0.28–2.5 for old and new data, respectively. Great majority in both data sets were subcritical panels. The new data concentrate more on pillar stress for gateroad chain pillar design.

The results are summarized in the following sections (note that all data will be referred to by panel number). It must be emphasized that the results presented here did not consider all factors that are known to influence abutment pressures and overburden movement. The intention is only to present an overall general trend:

3.3.2.1 Old data

A. ABUTMENT PRESSURE

(1) Front abutment pressure   Depending on local conditions, the front abutment pressure in the solid coal can first be detected approximately at a distance of 500 ft (152 m) outby the face. At this time, however, the pressure is very small, but it begins to increase rapidly when the face approaches to within 100 ft (30 m). It reaches the maximum value (i.e., maximum front abutment pressure) when the face is 3–20 ft (0.9–6.0 m) inby. Thereafter, the pressure drops drastically and vanishes at the faceline (Figs 3.3.2).

Figure 3.3.3 shows the width of the front abutment. The distance from the tailgate is normalized by the face width such that $N = 0$ is at the tailgate and $N = 1$ is at the headgate.

The width of the front abutment depends not only on the overburden depth but also on the position along the face. Figure 3.3.3 shows the width of the front abutment. Obviously, the front abutment width is not uniform across the panel width. It is wider at both ends of the face and decreases toward the center. The tail end is generally wider because of the effects of the adjacent mined-out panel. The width at the panel center ranges from 0.35 to 0.5 h (h = seam depth). It is smaller when the face is close to the bleeder pillars than when it is far away from them. Two curves in Figure 3.3.3 illustrate this point: curve 3B represents the front abutment width when the face is 200 ft (61 m) outby the bleeder pillars and curve 4 is about 80 ft (24 m) outby. There is quite a difference in magnitude between the two curves. The zone of front abutment shown by these two curves is relatively uniform, because they are near the bleeder pillars.

The maximum front abutment pressure is not uniformly distributed, and the peak front abutment pressure occurs either at the corner or at the center, depending greatly on the physical properties of the roof rock.

The maximum front abutment pressure is defined as the highest value in the front abutment profile as shown in Figures 3.3.1 A and 3.3.2. The peak front abutment pressure is defined as the highest pressure in the maximum front abutment pressures found in all cross-sections.

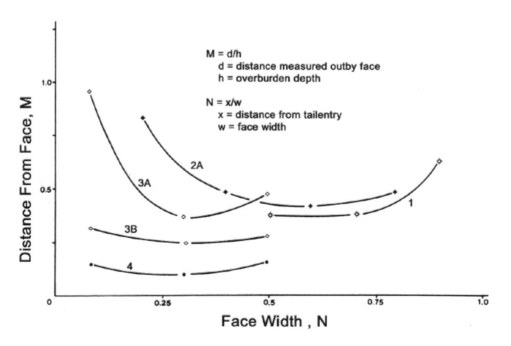

Figure 3.3.3 Measured width of front abutment pressures (the number on each curve is the panel number listed in Table 3.3.1)

The peak front abutment pressure occurs usually at the corners (or T-junctions) of the panel when the immediate roof rock is weak. A weak immediate roof does not impose a strong weighting effect on the front abutment, because it caves immediately behind the shield supports. The peak abutment pressure will remain at the corners as long as the weighting effects of the main roof do not come into play. In this respect, there are two possible conditions: first, the location of the main roof is high above the coal seam (i.e., thick immediate roof) so that its movements cannot be felt at the face area; and second, the face is located during the non-weighting period. In general, most geological conditions, coupled with the stress concentration around the ribs of the headgate and tailgate, satisfy the first condition, and the peak front abutment pressure occurs at the T-junctions. With the exception of panels 1 and 5A, all the cases in Figure 3.3.4 belong to the first condition. The immediate roof in panel 5A is composed of a strong sandstone stratum, 15 ft (4.6 m) thick, which exerts strong periodic weighting and the peak front abutment pressure at the T-junctions occurs only when the second condition is satisfied. Similarly, the instrumentation site where the data were obtained for panel 1 was very likely the place where the non-weighting period occurred.

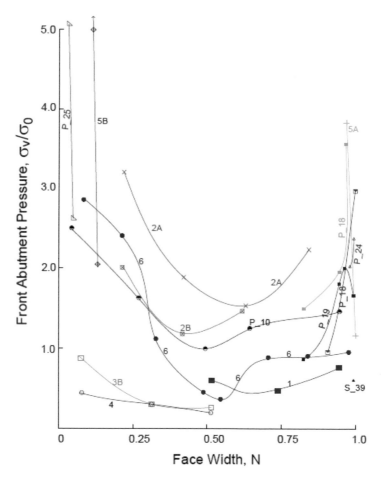

Figure 3.3.4 Measured maximum front abutment pressure profiles parallel to faceline (the number on each curve is the panel number listed in Tables 3.3.1 and 3.3.2)

In general, the maximum front abutment pressure ranges from 0.2 to 6.4 σ_o (where σ_o is the average *in-situ* overburden pressure) depending on the geological conditions, face location with respect to periodic roof weighting and the setup entry, and adjacent mined-out areas. As mentioned earlier, the front abutment pressure is smaller when the face is close to the bleeder pillars. If there is an adjacent mined-out panel, the maximum front abutment pressures are larger near that side. This is why in most cases the peak abutment pressure occurs at the tail-gate T-junction. The difference in maximum front abutment pressure between the headgate and the tailgate sides is $1\sigma_o$ for panel 2A and $1.7\ \sigma_o$ for panel 6 (Figure 3.3.4).

The location of a wider abutment pressure zone coincides with that of the maximum front abutment pressure (Figs 3.3.3. and 3.3.4).

The yield zone is the area between the faceline and the point where maximum front abutment pressure occurs. It is not uniform across the face width. The yield zone width ranges from 0.45 to 2.25 H (H = mining height) with the widest at the center of the panel. The yield zone is narrower when the face is near the bleeder pillars. If the yield zone is wide, there will be more roof weight to be borne by the shield supports at the face.

(2) Side abutment pressure   The side abutment pressure change is felt at the ribs of the headgate and tailgate at about the same time as the front abutment pressure. As the face advances farther, the side abutment gradually builds up as well as extends outward away from the ribs of the headgate and tailgate.

The process of the development of the side abutment pressure at the ribs can be classified into two extreme types based on the relation between pressure buildup and face location. The first type is that the side abutment pressure at the ribs has reached the maximum values before the face arrives (Fig. 3.3.5). After the face passes, the side abutment pressures in the

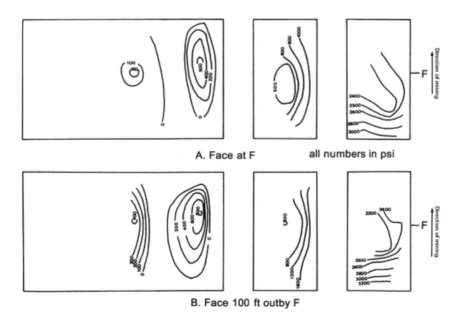

A. Face at F   all numbers in psi

B. Face 100 ft outby F

Figure 3.3.5 Development of measured side abutment pressures – type I (Peng, 1976)

first row of chain pillars change very little while those in the following rows of chain pillars increase considerably. The second type is just the opposite. Only a small increment of side abutment pressure occurs when the face arrives (Fig. 3.3.6). It reaches the maximum value only when the face has well passed.

The causes for these two extreme types may have something to do with the panel width, geology, and coal properties. The side abutment pressure at the ribs of the headgate and tailgate is the smaller, the better for purposes of entry stability and ease of maintenance. A fully developed side abutment at the ribs before the face arrives will cause more roof control problems.

Zones of high shear stress will be created in the immediate roof near both ribs of the entry due to the large difference between the side abutment pressure at the pillar rib and that at the edges of the panel. If the immediate roof is very weak, the roof control problems will spread over to the second and third entries. In this respect, the second type has the clear advantage; however most actual cases fall in between these two types.

The side abutment pressure is largest at the ribs of the headgate and the tailgate and drops exponentially away from the active panel (Figs 3.3.1 A and 3.3.5–3.3.7). The magnitudes of the side abutment pressure for the first row of chain pillars range from 0.4 to 3.5 σ_0 depending

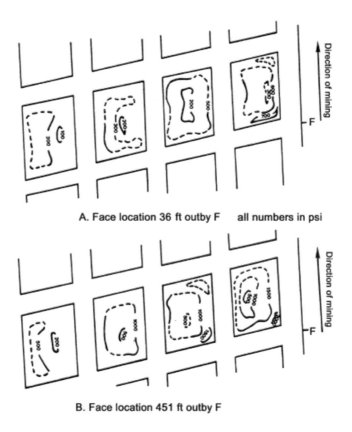

A. Face location 36 ft outby F   all numbers in psi

B. Face location 451 ft outby F

Figure 3.3.6 Development of measured side abutment pressures – type II

Source: Peng (1976)

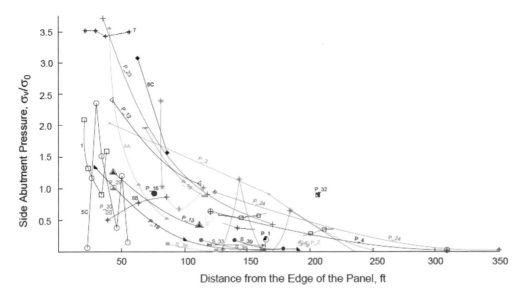

Figure 3.3.7 Measured side abutment distribution along cross-section parallel to faceline (the number on each curve is the panel number listed in Tables 3.3.1 and 3.3.2)

on the location inside the pillar. For most cases, depending on the size of pillars with respect to seam depth, there is a pillar core where the pressure increase is smaller than all edges of the pillar. The distance from the bleeder pillars will definitely affect the magnitude of the side abutment pressure (panel 8B in Fig. 3.3.7). However, this influence will diminish beyond a certain distance.

Although it is possible that the side abutment pressure at the ribs reaches the maximum value before the face arrives, the side abutment pressure continues to extend outward from the active panel until the face has passed far beyond. It has been found that the side abutment (or the influenced zone) increases with the overburden depth such that:

$$W_s = 9.3\sqrt{h} \tag{3.3.1}$$

Where W_s is the width of the side abutment (or influenced zone) in feet, and h is the overburden depth in feet. No specific relations between the side abutment and panel width or seam thickness can be found. Because the zone of side abutment is limited, the chain pillar system can be properly designed such that the next panel is essentially undisturbed using Equation (3.3.1).

Depending on the total pillar width and the influenced zone of the side abutment, the stress change (mainly stress increase) in the pillar, when the face of the second panel is passing by, may reach up to seven times that of when the first panel is being mined. The magnitude due to second panel mining ranges from 1.6 to 10 σ_o as compared to 0.4 to 3.5 σ_o due to the first panel mining.

(3) Gob pressure   Due to difficulties in measuring the gob pressure, no useful data were available for analysis except that shown in Figure 3.3.2. Therefore, the following statements rely more on computer modeling results calibrated by Figure 3.3.2.

When the roof rocks first cave in the gob, the weight of the caved fragments forms the gob pressure. As the caved fragments continue to pile up, so does the gob pressure. At some distance into the gob, the caved fragments start to take load from the upper strata. Figure 3.3.1 A shows a three-dimensional view of the gobs for two adjacent panels, one has been mined out and the other is being mined. Around the panel edges, the gob pressure is mainly due to the weight of the caved rock fragments. The gob pressure increases toward the center of the mined panel where the gob is more compacted (Fig. 3.3.1 C). The maximum gob pressure is the overburden weight that occurs when the gob takes the full load of the overburden weight. Whether the gob pressure reaches the overburden weight or not, it depends highly on the panel width. If the panel is too narrow, the upper unbroken strata will be bridged by the side abutments, resulting in gob pressure being more or less the weight of the rock fragments equivalent to the caving height.

B. HEIGHT OF CAVING ZONE

Because the un-caved roof strata in the gob are supported mainly by the abutments, the caving height determines the magnitude of abutment pressures. A higher caving height represents a smaller load to be supported by the abutments. The caving zone across the panel width is believed to assume more or less an arch shape with the highest points in or around the panel center, decreasing toward both sides. The caving height is highly dependent on geological conditions such as the location of the thick competent strata and the bulking factor of the caved strata. Caving stops when a self-supporting stratum is encountered or the overlying strata are fully supported by the caved fragments. Strong roof strata tend to delay the caving process and extend further into the gob. This condition will increase the front and side abutment pressures.

An example of the caving process is shown in Figure 3.3.8. After initial face advance of about 15 ft (4.92 m), the first 60 ft (18 m) of the immediate roof caved immediately after the

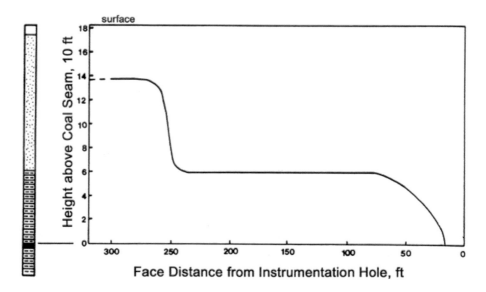

Figure 3.3.8 Development of caving height as face advance

Source: Wright et al. (1979)

powered supports were advanced. The overlying shale layer about 60 ft (18 m) thick tended to overhang for at least 50 ft (15.3 m) before it caved in. It was "at least" because the location of the borehole where the measurements were made was not necessarily in the starting position of a weighting cycle for that shale layer. The weighting effect of the shale layer was believed to have limited impact on the front abutment, because it was 60 ft (18.3 m) above the coal seam. The caving zone reached its maximum height when the face was at least 240–280 ft (73.2–85.3 m) outby. This indicates that the 50 ft (15 m) thick sandstone layer was strong enough to self-support for at least 220–260 ft (67.1–79.2 m) into the gob.

In addition to the sequences of weak and strong strata above the coal seam and the bulking factor of caved fragments, caving height is also affected by the seam thickness, panel width, and overburden depth. More caved fragments are needed to fill up the void created by mining in a thick seam than in a thin one. Also, the caved fragments in the thick seams are more compressible than those in thin ones because their piles will be higher. The wider the panel, the larger the roof deflection, especially at the panel center, thereby causing the strata to break sooner. A larger depth of overburden causes more caving due to its higher overburden pressure.

C. GATEROAD CONVERGENCE

Convergence is the reduction in the entry height due to the stress redistribution resulting from various mining activities. Traditionally, entry convergence is mostly assumed to be solely due to roof sag, while floor heave is negligible. Research and underground operational experienced have demonstrated that floor heave can be significant in many cases. Entry convergence can be first felt at the same time as the side abutment or front abutment pressure is first detected. Convergence at this time, however, is very small. It increases at an accelerating rate as the face approaches. After the face has passed, entry convergence continues to increase until it collapses

Entry convergence varies with the distance from the active panel. Convergence in the headgate is obviously larger than that in the second entry or the third entry. Convergence in the headgate is largest near the panel rib and decreases toward the first chain pillar. The amount of decrease depends on local geology, entry width, and pillar width. There are no obvious trends in the track entry (or second entry).

Convergence in the tailgate is much more severe than that in the headgate. The maximum convergence in the tailgate can be as high as three times that in the headgate. Maximum convergence in the headgate is toward the active panel but that in the tailgate is toward the previously mined-out panel.

A strong roof tends to resist deformation due to mining activities. Therefore, convergence with a strong immediate roof is less than that with a weak one.

3.3.2.2 New data

A. FRONT ABUTMENT PRESSURE

For convenience of comparison, the front abutment pressures for the new data are also plotted on Figure 3.3.4. where the new US data are prefixed with "P" and the Australian data with "S."

There are four new data lines in Figure 3.3.4 and they in general fit well with the old data, even though there are differences in panel width and depth between the old and new data.

Based on the new collected data set for the US mines, the front abutment pressure can be detected about 900 ft (274 m) outby the face. When the face distance is at the instrumentation site, the yield zone in the solid coal block do not exceed 10 ft (3 m).

B. SIDE ABUTMENT PRESSURE

Figure 3.3.9 shows an example of the pressure increase as the face advances when two panels on both sides of chain pillar are mined. The side abutment pressure could be detected about 750 ft (229 m) outby. During mining of the first panel, the side abutment pressure increases gradually as the face advances and accelerates when the face approaches the instrumented pillar. Pressure increase near the instrumented pillar is more obvious when the second panel approaches the instrumented pillar. Full side abutment pressure is the pressure increase experienced by a pillar system from the start of longwall mining to long after the face has passed it and the pressure on that pillar system has stabilized. As shown in Figure 3.3.9, the side abutment pressure on the instrumented pillar keeps increasing and stops when the face is 1000 ft (305 m) outby. However, the rate of increase is small once the face is 500 ft (152 m) outby.

The development of the side abutment pressure can be classified into three types vs. two types in the old data. In type 1, yielding occurs at the pressure sensor locations just after the face passes it. Most yield pillars in the US data belong to this type, where the entire pillar or most of the pillar yields except the pillar core. In type 2, yielding at the pressure sensor locations occurs before the face reaches the instrumented pillar. In type 3 group, the panel is completely mined out without any yielding at the pressure sensor locations (Figure 3.3.10). Basically, if the pressure sensor is located in high confinement zone closer to the pillar core, type 3 is more likely to occur even under high side abutment load. However if the pressure

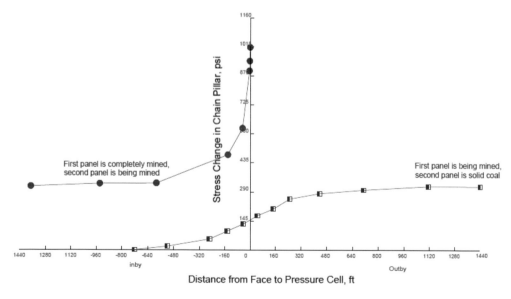

Figure 3.3.9 Side abutment pressure distribution with relative face position during mining of two longwall panels

Source: After Colwell (2006)

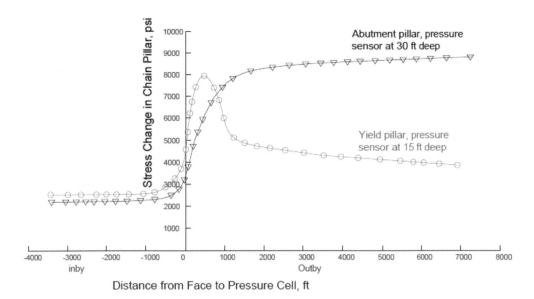

Figure 3.3.10 Side abutment pressure distribution for yield pillar and abutment pillar with pressure sensors at 15 ft and 30 ft, respectively

Source: Rashed (2019)

sensor is located closer to the ribs and the side abutment pressure is high, type 1 or 2 is more likely to occur.

The variation of the side abutment pressure across chain pillars for the old and the new data when the face is at the instrumentation site is shown in Figure 3.3.7. The old and new US data fit well within data range, while the Australia data appear to have smaller abutment pressures.

The common factors in all cases with small side abutment pressure in the new US data are that the roof caves well and that the ratio of panel width to panel length is more than 1.0. Conversely, the common characteristics in all cases with high side abutment pressure are that a strong massive sandstone or siltstone exists in the immediate and main roofs, plus the immediate roof is not easily cavable, and that the ratio of panel width to panel depth is less than 1.0. Most cases with high front and side abutment pressures are in the western US coalfields, while most data in Table 3.3.1 are from the eastern US coalfields.

For the new US data, the side abutment or influence zone is this:

$$W_s = C\sqrt{h} \tag{3.3.2}$$

Where $C = 5.6$–8.5.

Chapter 4

Panel development

4.1 Introduction

Panel development is an integral part of longwall planning. As the longwall becomes more productive, the possibility that development cannot catch up with longwall mining is very real. In fact, there are many cases where the next panel was not ready, and impromptu measures had to be adopted, such as cutting the next panel short or employing innovative ventilation methods to make the next panel ready, reducing the number of entry development, slowing down the longwall mining rate in the current panel, adding one more development unit on the opposite end, or simply parking or idling the longwall. With a quoted revenue loss of US$ 500 to US$ 2000 per minute of longwall idle time, inadequate panel development simply cannot be tolerated.

Since the longwall can outproduce gateroad development using room-and-pillar mining by up to 10 folds, the strategy for gateroad development is speed, not coal volume, as originally perceived in the 1970s. In fact, in a longwall mine, longwall accounts for 80 to 85% of total mine production while mains and gateroad developments account for the remainder.

Traditionally, coal mining in the United States employs the room-and-pillar method. In this mining system, the equipment usually consists of one continuous miner for cutting coal, two shuttle cars alternatively loading and transporting coal between the continuous miner and the conveyor belt feeder-breaker, and one twin-drill roof bolter for installing roof bolts. Since the continuous miner is designed to cut coal using a long cylindrical drum, the entry driven by it is an in-seam rectangular opening. An operator driven a continuous miner normally cuts 20 ft (6.1 m) deep, but a remote-controlled miner cuts up to 40 ft (12.2 m) deep. It then pulls out the newly cut opening and lets the roof bolter move in to install roof bolts.

Therefore, the so-called continuous mining using a continuous miner is absolutely not continuous, as its name implies, but requires place changes during which the continuous miner is not cutting coal. In addition, when the shuttle cars are moving coal between the continuous miner and the belt feeder-breaker, the continuous miner is idle and waiting for the next shuttle car to load.

The 1969 Coal Mine Health and Safety Act as amended in 1977 (US Congress, 1969) stipulates that anytime during coal mining, a separate intake and return escapeways must be provided, and the conveyor belt must maintain neutral with nominal air movement. Accordingly, the minimum number of entries required using the room-and-pillar mining method is three. The same law also says that no one can work under an unsupported roof, and because the CM operator's seat is normally located 20 ft (6.1 m) behind the cutting drum, a maximum of 20 ft (6.1 m) of deep cut is all that is permitted before the machine must pull out, although

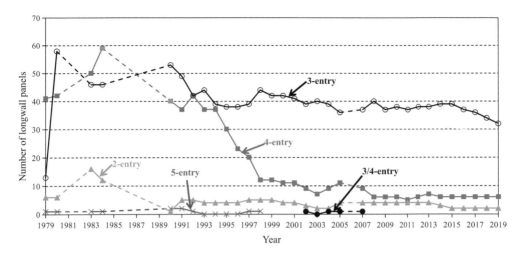

Figure 4.1.1 Historical trend of distribution of gateroad development system since 1979

in recent years the use of remote control miners with special permission can cut as deep as 40 ft (12.2 m).

When longwall mining was introduced in the 1960s and 1970s, US coal miners were not familiar with modern longwall equipment. So to reduce the training time and scope, the traditional continuous mining method was used to develop the panels.

This is why the multiple-entry gateroad system is used in US longwall mining as opposed to single entry in most countries. Experience has shown that this development system is highly efficient and allows for a fast longwall advancement rate.

In the 1970s and early 1980s, nearly all longwall mines were originally designed for room-and-pillar mining, converting later to longwall mining. So there was then gateroad development using the five-entry system. But, as experience increased over the last 30 years, the three-entry system has become the preferred choice (Fig. 4.1.1). Note, as stated earlier, the three-entry system is the minimum number of entry in compliance with the law.

For multiple entry development, the total chain pillar width ranges from 131 ft (40 m) to 442 ft (135 m). Under the prevailing seam depth and geological condition practiced in US longwall mining, this pillar width can isolate completely or reduce considerably the side abutment pressures created by the mining of a previous adjacent panel. Consequently, the tailgate of the next (or current) panel being mined is subject to less overlapped abutment pressure and can be handled easier. This is one of the major factors that allows for the fast advance of a retreating longwall mining face. The downside is obviously a much lower coal recovery rate in the panel because those chain pillars are left unmined.

4.2 Distribution and comparison of gateroad development systems

Figure 4.1.1 shows the historical change of the gateroad development system. There are two-, three-, four-, and five-entry and three- and four-entry combination systems. In the 1970s and early 1980s, the four-entry system was quite popular. But in the 1990s the three-entry system had gained favor, and today it accounts for 81% of all longwalls, while the four-entry system accounts for 14%. The reason is quite obvious. The three-entry system complies with the law, and it requires less development footage than the four-entry system. The two-entry system is not in compliance with the law, but upon request, MSHA (Mine Safety and Health Administration) has been approving it under special conditions.

4.2.1 Two-entry system

A special approval from MSHA is required. This system has been practiced in three western states (Colorado, Wyoming, and Utah), and in 2019 exclusively in Utah, where the coal seams are deep and the overburden consists of thick sandstone strata. The strata tend to overhang and induce very large abutment pressures in and around the longwall panels, resulting in coal bumps, or sudden outbursts of large volume of coal. In order to reduce this hazard, the two-entry system using a yield pillar of 26–30 ft (7.9–9.1 m) has been found helpful in reducing pillar bumps. Also the two-entry system contains half the number of pillars in the three-entry system and the probability of coal pillar bumps reduces accordingly. The rule of thumb that MSHA uses is that the seam must be larger than 1100 ft (335 m) deep for approval of the two-entry system.

The first entry is used for the conveyor belt (and track or diesel car road) while the second entry is the tailgate for the next or future panel.

4.2.2 Three-entry system

This system is in compliance with law in terms of the minimum number of entries: one intake and one return escapeway, and one neutral conveyor belt entry. All three entries must be isolated.

The first entry is for the conveyor belt (i.e., headgate) and can be used for intake if carbon monoxide (CO) monitors are installed; the second entry is the track for miners' travel and supplies, and the third entry is the tailgate for the next panel.

This system may require battery-operated man trips in the track entry and/or CO sensors in the conveyor belt entry.

In some cases, the conveyor belt and track or travel road are side by side in the first entry. In such cases, the second entry is used for ventilation and the mule train (i.e., power center, hydraulic pumps, etc.).

4.2.3 Four-entry system

This system is being used for coal seams of low thickness, high methane content, and large depth panels in Alabama and Virginia, mainly because the entry cross-section is relatively smaller and requires more entries to provide sufficient air to the face.

The first entry is for the conveyor belt (i.e., headgate); the second entry is for track or travel road; the third entry is for air intake; and the fourth entry is the tailgate for the next panel.

In terms of the amount of development work per unit longwall face advance, the two-entry system requires the least development work. But, the continuous miner development system consists of several pieces of large equipment near the face area, as mentioned earlier, the two-entry system is too crowded for efficient equipment place change. The complete opposite is true for the four-entry system. For long panels and/or gassy and/or thin seams, ventilation requirements may prevent the use of two-entry or even three-entry systems.

4.2.4 Summary

Theoretically, the required gateroad development footage is proportional to the number of entries developed. Everything being equal, the footage for a four-entry system is 1.4 times that for a three-entry or 2.3 times that for a two-entry system. But a 1995 survey (Peng, 1995) showed that this was not completely true. Some development crew were used to the four-entry development and when converted to three-entry, they gained little or even slowed. The same survey also found that crew's dedication and incentives are the most critical factors in the rate of entry development, regardless of equipment used.

4.3 Methods and equipment used for gateroad development

Essentially two methods have been used for gateroad development while using three types of cutting machine (Demichiei and Beck, 2001): the place-change method and the in-place method.

4.3.1 Place-change method

As mentioned earlier in Section 4.1, this is the traditional continuous mining method in which the continuous miner cuts 20 ft (6.1 m) or 40 ft (12.2 m) deep and then trams to another entry for cutting (Fig. 4.3.1). The roof bolter is moved in to bolt the roof. In this method, the continuous miner moves from entry to entry, followed by the roof bolter. Therefore the continuous miner wastes a large portion of its working time tramming from entry to entry because the newly cut roof needs to be bolted by the roof bolter. In other words, the equipment is not even close to utilizing its full potential. Since the width of a continuous

Figure 4.3.1 Conventional continuous miner 12CM
Source: Courtesy of Komatsu Mining

miner is usually 10–11 ft (3.05–3.35 m), the entry width developed by it normally ranges from 18 ft (5.49 m) to 20 ft (6.10 m) by making two cuts side by side. The roof bolting pattern, just like in room-and-pillar mining, is normally 4 × 4 ft (1.2 × 1.2 m).

The average advance rate using the place-change method is slower than that of the in-place method. But some mines employ the modified super-section method in which two sets of continuous mining systems are used and routinely advance 170–250 ft day (51.8–76.2 m day).

4.3.2 In-place method

This method is further divided into two types, depending on the use of equipment.

4.3.2.1 Miner bolter

In the mid-1980s, Joy developed a miner bolter in which two chassis-mounted roof drills were incorporated into Joy's popular continuous miner 12CM and 14CM. With this model, the continuous miner does not need to change places. Instead, it simply stops cutting after the desired depth of the cut. The roof drills then take over to do roof bolting. Obviously, in this model, the miner and roof bolter are combined, so it cannot do cutting and roof bolting at the same time. The two roof drills install two side bolts in a row, approximately 8 ft (2.4 m) apart. A center bolt, if needed, is installed later. Miner-bolter is no longer used.

4.3.2.2 Satellite bolters and bolter miners

Satellite miner is designed to do cutting and bolting simultaneously by utilizing a separate independent drill frame unit such that the bolter unit is fixed on a bridge structure and moves relative to the miner. The roof bolts can be installed 6 ft (1.8 m) behind the face. Note, just like the miner bolter, the two roof drills install two side bolts in a row, approximately 8 ft (2.4 m) apart. A center bolt, if needed, is installed later.

M451 (Fig. 4.3.2) is a bolter miners developed by Sandvik. Its cutter drum is mounted on a hydraulically actuated sliding frame, it can sump into the coal face independently of the mainframe and tracks. Its roof and rib bolters are mounted on the stationary mainframe and can be operated throughout the cutting cycle. The number of bolts in a row varies from two to eight, including rib bolts.

Figure 4.3.2 M 451 miner bolter

Source: Courtesy of Sandvik

Since the in-place method uses a machine that combines the continuous miner and roof bolter, it is much heavier and complicated, and consequently they are wider, normally 15.5 ft (4.73 m) wide. Therefore, for those entries developed by these machines, the entry width is the machine width. With the in-place method, the maximum rate of advance is estimated to be 250 ft d^{-1} (76.2 m d^{-1}).

4.3.2.3 Flexible conveyor train

In a continuous mining system, the miner-bolter concept integrates the coal cutting and roof bolting cycles into one subsystem, leaving coal haulage to remain as an independent subsystem. The flexible conveyor train (FCT) technique further integrates the so-called continuous mining system into a truly continuous one. In other words, coal cutting, roof bolting and coal transportation between continuous miner and section belt tailpiece operate simultaneously as an integrated system.

The first floor-mounted FCT was introduced in 1985. The current model is the fourth generation, or 4FCT.

The 4FCT is a remote-controlled, single operator, continuous haulage system that runs at low speed, and follows a fixed path of travel. Its length ranges from 200 ft (65 m) to 575 ft (175 m) depending on the mining panel layout, pillar size, and sequence of operations. It consists of a front end hopper complete with directional steering wheels (Fig. 4.3.3) and a lumpbreaker to handle the surge capacity of the continuous miner, and a single flexible conveyor belt that is driven from the return end and captivated by "high-hat" rollers. The whole conveyor belt structure is mounted on several matched Optidrive VFD-controlled haulage sections, with the number of haulage sections being determined by the length of the 4FCT (Fig. 4.3.4). Coal is discharged directly onto the panel belt or section belt through the dynamic move-up unit (DMU), which is the transition interface between the FCT and the outby belts.

Figure 4.3.3 4FCT underground – hopper and steering arrangement
Source: Courtesy of Komatsu Mining

When the 4FCT retreats, it accurately follows the same path on which the 4FCT comes in, without trying to cut the corners.

The full FCT system uses 14CM ED bolter miner that cut 25 tph with 4 ft (1.2 m) deep sumping with two roof bolts and two rib bolts on each side (Fig. 4.3.5). Underground coal mine trials for three-entry development consistently completed more than 200 ft/shift (61 m/shift) with maximum in the 364 ft/shift (111 m/shift).

FCT does require high maintenance standards.

Figure 4.3.4 FCT navigating a pillar underground

Source: Courtesy of Komatsu Mining

Figure 4.3.5 14CM ED bolter miner

Source: Courtesy of Komatsu Mining

4.4 Rapid development

In modern longwall mining, the bottleneck to consistent high production has always been attributed to the slow rate of development by the conventional methods, either place change or in-place. Consequently, gateroad development has received focus attention for improvement in the past decades, mostly concentrating on the development of new continuous miners and haulage system dedicated for entry development, including automation.

It is well known that a conventional continuous miner can do very well if the roof is strong (limestone or sandstone roofs) and methane emission is low. Both conditions allow deep cuts (up to 40 ft or 12.2 m deep), leading to more cutting time and less for tramming, resulting in fast development.

Rapid development has been a general terminology used by coal industry professionals without a common definition. So, in this book, rapid development refers to those rates of gateroad development (or development ratio) that exceeds the rate of face advance. Under this definition, development is always ahead of longwall retreat mining. And it has no minimum quantitative cutting distance per unit time. Consequently, under this definition, rapid development varies with mine production and mean different for different LW mines. For US longwalls with an average panel width of 1200 ft (366 m), the average face advancing rate is 25 ft (7.6 m) or seven cuts per shift. How much linear footage must the continuous miner cut and advance in the gateroad (or development ratio) in order to match this face advancing rate? It obviously varies with mines, number of entry development and total pillar width in the gateroad. For US longwalls, the threshold development ratios are about 2.5–2.8, 4.0–4.4, and 5.5–5.9 for two-, three-and four-entry development systems, respectively (see Section 4.9, p. 102). In other words, for three-entry development, a 25 ft per shift face advance will require at the minimum 25 × (4.0–4.4) = 100–110 ft linear foot per shift of continuous miner advance rate. Based on this definition, any continuous miner advance rate above 110 ft per shift is rapid development for this mine case.

The place change method, although not necessarily continuous in cutting, is highly mobile and flexible as compared to the in-place method. It is mobile because it is lighter and thus easier to move around. It is flexible because cutting and bolting use a separate machine, breakdown in one machine will not affect the other. Conversely, any part breakdown will shut down the whole operation for a bolter miner. Therefore, if a dedicated skilled work force can make use of those advantages under favorable geological conditions, the results will be formidable. In fact there are longwall mines routinely advancing 230 ft/shift (70 m/shift) in central and Appalachian coalfields.

4.5 Survey of types of pillar systems in gateroad development

In the two-entry system, there is only one row of chain pillars, which are yield pillars, 26–30 ft (7.9–9.1 m) wide regardless of seam depth. The yield pillars are designed to yield soon after development or when the longwall face is approaching such that they will not fail suddenly and violently.

In the three-entry system, there are two rows of chain pillars. Those two rows of pillars can be equal in size or unequal, i.e., one larger than the other. If they are unequal, they can be arranged such that either the large-sized pillar is next to the headgate or to the tailgate. Pillars are mostly rectangular in shape, with their length being much larger than their width in order to reduce number of crosscuts. They may be diamond-shaped with crosscuts sloping inby.

In the four-entry system, there are three rows of chain pillars. Theoretically, there can be many ways to arrange pillars. Experience has shown that the yield-stiff-yield system is the most popular. In this system the two side rows are yield pillars, 20–40 ft (6.1–12.2 m) wide, while the center row is stiff pillar, 100–220 ft (30.5–67.1 m) wide. Other types of pillar arrangements include the following: the first two rows are the same size and the third row is either smaller or bigger.

Figure 4.5.1 shows the entry width distribution (Tsang *et al.*, 1996). The width ranges from 15 ft (4.6 m) to 20 ft (6.1 m). The 17–20 ft (5.2–6.1 m) widths are cut by traditional place-change miners and are more popular, while the 15–16 ft (4.6–4.9 m) widths are cut by in-place miners.

Figure 4.5.2 shows the distribution of pillar width for the three-entry system (Tsang *et al.*, 1996). Widths range from 32 ft (10 m) to more than 130 ft (40 m). The most common ones are from 100 ft (30.5 m) to 115 ft (35.1 m).

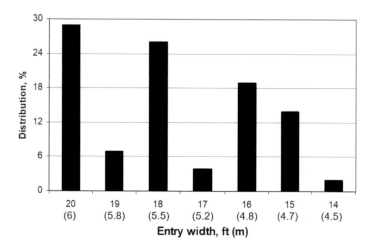

Figure 4.5.1 Distribution of entry width for US longwall gateroads

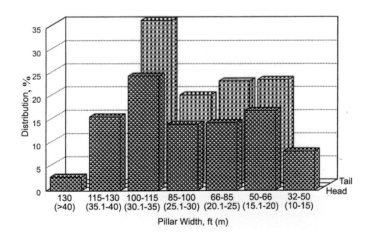

Figure 4.5.2 Distribution of pillar width for three-entry gateroad system

Figure 4.5.3 shows the distribution of total pillar width for the three-entry system. The total width ranges from 131 ft (40 m) to 279 ft (85 m) with the most common ones from 184 ft (56 m) to 230 ft (70 m).

Figure 4.5.4 shows the distribution of pillar width for the four-entry system (Tsang *et al.*, 1996). The width ranges from 32 ft (10 m) to larger than 130 ft (40 m). Figure 4.5.5 shows the distribution of total pillar width for the four-entry system, ranging from 197 ft (60 m) to 442 ft (135 m), with the most common ones ranging from 250 ft (76 m) to 295 ft (90 m).

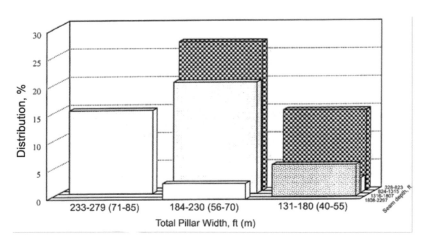

Figure 4.5.3 Distribution of total pillar width in three-entry gateroad system

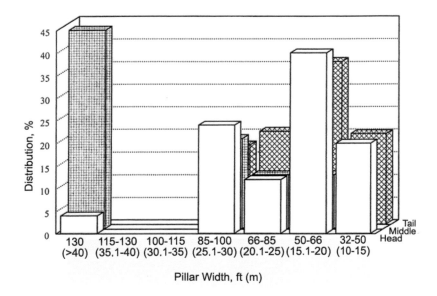

Figure 4.5.4 Distribution of pillar width in four-entry gateroad system

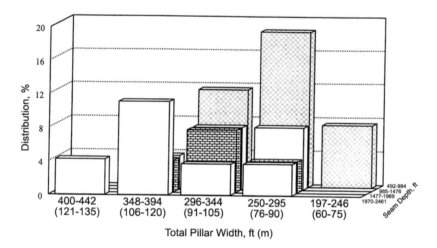

Figure 4.5.5 Distribution of total pillar width in four-entry gateroad system

It is obvious that longwall development by multiple entry with multiple rows of chain pillars decreases the overall coal recovery considerably. For example, the average mining height of current longwall panels is 8 ft (2.4 m) and average panel length is 1100 ft (3354 m). Every one-foot increment of pillar width contains 3740 tons of coal left unrecovered. Consequently, the importance of accurately determine the proper size of pillar cannot be over-emphasized.

4.6 Design of chain pillars and barrier pillars

4.6.1 Chain pillars in the gateroads and bleeder system

Currently there are two popular methods for logwall gateroad chain pillars design: the Analysis of Coal Pillar Stability (ACPS) method (Mark, 1990) and numerical modeling method. ACPS is for stiff pillar design. Stiff pillars refer to those with strength greater than the roof loading as opposed to yield pillars, which are designed to yield as roof loading increases. Stiff pillars are much larger than yield pillars. Stiff pillars are used in all US longwalls except Utah. Chain pillars in the gateroads are left behind without being mined, although pillar mining concurrent with longwall retreating mining had been done routinely in the past (Hendon, 1998).

4.6.1.1 The analysis of coal pillar stability

The longwall tailgate module of ACPS uses the following formulae to determine the stability factor (SF) for chain pillar systems subjected to abutment loads:

A. STRENGTH OF INDIVIDUAL PILLAR

$$Sp = S_1 \left(0.64 + 0.54 \, w/h - 0.18 \left(w^2 / h l_p \right) \right) \tag{4.6.1}$$

Where S_p = pillar strength (psi), S_1 = *in-situ* coal strength (psi), w = pillar width (ft), h = pillar height (ft), and ℓ_p = pillar length. The default value of S_1 = 900 psi.

B. BEARING CAPACITY OF THE PILLAR SYSTEM

The bearing capacity of an individual pillar is:

$$B_P = \frac{S_P w l_p (144)}{(l_p + w_e)} \text{(lbs per foot of gateroad)} \tag{4.6.2}$$

The load bearing capacity of the pillar system is:

$$B = \sum_i^n B_{pi} \tag{4.6.3}$$

Where i = 1 to n = number of pillars in the chain pillar system.

C. PILLAR LOADING

1   Headgate loading = loading at the headgate and tailgate T-junction of the first panel:

$$L_H = \left[L_d + (L_s)(F_h)R\right] * \left[(1+F_p)/2\right] \tag{4.6.4}$$

2   Bleeder loading = loading at the bleeder pillars:

$$L_B = \left[L_d + (L_s)(R)\right] * \left[(1+F_p)/2\right] \tag{4.6.5}$$

3   Tailgate loading = loading at the tailgate T-junction of the second panel onward:

$$L_T = \left[L_d + (1+F_t)(L_s)\right] \tag{4.6.6}$$

Where L_d = development load, L_s = total side abutment load, F_t = 0.7, F_h = 0.5, and:

$$R = 1 - \left[\frac{D - w_t}{D}\right]^3 \tag{4.6.7}$$

Where w_t = width of pillar system and:

$$D = 9.3\sqrt{H} \tag{4.6.8}$$

Where H = seam depth, and:

$$F_p = 1 - 0.28\left[\ln\left(H / W_{cps}\right)\right]$$

Where W_{cps} = width of the chain pillar system.

D. STABILITY FACTOR

$$SF = B / L$$

Where B = load bearing capacity of the pillar system and L = design loading.

Mark and Agioutantis (2018) recommend that for the tailgate loading condition, the SF should be about 1.4 where the roof is weak but could be as small as 0.8 where the roof is exceptionally strong. The ACPS program is available for download, and users can simply input design pillar dimension to check if SF meet the recommended values.

4.6.1.2 Yield pillar design

Yield pillars are used in longwalls located in Alabama and Utah where mining depth is more than 2000 ft (610 m). In Alabama where the floor strata are soft and tend to heave, the four-entry system with a yield-stiff-yield pillar arrangement was found to be applicable (Martin *et al.*, 1988). The yield pillars on both sides of the gateroad are 20 ft (6.1 m) wide and have been successful in dealing with floor heave under high overburden load. Yield pillars work best when both the roof and floor strata are strong or firm, because they have to sustain a roof span of more than 60 ft (18.3 m), i.e., yield pillar width + two entry widths (one on each side), when a yield pillar yields.

In Utah where the overburden is very thick with a great majority of the strata consisting of massive sandstone, coal bumps in pillars occur frequently and have caused injuries and fatalities to miners. Experience has demonstrated that a two-entry system with a yield pillar would prevent pillar bumps. But, in order to avoid bumps, the yield pillar must be in the range of 26–33 ft (7.93–10.1 m).

In the northern Appalachian coalfield, where nearly two thirds of US longwalls are located, yield pillars have never been systematically used. But in the late 1980s, underground experiments on yield pillars of 10 and 20 ft (3 and 6.1 m) had been conducted in the Pittsburgh seam where the cover was 450–600 ft (137.2–182.9 m). The 20 ft (6.1 m) yield pillars in a four-entry, yield-stiff-yield gateroad system ran for seven and one-half panels. They were designed to yield during longwall retreating. In practice, sometimes a pillar bumped outby the face, and others would not yield until they were behind the face in the gob. The 10 ft (3 m) yield pillars ran only one and one-half panels. They were designed to yield during development. In the worst case, the roof converged 5–6 ft (1.5–1.8 m) about one and a half pillar blocks outby the face during development. Thereafter further convergence was minimal. The ribs sloughed severely but non-violently during retreating. The practice was discontinued for safety reasons.

Yield pillar design has been based on trial and error. Morsy and Peng (2003) developed a computer model and yield pillar stability evaluation protocol for yield pillar design. The method was calibrated with four cases of longwall yield pillar designs in Utah.

4.6.1.3 Numerical modeling method

ACPS is a formula that, like most conventional pillar design formulae, does not consider the effect of interaction among roof, pillar, and floor. It is well known that coal strength increases, if confined, three to four times the amount of confining pressure. So if the roof and floor are harder rock such as sandy shale or sandstone, then their confinement will increase pillar strength. Conversely, if the roof and floor are of clayey material, pillar strength will be reduced. In addition, coal pillars may contain one or more partings, weaker or stronger than the coal. Numerical analysis can take all of those factors into consideration (Morsy and Peng, 2003). The disadvantage of numerical modeling method is that it is much more complicated and requires in-depth training to use.

The current industry practice for chain pillar design deserves some discussion. By looking at a longwall mine map, one normally finds that one pillar size or one pillar arrangement for simplicity and design convenience is used all throughout a panel or a section of panels or, in most cases, even the whole mine. (Note that pillar arrangement refers to the number and sizes of chain pillars in a row). Since surface topography changes from place to place, very often drastically especially in central Appalachia, pillar size should logically be changed with topography for better use of coal reserves. But, in practice, it is impractical to change pillar size within a gateroad development. So it is not easy to determine what part of the gateroad or which panel gateroad the selected pillars are designed for or vice versa if one is asked to design a chain pillar system. The answer probably lies in the method of how to determine the average seam depth and how and where to sample coal specimens for laboratory strength tests. The manner by which those two factors are determined will greatly affect pillar size determination.

4.6.2 Barrier pillar

A barrier pillar refers to that coal block left at the retreat mining end of the panel between the recovery room and the mains. In the past, trial and error had dictated its design to around 500–600 ft (152–183 m) wide. In recent years numerical finite element computer modeling has been used, and experience has validated its results. Barrier pillars have been reduced to 200–300 ft (61.0–91.5 m) in many cases. Using the numerical computer method for barrier pillar design can also in the meantime consider the stability of mains entries, which is useful because many longwall mines have laid out their panels on both sides of the mains. After panels on both sides are mined out, the permanent front abutments from both panels may superimpose on the mains and cause stability problems, especially on those outer entries closer to the mined out panels.

4.7 Roof support

Roof support can be divided into two types with respect to a continuous miner's method of mining: primary and secondary or supplementary supports. The primary support, mainly roof bolts, are installed in cycle immediately or soon after the continuous miner's cutting. Secondary supports, on the other hand, are installed long after the continuous miner's cutting. Secondary supports are usually longer bolts/cable bolts or standing supports that require more time and labor to install. For longwall mining, the secondary supports are additional supports installed to cope with the moving front and side abutment pressures. This way of installing supports, i.e., two-stage support installation, allows for the rapid development of gateroad.

4.7.1 Entry

Entries, 16–20 ft (4.9–6.1 m) wide, are developed by the conventional continuous miners. The primary support is 3/4 in. (19 mm) diameter by 6–8 ft (1.8–2.4 m) long in a 4 × 4 ft (1.2 × 1.2 m) pattern. They can be either fully grouted resin, combination, or resin assisted high-tension bolts (Peng, 2008). In the entry/crosscut intersections, one to four cable bolts, 0.6 in. (15.2 mm) diameter by 14–16 ft (4.3–4.9 m) or up to 24 ft (7.3 m) long may be added.

For those entries, 15–16 ft (4.6–4.9 m) wide, developed by the in-place method, two side bolts, 7–9 ft (2.1–2.7 m) apart and 3–4 ft (0.91–1.2 m) between rows, are installed in cycle.

A center bolt, if needed, is installed later. Bolt types are similar to those for the entries developed by the place-change method.

Depending on the severity of the front abutment pressure, supplementary supports in the form of cribs or props/engineered props, such as ACS and propsetter, or in recent years pumpable cribs may be added at the headgate T-junction, especially at the entry side of the headgate and crosscut intersection.

4.7.2 Tailgate

4.7.2.1 Type of tailgate support

Prior to 1985, when gateroad chain pillar design was in its infancy, there were many tailgate problems. Since then a combination of properly designed chain pillar system plus tailgate secondary support has resolved the problems, even though the tailgate has been designated as an emergency escapeway that must be kept open all time since 1985. Traditionally, tailgate support employs one or two rows of wood cribs, either side by side or staggered. Following Hurricane Andrew in 1992, high quality woods were in demand for re-construction work, and the price of mining timber increased dramatically. In addition, as the panel became longer, the amount of timber required for tailgate support also grew. Furthermore, crib blocks are bulky and require ample space for storage, plus they are not easily transported underground. On top of that, miners frequently are injured handling the crib blocks. Subsequently, various alternative tailgate supports were developed. According to Mucho *et al.* (1999) there are six types of secondary supports: (1) conventional wood cribbing, (2) engineered wood crib supports, (3) conventional and engineered wood post supports, (4) non-yielding concrete support, (5) deformable concrete supports, and (6) yielding steel supports. Conventional wood cribbing (Fig. 4.7.1) is the softest type and may not have consistent quality. The wood will

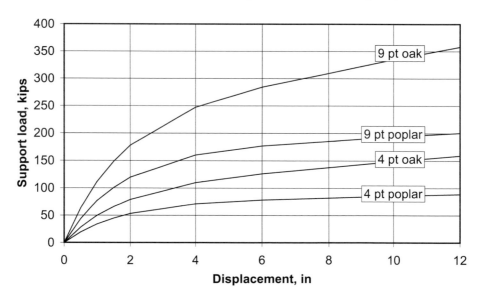

Figure 4.7.1 Load-displacement curves for conventional wood cribs

Source: Barczak (1999)

not take much load before it deforms substantially. Loads increase with deformation but at a decreasing rate. For engineered wood cribs (Fig. 4.7.2), quality is uniform, and the cribbing takes loads faster than the conventional type. It reaches the peak load at around 2 in. (52 mm) deformation. Thereafter, load increase is minor or becomes steady.

Conventional timber props (Fig. 4.7.3) take loads quickly but fails quickly soon after. The engineered timber props (propsetters) will also take a load quickly but maintain that peak load as it deforms for a considerable length. Non-yielding concrete support (Fig. 4.7.4) will also take a load fairly quickly until it reaches the peak load, which is the highest among all six types of supports. But these supports fail quickly after the peak load is reached. ACS props (Fig. 4.7.5) can be adjusted by using different head plates so that the peak load is reached at different rates and after that drops gradually to a residual load. Among the deformable concrete supports (Fig. 4.7.6), the cans take the load quickly and reach the peak load fairly quickly. After that they maintain the peak load for a considerable time. Pumpable cribs take the load very quickly also but drop quickly soon after the peak load.

Currently, the cans and pumpable cribs are most commonly used.

4.7.2.2 Selection of tailgate support

Based on the characteristics of load-displacement curves (Figs 4.7.1–4.7.6), two types of approaches can be used for tailgate secondary support design: stiff and soft. The stiff type refers to those that take a load quickly and fail soon after reaching the peak load (see Figs 4.7.3, 4.7.4, and 4.7.5). This type of support, since it loses its load-bearing capacity after failure, must be designed such that the supporting load density is sufficiently high so that it

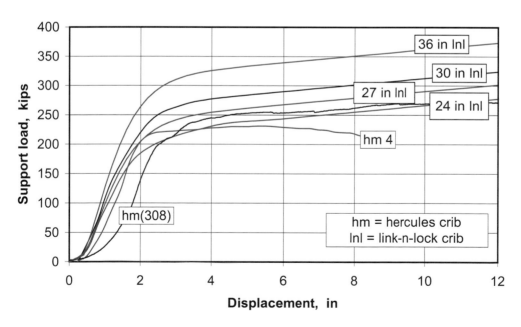

Figure 4.7.2 Load-displacement curves for engineered wood cribs
Source: Barczak (1999)

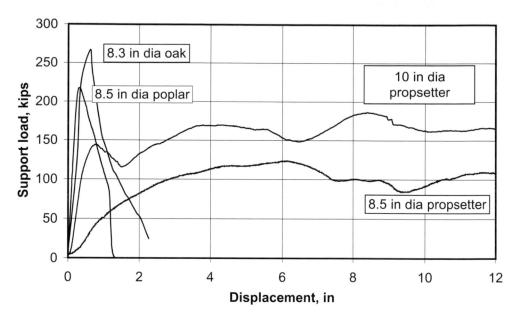

Figure 4.7.3 Load-displacement curves for conventional and engineered timber props
Source: Barczak (1999)

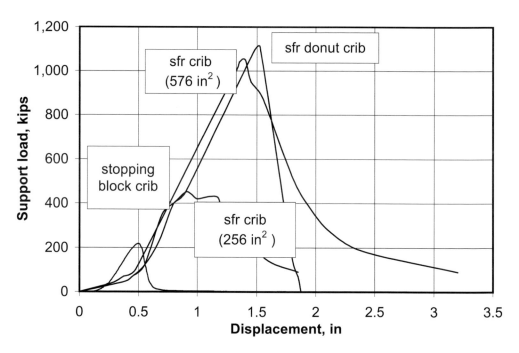

Figure 4.7.4 Load-displacement curves for non-yielding concrete supports
Source: Barczak (1999)

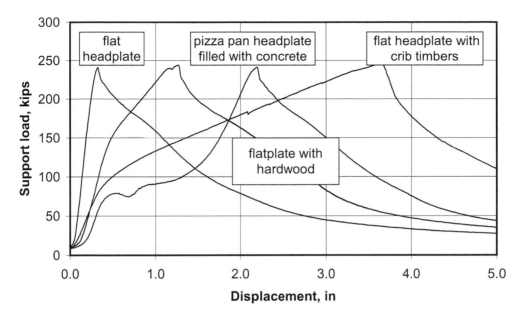

Figure 4.7.5 Load-displacement curves for deformable concrete supports

Source: Barczak (1999)

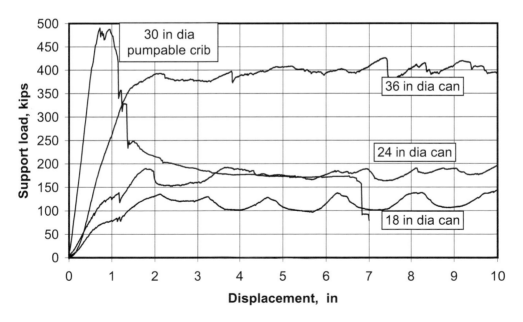

Figure 4.7.6 Load-displacement curves for yielding steel supports

Source: Barczak (1999)

would not fail when the longwall face, and thus, the front abutment pressure is approaching. Conversely, the soft type refers to those that take the load slowly, say after 10–12 in. (25–30 mm) of displacement before they reach the peak load and, after failure, still maintain a large proportion of the bearing capacity for considerable displacement (see Figs 4.7.1, 4.7.2, and 4.7.6). Since the post-failure bearing capacity of this type of support is still high, roof convergence of the tailgate continues until the roof, floor, and support reach equilibrium. For soft type of support, the post-failure supporting density must be sufficient so that the roof-to-floor convergence will not be excessive, leading to roof collapse.

Barczak (2001b), using the concept of ground reaction curve, developed a tailgate support design method, STOP (Support Technology Optimization Program). The three-dimensional finite element method has also been used for standing supports design for open entries or pre-driven recovery rooms (Tadolini *et al.*, 2002). The advantage of the finite element method is that a prior knowledge of the ground reaction curve is not needed.

4.7.3 Pumpable cribs

Pumpable cribs were developed in late 1990s and were first used to replace wood cribs or concrete cribs in longwall tailgates. Since their development, pumpable cribs have been increasingly used in longwall gate entries as well as longwall bleeders for standing supports under various conditions. Pumpable cribs are installed on site by pumping a two-component grout from the surface to underground into a fabric bag. Compared to the conventional standing support such as wood cribs and concrete cribs, pumpable cribs have several advantages including relatively higher support capacity, desirable stiffness, and most importantly, the pumpability of the crib material. The crib material can be pumped over a long distance of about 12,000–15,000 ft (3658.5–4573.2 m) from a surface borehole site to an underground installation location, which greatly reduces material handling for crib installation. Whether pumpable cribs are applicable to certain longwall gate entries and bleeders depends on their mechanical properties, and the geological and mining conditions under which they are installed. In using pumpable cribs, it is important to understand the operation and technical requirements and to design a support pattern which will securely maintain the stability of longwall gate entries and bleeders.

Pumpable cribs were first used in longwall gate entries to support the roof under abutment pressure and then in longwall bleeders to establish long-term support for reliable ventilation. Figure 4.7.7 shows the locations of the pumpable cribs installed in longwall gate entries with a three-entry chain pillar system. Figure 4.7.8 shows the pumpable cribs installed underground. New cribs are pumped in the headgate entries close to the face line as the longwall is being retreated. These cribs provide support to the roof in the headgate entries under the side abutment pressure as the current panel is being mined. As the next adjacent panel is mined, the previous headgate entries become the second entry and tailgate entry for the next panel. The cribs in the tailgate near the face now will not only provide support to the roof under the influence of both side and front abutment pressures, but it will also maintain an opening at the edge of the gob inby the face to allow the ventilation air to pass from the face to the second entry. The cribs in the second entry will help maintain it open for ventilation during and after the second panel is mined (Zhang *et al.*, 2012).

Pumpable crib materials are basically special types of cements or grouts made by mixing with certain additives and large proportion of water. Four main grouts are available on the market: (1) aerated cements, (2) Portland flyash cement, (3) Portland pozzolan cement,

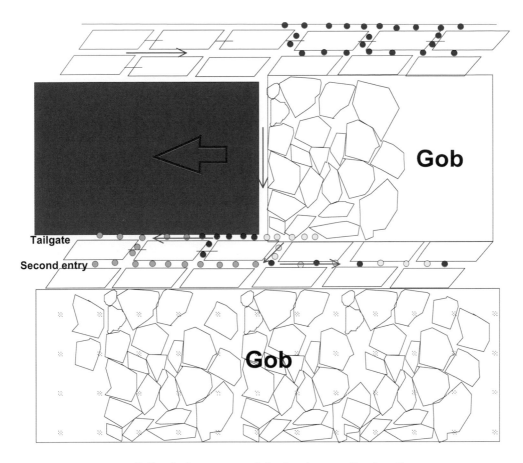

●   **Cribs being pumped during longwall retreating**

◉   **Cribs lightly loaded**

●   **Cribs heavily loaded**

○   **Cribs yielded**

——————▷   **Air flow**

Figure 4.7.7 Locations of pumpable cribs installed in longwall gate entries

and (4) ettringite-based cement (Barczak and Tadolini, 2008). Of these, the ettringite-based cements, calcium sulfoaluminate (or CSA) cements in particular, have several superior qualities such as high water-to-solid ratio of 1.75 to 1 or 2.0 to 1 and quicker setting time of 10 to 15 minutes. These qualities have not been matched by the other grouts.

When pumpable cribs are used for longwall roof support, it is important to first understand the operational requirements for crib installation. As the crib material is pumped from the surface to the underground over more than the length of a longwall panel, it is required that crib material be pumped to a distance of 12,000 ft (3658.5 m) at minimum, preferably

Figure 4.7.8 Pumpable cribs installed underground

15,000–18,000 ft (4573.2–5487.8 m). To ensure pumpability and to reduce cost, it is desirable for the material mix to have a high water-to-solid ratio. As crib material is pumped into a flexible bag hung under the roof, it is also beneficial for the material to have a short gel time for quick crib installation. For installation efficiency, low line maintenance and high pumping rate (cribs pumped per shift) are desired. For a high production longwall with an advancing rate of 50–100 ft d^{-1} (15.2–30.5 m d^{-1}), cribs should be installed at a rate of 20–25 cribs/shift at minimum.

The mechanical properties of a pumpable crib including its capacity and stiffness are determined by the strength of the cured crib material as well as the strength of the bag reinforced by steel wires. Lab testing shows that the uniaxial compressive strength of the cured pumpable material ranges from about 400 psi to 800 psi (2.76–5.52 MPa). The peak and residual capacity of the crib are controlled not only by the material strength but also by the strength of the bag as well as of the strength and spacing of the reinforcing steel wires. The crib's full capacity is achieved four weeks after being pumped. The crib capacities are determined by lab testing. Figure 4.7.9 shows a typical load-displacement curve for a 72 in. high by 30 in. (1.83 by 0.76 m) diameter pumpable crib. The load-displacement curve has a linear portion up to about 1 in. (25.4 mm) convergence with about 175 tons of peak load, and about 8–12 in. (203.2–304.8 mm) residual convergence with 100 tons of residual load. It should be pointed out that the load-displacement curves from lab testing only show crib performance

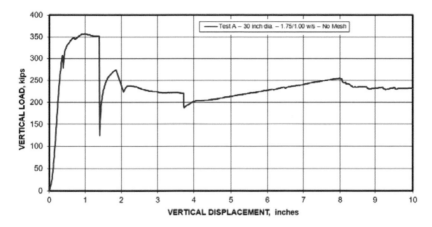

Figure 4.7.9 Typical load-displacement curve for pumpable cribs

under ideal conditions with flat roof and floor but *in-situ* performance could be different due to uneven roof and floor surfaces.

To fully utilize the capacity of pumpable cribs, one has to understand the technical requirements for crib properties in longwall gate entries and bleeders. The important properties of pumpable cribs are stiffness, peak load, residual load, and residual convergence. The stiffness of a crib is defined by how fast the crib can be loaded by roof and floor deformations. Though the Young's modulus of the crib material is a good indication of the crib's stiffness, practically the crib stiffness can be determined by the peak load and the maximum convergence at the peak load. The key point in crib selection is that the deformation of the crib should accommodate the normal deformations of the roof and floor so that the crib can use its peak capacity instead of residual capacity to support the roof as much as possible. It follows that the cribs in the tailgate should not undergo yielding before the face has advanced to nearby.

It should also be noted that cribs in bleeders should be stiffer than those in tailgates since roof-to-floor convergence in bleeders are less than in tailgates. In the entries where large roof and floor deformations have to occur, for instance, the second entry between two longwall gobs, it is desirable for the cribs to have large residual convergence with higher residual load. The crib stiffness requirement in the tailgate can be determined based on the measured convergence in the tailgate near the face.

Pumpable cribs are now widely used in longwall gate entries and bleeders either for resisting abutment pressure or for long-term support, and crib support design should take into consideration both geological conditions and safety requirements. Crib pattern and spacing are largely dependent on overburden depth, roof, floor and rib conditions, and ventilation and safety requirements. In the worst case, the total capacity of the pumpable cribs installed should be sufficient to support the dead weight of the roof strata within the highest potential roof fall height. The potential roof fall height can be determined by analyzing historical roof falls in the mine. It should also be pointed out that crib pattern should be designed not only to prevent roof falls but also to protect personnel walking in the cribbed entries from roof and rib hazards such as roof cutters, roof cavities between roof bolts, and rib sloughage.

4.8 Factors affecting the development rate

Since the gateroad consists of pillars, entries, and crosscuts, their dimensions and geometric relationships determine the amount of excavation needed. In particular the deciding factors include number of entries and pillar size.

4.8.1 Number of entries

Obviously, everything being equal, the linear development footage is proportional to the number of entries. The number of entries used is mostly three- and four-entry while 2 two-entry systems are being used in Utah.

4.8.2 Pillar size

Most pillars are square or rectangular, but some are diamond-shaped to facilitate the miner's movement. Pillar dimensions that affect development rate are width and length. Again, everything being equal, the wider the pillar, the more development work needed. Conversely, the longer the pillar, the less development work needed on a fixed panel length, because the number of crosscuts reduces. Since pillars are designed to support the roof load and protect the adjacent entries, the minimum total width of pillars depends solely on seam depth, coal strength, and roof and floor properties, so mine management has fewer options. Pillar length is different. Since crosscuts are used mainly to direct air between entries, crosscut spacing is important from a ventilation point of view. From a gateroad development point of view, the more crosscuts the slower the development rate. Conversely, the longer the pillar is, the faster the development rate. Crosscut spacing up to 300 ft (91.5 m) centers or pillar length 285 ft (86.9 m) has been used especially when the in-place method is used. The current maximum of 300 ft (91.5 m) crosscut spacing is set by MSHA for proper face ventilation.

Aside from the entry/crosscut and pillar geometrical factors mentioned earlier, the rate of development most critically depends on geological and mining conditions. Good roof/floor and minimum methane always produce a high rate of advance, whereas the reverse condition spells trouble for the development crew. Over the mine's life or even the panel life, these factors come and go and the development rate fluctuates up and down correspondingly. Other factors, including cutting sequence, work schedule, equipment maintenance, and management (Whipkey, 2005) are also factors contributing to development rate.

4.9 Determination of required daily panel development footage

4.9.1 Development ratio

Assuming the planned longwall advance (or mining) rate is V_{lw}, ft d^{-1}, the required footage on gateroad development per day can be determined as follows:

1   Along the entry direction:

$$[L_1 = N \times V_{1w}]$$

(4.9.1)

Where N is the number of entries in the chain pillar system, e.g., three-entry system, $N = 3$.

2   Along the crosscut direction:

$$L_2 = \frac{W_c}{\cos\alpha} \times \frac{V_{lw}}{L}$$

$$W_c = (N-1)W + (N-2)w$$

(4.9.2)

Where W_c is the total width of chain pillars, α is the angle of the crosscut with respect to the entry width direction, L is the pillar length, and W is the pillar width, and w is entry width.

3   Setup room, bleeder entries, and recovery chutes:

Normally the bleeders consist of three entries including the setup entry. The time required for development of the setup entry, which normally is 22–24 ft (6.71–7.32 m) wide, varies considerably, depending on geological condition and supporting requirements. The number and length of recovery chutes also vary depending on face move method adopted and local practice. The normal practice when one refers to the ratio of development rate to longwall mining rate is that this part of development is not considered. But, if it must be considered, and for simplicity, the extra footage is:

$$[L_3 = \eta(3L_b + W_b / S_1 \times P_w / P_1)]$$

(4.9.3)

Where $\eta > 0.11$–0.15 = a constant considering the development of setup entry and recovery chutes, L_b = bleeder pillar length (center-to-center), W_b = the total width of bleeder entries including the setup room, S_1 = crosscut center to center distance of the bleeder system, P_w = panel width, and P_1 = panel length.
    For simplicity, L_3 can be expressed as a percentage of $(L_1 + L_2)$ or:

$$L_3 = \delta(L_1 + L_2)$$

(4.9.4)

Where $\delta = 0.1$–0.4.

4   Minimum daily panel development footage:

$$L_d = L_1 + L_2 + L_3$$

(4.9.5)

Frequently the development ratio, b, is used to indicate the amount of development work needed per foot of longwall advance or:

$$\beta = L_d / V_{lw}$$

(4.9.6)

The development ratio normally runs between three and five, but less than four is considered desirable.

4.9.4.1 Example

Assume a longwall panel of 1000 ft (305 m) wide by 12,000 ft (3658 m) long is developed with a three-entry system having equal sized chain pillars of width 100 ft (30.5 m) center and crosscuts 184 ft (56.1 m) center and entry width 16 ft (4.9 m). The bleeder system is developed the same as the gateroad. Calculate the required panel development footage per foot of longwall face advance. It is assumed that the crosscut has been driven parallel to the entry width direction.

4.9.4.2 Solution

The following data are given for the longwall panel:

Longwall panel length $(P_l) = 12,000$ ft

Longwall panel width $(P_w) = 1000$ ft

Gateroad pillar width $(W) = 84$ ft

Gateroad pillar length $(L_g) = 168$ ft

Bleeder pillar width $(W_b) = 84$ ft

Bleeder pillar length $(L_b) = 168$ ft

Entry width (w) $= 16$ ft

Number of entries (N) $= 3$

$W_c = 2 \times 84 = 168$ ft

Development footage pet foot of face advance is:

$$L_1 = N \times V_{lw} = 3\,\text{ft}$$

Development footage along the crosscut per foot of face advance for rectangular pillars is:

$$L_2 = \frac{W_c}{\cos\alpha}\frac{V_{lw}}{L_g} = 168/184 = 0.913\ \text{ft}$$

For the three-entry gateroad system, the development footage per foot of face advance for the setup entry and bleeder entries, and recovery chutes is:

$$L_3 = \eta\left(3L_b + W_b\,/\,S_1 x P_w\,/\,P_l\right)$$

Substituting all given values and assuming $\eta = 0.11$:

$$L_3 = 0.11\left[3 + (216/184) \times (1000/12,000)\right] = 0.438\ \text{ft}$$

Minimum daily panel development footage per feet of longwall face advance is:

$$L_d = L_1 + L_2 + L_3 = 3 + 0.913 + 0.438 = 4.351\ \text{ft}$$

Therefore the development ratio $\beta = 4.351$.

Based on Equations (4.9.1) through (4.9.6), the effect of panel width and panel length on the development ratio can be plotted as shown in Figure 4.9.1. Development ratio decreases with increase in panel length and increases with increasing panel width. But when panel length gets longer, say greater than 8000 ft (2438 m), the effect of increasing panel width on the development ratio decreases.

4.9.2 Summary

From Figure 4.9.1, it is obvious that when only considering a single panel, panel width and panel length have limited effect on the development ratio. Total pillar width is the dominant factor. The smaller the total pillar width, the smaller the development ratio. For instance, in the same example cited in the previous section, if an individual pillar is reduced to 30 ft (9.1 m), development ratio is reduced to around 3.68.

Therefore, in order to reduce the development ratio effectively, the proper way is to reduce the individual pillar width and/or number of entry development. For the pillar sizes commonly used in US longwalls, development ratios for two-, three-, and four-entry development systems are on the order of 2.5, 4.4, and 5.8, respectively.

The most effective way of greatly reducing the development ratio is to increase the panel width in a fixed reserve block, thereby cutting down the number of panel. In a study by Trackemas and Peng (2013) for panel widening effect for a reserve block of 10,700 × 7700 ft (3262 × 2348 m), increasing panel width from 1050 ft (320 m) to 1600 ft (488 m) reduces the number of panels from six to four, a reduction of two gateroad systems and an increase of about 10% clean tons of coal.

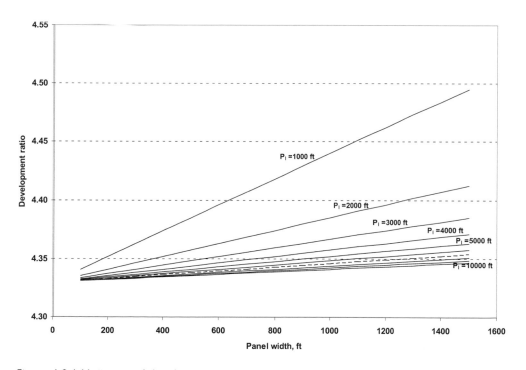

Figure 4.9.1 Variation of development ratio with change in longwall panel dimension

4.10 Examples of gateroad development systems

In this section, examples of the cutting sequence of gateroad development are cited for each of the gateroad development systems, including two-, three- and four-entry systems.

Figure 4.10.1 shows an example of the two-entry development system. Since the two-entry system is used exclusively in western states, such as Utah where diesel is permissible, diesel cars and no track are used. The cutting sequence reflects this advantage.

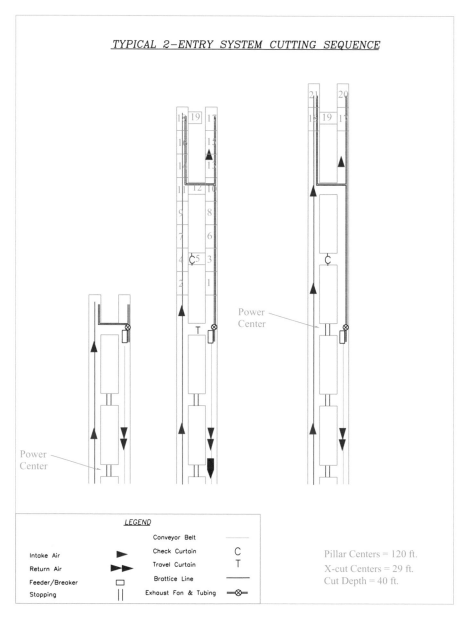

Figure 4.10.1 An example of cutting sequence for a two-entry gateroad system

The numerical numbers in Figures 4.10.1–4.10.6 denote continuous miner's cutting sequence.

Figure 4.10.2 shows a cutting sequence of a three-entry system using a modified super-section, i.e., two miners (left and right), one roof bolter, two shuttle cars, and one belt conveyor at the second (center) entry. The left miner is placed in and cuts number one entry only, counting from the left side. It cuts the first entry when the right miner that will be used for cutting the number two and number three entries is not cutting. So, the system is operated by one crew only. The key elements for this system are (1) a single right-hand turn per crosscut

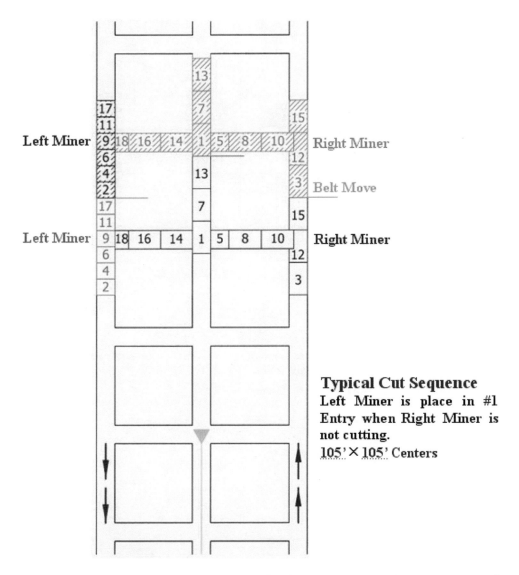

Figure 4.10.2 An example of cutting sequence for three-entry gateroad system using modified super-section method

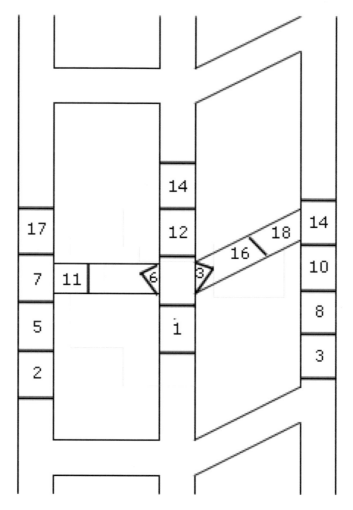

Figure 4.10.3 An example of cutting sequence for three-entry gateroad system using conventional single continuous miner

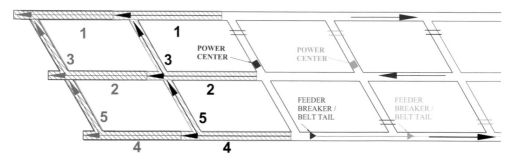

Figure 4.10.4 An example of cutting sequence for three-entry gateroad system (equal pillar dimension) using in-place miner

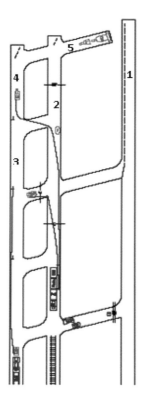

Figure 4.10.5 An example of cutting sequence for three-entry gateroad system (unequal pillar dimensions) using in-place miner

line, (2) the middle entry is farther advanced than the outside entries, forming a Christmas tree shape, (3) the belt conveyor is moved to minimize haul distance and to keep equal distance from the face to feeder while keeping the feeder as inby as possible, and (4) switch out is at the most inby intersection. Variation of the plan includes the belt in the third entry, using only one miner, and different belt move spacing. This system routinely advances 170–250 ft d^{-1} (51.8–76.2 m d^{-1}).

Figure 4.10.3 shows another example of a cutting sequence for the three-entry development system using the conventional single continuous miner.

Figure 4.10.4 shows the five-step three-entry development using the in-place method. In this system, the two rows of chain pillars are equal in width and length, but the crosscuts are inclined at 60° to the entry direction. The inclined crosscuts facilitate the turn negotiation of the miner, which is bulky and long. This system routinely advances 70–110 ft d^{-1} (21.3–33.5 m d^{-1}).

Figure 4.10.5 shows the five-step three-entry development using the in-place method with unequal-sized pillars both in width and length. Note the crosscuts are inclined at 70° to the entry direction.

Figure 4.10.6 shows an example of a four-entry development, while Figure 4.10.7 is a five-cut three-entry for FCT continuous haulage system.

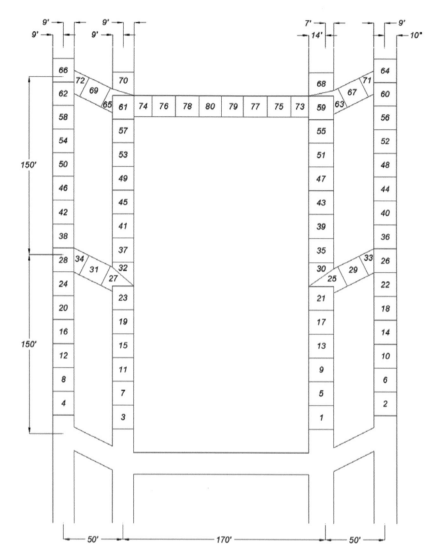

Figure 4.10.6 An example of cutting sequence for four-entry gateroad system

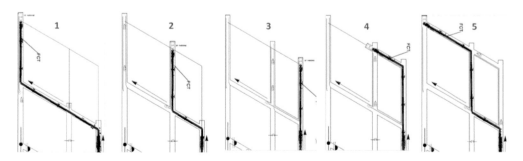

Figure 4.10.7 An example of FCT cutting sequence for three-entry development

4.11 Calculation of coal recovery rate in a longwall panel

P_l: panel length, ft
P_w: panel width, ft
P_{cw}:   panel center width, ft
W: entry width, ft
N: number of entries in the gateroad (normally 3)
M: number of entries in the bleeder (normally 3)
N_g: number of crosscuts in the gateroad
N_b: number of crosscuts in the bleeders
L: pillar length, ft
W: pillar width, ft

For L and W, subscripts g and b represent gateroad and bleeders pillars, respectively.

W_{bp}:   width of barrier pillar between mains and panel recovery line, ft (normally 200–500 ft or 61–152 m)

Calculation

Total area of the panel coal block is:

$$A = P_{cw} \times \left[P_1 + 2\left(W_b + W \right) \right]$$

Where $P_{cw} = P_w + 2\left(W_g + W \right)$

After extraction of the panel, the pillars left in the panel consist of three parts: (1) those in the gateroad, (2) those in the bleeders, and (3) the barrier pillar left to protect the mains.

1   Area of the left-out gateroad pillars, A_g:

$$A_g = N_g \sum_{i=1}^{n-1} L_{gi} \times W_{gi}$$
$$N_g = \left[P_1 + 2\left(W_b + W \right) \right] / \left(L_g + W \right)$$

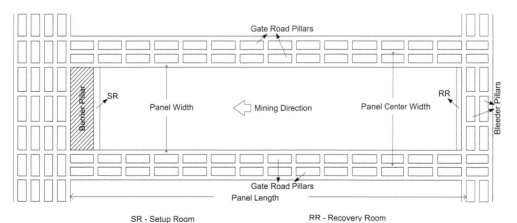

Figure 4.11.1 Panel terminology for coal recovery calculation

2   Area of the left-out bleeder pillars, A_b:

$$A_b = N_b \sum_{i=1}^{m-1} L_{bi} \times W_{bi}$$
$$N_b = P_W / (L_b + W)$$

3   Area of the barrier pillar, A_{bp}:

$$A_{bp} = W_{bp} \times (P_w - W)$$

Total coal recovery (U) in a longwall panel assuming uniform coal thickness is:

$$U = \left[A - \left(A_g + A_b + A_{bp} \right) \right] / A$$

Example

Assuming a longwall panel, 1000 ft (305 m) wide by 12,000 ft (3658 m) long, is developed with a three-entry system having equal sized chain pillars of width 100 ft (30.5 m) center and crosscuts 184 ft (56.1 m) center and entry width 16 ft (4.9 m). The bleeder system is developed the same as the gateroad.

$$
\begin{aligned}
A &= P_{cw} \times \left[P_l + 2(W_b + W) \right] \\
&= \left[1000 + 2(84 + 16) \right] \times \left[12,000 + 2(84 + 16) \right] = 14,640,000 \, \text{ft}^2 \\
N_g &= \left[P_l + 2(W_b + W) \right] / (L_g + W) \\
&= \left[12,000 + 2(84 + 16) \right] / (168 + 16) = 66.3 \\
N_b &= P_w / (L_b + W) \\
&= 1000 / (168 + 16) = 5.43 \\
A_g &= N_g \sum_{i=1}^{n-1} L_{gi} \times W_{gi} \\
&= 66.3 \left[84 \times 168 + 84 \times 168 \right] = 1,871,251.2 \, \text{ft}^2 \\
A_{bp} &= W_{bp} \times (P_W - W) \\
&= 200 \times (1000 - 16) = 196,800 \, \text{ft}^2
\end{aligned}
$$

Panel coal recovery ratio is:

$$
\begin{aligned}
U &= \left[A - \left(A_g + A_b + A_{bp} \right) \right] / A \\
&= \left[14.64 \times 10^6 - (1,871,251.2 + 153,256.32 + 196,800) \right] / 14.64 \times 10^6 \\
&= 84.38\%
\end{aligned}
$$

Panel coal recovery increases with panel width and length as shown in Figure 4.11.2. But the rate of increase decreases with increasing panel length. When panel length reaches

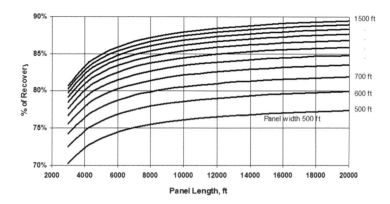

Figure 4.11.2 Coal recovery increases with increase in panel width and length

16,000 ft (4877 m), the rate of increase becomes insignificant. On the other hand, the effect of increasing panel width is more significant when panel width is less than 1000 ft (305 m) and panel length is less than 6000 ft (1829 m). Beyond that, the effect of increasing panel width and length on coal recovery becomes insignificant.

Chapter 5

Shield support – general

5.1 Introduction

Modern longwall mining employs self-advancing hydraulic powered supports (which will simply be called powered supports or support or shield in this book) at the face area. The powered support has, as it has evolved over the past few decades, not only holds up the roof providing a secure area for miners and face equipment and all associated mining activities, but also pushes the armored face chain conveyor (AFC), and advances itself. In recent years, it also provides a delivery platform for dust-suppressing water sprays and a mounting platform for workplace lighting. Its successful selection and application are the prerequisite for successful longwall mining. Furthermore, due to the large number of units required, the capital invested for the powered support usually accounts for more than 60–80% of the initial investment for a longwall face depending on the panel width. From both technical and economic points of view, the powered supports are very important pieces of equipment in a longwall face.

The application of modern powered supports can be traced to the early 1950s. Since then, following their adoption in every part of the world, countless models have been designed, manufactured and used in various countries. Figure 1.5.9 shows the year by year change in number and type of powered supports used in US longwall mines since 1976. Four distinct types: the frame, chock, shield, and chock shield are described in the order of their evolution of development.

5.1.1 Frames

The frame support (Fig. 5.1.1) is an extension of the single hydraulic props conventionally used underground, and the first type developed in modern self-advancing hydraulic powered supports. Its application involves setting up two hydraulic props or legs vertically in tandem that are connected at the top by one- or two-segmented canopies. The two-segmented canopies can be hinge-jointed at any point between the legs or in front of the front leg. The base of the two hydraulic legs may be a circular steel shoe welded at the bottom of each leg or a solid base connecting both legs. If the steel shoes are used, spring plates connecting the legs are used to increase the stability. This frame support is very simple but also flexible and less stable structurally. There are considerable uncovered spaces between the two pieces of canopy that allow broken roof rocks to fall through (Peng and Chiang, 1984). For this reason and due to its flexible structure, the frame support was quickly abandoned in the late 1970s, and its application completely eliminated in 1985 (see Fig. 1.5.9).

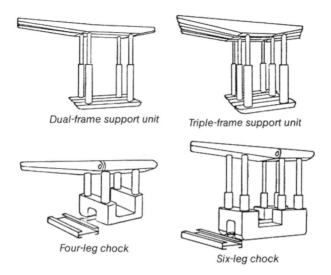

Dual-frame support unit   Triple-frame support unit

Four-leg chock

Six-leg chock

Figure 5.1.1 Frames and chocks

5.1.2 Chocks

In a chock support (Fig. 5.1.1), the canopy is a solid piece and the base is two separate parts connected by steel bars at the rear and/or the front ends. A large open space is left at the center for locating the double-acting hydraulic ram (DA ram), which is used to push the chain conveyor and to pull the chock unit forward, respectively, which is a distinctive difference from the frame support. This setup is also used for shields and chock shields. All hydraulic legs were installed vertically between the base and the canopy. The number of legs ranged from three to six, but four-leg chocks were by far the most popular ones. In most chock supports, hinge joint connections ran between the legs and the canopy and between the legs and the base. In order to increase the longitudinal stability, the chock support was reinforced mostly with a box-shaped steel frame between the base and each leg. A leg restoring device was installed around each leg at the top of the box-shaped steel frame. Chocks were suitable for medium to hard roof. When the roof overhung well into the gob and required induced caving, chocks provided access to the gob (Peng and Chiang, 1984). In the United States, chocks were abandoned and ceased to be used in 1994 (see Fig. 1.5.9).

5.1.3 Shields

Shields (Fig. 5.1.2) were first introduced in a northern West Virginia mine in 1975. Its acceptance by the industry largely established longwall mining as a viable coal mining method. Shields are characterized by the addition of a caving shield-lemniscate-link system at the rear end between the base and the canopy. The caving shields, which in general are inclined, are hinge-jointed to the canopy and, together with the lemniscate assembly, to the base making the shield a kinematically stable support, a major advantage over the frames and the chocks.

It also seals off the gob and prevents rock debris from getting into the face side of the support. Thus, the shield-supported face is generally clean.

The hydraulic legs in shields are generally inclined to provide a larger working height range and more open space for traffic. Because the canopy, caving shield, and base are interconnected, the shield can well resist the horizontal force without bending the legs. Unlike the solid constraint in the frame/chock supports, the spherical head connections between the legs and the canopy and between the legs and the base plate in a shield support make it possible for the angle of inclination of the hydraulic legs to vary with mining heights. Since only the vertical component of the hydraulic leg pressure is available for supporting the roof, the roof support capacity of the shield also varies with the mining heights (Fig. 5.1.3).

In a two-leg shield, both legs are connected between the canopy and base plates. Two-leg shields are further divided into two types: the first shield type, introduced in 1975, was the caliper shield, in which the canopy tip moved away from the faceline as the shield is raised up, resulting in a larger unsupported distance between canopy tip and faceline. The caliper shield was abandoned as fast as it was introduced. The second type is the lemniscate shield, which remains the most popular type.

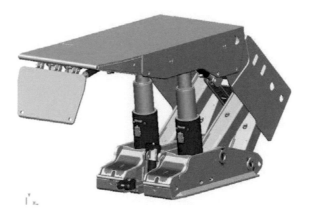

Figure 5.1.2 Shields
Source: Courtesy of Caterpillar

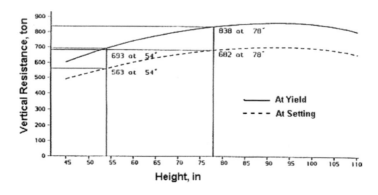

Figure 5.1.3 An example of shield load varying with its operating height

The caving shield and the base are joined by two lemniscate bars on each side that have a total of four hinges. As the hydraulic legs are raised and lowered, the kinematics of the lemniscate assembly is such that the tip of the canopy moves up and down nearly vertically, thus maintaining a nearly constant unsupported distance between the faceline and the tip of canopy. This feature is widely considered to be most desirable for good roof control (Peng and Chiang, 1984). There are clear limits of mining height, within which the leading edge of the canopy moves nearly vertically. These limits are strictly controlled by the dimensional and positional arrangements of the canopy, caving shield, lemniscate bars, and the base. Beyond these limits, the canopy will move rapidly away from the faceline, creating a large unsupported area. All shields in the United States now employ lemniscate links that have also become standard worldwide.

Shields are equipped with a stabilizing ram (or tilting cylinder) between the canopy and caving shield. Its major function is to maintain the canopy position during support advance so that the canopy and the caving shield will not collapse at the hinge. To some extent, the stabilizing ram can be used to change the application point of the resultant force on the canopy. It could be considered as an additional leg, especially in recent models of high capacity shields; the capacity of the stabilizing cylinder can reach up to 200 tons.

Research in the 1980s demonstrated (Peng et al., 1989; Peng, 1990) that the two-leg lemniscate shield is the most desirable one. So, since 1990, all new shields purchased are the two-leg type and it is now the only type used in US longwall mining.

Shields are made in 1.5, 1.75, and 2.05 m wide in order of development with yield capacity up to 1925 tons (1750 metric tons) and working height range 2.6–25.6 ft (0.8–7.8 m) and prop bore size 125–480 mm. Currently most shields are 1.75 m wide and 1.5 m wide is phasing out.

5.1.4 Chock shields

The chock shield (Fig. 5.1.4) possesses the advantages of both the chock and shield because it combines the features of both of them, i.e., it is a shield with four legs between the canopy and base plates. All of the four legs are installed vertically or in a V shape between the

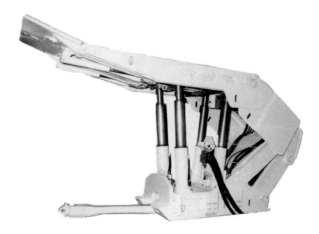

Figure 5.1.4 Chock shield

canopy and the base. This is why it is called a chock shield. A chock shield has the highest supporting efficiency and is suitable for a hard roof. However, research showed that the rear two legs were in most cases underused. Chock shields were totally abandoned in the United States in 2002 (see Fig. 1.5.9).

5.2 Two-leg shields

As mentioned earlier, two-leg shields are now standard for longwall mining in the United States (see Fig. 1.5.9) and note that the decrease in the number of faces using two-leg shields since 1992 reflects the decreasing number of longwalls, not the percentage of two-leg shield. After nearly three decades' evolution in the late 20th century, the structural design and component features and characteristics of two-leg shields have more or less been standardized and adopted for use in modern high-production, high-efficiency longwalls all over the world. Figure 5.2.1 shows two typical two-leg shields manufactured by the two current major original equipment manufacturers (OEM) for longwall face equipment.

5.2.1 Why two-leg shields?

The reasons that two-leg shields became the choice for modern high production longwalls are discussed here.

5.2.1.1 Simplicity

Two-leg shields have only half as many hydraulic components (e.g., legs, cables, hoses, and valves) as those for four-leg shields or chock-shields. There are only three hoses (one each of high pressure and return hydraulic hose, water) and one electrical cable hanging between shields, making the walkway spacious and the face cleaner. They are much simpler to operate and require less maintenance due to their fewer hydraulic components.

5.2.1.2 High capacity

One of the major reasons for the introduction of four-leg powered supports in the late 1970s was to increase support capacity by increasing the number of legs. As the technology to make a larger leg prop became reality, a two-leg shield could have a large load-bearing capacity by using large diameter leg props. With current leg diameter exceeding 480 mm, the maximum capacity of a two-leg shield can easily be more than 1800 tons (1636 metric tons).

5.2.1.3 Improved design features

In two-leg shields, the maximum floor pressure is located under and near the toe of the base plate (see Fig. 10.14.3), which tends to push the toe into the floor if the floor material is soft. The universal use of a positive base-lifting device has already corrected/reduced the problem. Note that "positive" refers to the action that the toes of the base plates are lifted right before the DA ram is engaged to pull the shield forward, thereby facilitating the advancement of the shield in soft bottom conditions.

  The stabilizing ram or tilt cylinder between the canopy and caving shield has its own hydraulic supply and is independent of all other hydraulic supply, such that in some cases, it can be used to adjust the location of the roof resultant load.

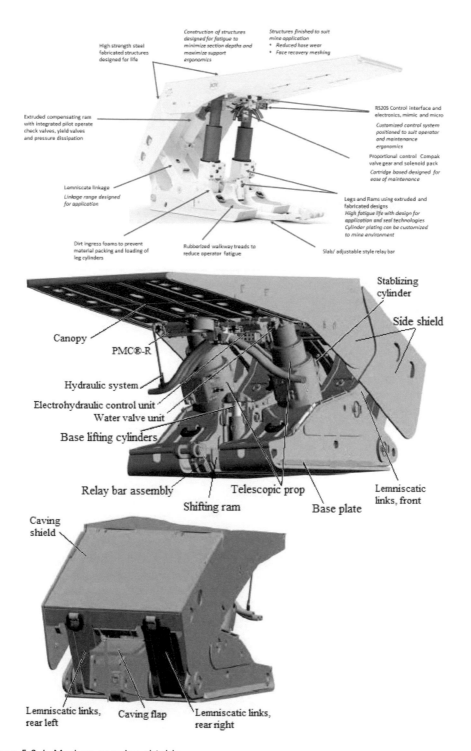

Figure 5.2.1 Modern two-leg shields

Source: Upper, courtesy of Komatsu Mining; lower, courtesy of Caterpillar

5.2.1.4 More efficient roof support

Another significant advantage of the two-leg shield design is that it has a single load path, resulting in a more efficient support design. With only two leg cylinders, all of the load must be transferred from the mine roof or floor through these two legs. This design ensures that the full support capacity will be achieved, regardless of the roof or floor contact configurations, and that the resultant roof force will be consistently located near the leg location. In comparison, the load is generally shared between the front and rear legs in four-leg shields, making the resultant force location more variable and generally further from the face. Imbalanced loading between the front and rear set of legs can occur, and one sets of legs may yield, and thereby lowering the overall capacity of the shield. In extreme situations where cavities are formed above the canopy, one set of legs may not carry any load and thereby reduce the rated shield capacity by 50%.

5.2.1.5 Active horizontal force

Since the two hydraulic legs in a two-leg shield are inclined toward the face, the horizontal component of the leg force when the shield is being set against the roof provides an active horizontal force to the immediate roof including the unsupported distance between canopy tip and faceline. For a medium and/or weak roof where pre-mature roof falls frequently occur in the unsupported distance between the canopy tip and faceline, this is a very important feature. Rocks are weak in tension and the active compressive horizontal force can reduce or eliminate the tensile stress in the immediate roof (see Section 10.6 for details).

5.2.2 General features of two-leg shields

Modern two-leg shields have the following structural and operational features.

5.2.2.1 Solid canopy

In the 1970s and early 1980s, the canopies of many models of powered supports were made of two parts: the main and extension canopies. The idea was that a shorter canopy can better accommodate roof configurations resulting from undulations or cavities and that an extension canopy or a forepole capsule in the front end could deliver a larger tip load to the areas around the tip of the canopy. However, underground application showed that this type of design was not effective.

Modern two-leg shield employs a one-pieced solid canopy, 12–16 ft (3.7–4.9 m) long using high-strength steel with uni-axial tensile strength, 120,000 psi (827.4 MPa), and the canopy tip is raised up 2–3 in. (51–76 mm) higher than the rear edge in order to ensure the designed tip loading is realized in practice. The canopy thickness is either uniform or mostly not uniform throughout the canopy length. It tapers toward the front with the front end has a thinner cross-section since the tip loading is considerably less than other sections of the canopy.

5.2.2.2 Split base plate with rigid connection or catamaran

The base consists of two sections separated by a gap where the double-acting cylinder and relay bar are placed for pushing the pans and pulling the shields forward. The two sections are commonly connected in the front end by a rigid bridge shaped like a tomb stone. The rigid base design can negotiate soft floor better than split base designs as a whole unit.

Split base can better negotiate uneven floor, while rigid connection between the base plates provide better stability.

5.2.2.3 Base-lifting device

The base-lifting device is located between the two base plates either in the front (more popular) or the back side of the rigid connection bridge. As mentioned earlier, the shield is programmed to lift the base plates right before the shield is pulled forward to prevent blockage by the floor steps in hard floor conditions or hooved floor material in soft bottom conditions.

5.2.2.4 Reversed-mounted double-acting ram (DA ram or cylinder)

The DA Ram is used to advance the pans and pull the shields forward a pre-determined distance (i.e., normally equivalent to the shearer's cutting web) and to keep the pans and shields in two separate straight lines parallel to the faceline. It is mounted in reverse manner so that the force that pulls the shield is that generated by the front side of the piston, the area of which is larger, and thus, the force produced is larger; whereas, that pushing the pans is from the rear side of the piston, the area of which is smaller with smaller force (Fig. 5.2.2).

5.2.2.5 Full side shield

A side shield is installed on either the head or tail side of the canopy, caving shield, and lemniscate links. It consists of two overlapping sections, the canopy and caving shield, thereby

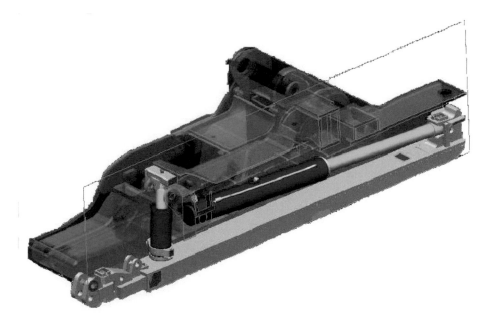

Figure 5.2.2 Reversed-mounted double-acting (DA) ram
Source: Courtesy of Caterpillar

extending from the rear edge of the canopy to the leg socket location or up to the AFC furniture. The side shield is also controlled by DA rams and normally is used to cover the gap between shields to prevent roof material from falling off between them, preventing injury to the face crew and/or debris from falling on the shield walkway, thereby keeping the face clean.

5.2.2.6 Long life

Modern shields are made of high-strength corrosion resistant material using state-of-the-art welding technology. They are rigorously tested to last at least 60,000 (mostly)–75,000 mining cycles.

5.2.2.7 Automation

All shields use electrohydraulic control to enable fast cycle time, positive setting capability, and automated support advance.

5.3 Elements of shield supports

The modern high capacity shield supports have the following components (Fig. 5.2.1):

1   Load-bearing units – these include the canopy, base, caving shield, lemniscate bars, and hinge pins (Peng and Chiang, 1984).
2   Hydraulic rams or cylinders – these include the hydraulic rams for (1) pushing the chain conveyor and advancing the shield (DA Ram), (2) operating the flippers, (3) stabilizing or limiting the position of the canopy with respect to caving shield, and (4) operating other auxiliary equipment, such as the base-lifting device, side shields, and, above all, (5) operating the shield legs or props.
3   Hydraulic control and operating units – these include internal control valves, such as check and yield valves, in the hydraulic legs, unit control valves, yield valves, high pressure hydraulic hoses, and water valves.
4   Auxiliary devices – these include flippers, base-lifting, lighting, water sprays, and specialty sensors.
5   Electrohydraulic control unit – each shield is equipped with a microprocessor with a keyboard that controls a group of solenoid valves, which in turn control the hydraulic functions of the shield, either manually or automatically or remotely.
6   Hydraulic power supply – this is the pump system and emulsion for operating the shield hydraulics.

Each of the six components is discussed in detail in the following sections.

5.3.1 Load-bearing units

5.3.1.1 Canopy

The canopy is made of top and bottom steel plates forming a deck structure that is strengthened by internal grids. The steel grids are welded in transverse and longitudinal directions. The longitudinal grids are really the main load bearing sections. They act as a beam and are designed to prevent bending of the canopy from the leading end (tip) to the rear end. The transverse grids, on the other hand, control bending across the canopy width and torsional loading (twisting) of the canopy. This arrangement employs continuous butt weld and is more tolerant to fatigue motion.

Since the canopy is the only component that makes direct contact with the roof, and the roof is usually stepped and/or undulated, it must be strong and stiff so that it can convey and resist loading to and from the roof.

As mentioned earlier, the canopy is solid and single-pieced. It may be uniform in thickness along the whole length. On the other hand, the portion in front of the leg-connection point is usually tapered toward the face due to lower stress acting toward the canopy tip. The canopy is made of high-strength steel in order to carry the load and, in the meantime, to reduce the weight of the component to a minimum due to the ever-increasing shield size. This design produces a stiff canopy and usually makes less contact with the roof due to inflexibility over the whole canopy area. To reduce this problem, the tip is raised up 2–3 in. (51–76 mm) in order to ensure good tip contact, although studies show this is not necessarily effective (Pothini *et al.*, 1992). The canopy is connected to the caving shield by two hinge assemblies on both sides.

The proportion of canopy length is an important point in shield design. A ratio of two to one was generally employed to represent the ratio of the canopy length in front of a fixed point to that behind it. The fixed point can be either the hinge point where the legs are anchored or the point where the resultant force acts. Such a design produces an idealized triangular load distribution with the largest component at the rear end and decreasing linearly to zero at the leading end. This arrangement ensures that the whole canopy will provide roof support. However, in practice, the canopy length ratio is mostly between 2.5:1 and 3.0:1 so that the shield can provide a prop-free front of 12–16 ft (3.66–4.88 m) in length and minimize the unsupported area in front of the shield.

5.3.1.2 Base

The base consists of two separate equal sections forming a split base. The structural design of the base is similar to that of canopy. Split bases with a rigid connection or catamoran are used exclusively because they are better able to control loading in undulated floor but are sufficiently strong to resist large differential displacement

The base is the unit that receives and transfers the roof pressure to the floor. Just like the canopy, it must be strong and stiff so that it is stable for the uneven floor. It must control the pressure distribution, especially near the front end, in order not to penetrate into the floor. It must also provide sufficient space for hydraulic legs, the double-acting ram, and other auxiliary devices.

Similarly, the bases are made of steel plates that are welded into grid deck structure. The lower front edge of the base is beveled (or has a swiveled front toe) so that it not only can avoid scooping up bottom coal/rock debris, but also prevents it from digging into the floor during shield advance. In order to avoid bias loading, the connection between the base and the legs is such that it has a spherical contact surface. The top surface of the front toe is covered with an anti-slip plate for safe walking.

5.3.1.3 Caving shield, lemniscate links, and pins

The caving shield is also a grid-deck structure. It makes the shield a kinematically stable unit. The caving shield together with lemniscate links resist horizontal forces, both transversely and longitudinally, and biaxial loadings. It dips about 20–25° from the horizontal. It is also used for preventing the caved rock fragments from getting into the face area. As such, it is frequently required to bear the load from the caved rock. The straight solid caving shield is exclusively used.

The lemniscate links are formed from strong steel plates welded into a box structure. There are two lemniscate links symmetrically placed on both sides of the shield. One end of each of the lemniscate links is hinged to the bottom portion of, and moves with, the caving shield while the other end is hinged to the rear end of the base. Such an arrangement enables the tip of the canopy to move up and down more or less vertically within a certain range of mining height. The lemniscate links must be strong because, together with the caving shield, they are subjected to the lateral forces toward the gob and the external force from the caved rock fragments. In general, the upper or front lemniscate link is mostly under compression while the lower or rear one is under tension.

Round steel pins are used for the hinge between the canopy and the caving shield and between the lemniscate links and the caving shield and the base. They are undoubtedly subjected to high loadings. The pins are usually 3–6 in. (76–152 mm) in diameter. The larger pins are used to withstand extreme pressures. The pin diameter must be tightly fitted into the hinge hole so that the canopy tip will follow the theoretically vertical up and down trace of the lemniscate links.

5.3.2 Hydraulic props or cylinders or rams

Figure 5.3.1 shows the full range of hydraulic props or cylinders or rams used in a shield.

Hydraulic rams are the power source that converts fluid power into mechanical power. They not only have to resist long-term high pressure generated within the shield structure both in tension and in compression but also in bending. They must have good sealing characteristics to maintain long-term high pressure with no or minimal leakage during their service life.

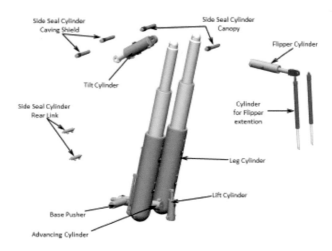

Figure 5.3.1 Full range of hydraulic cylinders used in shield

Source: Courtesy of Caterpillar

The leg prop size ranges from 125–480 mm in diameter. There are single- and double-acting and single-, double-, and triple-stage telescopic hydraulic cylinders. In a single-acting cylinder, the piston is raised by hydraulic pressure but lowered by the weight of the structure it is supporting. Since retraction of the piston by gravity loading is very slow, the single-acting cylinder is no longer in use. On the other hand, in a double-acting cylinder, the piston is raised and lowered by hydraulic pressure such that piston retraction is much faster, thereby reducing shield cycle time. Pistons are either solid or hollow tubing to reduce shield weight.

The extension of a single-stage (single-telescopic) hydraulic cylinder is adequate, but the collapsed height limits its operation range, making it unsuitable for coal mining, because coal seams vary considerably from panel to panel or even within a panel. Due to the "overlap" or the designed length for the guidance of pistons and bushings, the extended length for a single-stage cylinder is less than twice the cylinder length. Therefore, if a single-stage prop is mounted vertically, the ratio of maximum height to collapsed height is less than two. The major advantage of the two-stage (double-telescopic) cylinder is its small collapsed height. With the use of a double-telescopic hydraulic cylinder, the ratio of the maximum extended height to the closed height has increased to two to one. In order to further increase the ratio of extended to collapsed height, the hydraulic legs in a shield are mounted in inclined positions. The larger the inclination from the vertical, the larger is the height adjustment ratio (Fig. 5.3.2).

Figure 5.3.2 shows that for a double-telescopic prop or cylinder, in addition to inclination effect, the smaller the closed (or collapsed) height, the smaller the maximum attainable ratio of extended to closed height. Note the steeper the prop angle the more uniform the support resistance over the entire range of vertical adjustment of the shield. The characteristics curves shown in Figure 5.3.2 vary with the design of prop including prop diameter and piston width. For the same angle of inclination, a triple-telescopic prop or leg has the largest height adjustment ratio followed by the double-telescopic and then the single-stage prop. But the triple-telescopic props are not used due to their complexity, even though they are designed to improve the characteristic curve of the shield and increase the range of vertical adjustment.

In a double-telescopic hydraulic leg prop, there are two pistons, one inside the other, forming a bottom and a top stage. The piston configurations and fluid inlets and outlets are arranged such that, during both extension and retraction, the bottom piston always moves first followed by the top piston at the end of the stroke. An example is shown in Figures 5.3.3 and 5.3.4. When the leg prop is to be raised (Fig. 5.3.3), the high-pressure fluid enters the front chamber of the bottom piston through the fluid inlet port causing the bottom piston to rise. When the bottom piston touches the distance-control ring at the cylinder top, and the bottom stage is fully stroked, the high-pressure fluid forces the check valve to open and enters the front chamber of the top piston, thereby causing the top piston to begin to rise. When the leg is to be lowered (Fig. 5.3.4), the high-pressure fluid enters the retract annulus of the bottom piston through its inlet port. The bottom piston is lowered first. When it reaches the bottom of the bottom cylinder, the check valve hits the stage rod at the center of the bottom cylinder causing the port between the front chambers of the bottom and top cylinders to open. At this time, the high-pressure fluid is switched to the inlet port leading to the retracting annulus of the top piston causing the top piston to begin to lower down.

Other auxiliary hydraulic rams, such as the DA ram for advancing the support or the stabilizing ram for controlling the canopy tip's up and down and balancing the canopy, are also double-acting single-telescopic cylinders.

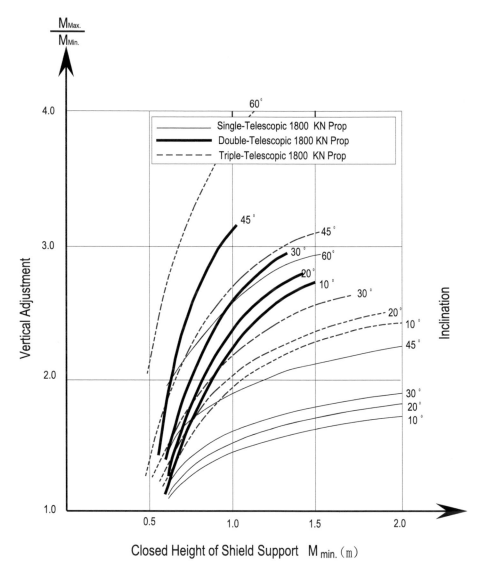

$$\frac{M_{Max.}}{M_{Min.}}$$

4.0

60°

— Single-Telescopic 1800 KN Prop
— Double-Telescopic 1800 KN Prop
- - - Triple-Telescopic 1800 KN Prop

45°

45°

3.0

30°   60°

20°

10°

30°

20°

10°

45°

2.0

30°

20°

10°

1.0

Vertical Adjustment

Inclination

0.5   1.0   1.5   2.0

Closed Height of Shield Support   M min. (m)

Figure 5.3.2 Vertical height adjustment of hydraulic legs varies with number of telescopes and angle of inclination

Longwall mining is operated under particularly demanding circumstances, due to the combination of high pressure, excessive lateral loadings, and low lubricity fluid in a dirty and abrasive working environment. As such, prevention of fluid leakage over the service life of prop or cylinder or ram is extremely important. Figure 5.3.5 shows the seal locations for a shield leg prop. The piston seal is the most critical, because it has to withstand high pressure and the effect of additional friction and force. The top stage working pressure is greater than that of the lower stage due to the difference in piston areas. Therefore, the top stage piston is the most susceptible to leakage. Seal material must match the interior surface of the prop.

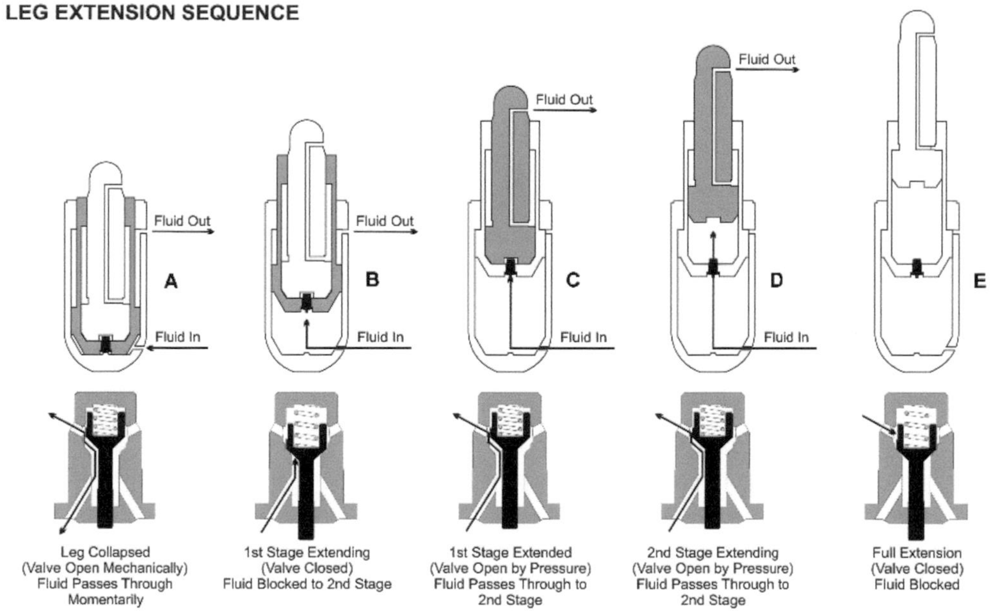

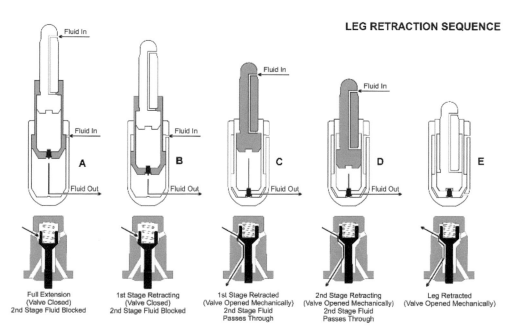

LEG EXTENSION SEQUENCE

Fluid Out

Fluid Out

Fluid Out

Fluid Out

A

B

C

D

E

Fluid In

Fluid In

Fluid In

Fluid In

Leg Collapsed
(Valve Open Mechanically)
Fluid Passes Through
Momentarily

1st Stage Extending
(Valve Closed)
Fluid Blocked to 2nd Stage

1st Stage Extended
(Valve Open by Pressure)
Fluid Passes Through to
2nd Stage

2nd Stage Extending
(Valve Open by Pressure)
Fluid Passes Through to
2nd Stage

Full Extension
(Valve Closed)
Fluid Blocked

Figure 5.3.3 Sequence of leg extension

Source: Courtesy of Swanson Industries

LEG RETRACTION SEQUENCE

Fluid In

Fluid In

Fluid In

Fluid In

Fluid In

A

B

C

D

E

Fluid Out

Fluid Out

Fluid Out

Fluid Out

Full Extension
(Valve Closed)
2nd Stage Fluid Blocked

1st Stage Retracting
(Valve Closed)
2nd Stage Fluid Blocked

1st Stage Retracted
(Valve Opened Mechanically)
2nd Stage Fluid
Passes Through

2nd Stage Retracting
(Valve Opened Mechanically)
2nd Stage Fluid
Passes Through

Leg Retracted
(Valve Opened Mechanically)

Figure 5.3.4 Sequence of leg retraction

Source: Courtesy of Swanson Industries

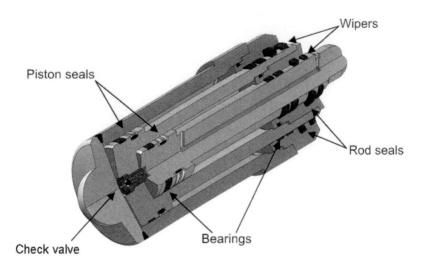

Figure 5.3.5 Type and location of hydraulic leg seals
Source: Courtesy of Caterpillar

Both the top and bottom of the prop are spherical in shape for easy connection to the canopy and base plate, respectively, and robust load transmission from canopy to leg and leg to base plate. Flexible inserts filled around leg/canopy, and leg/base plate pockets keep them free from debris buildup that can lead to bending loads (Lyman, 2017).

The pistons in the hydraulic legs and rams are coated with a thin layer of chrome or zinc to protect them from rust or corrosion. In addition, protective covers are used to cover the exposed piston surface to prevent it from contacting mine water, many of which are very corrosive. Corrosion has been identified as a key problem in longwall mining machines.

5.3.3 Control and operating units

The control and operating units are used to direct and control the hydraulic fluid flow. They must be well sealed from long-term high-pressure application and be sensitive to pressure change.

Bearings are subject to and must be able to withstand lateral loadings.

5.3.3.1 Internal control units

These include the check valve and the yield valve. The check valve is used to control the flow direction of the pressurized fluid, and the yield valve maintains a constant maximum fluid pressure in the hydraulic legs. When the fluid pressure in the piston chambers exceeds the preset value in the yield valve, the port opens and allows the fluid to exit, which reduces the fluid pressure to below the preset value.

A. CHECK VALVE

This controls the flow direction of high-pressure fluid inside the hydraulic legs. The bottom figures in Figures 5.3.3 and 5.3.4 show how the check valve works. During leg extension, when the shield is at the collapsed height, the valve is opened mechanically (Fig. 5.3.3A). As the high-pressure fluid enters the front chamber of the bottom cylinder, it passes through the valve momentarily. When the bottom piston begins to extend, the valve is closed, and the fluid is blocked to the top stage (Fig. 5.3.3 B). When the bottom stage reaches the end of the stroke, the valve is opened by pressure (Fig. 5.3.3 C) and the fluid passes through the valve to the front chamber of the top stage. The top stage begins and continues to extend (Fig. 5.3.3 D). When the top stage reaches the end of the stroke, the valve is closed, and the fluid is blocked (Fig. 5.3.3 E). During leg retraction, the sequence of valve action is just the opposite, beginning from when the top stage is fully stroked (Figs 5.3.3 E and 5.3.4 A). As the fluid enters the retraction annulus of the bottom stage, the bottom stage begins to retract, the valve is closed, and the top stage fluid is blocked (Fig. 5.3.4 B). When the bottom stage is fully retracted (Fig. 5.3.4 C), the fluid is switched to the retracting annulus of the top stage, the valve opens mechanically allowing the fluid in the front chamber of the top cylinder to pass through and the top stage to lower (Fig. 5.3.4 D) until the top stage is fully retracted (Fig. 5.3.4 E).

If the check valve has a poor seal, it cannot lock up the fluid in the front chamber of the cylinder. As a result, when the roof pressure increases, the legs will automatically lower. If the motional reaction of the check valve is not sensitive, the sealing pressure will be small. Consequently, the designed setting pressure will seldom be realized.

B. YIELD VALVE OR PRESSURE RELIEF VALVE

This very critical device is for protecting the integrity of leg props. Based on its capacity, a yield valve can be classified into two types: yield valve and burst (bump) valve. Yield valves limit the maximum pressure allowed, and burst valves expel a much higher volume of fluid to limit pressure during bump-prone mining conditions. In a bump-prone face, both yield valve and burst valve are installed on a shield leg.

Very often, in a longwall, hydraulic pressure in the leg suddenly increases to exceed the designed yield pressure of the legs, for example, when outbursts or bumps occur. When this happens the pressurized fluids in the legs must leak out rapidly allowing the pressure in the prop to drop instantaneously so that the prop will not be damaged. A burst valve is designed to let the pressurized fluids exit rapidly in large volume. The capacity of a burst valve has improved, just like all other longwall face equipment, and has increased from 52.6–131.6 gallons min^{-1} (200–500 liters min^{-1}) in the 1970s to more than 1316 gallons min^{-1} (5000 liters min^{-1}) in recent years.

In regard to pressure limitation, there are two types of yield valves: coil spring and gas pressure. The former is more popular. Fig. 5.3.6 shows a yield valve of the coil spring type. The limiting (or yield) pressure is determined by the stiffness of the coil spring. When the fluid pressure is less than the yield pressure, the fluid exit path is locked, and the valve is closed (Fig. 5.3.6 A). But once the fluid pressure exceeds the yield pressure, the spring is compressed, causing the fluid path to open and allow the fluid to exit (Fig. 5.3.6 B). As a result, the pressure in the fluid chamber of the bottom cylinder is reduced below the yield pressure. The common design of a yield valve is that each time a yield valve opens, and the fluid exits, the fluid pressure in the chamber will drop 10–15%.

Figure 5.3.7 shows a yield valve of the gas pressure type. The gas is an inert gas such as nitrogen. When the pressurized fluid in the front chamber of the leg piston exceeds the preset air pressure, the valve rod is moved upward. When the fluid exit port in the valve rod passes

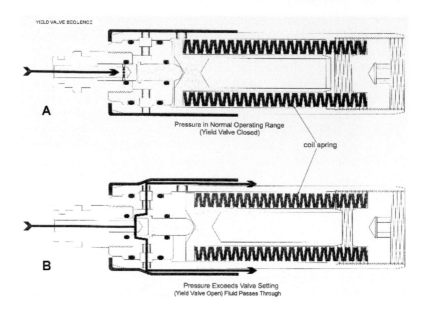

YIELD VALVE SEQUENCE

A

Pressure in Normal Operating Range
(Yield Valve Closed)

coil spring

B

Pressure Exceeds Valve Setting
(Yield Valve Open) Fluid Passes Through

Figure 5.3.6 Sequence showing the working principle of yield valve
Source: Courtesy of Swanson Industries

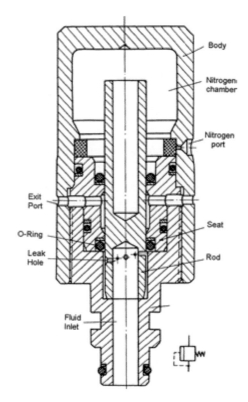

Body

Nitrogen
chamber

Nitrogen
port

Exit
Port

Seat

O-Ring

Leak
Hole

Rod

Fluid
Inlet

Figure 5.3.7 Gas-operated yield valve

the O-ring, the pressurized fluid leaks out. Similarly, a yield valve must have a good seal and be very sensitive to pressure change.

Two major types of yield valves are used in a shield: high flow rate for leg prop and low flow rate for all others. Defective yield valves can be identified by leaking fluid and pressure drop of the gauge.

5.3.3.2 Unit control valve or valve bank

This is a bank of directional hydraulic valves, each of which directs the hydraulic fluid to perform a predetermined function of the shield support including leg lowering, leg raising, support advance, conveyor advance, extension and retraction of side shield, stabilizing ram, and flipper. See Fig. 5.4.2 for an example of a valve bank or unit control valve.

5.3.4 Auxiliary devices

Depending on the support types and their applications, a shield support is equipped with one or more of the following auxiliary devices.

5.3.4.1 Push and pull device

A double-acting hydraulic advancing ram or DA ram (Fig. 5.2.2) is installed at the center between the two base plates of each shield. It is used to push the conveyor and pull (or advance) the shield support forward. Shields are set at one-web back from the AFC in the longwall face (see Fig. 1.4.2). In this system, after the shearer cuts and passes by, the shield is pulled forward first, followed by pushing the conveyor, thereby providing an immediate support to the newly exposed roof. In the sequence, the shield support is pulled (or advanced) forward immediately following the shearer's cutting. After the trailing drum of the shearer has passed the shield by 28.7–45.9 ft (8.75–14 m), the conveyor is pushed forward.

In the double-acting advancing ram, the force generated due to piston extension is larger than that due to piston retraction because of the larger area on the cylinder side without the rod (advancing ram). In order to utilize this property, the hydraulic ram is mounted on a relay bar (Fig. 5.2.2) in a reverse manner such that the force due to piston extension can be used to pull the support.

5.3.4.2 Base-lifting device

If the shield is to be used on a soft floor during its service life, it must be equipped with a base-lifting device. The base-lifting device is a double-acting ram and is generally located in the front-end space between the base plates (Fig. 5.3.8). They are usually mounted on the rigid bridge or rectangular arch connecting the two base plates (Figs 5.2.1 and 5.3.8).

The base-lifting ram is programmed to extend its piston against the relay bar and lift, at the minimum, the toes of base plates before the shield is advanced, thereby making sure the toe of base plate is lifted up above the floor for unobstructed advancement. The ram capacity must be higher than the weight of the shield, and the piston stroke must be larger than the maximum base plate penetration into the floor or thickness of any floor extrusion.

Figure 5.3.8 An example of base lifting device
Source: Courtesy of Caterpillar

5.3.4.3 Flipper or face guard or sprag

In thick seam mining, 10 ft (3 m) or thicker, the upper portion of the coal face tends to spall off (see Section 10.14), increasing the unsupported roof span and creating roof fall problems. During periodic weighting, coal spalling may be sudden and violent so that it ejects over the walkway and poses safety hazards to the face crew. A face guard, which consists of a steel plate and a double-acting hydraulic ram, is installed under the front canopy. The face guard is designed to prevent the spalling and/or the falling coal fragments from hitting the working crew. Under normal conditions, the face guard is activated to press against the face (Fig. 5.3.9). When the cutting machine is approaching, it is retracted. It is re-activated as soon as the support is advanced.

5.3.5 Electrohydraulic control

The unit control valve is a bank of directional hydraulic valves, each of which controls a shield function. Traditionally, each valve was manually controlled, and a shield cycle required three separate but consecutive manual engagements of valves, i.e., leg lowering, shield advance, and leg rising. Consequently, the shield cycle was slow. As a result, it was not unusual for the shield cycle to take 30 to 60 seconds or more, and the leg setting pressure seldom reached the designated value and usually was not uniform across the whole face.

In order to correct the problems associated with manually operated hydraulic valves, the electrohydraulic control was first introduced in the Pittsburgh seam in northern West Virginia in 1984. In this system, a solenoid valve is used to convert electrical impulses into

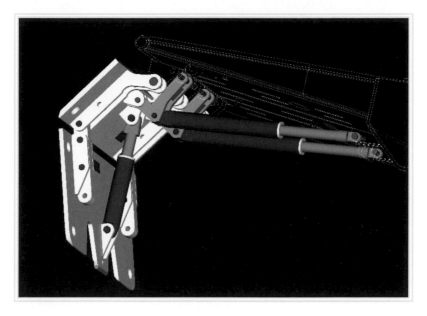

Figure 5.3.9 Shield flipper or sprag or face guard
Source: Courtesy of Caterpillar

hydraulic signals to operate the pilot valve that in turn operate the directional hydraulic valve, as opposed to being manually operated directly function by function. With a solenoid valve control, the directional hydraulic valve can be operated and perform its designated function quickly and remotely and in rapid sequence.

All shields are now equipped with an electrohydraulic control system. See Section 5.4 "Electrohydraulic Control System" for more details regarding its operation principles and shield control units.

5.3.6 Hydraulic fluid and its supply system

5.3.6.1 Hydraulic fluid

The hydraulic fluid used to operate shield support is critical to the operational life of the equipment. All longwalls operating in the United States use an ISO HFA Fire Resistant Fluid (Table 5.3.1). The hydraulic fluid is a water emulsion consisting of 95–98% water and 2–5% oil concentrates depending on the manufacturer of the fluid, the original equipment manufacturers (OEM) of the shield supports, and the quality of water used to mix the emulsions. The majority of longwalls use the water emulsion of 95% water and 5% concentrate. The reason that water-based fluid is used is that petroleum hydraulic fluids are highly flammable and not suitable for underground coal mines, and water is the best fire-resistant material. Plus it is inexpensive. However, the downside for using water is that it is a poor lubricant for pump gears and valves; it does not protect the metal parts in contact; and it leaks easier due to its low viscosity. Therefore, oil concentrates are added to make up these deficiencies. Today's

Table 5.3.1 Types of fire-resistant fluids

Type	Description
HFA	Emulsions of oil in water, also called high water-containing fluids (HWCFs), 95/5 fluids, or oil-in-water emulsions
HFB	Emulsions of water in oil, also called invert emulsions
HFC	Solutions of water and poly-glycol, also called water glycols
HFD	Fire-resistant fluids that do not contain water. Many types are available including polyester, phosphate ester, water-free polyalkylene glycol, and polyester polyol.

hydraulic fluids must meet the requirements for mine operators, OEM, and regulatory bodies. For the operator, the fluid must be stable, safe, and inexpensive; for OEM, it must be clean and not damaging to pump, gears, and valves; and for regulatory bodies, it must be nontoxic (CFR, Title 30, Part 35 gives a list of approved fluids).

There are two types of emulsions: micro- and macro-emulsions. Microemulsions are transparent with smaller particles 0.1–0.2 microns in diameter, whereas macroemulsions are milky with larger particles up to 50 microns in diameter. In the past, macroemulsions were prevalent. But as electrohydraulic systems became more sophisticated, the required cleanliness level increased and the deficiencies of macroemulsions soon became apparent. Microemulsions are more stable, contain less oil, and are more consistent in delivery of lubricant to pumps and valves than macroemulsions, and they allow the use of finer filters. Therefore, microemulsions are more popular (Sanda, 1996) and used exclusively.

A. CONCENTRATES

Oil concentrate itself is a concentrated water emulsion, because it contains 50–55% water droplets that are uniformly distributed within the matrix of oil. It consists of basic oil and an emulsifier. Water and oil are mutually insoluble. But if an emulsifier is added to an oil droplet, the emulsifier will form a sticky layer around the surface of the droplet. This is the emulsified oil or concentrate. Mixed with the water, the emulsified droplets will not congregate but uniformly distribute throughout the water body.

Some concentrates contain a proper amount of additives for various functions, such as, emulsifying agent, coupling agent, rust inhibitors, antifoaming agent, and bacteriological control additives. In selecting those products, toxicological effects of human contact and environmental disposal of the fluids must be considered.

B. WATER

The stability or performance characteristics of water emulsion depend on the quality of the water and the emulsified oil. The water hardness, pH, total suspended solids, specific ion concentrations, and biological activity all have an impact on the ability of the fluid to function properly. Most water sources are suitable for emulsion production.

The hardness of water is due to the presence of dissolved alkaline earths with calcium and magnesium. Generally speaking the chloride and sulphate content up to 200 mg L^{-1} of chloride and 400 mg L^{-1} SO_4 is acceptable. Water hardness is expressed as an equivalent amount of $CaCO_3$ it contains. It is soft if $CaCO_3$ is less than 250 mL, but it is hard if $CaCO_3$ ranges from 250 to 750 mL. Different hardness levels require different concentrates.

If water is low in pH (i.e., acidic) it is difficult to form an emulsion. But water with a high pH may have a buffer effect that reduces the efficiency to form a stable emulsion due to its excess carbonate. A pH level of six to nine is most desirable. When the pH of the concentrates for hard water runs from 9.5 to 11, the pH of the emulsions it produces will be from 8.5 to 10. Similarly, when the pH of concentrates for soft water runs from 8.5 to 9.5, the pH of the emulsions it produced will be from 7.8 to 9.0. If the pH level of an emulsion is too high, it will adversely impact the metals it contacts. On the other hand, if it is too low, bacteria may grow and increase corrosion. For instance, when the pH level of a soft water emulsion decreases to 7.0–7.5, it suggests that a significant level of bacteria activity has occurred. Therefore, the pH level of a water source has significant effect on water emulsion. Water alkalinity can also be influenced by water additives such as dust suppressants and water conditioners.

Suspended solids must be kept at less than 0.03% in order to maintain emulsion stability and avoid causing filter plugging and wearing of hydraulic components. Very hard water ($CaCO_3 > 750$ mL) must be softened, because it contains too much calcium and magnesium ions which, just like the chlorine ions, have a destabilizing effect on oil emulsification.

Table 5.3.2 summarizes the water quality standard for hydraulic emulsion fluid.

C. BACTERIA

Bacteria is found in all waters at some level whether naturally occurring or from a portable water source. The water used to mix emulsions is the primary route that bacteria/fungi enter the emulsion system, although airborne contaminants can enter the system through breathers on the longwall emulsion reservoir. The ability of the fluid to control bacteria and fungi growth is key to longer fluid and filter life. Left unchecked, bacteria/fungi (B/F) growth will lead to filter blockages, instability in the fluid, and potential corrosion problems. A by-product of bacteria growth is an acid. The pH of the emulsion will begin to drop as bacteria continue to multiply. The lower the pH the better the environment becomes for bacteria growth. The majority of longwall fluids available today contain an additive for controlling the growth of B/F, commonly referred to as a biocide (Russell, 2006).

Bacteria and fungi infestations usually occur following an oil separation problem. Where an oil film is allowed to develop on the surface of the emulsion to exclude oxygen from the system, the bacteria can breed quickly to infest the emulsion. Bacteria cause the deterioration of emulsion. It eats emulsifier components and produces biomass that can plug filters.

Table 5.3.2 The water quality standard

Item	Criteria
Appearance	Clear, colorless, no smell
Electrical conductivity	< 1800 ms cm^{-1}
pH	6.5–8.5
Water hardness	< 750 mg L^{-1} C$_a$CO$_3$
Cl	< 201 mg L^{-1}
SO$_4$	< 145 mg L^{-1}
NO$_3$	< 20 mg L^{-1}
Fe	< 1 mg L^{-1}
Bacteria load	< 15^5 col mL^{-1}

Abnormal bacteria growth can gradually split the emulsion, causing the formation of free oil and cream which can severely plug hydraulic cylinders, control valves, and filters. Depending on the severity, various bactericides and cleaning methods must be used (Pack, 1991). Bacteria growth can be monitored by enzymatic instruments or an aerobic plate count dip slide. The former is much more precise, while the latter is much simpler and most commonly used. The slides are dipped into the emulsion, stored at room temperature for 48 hours in a dark closet. The contaminant level is then compared to a chart for known concentrations of bacteria.

D. EMULSION STRENGTH

Emulsion strength is the total oil percentage in the emulsion. It can be determined by a refractometer. A simple test to determine the stability of the emulsion, i.e., the layering tendency of oil and water, is to put the emulsion in the oven and heat it for 24 hours at 150° F and watch for its layering intensity. The conductance of water or emulsion reflects the amount of ionized solutes dissolved. A sharp increase normally denotes impending emulsion instability and corrosion.

All emulsions are designed to perform at a certain ratio of water to oil concentrate, and it is important to maintain this ratio at all times. When the ratio changes to exceed the limits of tolerance, oil and water will segregate and become unstable. Concentrates are also designed to match water hardness. Thus, inconsistent water quality may also cause the emulsion to become unstable or layering to develop. Once the oil has separated, other problems set in, such as corrosion, filter and valve blockage, and erosion. Therefore, constant monitoring of water quality is required.

E. MIXING OF WATER AND CONCENTRATES

The mixing device for water and oil is crucial in maintaining a stable emulsion and for the proper water-to-oil ratio. Daily checkups and record keeping are recommended.

A closed-loop mixing system may be used. In this system, the designated proportion of water and concentrate is metered in by a separate flow meter. A refractometer is installed on-line to analyze the stability of the mixture continuously. Any deviations beyond the tolerance limits from the designated strength can be corrected automatically.

F. EMULSION FILTRATION

The filtering system is one of the most important features of the shields. Without good filtering, many problems will exist. It is very important to filter as much as possible prior to allowing the fluid to reach the shields. In fact, the ideal condition is to filter the fluid with 25 μm or less before entering each and every valve.

The newest shields are equipped with a high-precision valve design that enables them to have a cycle time of less than 10 seconds. These valves feature tiny bores to direct hydraulic fluid activity, and therefore are easily susceptible to blockage when sufficient suspended solids, including bacteria, exist in the emulsion. It is important to keep the emulsion free of impurities, and ordinarily multi-stage filtering is used to achieve this end. Consequently, filterability of the emulsion has become a critical issue in modern longwall mining.

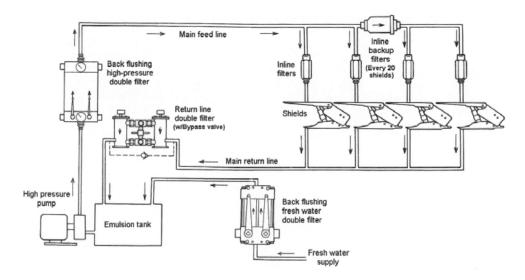

Figure 5.3.10 An example of shield hydraulic filtering system

Microemulsions being smaller in particle size are suitable for filters with smaller mesh size without compromising the fluid. Factors that can negatively affect fluid filterability include particles from mixing water, actual emulsion particle size, the presence of biomass in the aged fluid, fluid decomposition products, and the creation of water insoluble organic materials from the fluids. Figure 5.3.10 shows an example of a high-performance filtration system for hydraulic supply that features replaceable fine mesh stainless steel filter elements, high flow rates, and the strength to withstand high differential pressure. The double-filter assemblies feature an automatic back-flushing design that cleans the elements in place quickly and easily without system shutdown. Two pressure differential gages indicate the degree of filter loading by the pressure differential between the filter's inlet and outlet ports. When a significant pressure drop occurs at the outlet, part of the hydraulic system's flow is diverted briefly to back flush the accumulated dirts from the filter elements until their original filtering capacity is fully restored.

The recommended filter mesh size for the pressure lines is 10 μm while that for return lines is 7 μm.

G. EMULSION LEAKAGE

Emulsion leakage occurs in all systems, the severity of which increases with the age of the system. Replacement of concentrates normally runs 1000–2000 gallons (3785–7570 liters) per month, with the worst case being up to 5000–7500 gallons (18,927–28,391 liters) or more. Leakage occurs mostly in hoses and hose connections and DA rams. Due to its diluted nature, leakage may not be easily identifiable when mixed with spray water from the shearer dust suppression system. Tracer dye can be used to positively identify the water emulsion.

H. OEM SPECIFICATIONS

Both Caterpillar Global Mining and Komatsu Mining (Joy) have their own lists of approved concentrates. Both OEMs require fluid manufacturers to submit new products to their designated laboratories for specified tests. Those tests are designed to clarify the stability, sealing material, and corrosion of the hydraulic fluid. If the test results meet the OEM's criteria, a certificate of approval is issued. Only those concentrates that meet their requirements will be permitted for use with their individual shield control systems.

5.3.6.2 Maintenance of hydraulic fluid

Operational reliability of hydraulic fluid depends on supply of consistent good quality fluid. Consequently, a fluid maintenance program must be developed and maintained. At the minimum, weekly checkups of the following quality factors must be performed: pH value = 7.0–9.5, microbial load < 10^5 Kol mL^{-1}, and concentrate content as specified. Fluid cleanliness should maintain no more than 2500 particles less than 5 μm and no more than 160 particles larger than 15 μm in one mole of fluid.

5.3.6.3 Hydraulic supply system

As panel width gets wider and shield capacity becomes bigger, shield cycle time will have to be as short as possible (or the shield can move as fast as possible) to achieve rapid advance and high production. In order to achieve a short shield cycle time, the design of a hydraulic supply system becomes a very important task. A larger capacity and faster moving shield require a much larger volume of fluid supply, while a wider panel causes a larger pressure drop from headgate to tailgate. Therefore, if the hydraulic supply system is not properly designed, shields near the tailgate end may not get the desired pressure and/or volume for desired advance speed and setting pressure against the roof. A properly designed hydraulic supply system should supply a sufficient quantity of fluid at all time to ensure the designed setting pressure is reached easily at all shields across the face. Hydraulic pressure drops as it travels from pump station to headgate and from headgate to tailgate due to internal friction as fluid flows through hydraulic components such as hoses and valves. It can be minimized by sufficiently dimensioned components and avoiding dynamic pressure peaks in returns and short circuits.

The hydraulic fluid supply system consists of hydraulic pumps for generating high-pressure fluid and hydraulic hoses for fluid delivery to shields at the longwall face. There are two types of pump stations, one central and the other decentral. The central station is either located on the surface or in the mains and serves a section or sections each of which consists of several panels. Some pumps are installed permanently in a specially constructed pump station located in the mains or submains and serve a section of several longwall panels. In this case, due to long feed lines, a large pipe, up to 4 in. diameter (20.2 mm), is used to deliver the high pressure fluid from the pump station to the face, where it then steps down to 1–1/4 or 1–1/2 in. (32 or 38.1 mm) hoses. The central station requires less maintenance and can be controlled better. But it needs more capital investment and, consequently, is less used.

Conversely, the decentralized station is almost exclusively used in US longwalls. The hydraulic supply system consists of a pump station and an emulsion tank. In the pump station, three to five pumps are mounted on a flat rail car or skid and located, together with the power center, 1000–2000 ft (305–610 m) outby the retreating longwall face. Most of the systems are using three pumps to supply sufficient pressure and volume to the face. Usually two of these pumps will be active while the third pump is mostly idling. The pump capacity runs from 80 to 125 gpm (gallons per minute) per unit, normally 100 gpm, at pressure ranging from 4200 to 5000 psi (29.0–34.5 MPa) with 4160 volt power supply. The stainless steel emulsion tank, the size of which is 2500–3000 gallons (9.47–11.36 m³), is mounted in another car or skid (Fig. 5.3.11). The pump car in Figure. 5.3.11 contains four pumps. A pump car may contain two to four pumps.

Mixing of concentrates and water is performed either underground on the emulsion tank or on the surface. Surface mixing stations for preparation and distribution provide a stable source of clean emulsion with uniform oil concentration. Portable water is more easily available for emulsion preparation. Use of surface and underground mixing is approximately half and half in US longwalls.

Shield hydraulic system is designed for a nominal operating pressure of 4641 psi (320 bar) and face layout is designed to maintain at least 4060 psi (280 bar) at the shield. Typically, a shield requires 50–80 gpm to operate depending on cylinder size and shield cycle time.

The high pressure hydraulic hoses, 1.25–3 in. (31.75–76.20 mm) in diameter with rated strength of 5000 psi (34.5 MPa), are protected with braided steel wires on the outside and are either steel or zinc-coated at the ends. The effective ultimate strength (i.e., actual ultimate strength or burst pressure divided by a safety factor of 4, i.e., the burst pressure is 20,000 psi or 138 MPa) of hoses and connectors is the limiting factor for pump pressure. The hoses for the high pressure supply lines are normally smaller than those for the return lines. The rules of thumb for hose dimension and arrangement are that hydraulic pressure drop across the whole panel width or face length must be less than 725 psi (50 bars), and that the same

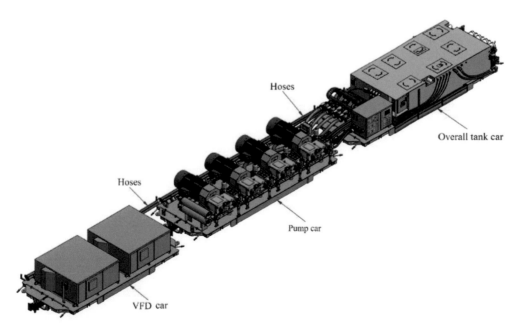

Figure 5.3.11 Hydraulic pump station and emulsion tank car
Source: Courtesy of Hutchinson

pressure drop applies to that between pump station and headgate T-junction. The return hoses do not need to be rated as high. They have a 1160 psi (80 bar) working pressure requirement with a 4640 psi (320 bar) burst pressure.

The hose layout depends on three factors: total flow, velocity and total drop to last shield in the longwall. There are three types of hoses arrangements (Bassier and Migenda, 2003; Fusser, 2005): (1) garland-shaped hoses from shield to shield, (2) a straight hose line along the AFC with a cross-over connection to each shield, and (3) two hose lines in a ring main system. The ring main system is exclusively used for modern high production longwalls. It consists of two subsystems, one for the high pressure line and the other for the return line (Fig. 5.3.12).

For each subsystem, one or more hose lines in a curtain arrangement between neighboring shields, run from shield to shield. The other line is straight along the AFC panline from headgate to tailgate. Both lines are connected by cross-over lines several times at selected shields. The number of cross-overs depends on the hose size, which in turn depends on bending radius and space constraints. In the return line, the hose size is very important, because the ratio of the retracted annular area to the loading piston area is 1:7–1:10 in shield legs such that the pressure available for leg lowering is only up to 10% of the pressure which is insufficient for operating several shields simultaneously. As a result, back pressure builds up and slows down leg lowering, or in some cases, triggers support functions. Therefore, the return line is generally at least one size larger than the high pressure line. The rules of thumb are that the back pressure should be less than 435 psi (30 bars). For modern longwalls, shield cycle time is about 10 seconds (3.5 seconds for leg lowering of 4 in. or 100 mm; 3.5 seconds for shield advance of 40 in. or 1000 mm; and 3 seconds for leg setting of 4 in. or 100 mm)

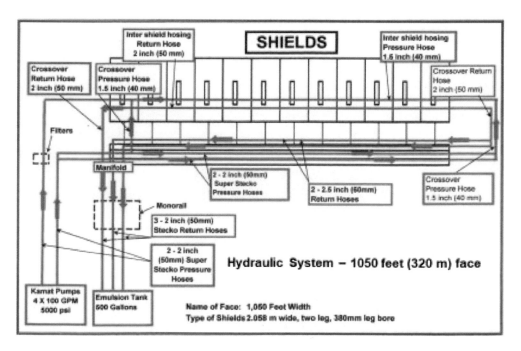

Figure 5.3.12 Schematic drawing showing an example layout of hydraulic supply system
Source: Trackemas and Peng (2013)

and requires 9.08–18.16 gallons/cycle (40–80 L/cycle) per shield. In automatic operation, several shields can be moving at the same time.

For modern, high-production longwalls in which wide panels or long faces are routinely used, minimum pressure loss and fluid volume across the whole face must be maintained all times. Fusser (2005) demonstrated that the desired minimum pressure drop and maximum fluid volume can be achieved through a combination of the use of properly dimensioned steel pipe and number of cross-over lines along the face. A steel pipe 2–3 in. (50.8–76.2 mm) in diameter replaces the hoses from pump station to headgate T-junction and from headgate to tailgate along the panline. Use of steel pipe in this arrangement can reduce the pressure drop considerably because (1) there are many hose couplings along the lines, and a hose coupling has a much smaller bore than that of the hose itself, (2) the friction on rubber is higher than that on steel, (3) hoses expand under high pressure, and (4) hoses tend to lay snaked, not straight like a steel pipe. The desired fluid volume for any shield location can be achieved by introduction of a cross-over line from the panline (the main) to the shield line (the ring) at or around that location. So, the number of cross-over lines depends on the desired fluid volume and the width of panel. The larger the desired fluid volume and the wider the panel are, the greater number of cross-over line is needed. Therefore, with a properly dimensioned pipe and number of cross-over line, the desired minimum pressure drop and fluid volume can be maintained at any point across the whole face.

Figure 5.3.13 shows an example of a detailed hydraulic supply ring main system for modern high-production longwalls. The longwall face is 1450 ft (442 m) wide and equipped with

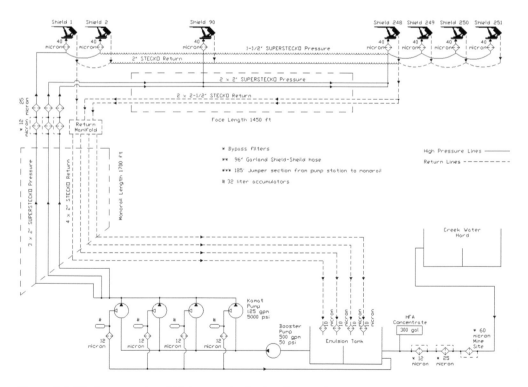

Figure 5.3.13 An example of high capacity hydraulic supply system

251 units of 1.75 m wide two-legged, 924-ton shields. The first three shields (#1–#3) are headgate gateend shields, while the last three (#249–#251) are tailgate gateend shields. Clean creek water on the surface or a city water supply, if needed, is piped underground and filtered in three stages in sequence: 60 micrometers → 25 micrometers → 12 micrometers before it is mixed with HFA concentrates in the 2500-gallon (9500 liters) emulsion tank, which is a part of the power train outby the longwall face. The mixing of water and concentrate takes place approximately 3 ft (0.91 m) prior to entering the emulsion tank at the pump station.

The creek water has a booster pump which is adjustable to maintain the proper ratios of the mixture. The concentrate is stored in a 300-gallon (1140 liters) container and pumped also by an adjustable booster pump. The lines for the creek water and concentrates are then teed together after leaving their respective booster pump for mixing. The mixture is measured once about every one or two weeks with a refractometer. The readings typically stay within ± 0.5% of the 5% concentrates. If the readings are too far off, the booster pumps are then adjusted to insure the proper flow.

The emulsion in the tank is first pumped by a booster pump to the four main pumps, each through a 12-micrometer filter when necessary. A booster pump is used to ensure that fluid from the emulsion tank supplying the main pumps is under positive hydraulic head, because axial piston pumps require positive pressure to operate. The number of pumps in operation depends on fluid demand from the shield operation. Normally, the first two pumps are on at 4800 psi (33.1 MPa). When the pressure drops down to 4600 psi (31.7 MPa), the third pump will turn on, and when the pressure drops down to 4400 psi (30.3 MPa), all four pumps will operate.

The maximum distance between the main pumps and the headgate T-junction is 1700 ft (518 m). There is a 12 micrometer filter immediately after the main pumps. In order to reduce pressure drop and supply the face with sufficient quantity of fluid at any location, the high-pressure fluid is supplied by three, 2 in. (50.8 mm) diameter lines, one going to the number one shield and then going from shield to shield until it reaches shield number 248. The other two high-pressure lines run through the panline with a cross-over at shield number 10. Note that in each shield, there is a 10 micrometer filter. Similarly, in order to reduce back pressure in the return lines, the return lines consist of three, 2–1/2 in. (63.5 mm) diameter lines, one going from shield to shield and the other two following the panline. The three lines are combined and then divided into four, 2 in. (50.8 mm) diameter lines at the stage loader and then on to the monorail. Right before the fluid is dumped back to the emulsion tank, there is a 10-micrometer filter in each line. The reason the ring main system does not cover shield number 249 through number 251 is to avoid any potential accidental damage of the hose when roof falls occur at the tailgate T-junction

In order to maintain the pressure and flow in the system with various flow requirements throughout the mining cycle, four 0.845 gallon (32 liters) accumulators, one for each pump, are installed at the pump station. These accumulators return the fluid to the emulsion tank once the pressure inside them reaches 4700 psi (32.4 MPa). An unloader valve may be installed in place of an accumulator at the pump to dump high pressure fluid back to the reservoir when and if a sudden pressure buildup occurs.

According to Hutchinson (2006), there are three longwalls in the United States that employ steel pipes rather than the traditional hoses for hydraulic supply from the pump station to the stage loader and then the ring-main from headgate to tailgate along the AFC. The steel pipes are 2–4 in. (50–100 mm) in diameter, 16.4 ft (5 m) long with either clamp or screw-type connectors. The pipes can either be hung on a separate monorail from that for the cables, water hoses, etc., or on a specially designed monorail in which cables, water hoses, etc., are on the top and the pipes are on the bottom.

Figure 5.3.14 A container on monorail that houses two hydraulic pumps
Source: Courtesy of Swanson Industries

Figure 5.3.14 shows an innovative hydraulic supply system in which the pump station is hung on the monorail and the emulsion tank is installed in the stage loader.

The pump station consists of five containers in a series, each of which houses two 100 gallon pumps with a 4 in. (100 mm) diameter hydraulic delivering steel pipe. Note that each container is covered with doors, the inside of which is padded with sound absorbing panels so that when the doors are closed, they are much quieter than open pump stations. With this design, the entire hydraulic supply system, including pumps and emulsion tank, is moved with the retreating longwall face, cut by cut.

5.4 Electrohydraulic control system

5.4.1 Shield control system

5.4.1.1 Definition and solenoid valve

Electrohydraulic shield control system, consisting of electrohydraulic control unit (hydraulic block valve + pilot valves) and electronic control (i.e., PMC-R or RS20s), is designed for control of shield support functions. Shield functions may not be directly operated by electrohydraulic control unit but rather by an electronic control in neighboring shield. The pilot valves on the electrohydraulic control unit are activated manually or via a program on control unit. Shield's individual functions are actuated directly by pilot valve on electrohydraulic control unit.

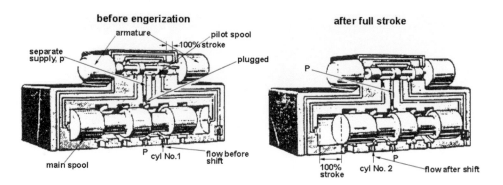

Figure 5.4.1 Operation principle of solenoid valve

In the electrohydraulic control system, solenoid valves are used to direct the pilot fluids and drive the hydraulic valves for various functions. The arrangement enables the hydraulic control of the shield to be automated and remotely controlled and thereby run much faster.

Figure 5.4.1 shows the principle of a solenoid valve. Before the solenoid is energized, the armature pushes the pilot spool to the right so that in the main spool the pressurized fluid, p, flows through cylinder 1. Once energized and the pilot spool reaches full stroke, the pilot fluid pushes the main spool to the far right such that the pressurized fluid, p, flows through cylinder 2. When de-energized, both pilot and main spools return to their original position.

The purpose of utilizing solenoid valve is that its armatures maintain the pilot spool ready to complete the engagement all time, thereby saving time and enabling automation.

5.4.1.2 System components and operations principles

There are two major shield control systems: one is a PMC-R by Caterpillar Global Mining and the other is a RS20s Faceboss by Komatsu Mining (Joy).

System configuration and components and, thus, operation procedures are different between the two systems, but, in general, they both consist of the following four components:

1 Hydraulic functional control valves typically found in conventional powered supports (Fig. 5.4.2).
2 Solenoid valves that convert electrical impulses into hydraulic signals for the operation of the hydraulic functional control valves. The application of solenoid valves make shield moving much faster and enable it to control the hydraulic valves with push button remote control (Fig. 5.4.2).
3 An electronic control unit (SCU) with microprocessor is equipped with a push button panel that issues commands via the microprocessor to control all support functions plus conveyor push. No central computer at the headgate T-junction or on the surface is needed and all parameters can be changed anywhere on the face (Fig. 5.4.3);
4 The microprocessor in each electronic control unit is designed to issue, compare, and relay commands directly to any support in sequence in batch control without the need for routing through the headgate master computer. The parameters, such as batch size, etc., can be programmed easily at any electronic unit (Fig. 5.4.3).

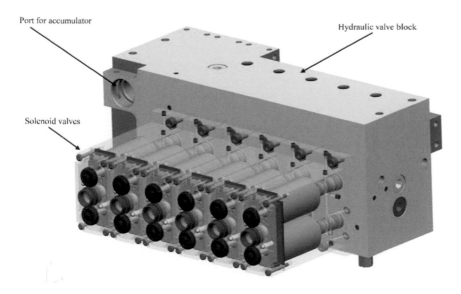

Figure 5.4.2 Twelve-function distributor block

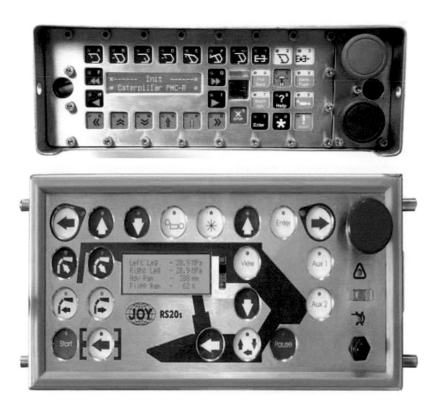

Figure 5.4.3 Shield control unit: upper: PMC-R and lower: RS20s
Source: Upper – courtesy of Caterpillar; lower – courtesy of Kamatsu Mining

SCU has two models, PMC-R and RS20s with a total number of key buttons 30 and 28, respectively, including infrared receiver and red and black push pull buttons for system and individual shield emergency stop, respectively.

SCU functions just like a personal computer (PC), except that almost all of the 84-button keyboard in a PC are single action command, while SCU's key stroke functions are so designed that all individual and sequential auto function can be activated either in a two– or three–push button operation.

For each shield, there are four action commands, leg lower, shield advance, leg rise, and pan push, all of which are controlled by hydraulic valves. In normal production operations, the first three actions are in sequence, while the fourth action is delayed depending on how far the shearer has cut and passed by. However, there are up to 170/198 units of shields for an average panel width of 1100 ft (335 m) with four actions each executed following shearer's passage. Consequently, to automate those vast quantity of shields in a longwall face, there are two types of factors of action functions: for individual shield there are four basic action functions and for the whole panel of shields, there are two basic action functions, i.e., direction and single shield or group of shields. The two SCU keyboards are designed to execute the aforementioned basic functions of individual and group of shields in various combinations through specialized software menu, plus for obvious reasons, safety features.

Sensors – for automatic and remote control in a longwall face, sensors are needed for hydraulic leg pressure monitoring, DA ram extension distance monitoring, detection of infrared emission from the shearer for shearer location, chain tensioning, etc. The most commonly used sensors for the first two types of monitoring are discussed here.

A. HYDRAULIC PRESSURE TRANSDUCERS

Hydraulic pressure enters the open end of the transducer and causes deflection of the diaphragm. The strain gages mounted on the opposite side of the diaphragm will measure the amount of deformation and convert it to an electrical voltage change, which can be monitored.

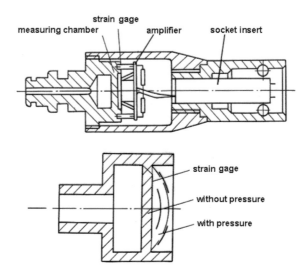

Figure 5.4.4 Schematic of a hydraulic pressure transducer

(Fig. 5.4.4) The measured deformation can be calibrated to closely match the magnitude of pressure applied. With this sensor installed in each leg, the magnitude of leg pressure can be determined at any instant. This capability is used to implement the positive set of shield leg pressure. In this system after the introduction of high pressure fluid into the leg cylinder, its pressure continues to increase to a predetermined level, e.g., 110 bars, then the pilot signal is shut off automatically. But the high pressure fluid continues to build up until the setting load is reached. Therefore, in an electrohydraulically controlled shield face, leg pressures are uniformly set all across the face, provided the cylinder and the hydraulic supply lines are not leaking.

B. REED ROD SENSOR FOR DA RAM POSITION

Reed rod contact switch is the most popular sensor for DA ram position measurement. Figure 5.4.5 shows the Reed rod position sensor. The rod consists of a multitude of serial impedances of identical value. The Reed contacts are soldered by one contact side to the connection points of the different impedances very much like the rungs of a ladder; the other side of the Reed contact is electrically interconnected to form the sliding contact of a linear potentiometer.

The impedance/Reed combination is located within a steel tube and sealed against the internal hydraulic pressure. The measuring rod of the Reed contacts are located inside and surrounded by an annular magnet that is built into the piston. As the piston moves, the Reed contacts are switched on and off by the annular magnet as it follows the movement of the piston rod over the Reed rod. This will change the resistance of the circuit. By monitoring the resistance of the circuit, the exact location of the cylinder and, thus, the ram stroke can be determined. With this system, it can be assured that the push-pull of DA ram is always to the full or predetermined length of stroke and ensure that shields and the panline are always parallel in straight lines. With a Reed rod sensor, the ram stroke is available in absolute value at any moment. The resolution and the measuring accuracy are better than 0.14 in. (3.6 mm). This, plus its precision, provides the basis for panline alignment, which in turn, provides the basis for automation for face alignment. Reed rod sensors are extremely reliable due to the absence of any moving electronic parts.

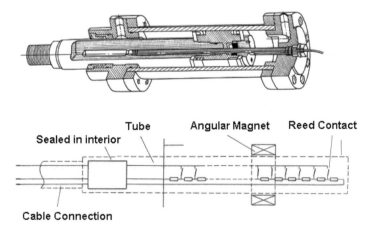

Figure 5.4.5 Reed rod for measuring the stroke of DA ram: upper, cutaway view of DA ram and lower, principle of Reed rod

C. OTHER SENSORS

Other sensors such as an inclinometer, for detecting the pitch and roll for AFC, proximity switches, and infrared receiver for shearer position, etc., are also used.

5.4.1.3 Two shield control (SCU) models – PMC-R and RS20s fireboss

An SCU is the heart of the electrohydraulic control system. It is a hardware that controls hydraulic function either manually or digitally.

SCU can control and display all functions of a shield and is simultaneously the interface between operator and machine. SCU can control all individual shield functions. As an interactive system, it allows the user to execute single shield functions as well as automatic sequence functions. SCU is mostly used for controlling the hydraulic functions of the shields by the operation of electrohydraulic solenoids. Since these units are in a network, it allows the operator to control a shield while standing on another shield, keeping him/her safe from falling debris and moving equipment. The operator is never allowed to stand on the shield he/she is controlling and move it with SCU.

SCU can be used to control a stageloader, a face conveyor tensioner, a filter station, and the dump valve, and backflushing the filter system. It can be part of the network or a standalone station.

As mentioned earlier there are two major shield control systems, PMC-R (formerly PM4) and RS20s (formerly RS20) (Fig. 5.4.3). Their designs are different, but the overall functional objectives and their operational principles are similar. Both systems are highly reliable and functionally flexible.

Figure 5.4.6 shows the components for the PMC-R roof control system. Each system is installed in every shield and consists of a shield control unit (SCU), a solenoid driver module, a valve block with solenoids, and various sensors. The SCU has 30 touch keys each of which is shown with a functional symbol for easy identification. Up to 30 functions can be performed by each SCU.

The buttons in SCU keyboards consist of four basic functions of individual shield, individual or group of shields, direction of action, and automation (a series of functions of different types), a group of menu selection/scroll keys and an LED screen for shield status, plus emergency stop of whole face and lockout for individual shield and cancellation of functions buttons.

In particular, many software programs have been included for every key stroke such that when a specific command button is activated, a string of menu programs including diagnostic are available by scrolling up and down on LED. When problems occur, error messages and remedial measures will show up.

The microprocessor is computer-based and employs the latest enhanced technology. Connection ports are provided for up to eight sensors. But for the shearer faces, the most common one includes two sensors for the left and right leg pressures, one for the Reed rod in the DA ram, and one for the infrared receiver.

The user interface menu can access all data available in the shield, including all transducer values, status information, and local parameters. Each SCU has a quick stop and a local lockout button. The former is to de-energize the entire face in an emergency situation while the latter is for local maintenance or diagnostic purposes (Fig. 5.4.3 upper).

The SCU has a one-piece enclosure made of stainless steel. It is sealed against moisture and high pressure water sprays, vibration, and shock. The keyboard is hermetically sealed

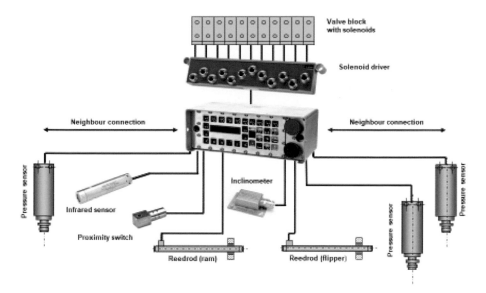

Figure 5.4.6 Components of PMC-R

Source: Courtesy of Caterpillar

with an O-ring and covered by a plate made of stainless steel. The SCU can be configured with integrated face lighting, eliminating a separate power supply for face lighting as commonly used today. To ensure the operator's safety, any manual activation of shield functions via the keyboard requires at least two buttons to be pushed, and the safety feature prohibits malfunctions by mistakenly pushing the wrong button.

The cables between the SCU and the Reed rod sensors, leg pressure sensors, and infrared receiver are steel-braided hose cables with waterproof connectors. These are four-wire cables, two for power supply and two for bi-directional communication. The same four-wire hose cables are used for the connection between neighboring shields. All sensors allow the detection of cable failures. These failures are displayed on the SCU and are sent to the headgate computer if installed.

The RS20s shield control system (Fig. 5.4.3 lower) is very powerful and extremely fast. It consists of a control system and electronics. The Compak valve assembly control system is operated by a set of 14 solenoid valves and is coupled to the RS20s system. The electronics for the RS20s consists of three units – a mimic, microprocessor, and a solenoid transducer unit (Fig. 5.4.7).

The mimic provides the man-to-machine interface. The front panel of the mimic unit shows the outline of a shield. On it, buttons for activation of individual shield functions are positioned so that each of them shows the individual function of the shield to be activated. Active shield indication LEDs surrounding the mimic illuminate to indicate clearly what is going to happen when the operator keys in according to prime type, direction of travel, and shield functions. The structural design of the mimic and microprocessor allows these two units to be mounted separately for maximum protection while accommodating the operator for maximum operation convenience (Fig. 5.4.8). All units are fully encapsulated against moisture and dirt. The system allows integration of video and voice communication in the future.

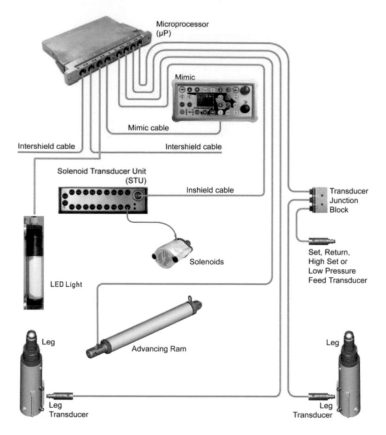

Figure 5.4.7 Components of RS20s systems

Source: Courtesy of Komatsu Mining

Figure 5.4.8 In RS20S, mimic and microprocessor are mounted separately

Source: Courtesy of Komatsu Mining

5.4.2 Shield control system layout

Figure 5.4.9 shows the PM^R-C shield control system layout for the longwall faces. The system consists of an SCU in every shield and an optional SCU to interface with the gateend computer for face visualization. The communication among SCUs in the entire face, via a four-core cable is very fast and reliable. Power supply is partitioned into several groups, and each group has its own power supply and is separated from the other groups by an isolation adaptor. If any group is faulty, the communication still can reach the entire face.

Server connections

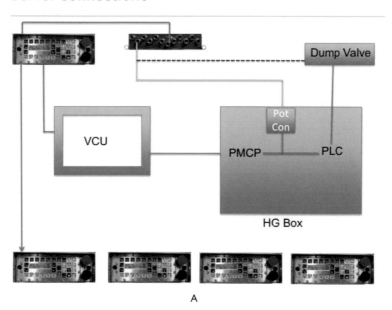

A

A brief view of the working network

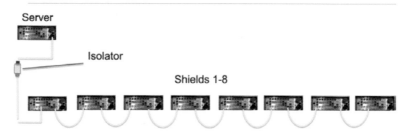

Network Parameter
"Number of Shields" = 8

B

Figure 5.4.9 System layout for PMR-C shield control system

Source: Courtesy of Caterpillar

Figure 5.4.10 shows the basic system layout for the RS20s shield control system. All cables use hose conduit with staple lock connections. Its modular construction simplifies fault finding and routine maintenance. All shields in the system can operate on full automation mode without the need for a master control computer at the headgate. The system can accommodate up to 350 units.

The system monitors and can display the following shield information:

1   Voltage at microprocessor.
2   Leg pressure.
3   DA ram position.
4   Shield cycle time analysis.
5   Position and direction of shearer.
6   Location of all prime sequences.
7   Location of lowered shields, isolated, faulty, or marked RS20s units.

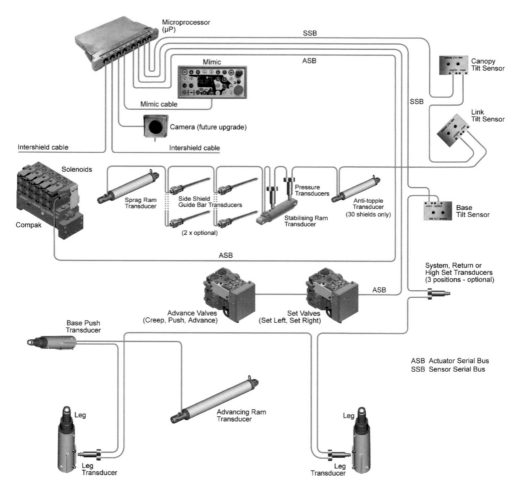

Figure 5.4.10 System layout of RS20s shield control system

Source: Courtesy of Komatsu Mining

5.4.3 Shield operation modes

Shields can be run in either manual mode or remote control (or batch) mode. In manual mode, each shield is operated separately, and the operator moves from shield to shield to initiate its movement. In the batch mode, since both PMC-R and RS20s are computer-based, software can be programmed to operate the shields in almost any imaginable modes of automation. RS20s has a complete library of every major automation sequence with more than 100 different face cutting methods. PMC-R also contains a full library of different cutting methods. Both systems allow the operator to quickly configure new cutting sequences. The menu also allows the operator to select and change parameters at any unit across the face, displays the system's status, displays error messages, and activates lockout functions. Programs can be loaded from external sources. In other words, the system functions very much like a desk top computer with a broad range of capabilities. Some examples of the more commonly used operation modes and features are highlighted in the following sections.

5.4.3.1 Common operation mode

A. BI-DIRECTIONAL ELECTRIC ADJACENT CONTROL

In this operation mode, a shield will automatically lower, advance, and reset by pushing a button on the electric control unit on either side of the shield. Bank (conveyor) push is activated by pressing a button on each shield or a remote button to initiate bank push at a preset offset distance. This mode is most frequently used.

B. FLOATING BATCH CONTROL

The batch size is adjustable from two units to full face, but usually 8 to 30 units are used, because those shields are within the sight of the operator.

There are different patterns of advancing shields and pushing AFC in order to match changing conditions. Following are three examples:

1   Standard – every other shield is selected to move simultaneously, beginning with the "even" numbered shields.
2   Rabbit – same as standard, but staggers the functions, meaning multiple shields are not lowering, advancing, or setting at the same time. Instead, when the first "even" shield lowers and begins its advance, the next "even" shield starts lowering. This procedure requires less volume from the pumps.
3   Sequential – shields are automatically advanced one at a time, moving either direction as desired.

For pushing the AFC, there are a few different ways:

1   Selecting a shield and using the dedicated key will activate the push circuit only until the key is released.
2   Pressing the locking push key on the shield or from the master shield from which it is controlled.
3   Backpushing shields from a certain point to a desired stroke.
4   Allowing the automation to push the pan based off the shearer travel.

5.4.3.2 Shearer initiation

Shearer initiation refers to the automatic initiation of support movement after passage of the shearer at a preset offset distance.

Shearer initiated shield advance (SISA) requires the instantaneous position of the shearer at all time. The pioneer approach introduced in mid 1980s used infrared technique (Peng, 1987). An infrared emitter is mounted on the gob side of the shearer (Fig. 5.4.11). A receiver is incorporated into the electronic control unit of each support. The emitted signals are received constantly by up to three receivers in three adjoining supports. These signals are analyzed to determine the location and the movement direction of the shearer and issues commands of shield cycling and conveyor push at preset offset distances behind the shearer. Under most favorable conditions, a shearer-initiated face requires no shield operators and only one or two shearer operators depending on cutting requirements present at the face. Since the infrared emission rays spread over three shields, the resolution of the shearer's location is not as accurate. In addition, thick dust or clouds of water sprays could obscure the emission rays and make it fail to work.

Since then, more sophisticated sensors have been developed and used for shearer position determination greatly improving shearer automation (see Sections 9.2, p. 312)

5.4.4 Face visualization

If a gateend computer or master control unit (MCU) is available, the face data information can be displayed on its screen for either the full face or local areas of interest. Some examples of displays are these:

1   Face layout, including shield number and position, AFC, shearer, and its location, and when cutting, the moving direction.
2   Leg pressures for and status of each shield.
3   DA ram position for each shield.

Several face data, such as face layout, lockout functions, parameters, etc., may be changed through MCU depending on the level of automation.

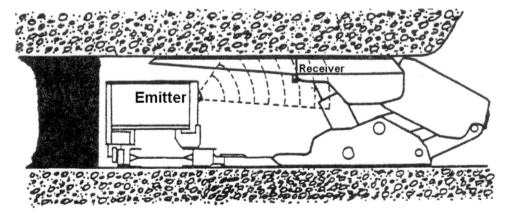

Figure 5.4.11 Principle of infrared system for determination of the location of the shearer
Source: Peng (1987)

5.5 Performance of shield supports

5.5.1 Supporting cycles and loading characteristic parameters of shield supports

After a shield is set against the roof and floor, its load (or actually resistance to roof convergence and/or floor heave) does not remain constant at the setting load level. Rather it varies from period to period as a result of interaction among roof strata, shield, and floor strata. There are distinct patterns of variations in shield resistance (or pressure) that can be used to evaluate the working conditions of the shield supports and the strata behavior.

Figure 5.5.1 shows a typical form of variation of support resistance in a supporting (mining) cycle (Peng and Chiang, 1984). Once the support has been advanced and reset, an initial setting load, $p_{s,}$ is rapidly achieved within a short time period, t_s, (10 sec–4 min). Thereafter, some may drop quickly various degrees depending on contact conditions (Fig. 5.5.2). This is the initial setting period (o–s). As the support starts to interact with the roof, the support resistance increases rapidly during time t_a until it reaches a relative equilibrium. This is the

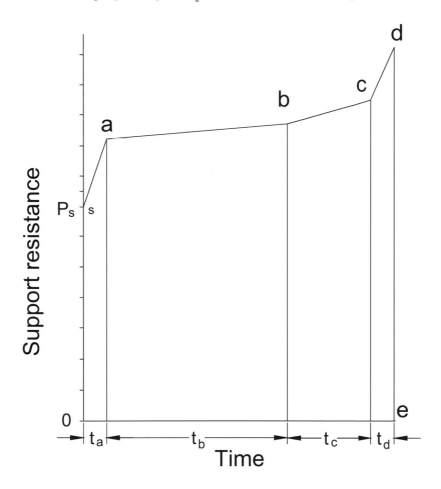

Figure 5.5.1 Typical pressure change in a shield supporting cycle

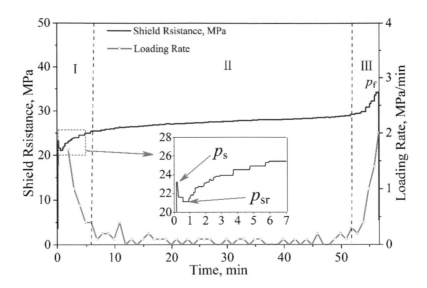

Figure 5.5.2 Variation of shield resistance and loading rate in a typical shield supporting cycle without yielding. The three periods are I initial period, II relatively stable period, III cutting influenced and neighboring shield advance period. ps, psr, and pf denotes setting pressure, revised setting pressure, and final pressure, respectively

Source: Cheng et al. (2018)

rapidly increasing resistance period (s–a). It is followed by a relatively stable loading period, a–b, during time t_b the shearer is not in the immediate vicinity of the support. In this relatively longer period of time, the roof loading is relatively stable. Notice this period is generally the longest in a mining cycle. When the drum cutting action of the shearer approaches and passes the support during time t_c, the roof loading increases as a result of the additional area of roof equivalent to the web width exposed. This is the cutting-influenced period, b–c.

Finally, as the neighboring support is disengaged and advanced during time t_d, the roof load is suddenly transferred to the support, thereby rapidly increasing the support resistance in a short time period (generally less than 10 sec). This is another rapidly increasing resistance period, c–d. As the support is released for advancing, its resistance drops almost instantaneously to zero (d–e). Thus, a supporting (mining) cycle causes the support resistance or load (or pressure) variation from o through s, a, b, c, d, to e. Note P_s at s is setting pressure and d is final pressure. The difference between setting and final pressures is load increment.

Prior to 1990, when support capacity was smaller, a typical normal supporting cycle took more or less the form shown by Figure 5.5.1. Even after 1990 when the support setting (partially due to positive set) and yield load began to increase significantly, the initial load increase (sa in Fig. 5.5.1) and load increase (bc in Fig. 5.5.1) due to shearer's cutting and passing remain significant parts of the supporting cycle. Peng (1998) reported that the pressure change within a supporting (or mining) cycle varies just like before from cycle to cycle and shield to shield. But in general pressure change can be divided into three major types: increasing, steady, and decreasing types depending on whether the leg pressure, after setting, increases substantially, stays more or less the same, or decreases substantially, respectively (Fig. 5.5.3). Shields in the increasing pressure type receive the largest pressure (or load) increase in a mining cycle (Fig. 5.5.3 A). The increasing type has many variations depending on mining and geological conditions. Figure 5.5.2 is one of the more significant ones. Typically, the increasing type of pressure curve can be divided into three segments according to its rate of pressure change. The first segment is the initial rapid increasing segment and usually lasts less than 10 minutes just after setting. The second segment covers the portion of relatively steady increase after the initial rapid increase and accounts for most of the cycle time. The third segment is the final rapid increase portion and occurs usually within five minutes just before the shield is released for advance. This type of pressure curve denotes a relatively intense roof loading.

If the roof movement is very intense, the leg pressure may reach the yield load before and cease to increase further until the end of the mining cycle. Shields with the steady type of pressure curves (Fig. 5.5.3 B) receive less load than those with the increasing type. This type of pressure change also can be divided into three segments. But the first two segments are quite different from those of the increasing type. The pressure in the first segment may remain constant, decrease, or increase, while the second segment is normally flat. This type of pressure curve denotes a relatively weak roof loading. The decreasing type of pressure curve (Fig. 5.5.3 C) denotes that either the roof is extremely weak, or it has too much rock/coal debris between the canopy and roof strata and/or under the base plates or a leg leakage. The leg pressure decreases rapidly after setting and continues to decrease at a decreasing rate. In some cases it increases suddenly at the end of the cycle, while in others cases it finishes at the same decreasing rate.

The increasing type of pressure distribution is a result of roof deflection and can be used to reflect the behavior of roof strata movement, whereas the decreasing type is mainly caused by

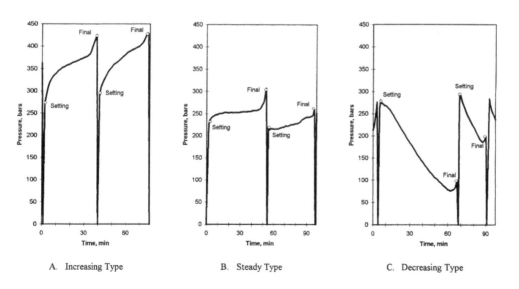

A. Increasing Type   B. Steady Type   C. Decreasing Type

Figure 5.5.3 Three types of pressure changes in a shield supporting cycle
Source: Peng (1998)

either continuous deterioration of roof/floor strata, the presence of too much rock/coal debris between supports and roof/floor strata, or leakage of hydraulic legs in a supporting cycle. To analyze and predict roof activity above a longwall face, the characteristics of pressure changes in a normal supporting cycle must be well defined and analyzed. A normal supporting cycle can be divided into three periods (Fig. 5.5.2) (Chen and Peng, 2003; Cheng et al., 2018):

1   Setting period – the time duration from when the shield canopy initially set against the roof to when the setting valve in the shield leg is closed. It lasts from several seconds to 35 seconds, depending on the effective pump pressure, which is the actual pump pressure minus pressure loss along the hydraulic supply system, the designed setting pressure, the setting method (positive or manual set), and contact condition between shield and roof/floor.

2   Roof deformation period – leg pressure develops gradually as the shield starts to react to roof deflection induced by the overburden strata movement. Depending on the longwall system's availability, shearer's travel speed, face width and location of the shield of interest, this period lasts several minutes to several hours.

3   Supporting area change period – variation of shield leg pressure mainly results from change in the total supporting area caused by the shearer's cutting and approaching as well as the adjacent shield's release from the roof for advance, the effect of which lasts 5–20 min immediately after setting and 1–5 min before adjacent shield advance.

Traditionally, the factors for analyzing the shield leg pressure in a working cycle include setting pressure, final pressure, load increment and time-weighted average resistance or pressure (TWAR or TWAP). These factors have been found insufficient to explain the implication of the pressure-time history regarding the complicated interaction between support and rock. Cheng and Peng (2017), Cheng et al. (2015), and Cheng et al. (2018) had collected and analyzed massive data on leg pressure, leg closure, and shearer's location and defined more than 20 factors such as setting pressure, revised setting pressure, TWAP, final pressure, rate of load increase, rate of leg closure, duration of working cycle, characteristics of yield valve opening, etc. and investigated their influence on the shield's loading characteristics and interaction between shield and surrounding rocks. Based on the results of the findings, they have developed a software program "Status of Shield and Roof IntelliSense System" (SSRI). The results of the effects of those factors are summarized in the following subsections.

5.5.1.1 Setting pressure and revised setting pressure

When setting the support using the electrohydraulic control, the fluid is fed into the bottom stage of the cylinder. As the fluid continues to flow, the leg continues to rise until the canopy touches the roof. Thereafter, the fluid in the bottom stage chamber rapidly increases to the working pressure (actually in practice, the working pressure equals pump pressure minus pressure loss due to friction in traveling through hoses and valves) of the hydraulic pump. At this time, the control valve is turned off and the fluid is locked in. The fluid is now operating at the working pressure of the pump. This is the setting pressure of the support. Assuming the bottom stage is not at full stroke, the force exerted by each leg under the setting pressure, P_s, is (Peng and Chiang, 1984):

$$P_s = \frac{\pi d^2}{4 \times 2000} \sigma_s \, (\text{tons})$$

(5.5.1)

Where d = inner diameter of the bottom stage of the leg cylinder, in.:

$\quad \sigma_s$ = pump pressure, psi

If there are n legs in the support, the support setting force, P, is

$$P = n\eta P_s \qquad (5.5.2)$$

Where η is the support efficiency, which is the ratio of the actual load available for supporting the roof to the total force exerted by all legs under the setting pressure. For two-leg shields, η varies primarily with the leg inclination of the roof support and only the vertical component of the force is considered in the context. In practice, the pump pressure ranges from 4200 to 5000 psi (28.96–34.48 MPa), which is limited by the maximum allowable pressure for the hydraulic hoses and connectors. Normally, a safety factor of four is used for hoses and connectors.

In an electrohydraulic control system, there are two levels of setting a shield: a positive set is to ensure that the hydraulic leg pressure reaches the rated setting pressure when the shield canopy was first raised to contact the roof, whereas a guaranteed set is that after a positive set is implemented at the beginning of the cycle, a pilot signal will scan all shields at a selected time interval. If the setting pressure in any shield is found to be below the programmed setting pressure threshold, it will be raised to the rated setting pressure. Therefore, for a weak roof, a guaranteed set may not be desirable because many shields may be under constant adjustments posing safety concerns. Chen and Peng (2003) described the setting features for the PMC^R-C control system in which there are three preset pressure thresholds: P_0 = 1450 psi (100 bars), P_1 = 4060 psi (280 bars), and P_2 = 4640 psi (320 bars). During the setting operation, the setting valve is held open until the hydraulic pressure reaches the preset value of P_0 = 1450 psi (100 bars). At this moment, the automatic setting system will take over the artificial control for a specified time period (i.e., 15 seconds). The setting valve will be turned off if the designated setting pressure threshold, P_2 = 4640 psi (320 bars), is reached, or the designated setting pressure is not reached, but the time exceeds the specified time period. A pilot signal will scan all shields at a selected time interval (i.e., 10 seconds) after the initial positive setting. The setting valve will be re-opened again if the actual leg pressure is less than the minimum setting pressure threshold, P_1 = 4060 psi (280 bars). This process usually repeats three times. In other words, the designated setting pressure (P_2) or at least the minimum setting pressure (P_1) should be achieved across the face if the shield legs are not losing pressure due to leaks or extension due to soft ground conditions, provided adequate pump pressure is provided.

Positive set can be turned on or off any time. Normally, when encountering a roof fall or very weak roof, it is turned off.

The most commonly ignored feature in analyzing the setting pressure is that after the leg has reached the setting pressure (p_s in Fig. 5.5.2), leg pressure will drop quickly from less than 1 min to no more than 3 min (p_{sr} in Fig. 5.5.2). The amount of pressure drop is 72.5–725 psi (0.5~5 MPa), followed by slow increase or stabilize. Figures 5.5.4 and 5.5.5 show the frequency distribution of the amount of pressure drop and their duration for a US and a Chinese mine.

Pressure drop is due to uneven floor or there are float coal and/or debris between canopy and roof, and base plates and floor. When shield reaches the setting pressure and after the high pressure inlet ports of the legs are shut off, those float coal and debris are compressed and crushed further causing the leg to extend slightly, thereby lowering the leg pressure.

Consequently as the legs extend, their pressures decrease; In addition, since the leg pistons extend upward quickly under the impact of high pressure fluid, the extension moment of inertia continues even after the inlet ports are shut off, causing the leg pressure to drop within a short period of time until it reaches equilibrium. Therefore, the actual setting pressure is

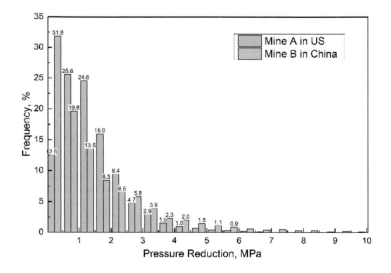

Figure 5.5.4 Distribution of pressure decrease after shield setting

Source: Cheng and Peng (2017)

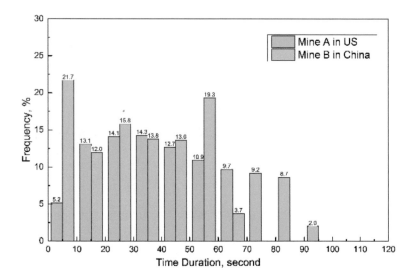

Figure 5.5.5 Frequency distribution of duration of decrease in setting load

Source: Cheng and Peng (2017)

that when shield reaches the new equilibrium, because this is the actual pressure available for supporting the roof at shield setting. The leg pressure at the new equilibrium condition is defined as the revised setting pressure.

5.5.1.2 Final pressure and TWAR or TWAP

Normally, the final pressure of shield legs (p_f in Fig. 5.5.2) occurs when the adjacent shield is advancing and lasts less than 1 min (about 3% of total shield cycle time). It is the highest pressure in a normal shield supporting cycle. For panels without or with a minimum amount of yielding, the TWAR or TWAP reflects better than the final pressure on the shield's loading characteristics and roof activities.

Since the roof pressure usually varies continuously with time, the time-weighted average resistance is more appropriate to illustrate the overall conditions of the support in a whole cycle. The time-weighted average resistance, P_t, is defined by the following equation:

$$P_t = \frac{\sum P_i t_i}{t_t} \tag{5.5.3}$$

Where P_i is the average support resistance during the period of t_i. In other words, P_t is the ratio of the area under the pressure variation curve to the total time in a mining (supporting) cycle. As an example, P_t in Figure 5.5.1 is:

$$P_t = \frac{\frac{1}{2}\left(P_a + P_s\right)t_u + \frac{1}{2}\left(P_b + P_a\right)t_b + \frac{1}{2}\left(P_c + P_b\right)t_c + \frac{1}{2}\left(P_d + P_c\right)t_d}{t_a + t_b + t_c + t_d} \tag{5.5.4}$$

A high maximum loading (or final pressure) sustained by a support does not mean it has efficiently supported the roof strata throughout the entire period of the supporting cycle. The maximum loading of the support is only an instant event that generally occurs under a series of operation-related factors. Therefore, the time-weighted average resistance of the support can better denote the roof pressure during the whole period of the cycle.

There are high and low of TWAP as the face advances. The high peaks denote a roof weighting and appear periodically at various intervals. This is the periodic weighting due to periodic adjustment of the roof structure. But as can be seen in Figure 5.5.6 not all weightings are serious (Peng, 1998). A general measure of the intensity of the periodic weighting is the mean and standard deviation of TWAP. The larger the mean and standard deviation, the more intense is the roof movement. By comparing the TWAP in an individual web cut with the mean or average TWAP for that panel, the intensity of roof loading or periodic weighting in a web cut can be evaluated. Mine A with a larger mean and standard deviation of TWAP, 342 and 38 bars respectively, had a conspicuous periodic weighting but in less frequency. Mine B with a small mean and standard deviation of TWAP, 218 and 22 bars, respectively, had a mild periodic weighting but occurring more frequently. Mine C with a large mean but a small standard deviation of TWAP, 331 and 23 bars, respectively, had a relatively small periodic weighting interval or load change but a persistent large roof loading.

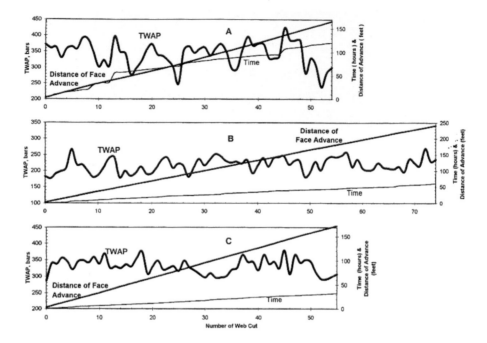

Figure 5.5.6 TWAP, time and distance of face advance vs. number of cut for three mines (Mine A, B, and C)

Thus, a real-time construction of a figure like Figures 3.2.10 and 5.5.6 can predict the arrival of an incoming periodic weighting if the shield supporting cycle has no or with minor yielding.

5.5.1.3 Loading rate in each period

The load increase rate in the initial load increase period is that occurs within a short period after setting, generally 5–20 min, depending on shield's load increase characteristics and geological conditions (period I in Fig. 5.5.2). It is an important index for evaluating the intensity of roof activity. Relative stable load increase period (period II in Fig. 5.5.2) is that located between the initial load increase period and the increasing load increase before the shield advance period (period III in Fig. 5.5.2). In this period, the shearer has passed and moved far away. The roof activity stabilizes. The load increase of shield in general is low. For most shields' loading cycles, relative stable load increase period account for the majority of the total cycle time. Therefore, the load increment rate in this period can reflect the loading condition of shield. When shield is advanced sequentially following the shearer's cutting and passing by, the load increase on each shield is equivalent to the rock weight of one web depth area exposed in front of it. Afterward, when the adjacent shield is lowered and advanced, the original rock weight on the adjacent shield is immediately transferred to it, causing a rapid shield load increase in a very short period. This is the load increase before shield advance period.

5.5.1.4 Parameters associated with cutting influence and neighboring shield advance period

As shown in Figure 5.5.2, the load increase in the last few minutes before shield advance is due mainly to shearer's cutting and neighboring shields' advance. For convenience, the last five minutes before shield advance is considered as the starting time of this period. Three parameters are proposed to reflect the leg pressure changes within this period of shearer's cutting and neighboring shields' advance before the shield advances itself (Cheng et al., 2015), i.e., leg pressure increase before shield advance (LPI), leg pressure increasing rate before shield advance (LPIR), and the ratio of leg pressure increase to the pressure five minutes before shield advance (RLPI). The rate of increase in leg pressure before shield advance is the average increasing rate during this five-minute period.

5.5.1.5 Parameters associated with yielding

When heavy weighting, or a long idle time period, or an insufficient yield load is encountered, shield load may reach or exceed the yield load. In order to prevent the shield legs from being damaged by the high pressure emulsion fluid, the yield valve opens immediately and the fluid exits to lower the leg pressure. The yield valve then closes and the leg pressure increases again under the roof loading. The whole process from valve open to close is a yield cycle (yield valve opens once). Fig. 5.5.7 shows the variation of leg closure and leg pressure

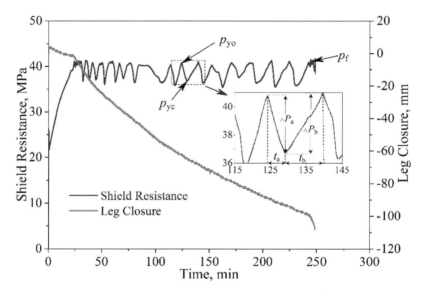

Figure 5.5.7 Variation of shield resistance and leg closure in a typical shield supporting cycle with continuous yielding. Negative leg closure means shortening of shield leg. p_{yo} – pressure at which yield valve opens, p_{yc} – pressure at which yield valve closes, p_f – final pressure, Δp_a – pressure drop during a yielding cycle, Δp_b – pressure increase in between two consecutive yielding events, t_a – duration of a yielding cycle, t_b – pressure increasing period in between two consecutive yielding events

Source: Cheng (2015); Cheng and Peng (2017); Cheng et al. (2018)

in a typical shield loading cycle in which yield valve opens frequently. In addition to yield frequency, the factors that are related to the duration of yield valve opening are as follows (Cheng *et al.*, 2018; Cheng and Peng, 2017):

1   Shield leg pressure when the yield valve opens (p_{yo} in Fig. 5.5.7).
2   Shield leg pressure when yield valve closes, i.e., the leg pressure when yielding ends (p_{yc} in Fig. 5.5.7).
3   Amount of pressure decrease when yield valve opens which is equal to the leg pressure when the yield valve starts to open less the leg pressure when yielding ends (Δp_a in Fig. 5.5.7).
4   The accumulated pressure increase when yield valve is closed, i.e., the pressure increase occurred from the time when the yield valve closes to that when the yield valve opens in the immediate following cycle (Δp_b in Fig. 5.5.7).
5   Duration of yield valve open, i.e., time period from yield valve opens to yield valve closes (t_a in Fig. 5.5.7).
6   Duration of pressure increase before yield valve opens again, i.e., the time period from yield ends to starts of next yielding cycle (t_b in Fig. 5.5.7).

In addition, similar factors can be defined for leg closure.

5.5.1.6 Leg closure of the shield

Shield supports the roof by introducing the high pressure fluid into the leg cylinder causing the leg piston to rise until it make contact with the roof. When the leg pressure reaches the setting pressure the inlet port is closed and the fluid inside the cylinder is sealed completely. During roof loading, leg closure of various amount occurs due to elastic compression of the hydraulic fluid, frequent loading and unloading as the yield valve opens and closes or leaking from damaged seals. For example, as shown in Fig. 5.5.7, different rates of leg closure occur at different periods of shield loading. Consequently, just like the leg pressure factors stated in (1) through (6), the amount of leg closure and leg closure rate occurring in the whole or each supporting period of shield supporting cycle can be determined for analyzing the shield loading characteristics and the interaction between shield and rock.

5.5.2 Factors influencing the supporting characteristics of shield

Since the leg pressure denotes the results of interaction among roof, shield, and floor, each and every pressure curve in a mining cycle represents the results of a specific interaction among roof, shield, and floor. It follows that all other things being equal, if the pressure curve in a mining cycle varies from cycle to cycle, then the interaction among roof, shield, and floor is not the same from cycle to cycle and, similarly, if the pressure curve in a mining cycle varies from shield to shield, then the interaction must be different from shield to shield. Does this mean the roof condition changes from cycle to cycle and from shield to shield? Obviously, unless the roof strata encounter a localized geological anomaly (e.g., slickensides) or are badly broken and totally lose their "beam" characteristics, the varying nature of the pressure curves in a mining cycle is unlikely to be due purely to changes in roof structures. Therefore, it is most likely, and indeed underground observations show, that contact conditions between shield canopy and the immediate roof strata and/or shield base plate and floor are the key factors accounting for the varying shield leg pressures. In fact, the shield canopy rarely has a full and smooth contact with the immediate roof strata due to the ever presence of cutting steps within and between the shearer's web cuts, roof cavities, rock/coal debris, localized geological anomalies, partings, and their sudden change in thickness. Unless these complicated factors are known and well-defined, the roof strata must be assumed to be more or less continuous beams and, as such, the interaction among roof, shield, and floor should consider the interaction between adjacent shields. There are several methods available to handle this type of data and its controlling factors. But due to the massive quantity of data, the statistical method is the most appropriate. It must be noted however that, due to advance in automation in shearer's roof and floor controls in recent years, the roof and floor horizons have been improved considerably.

Other factors that affect the interaction of roof, shield, and floor include major geological anomalies: setting pressure, time, yielding characteristics of yield valve, and mining techniques. Those factors are selectively discussed in the following subsections.

5.5.2.1 Effect of setting load on the loading characteristics of shield

Since the commonly accepted practice in the coal industry is to use yield capacity to represent the capacity of shield support, and yet when the shield is set against the roof, it is the setting load, not yield load, that is used to support the roof. In this respect, the ratio of setting-to-yield load (or pressure) ratio is important. Shields with the same yield load capacity but different setting loads may produce different results against the same roof, because the roof may react one way with a lower setting load but another way with a higher setting load. Knowing the yield load only is not enough. Both yield load (or pressure) and setting load (or pressure) or yield load and setting-to-yield load ratios must be known.

Since the total roof-to-floor convergence at the face area is a measure of the intensity of the roof activity and thus the roof stability, it has been used to correlate with the adequacy of the setting load. The relationship between the roof-to-floor convergence and the setting load density is a hyperbolic curve, as shown in Fig. 5.5.8. Point a is the infection point, which represents the point at which the curvature of the curve changes from left to right and vice versa. To the left of point a, the setting load density varies rapidly with small change in convergence. In other words, a large increase in setting load density is necessary to produce a small decrease in convergence. In fact, for setting load densities greater than P_{s2}, convergence is not reduced to any significant amount less than C_{s2}. The situation is just the opposite for the area to the right of point a. In this region, a small change in setting load density induces a large amount of convergence change. In other words, decreasing the setting load density a small amount will increase the convergence significantly. When the setting load densities are below $\overline{P}_{s1}$, the roof cannot be controlled. It appears from Figure 5.5.8 that there is an effective range of setting load density, within which the roof can be effectively controlled. The minimum load density, $\overline{P}_{s1}$, can be considered as the setting load density (SLD), while the yield load density (YLD) can be anywhere from $\overline{P}_{s1}$ to $\overline{P}_{s2}$, depending on the setting-to-yield load ratio. In practice, a setting-to-yield ratio of 0.6–0.9 has been used.

Both theoretical (Peng and Chiang, 1984) and field experiments (Graham, 1978; Price and Pickering, 1981; Bates, 1978) have shown that if the setting pressure is too low to effectively control the roof convergence, increasing the setting pressure has clear advantages. In

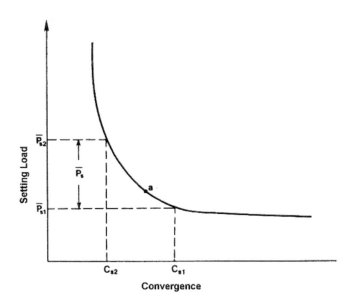

Figure 5.5.8 Relationship between roof-to-floor convergence and shield setting load
Source: Peng *et al.* (1987)

addition to reducing the roof-to-floor convergence, it also reduces roof fractures and face sloughing.

The conventional wisdom is to raise the setting pressure as much as possible such that it not only reduces bed separation and increases roof stability, but also reduces face spall and roof falls in the unsupported distance. The recommended setting to yield pressure is 0.6~0.9 or 0.6~0.8. This conclusion was obtained during the 1980s–1990s when the capacity of powered support was low. As the yield load of powered supports continued to increase, underground monitoring had shown that the average setting pressure lied between 2646.3 psi (18 MPa) and 3744 psi (25.82 MPa) (Table 5.5.1). The measured setting-to-yield ratios were mostly around 0.5, far lower than the traditionally recommended range of 0.6–0.9. In spite of this, the measured support density in most mines already exceeded 92.8 psi (0.6 MPa), being sufficient to keep the roof stable, according to all underground monitoring results. Therefore, as the shield's yield load continues to increase, it is the setting pressure, rather than the setting-to-yield ratio, is more representative of support density (Cheng and Peng, 2017). Figure 5.5.9 shows the current setting pressure already reaches or nearly reaches P_{s2}, further increase in setting pressure will not reduce any significant amount of roof-to-floor convergence.

Recent research shows that the effects of setting pressure (load) on shield load can be divided into two types (Cheng *et al.*, 2018; Cheng and Peng, 2017; Peng *et al.*, 2019): the first type refers to those shields with adequate shield capacity, for which there will be no or few yielding events at the face. Consequently, there will be little effect on the pressure increase after shield setting as shown in Figure 5.5.9. Within the same time period, although the setting pressure is different, the trends of pressure increase after setting are the same and remain so until the shield advance. The magnitude of setting pressure (load) has no effect

Table 5.5.1 The measured averaged setting pressure in 10 longwall panels

Mine	Average measured setting pressure, MPa	Average measured setting load/ shield load capacity	Average measured setting load density, MPa
A	23.74	0.5	0.87
B	25.82	0.56	0.87
C	23.77	0.48	0.85
D	23.49	0.51	0.79
E	23.42	0.51	0.79
F	22.98	0.49	0.71
G	25.03	0.56	0.63
H	21.85	0.48	0.59
I	20.18	0.47	0.57
J	22.66	0.53	0.44

Source: Cheng and Peng (2017)

* A–I: Chinese mines, J: US mine

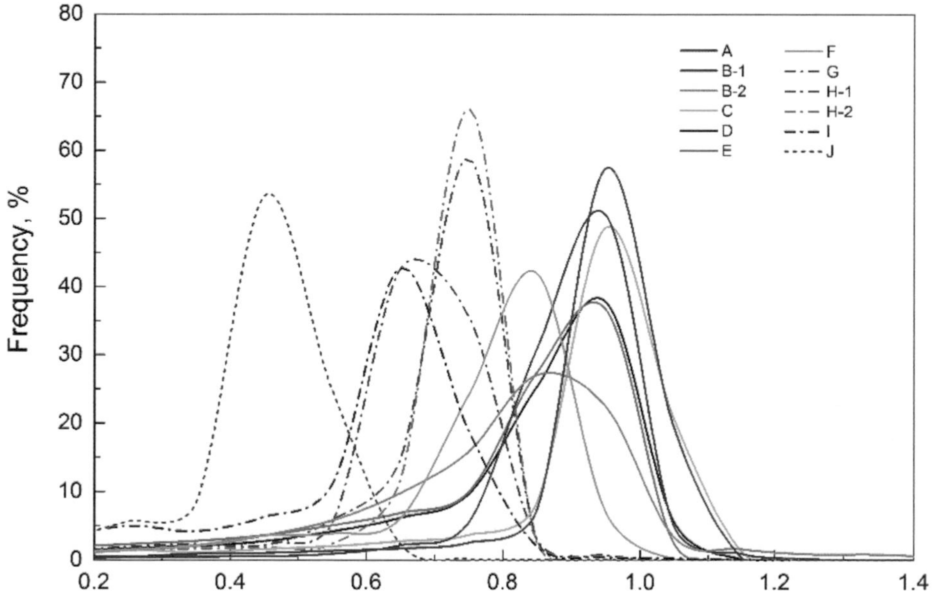

Figure 5.5.9 Shield support density distribution for mines shown in Table 5.5.1
Source: Cheng and Peng (2017)

on the pressure increase (Fig. 5.5.10). However, setting pressure (load) is somewhat linearly proportional to TWAR (Fig. 5.5.10).

The second type refers to those shields with inadequate shield capacity for strong roof pressure and frequent yielding phenomena. A larger setting pressure tends to shorten the time

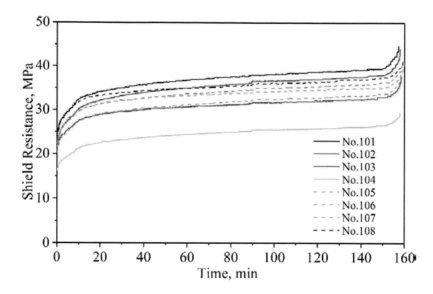

Figure 5.5.10 Load increase characteristics for neighboring shield under different setting pressure in the same shield cycle

Source: Cheng and Peng (2017)

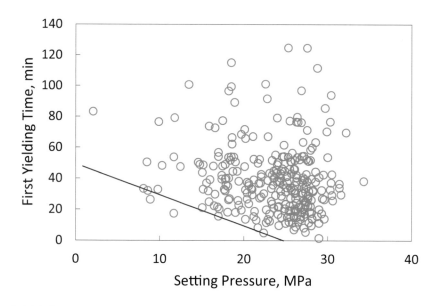

Figure 5.5.11 Setting pressure vs. the time to first yielding

Source: Cheng and Peng (2017)

period between setting and initial yielding (Fig. 5.5.11), thereby increasing the frequency of yielding and the magnitude of leg closure within a shield cycle (Figs 5.5.12 and 5.5.13). Shields with a larger setting pressure will reach quicker the onset of period in which yield valves begin to open. During yielding, shield stiffness and its capacity to resist deformation

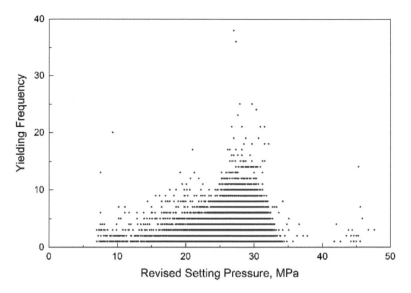

Figure 5.5.12 Number of yielding vs. setting pressure

Source: Cheng and Peng (2017)

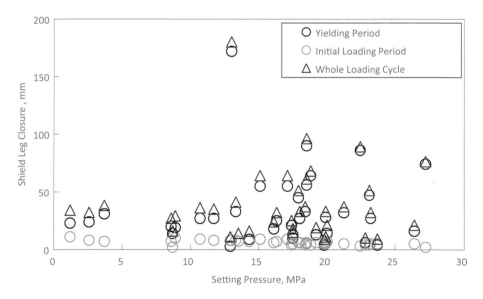

Figure 5.5.13 Setting pressure vs. leg closure before and during yielding

Source: Cheng and Peng (2017)

reduces greatly. Consequently, the longer the yield valve opens, the larger the hydraulic leg closure.

If the setting pressure is too low, it will lead to a larger pressure increase, resulting in roof convergence being too much in a shield cycle. Conversely, a large setting pressure leads to frequent opening of yield valves resulting in increasing roof convergence. Therefore, from controlling the roof convergence point of view, a proper setting pressure means it will not lead to too much load increase in the setting period, nor will it cause frequent opening of the yield valves.

5.5.2.2 Effect of shearer's cutting and advance of neighboring shield on shield load

During a shearer's coal cutting and shield advance, the equilibrium loading condition between support and roof is destroyed, causing rapid increase in leg pressure before shield advance and right after setting until a new state of relative equilibrium is reached. Underground monitoring data show that the duration of influence of coal cutting and shield advance lies within 20 min after setting and 10 min before shield advance. During this period, the rate of load increase 29–435 psi min⁻¹ (0.2~3MPa min⁻¹) is several times that of relative stable loading period (less than 14.5 psi min⁻¹ or 0.1 MPa min⁻¹).

Recent research demonstrates (Cheng *et al.*, 2015; Cheng and Peng, 2017) that under the combined influence of coal cutting and shield advance, pressure increase ΔP_R during this short period is related to the distance to the front drum D by (Figs 5.5.14 and 5.5.15):

$$[\Delta P_R = a\mathrm{D}^2 + b\mathrm{D} + c] \tag{5.5.5}$$

where a, b, and c are constants.

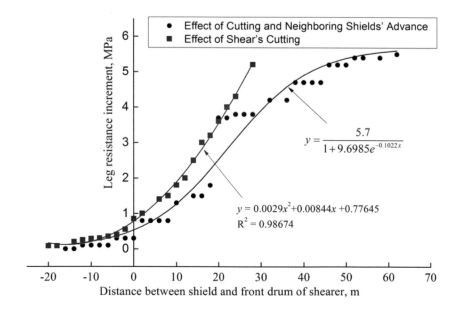

Figure 5.5.14 Relationship between leg pressure increase and the distance between shield and front drum of shearer under the independent influence of drum cutting and the combined influence of shearer's cutting and neighboring shields' advance

Source: Cheng *et al.* (2015); Cheng and Peng (2017); Cheng *et al.* (2018)

During shield advance, it is only subjected to cutting influence, the load increase ΔP_R is related to the distance to front drum by (Fig. 5.5.14):

$$R\Delta P_R = \frac{\Delta L}{1 + \Delta L pe^{D \ln q}} \tag{5.5.6}$$

Where ΔL is the finial pressure after pressure increase and p and q are constants.

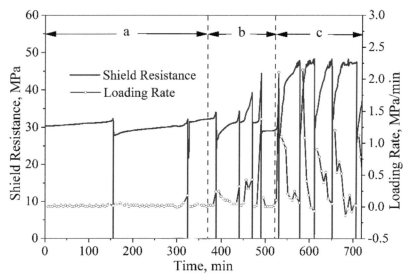

(A) Strong roof and inadequate shield capacity

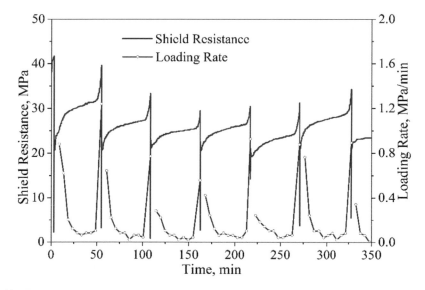

(B) Medium-weak roof and adequate shield capacity

Figure 5.5.15 Shield pressure and loading rate of different conditions

Source: Cheng and Peng (2017); Cheng et al. (2018)

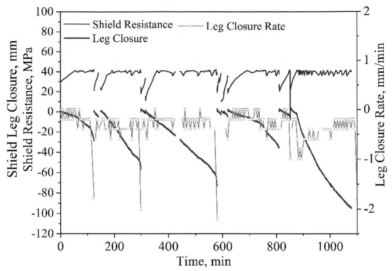

(C) Weak roof and inadequate shield capacity

Figure 5.5.15 (Continued)

   In addition, geological conditions and whether or not shield capacity is sufficient have great influence on the characteristics of load increase during coal cutting and shield advance. For panels with hard roof, the development of stage of fracture in the roof affects the characteristics of pressure increase during coal cutting and shield advance. As shown in Figure 5.5.15 A, when the roof overhang is short, the load increase is insignificant after setting and before shield advance (stage a in Fig. 5.5.15 A), as the overhanged length increases, coal cutting breaks the roof's equilibrium condition, resulting in first increasing rate of pressure increase before shield advance (stage b in Fig. 5.5.15 A), followed by the roof's entering into severe active period due to roof subsidence, fracture, and rotation and leading to large increase in the rate of pressure increase after setting and before shield advance (stage c in Fig. 5.5.15 A) until the yield valve is opened and closed repeatedly.
   For panels with medium weak roofs where shield capacity is sufficient (Fig. 5.5.15 B), shield load increase is larger within the short period after setting and before shield advance (190–1450 psi or 2–10 MPa or 5–25% of the shield capacity in the case of Fig. 5.5.15 B). Furthermore the rates of pressure increases are similar after setting and before shield advance. This may be due to the fact that overhang condition is insignificant for weak roof as compared to hard roof. When the supporting area changes, the overburden load can be quickly transferred.
   When shield capacity is sufficient, shield load increase is larger (Fig. 5.5.15 B); when shield capacity is insufficient, the influence of coal cutting and shield advance is expressed by frequent opening of yield valve (Fig. 5.5.15 A and C). However, the difference is difficult to be expressed by factors related to shield leg pressure and rate of load increase. In comparison, leg closure is much better. For example, as shown in Figure 5.5.15 C, leg closure is rapid within the short period. The rate of closure being one digit larger, generally exceeding 1mm min[-1].

5.5.2.3 Time dependent effect of support pressure

It is well known that shield pressure increases with time. Figure 5.5.16 Mine A shows that in a medium-hard roof condition the shield pressure in Friday's production cycles reached the yield pressure only in 1.5% of the shield operation time (Peng, 1998). But, on Saturday's idle shifts, the pressure increased considerably and reached the yield pressure in 14% of the idle time. This trend continued and ultimately 28% of the time reached yielding on Sunday's idle shifts. Similar trends are shown in Figure 5.5.16 Mine B for a hard roof condition. The percentage of time shields yielded 49% during non-production shifts, which is significantly larger than the 26% found during production shifts.

Analysis of shield pressure increase in every cycle shows that when the cycle time is less than one hour, the rate of average pressure increase is large, about one bar per minute. When the cycle time is more than one hour, the average rate decreases to 0.3–0.05 bar per minute. Therefore, roof movement above and around the face area in response to longwall mining is a time-dependent process, and the harder the roof is, the longer it lasts, and the larger the time-dependent incremental load.

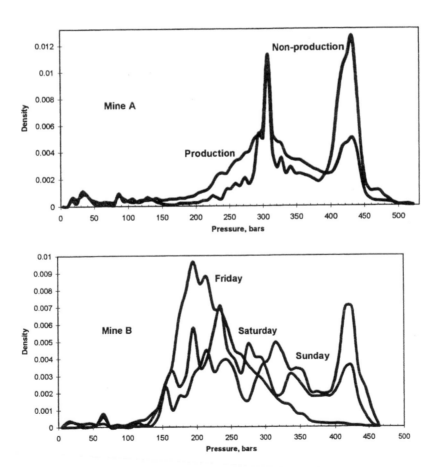

Figure 5.5.16 Comparison of shield pressure distribution between production and non-production shifts

Time factor including face advancing rate, cutting speed and shield advance rate has great influence on roof activity, and characteristics of load increase and load transfer between shields. Laboratory experiments show rock deformation is time dependent. The effects of time factor is more significant on the deformation and failure of *in-situ* rocks during underground mining operation (Peng, 2008). Since the normal production cutting cycle in longwall mining ranges from less than an hour to several hours, the shield leg pressure change during idle time is more representative of time effect on overburden movement.

Research has demonstrated (Cheng, 2015; Cheng *et al.*, 2018) that shield's leg pressure changes are related to face geological conditions. Figure 5.5.17 A is a 3D shield pressure distribution map for 130-hour continuous operation of a longwall panel in northwestern China. During this period, there were two stoppages lasting for 40 and 25 hours. The roof strata were hard. Periodic weighting was conspicuous and severe, causing the yield valve to open frequently; Due to difference in the arrival of periodic weighting, the leg pressure change exhibited two very different characteristics, one is a rapid pressure increase after setting (20th–60th shields in Fig. 5.5.17 A). It reached or approached the yield load within a short period of time (0.5–5.0 hr after setting). Yield valves opened frequently and the accumulated roof subsidence was large. Conversely, in the area from the 70th to the 150th shield, increase in shield pressure was minor within 20–40 hr after setting with shield pressure density mostly less than 4350 psi (30 MPa).

Figure 5.5.17 B is another 3D shield pressure distribution map for 70-hour continuous operation of a longwall panel. During this period there was a stoppage lasting for 35 hours. The roof strata were medium weak. Periodic weighting was not conspicuous, shield capacity was adequate and yield valve did not open. During stoppage, all shield pressure increased continuously to various degrees and its rate of pressure increase was smaller than the hard roof panel. But it lasted longer and stabilized 10–25 hrs after setting.

Overall, during long stoppage, shield pressure change after setting for the hard roof panels depends on what time section of the shield is in the periodic weighting period (or stage of the roof fracture process). If it is on the stage that the main roof has already broken and begin to rotate, shield pressure will increase rapidly because the main roof needs to rotate fairly large amount before stabilization. It maintains high shield pressure during the whole stoppage period. The longer is the stoppage, the larger are the amount of load increase and roof subsidence. If production stoppage occurs at the end of periodic weighting (before the main roof breaks), shield load is the weight of immediate roof and will maintain smaller loading during the stoppage. Load increase is not obvious. For weak roof panels, no clear periodic weighting occurs during the long stoppage. Shield load is the weight of immediate roof. Pressure increase along the faceline direction does not exhibit localized section model, unlike for the hard roof panel where the roof is in blocks, reflecting more of the slow roof subsidence and slow increase of load increase under the influence of mining pressure.

5.5.2.4 Effect of yielding on the loading characteristics of shield

For panels with frequent yielding, yielding is a much more important characteristics than final load and TWAP. However, both management and technical staff in the coal industry use shield capacity as the only criterion for evaluating shield's loading capacity, i.e., shield's capacity is insufficient when continuous yielding in shield occurs. If yielding occurs frequently or continuously in operating shields, next time when opportunity comes for designing

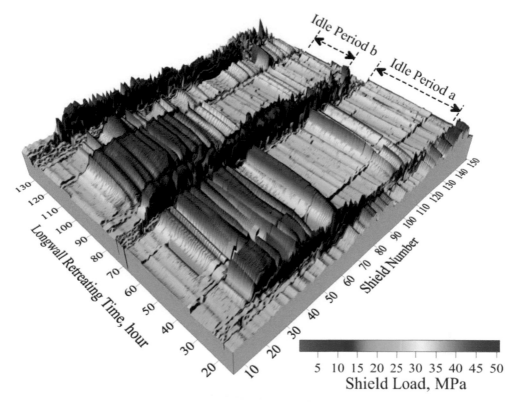

(A) 3D view of the shield leg pressures including two idle shifts of long idle periods in longwall panel with hard roof

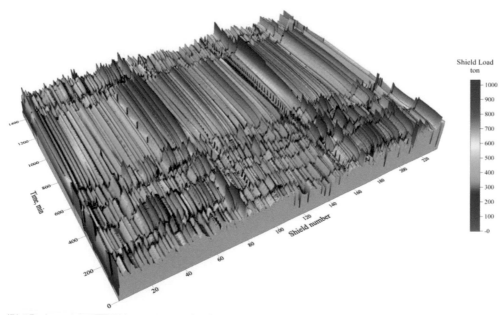

(B) 3D view of the shield leg pressures for the panel with medium weak roof after advancing more than 1400 min that included an idle shift of about 800 min

Figure 5.5.17 Change in shield pressure with longwall retreating time

Source: Peng *et al.* (2019); Cheng *et al.* (2018); Cheng (2015)

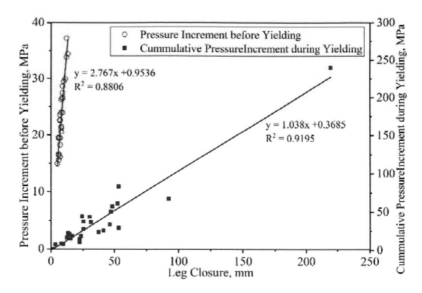

Figure 5.5.18 Leg pressure increase vs. leg closure during and before yielding

Source: Cheng and Peng (2017); Cheng et al. (2018)

a new shield, a higher capacity shield is designed and selected. This logic cycle repeats again and again, resulting in the ever-increasing load capacity of new shield ordered. The irony is that shields of 2473 tons (22,000 KN) are still yielding in western Chinese coal mines.

When severe roof weighting occurs or shield's yield load capacity is insufficient, the supporting load provided by the shield is unable to control the subsidence of the overburden. Once shield takes up the amount of leg closure occurred during the pressure increasing period, it enters into the yielding period. Underground monitoring shows that the amount of leg closure during pressure increase period is very limited (normally not exceeding 0.39 in. (10 mm) in Fig. 5.5.18 except when the setting pressure is very low). Furthermore, shield's stiffness is far smaller during continuous yielding than before yielding. In fact, the latter is 2.8 times that of the former (Fig. 5.5.18): due to a much smaller stiffness during yielding leg closure accumulated during yielding is larger.

Therefore, if continuous yielding is unavoidable, the ideal yield working characteristics is that the yield valve close up rapidly as soon as it discharges a fixed amount of high pressure fluid in order to reduce leg closure as much as possible, thereby reducing the risk of running the leg cylinder into solid. This way shield can maintain sufficient height for shearer's passage even during long stoppage.

5.5.3 Application of shield leg pressures

5.5.3.1 Early warning of severe roof weightings

It is well known that a thick and hard roof in the overburden induces periodic weighting that in turn may cause roof stability problems at the face. The conventional method of utilizing shield leg pressure to determine if periodic weighting arrives is to use the final pressure

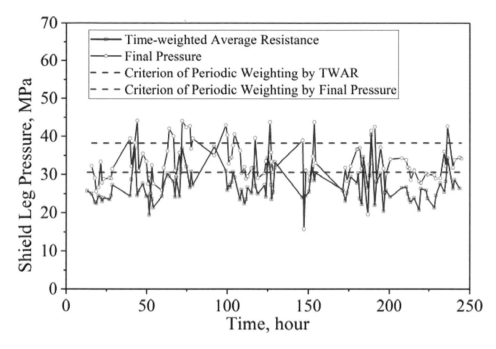

Figure 5.5.19 Comparison of roof weighting determined by final load and TWAR

or TWAP plus root mean square on one or a group of shields (Fig. 5.5.19). However, this method of analyzing the roof weighting characteristics is not realistic because shields leg pressures exhibit different characteristics at different panels. The roof weighting arrival as determined by the TWAP of a shield often does not match what actually occurs at the face. In addition, it ignores the yielding characteristics of yield valve during shield loading. In panels where yield valves open frequently, the final pressure and TWAP commonly employed cannot reflect the actual shield loading characteristics. Under these conditions the final pressure and TWAP + root mean square cannot be used to determine the arrival of periodic weighting.

Figure 5.5.19 also shows that the magnitude of periodic weighting and periodic weighting interval determined by TWAP and final pressure are different, some of which are significant.

The factors related to weighting arrival include periodic weighting interval, periodic weighting intensity factor, average TWAP support density (TWASD), and number shield supporting cycles occurs during periodic weighting. Analysis those factors by SSRI can show the characteristics of pressure distribution of the panel at different location and time as the face advances, including impending roof weightings.

Periodic weighting develops with time. In the process, shield load and loading characteristics vary with time passed. In addition, periodic weighting occurs at different time across the whole face with some shields exhibiting weighting arrival first. Therefore, identifying the weighting characteristics of those few shields can predict accurately the weighting arrival of the whole face.

Using the developed pressure prediction model, SSRI employs the 20+ factors from the shield leg pressure cycle to identify the loading characteristics of the current cycle as well as the following cycle, which are then used to predict the pressure-increasing trends of the current cycle. The same method is applied to all shield cycles on the face. When the shield pressures across the whole face reach the threshold value, SSRI will give warning (Fig. 5.5.20).

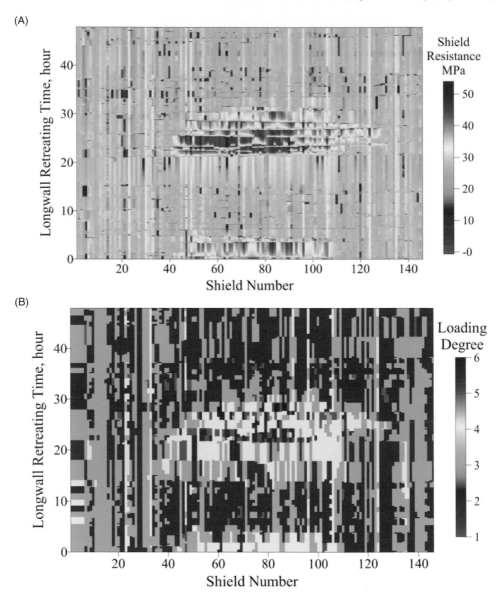

Figure 5.5.20 Changes in shield and mining pressure level before and after weighting. A: Shield pressure changes within 48 hrs before and after weighting arrival. Red area denotes weighting arrives at the center at the 23rd hr and lasts for five cycles for about 9 hr. B: Change in mining pressure level for all shield before and after weighting

Source: Cheng (2015); Cheng et al. (2018)

5.5.3.2 Assessment of the supporting quality of shield

A. EVALUATION OF YIELDING OF YIELD VALVE

A yield valve is designed to let the high pressure fluid to leak out a fixed amount and closed out quickly when the pressure reaches the preset value. In other words, it ensures that the cylinder is not damaged and that the leg closure is reduced to the minimum possible, so as not to push the piston to solid. Analysis of the shield pressure and leg closure data will show the working condition of the yield valves and find the problems if it exists. Underground monitoring shows the following major problems due to lack of maintenance exist (Cheng and Peng, 2017; Cheng *et al.*, 2018):

1   Yield valve yields before reaching yield pressure or does not yield after exceeding the yield pressure.

A shield's designed capacity cannot be fully utilized if the yield valve yields before reaching the yield pressure. On the other hand, if it does not yield after exceeding the yield pressure, the cylinder may be exploded.

2   After yielding, the yield valve cannot close quickly and let yielding to last too long.

The amount of leg closure is related to yielding time of the yield valve. The longer is the yielding the smaller are the shield's stiffness and load capacity leading to excessive roof convergence. When the roof pressure is heavy or the cycle time is long, the risk of pushing the piston into solid increases significantly.

B. DIAGNOSTICS OF SHIELD LEG FLUID LEAKAGE

Aside from debris, soft roof and floor, fluid leakage on shield legs is the major reason for a decrease in shield leg pressure due to valve damage or poor seal on cylinders (Cheng *et al.*, 2018). Figure 5.5.21 shows the characteristics of decrease in shield leg pressure due to two types of fluid leakages. In Figure 5.5.21 A, pressure decrease is due to minor fluid leakage with the rate of pressure decrease around 1.45–14.5 psi min^{-1} (0.01–0.1 MPa min^{-1}) and is constant during the shield supporting cycle. There is residual loading capacity, depending on the time duration of the cycle, before shield advance. This type of minor leakage is not easy to detect due to the slow decrease in pressure. The pressure decrease shown in Figure 5.5.21 B is caused by rapid leakage of fluid with the rate of leakage around 72.5–145 psi min^{-1} (0.5–1 MPa min^{-1}) or more. A shield loses its loading capacity soon after setting; if the shield activates its positive setting option automatically, leaking will continue, and the fluid supply system will pump the fluid automatically several times within a short period of time (Fig. 5.5.21 B) until it reaches the preset number of repeat pumpings of fluid or the duration of fluid pumping, thereby overloading the fluid supply system.

Fluid leakage of shield legs starts with minor leakage. It grows gradually but turns to rapid leakage due to repeat loading of shield. The result is that roof falls may occur.

(A)

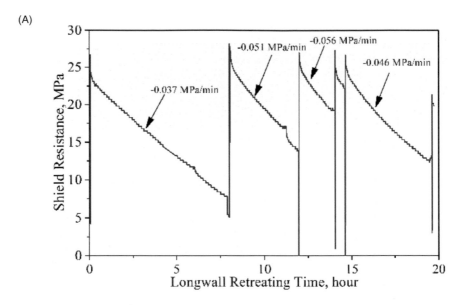

(B)

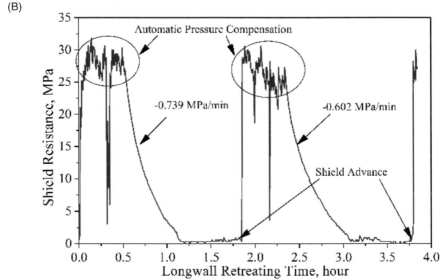

Figure 5.5.21 Pressure decrease in shield legs caused by fluid leakage. (A) slow decrease and (B) rapid decrease

Source: Cheng *et al.* (2018)

C. OTHER DIAGNOSTICS APPLICATION

Changes in shield's leg pressure characteristics can be used to evaluate roof geology (Peng *et al.*, 2019) as well as prediction of a roof fall or roof cavity (see Fig. 10.3.1).

Chapter 6

Shield support – design/selection

6.1 Introduction

Shields installed in the United States since the early 1990s have been of the two-leg design (see Fig. 1.5.9). Prior to that period, several types of powered supports were in use, and there were research efforts devoted to determining which types of powered support is most compatible with various geological conditions (Peng and Chiang, 2000). All of the current 43 longwalls operating in the United States employ two-leg shields (for reasons why, see Section 5.2.1). Therefore, based on current industry practice, there is no need to choose the type of powered supports, greatly simplifying the whole design process. This chapter will concentrate on the design and selection of two-leg shields.

As shown in Figure 1.5.10, the average capacity of powered supports for all US longwalls increases year after year, although there seems to be a stabilizing trend in recent years. One logical explanation for this trend is that in the 1970s and 1980s, there were quite a few cases of the collapse of longwalls that buried all of the face equipment. Since the capital investment for a longwall is so large, losing a set of longwall equipment is a huge loss (equipment and production loss). So when a new longwall was ordered, support capacity was increased and the support manufacturers were able to produce it. Several reasons the support capacity seems to have stabilized are the restriction in shield width to accommodate the two large legs (maximum 480 mm), shield control unit, etc., for ergonomically safe operation, and shield weight for floor stability and safe handling. To state this differently, many supports built prior to the late 1980s were under capacity. Conversely, most supports currently being used are overrated. Against this background, the design of a powered support is less related to capacity and more related to enhanced reliability, service life, and diagnostic monitoring capabilities. With the trend toward two-meter width in recent years, the one remaining structural design is shield width. A shield wider than two meters may need to be further strengthened for torsional and asymmetric loading.

There are many positive advantages for employing an over-rated shield support. For one thing, all theories developed for support capacity determination relate to static loading and do not account for the fact that shields are moving with the longwall face. As such, they are loaded and unloaded cyclically, which causes wear and tear cycle after cycle. This normal wear accumulates over years, and with an expected service life of 12 to 15 years, it has considerable impact on the integrity of the shield and, thus, performance. In this respect, in addition to improved structural design over the past two decades, a larger capacity shield can withstand cyclic loading and unloading much better than a smaller one. Consequently "fatigue life," not "static life," should be used in shield support design.

6.2 Elements of shield design

Traditionally, support design meant the determination of support capacity. This concept ignores certain elements of the supporting system, including support itself and the foundation floor on which the support rests. The traditional method is perfectly adequate when the support itself is simple in structure, such as a coal pillar, but is completely inadequate for standing supports that must transmit the roof load to the floor through itself. For the case of shield supports in longwall mining, a shield has a complicated structure, the design of which must be compatible with the requirements of controlling roof and floor, and it must also have a long service life. Consequently, a complete shield support design consists of four elements: the amount of roof and lateral loads to be supported, structural design and testing of the shield itself, determination of floor load distribution, and fatigue life span.

6.2.1 Roof load or support capacity

As the roof is undermined, some roof collapses immediately, some gradually deflects until it touches the floor or gob piles, and other roof overhangs for various lengths of time. All of these will induce various degrees of roof loadings and movements in the face area. Therefore, the behavioral characteristics of the immediate and main roofs are paramount in determining the loading conditions on the supports.

6.2.2 Structural integrity and stability of the powered support

The dimension of each component of the support and its interconnection must be adequate for proper functioning at the face area. The size and strength of each component and welds must also be adequate to withstand the expected loadings, static, and fatigue.

6.2.3 Interactions among support, roof, and floor

Certain types of powered supports are best suited for certain types of roof and/or floor behaviors and vice versa. For two-leg shields, the selection of the dimensions of shield elements should take into account the expected interactions among roof, support, and floor.

6.2.4 Service life

Nearly all supports used in underground coal mines are stationary, i.e., once installed they stay in place long after the supported areas have been abandoned. Conversely, shield supports are moving, cut by cut or supporting cycle by supporting cycle, with the retreating longwall face. Therefore, it is loaded and un-loaded cyclically. This is "fatigue" as opposed to static loading throughout the service life of the support. Fatigue life is shorter than static life.

In order to cover these four factors, a procedure consisting of six sequential steps as shown in Table 6.2.1 is necessary. The recommended shield design procedures represent the elements that need to be considered for a complete shield design. In practice, however, due to lack of qualified personnel and the required length of time involved in the analysis, mine personnel, when ordering new shields, exercise only steps one and six, leaving step two through five for the shield manufacturers. For completeness, this chapter will discuss the salient features in each step.

Table 6.2.1 Procedures for designing a shield support

1 Preliminary determination of component sizes by considering:
 (a) Mining height
 (b) Prop-free front and space for crew travel
 (c) Ventilation requirements
 (d) Other considerations
2 Determination of the roof and lateral loadings by considering:
 (a) Characteristics of the immediate roof thickness and caving conditions
 (b) Gob loading on the caving shield
 (c) Characteristics of main roof movements
3 Determination of the bearing capacity of floor rocks and water conditions.
4 Determination of geometrical configuration based on preliminary analyses of:
 (a) Stability based on external loadings and geometrical configurations
 (b) Supporting efficiency
 If necessary, revise the geometrical configuration and repeat the analyses until the
 desired stability and supporting efficiency are obtained.
5 Structural design: determine the actual size and stiffness of each element by performing
 static stress analyses of the support, including simulation of critical loading conditions
 for each element. Revise the design and repeat the whole procedure, if necessary.
6 Testing of the full-sized prototype shield model and electronic and hydraulic units.

6.3 Preliminary determination of the overall dimensions of shield components

Initially, a rough estimate of the dimensional requirements for each component of the proposed shield support can be achieved by considering the factors in the following sections.

6.3.1 Mining height

Since the seam thickness usually varies from place to place, the hydraulic legs of the support must have a sufficient stroke to accommodate the changing mining height. The maximum (H_{max}) and minimum (H_{min}) working height of the support can be determined by:

$$H_{max} = H + \Delta H - C - C_r \qquad (6.3.1)$$
$$H_{min} = H - \Delta H - C - C_r \qquad (6.3.2)$$

Where H = the average seam thickness, ΔH = one-half of the maximum change in seam thickness, C = roof convergence at the face area, and C_r = the height of rock debris and asperities on the canopy and the floor underneath the base plate. C = usually 1–3 in. (25.4–76.2 mm) per mining cycle, but C_r varies considerably from 0 to 6 in. (0–152.4 mm) or more, depending on the immediate roof and floor condition.

For two-leg shields, the legs are inclined, and the length of each component must be chosen so that the final kinematic configuration of the shield will produce a maximum travel distance from H_{max} to H_{min} and vice versa. Within this travel distance, as the support changes to suit the mining height, the canopy tip moves vertically up and down, maintaining a near constant distance between the faceline and the canopy tip (Peng and Chiang, 1984). In reality, since the piston stroke is smaller than its closed length, a single telescopic leg in vertical orientation can only have a vertical adjustment of $H_{max}/H_{min} = 1.72$. But as the leg angle of inclination from the vertical increases, the ratio of H_{max}/H_{min} also increases (see Fig. 5.3.2). However, the larger the leg inclination, the smaller the vertical component of leg resistance for roof support. As a compromise, most support legs are installed with a leg angle from 70 to 80 degrees from the horizontal. Under this design, a double-telescopic leg has a maximum vertical adjustment of $H_{max}/H_{min} = 2.5$. In addition, within this vertical adjustment range the maximum allowable horizontal movement of the canopy tip is 2–4 in. (5–10 mm).

6.3.2 Prop-free front

The prop-free front refers to the distance from the face to the shield legs at the prevailing mining height, roughly equals to AFC/shearer width + web width + travelway width before shear's cutting, or + web width after shearer's cutting (Fig. 6.3.1). It usually ranges from 12 to 16 ft (3.66–4.88 m) long. The size of the prop-free front depends on the width required to install the cutting machine and the face conveyor and the location of space for crew travel. A large, prop-free front is not desirable for the weak immediate roof, which requires proper immediate support near the canopy tip right after undermining, because the supporting loads transmitted in this area are usually very small. For a two-leg shield, the crew usually travels in front of the legs. The optimal width for crew travel is 2.0–3 ft (0.61–0.9 m).

6.3.3 Ventilation requirements

If the coal seam contains a high volume of methane, a large volume of air ventilated through the face area is a common way of diluting the methane and carrying it away immediately after emission. If this is necessary, a shield support is required with a larger longitudinal cross-section in order to provide sufficient air to dilute the methane content to the safe level required by law. The maximum methane content allowed at the face is 1%, while the minimum air velocity varies with MSHA districts [most commonly 400 fpm (2 m s^{-1}) at the 10th shield from the headgate T-junction].

6.3.4 Other considerations

Other considerations include, for example, support transportation during first-time installation and subsequent face moves. The shield is transported in a whole unit. Its cross-sectional area (height and length) in the collapsed position must be smaller than that of the entries (height and turn radius) en route in order to save the time and expense of enlarging the entry width (correcting the curvature) and/or height or dismantling the equipment.

6.3.5 Summary of preliminary design

Figure 6.3.1 shows the widths of major longitudinal cross-sections at the face that influence shield design, including shield's total width, collapsed and extended heights, canopy tip to faceline distance, web width, travelway width, and AFC/Shearer width.

6.4 Determination of external loadings

External loadings on shield supports consist of those applied on the canopy and on the caving shield. The former includes the vertical load due to the weight or movement of overburden

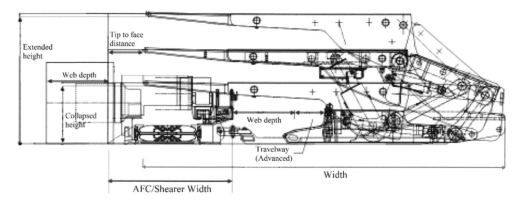

Figure 6.3.1 Dimensions in major cross-sections that influence shield design

Source: Courtesy of Caterpillar

strata and lateral forces, both parallel and perpendicular to the face, due to the horizontal movements of roof strata. The latter is the weight of caved rock fragments that are piled up on the caving shield.

Most researchers (Barry *et al.*, 1969; Jacobi, 1976; Wade, 1977; Wilson, 1975) in past years have concentrated on the derivations of the vertical load due to the weight of the roof strata. Although there are differences in the roof load models, they can be summarized by a model as shown in Figure 6.4.1. The immediate roof breaks along the faceline and at the gob edge of the support either vertically or sloping upward at an angle θ toward the gob. Depending on the roof rock property and geological conditions, sometimes the roof over-hangs and sometimes it breaks along or in front of the gob edge of support. Because the roof convergence increases from the faceline toward the gob and due to the structural stiffness of the canopy and location of the leg cylinders, the force applied on the support canopy is not uniform. Therefore, the analysis of canopy loading should consist of two components: the location and the magnitude of the resultant force.

6.4.1 Location of roof loading resultant force

The weight of the immediate roof (Fig. 6.4.1) on the support, W_1, is:

$$W_1 = \text{weight of Block A} + \text{weight of Block B}$$

$$= h_{im}\gamma_{im}BL\cos\alpha + \frac{h_{im}}{2}h_{im}\gamma_{im}B\tan\theta\cos\alpha$$

$$= h_{im}\gamma_{im}B\left(L + \frac{h_{im}}{2}\tan\theta\right)\cos\alpha \tag{6.4.1}$$

Where B is the center-to-center distance of the shield supports, α is the angle of seam inclination, and:

$$L = L_w + L_u + L_c + L_o \quad \text{after shearer's cutting or}$$
$$L = L_u + L_c + L_o \qquad \text{before shearer's cutting}$$

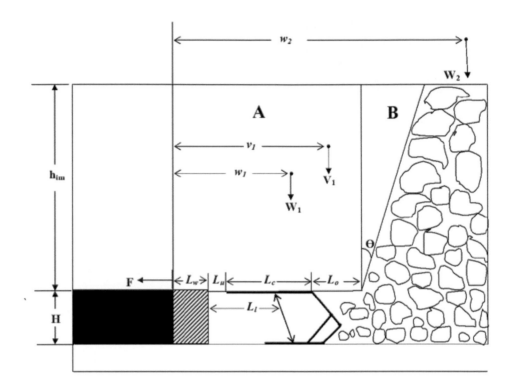

H : mining height L_w : width of cut(web) L_o: rear overhang
h_{im}: caving height L_u : unsupported distance L_e: distance from face line to leg
                                                       before cutting
F : panel coal block L_c : shield canopy length Θ : caving angle

Figure 6.4.1 A generalized roof loading model

The location where W_1 acts on the canopy is:

$$w_1 = \frac{L + h_{im}\tan\theta}{2} \qquad\qquad (6.4.2)$$

When the roof is strong and overhangs, then $w_1 > L_1 =$ the distance from the faceline to the shield legs, and the rear portion of the canopy will be subjected to larger loadings. Conversely, when the immediate roof is weak, it breaks before it passes beyond the gob edge of the canopy, $w_1 < L_1$, which generally means that a larger loading is on the front portion of shield support.

When periodic weighting arrives, the additional weight, W_2, will be added. The location where W_2 acts on the immediate roof is very important.

Because:

$$v_1 V_1 = w_1 W_1 + w_2 W_2 \qquad\qquad (6.4.3)$$

and:

$$V_1 = W_1 + W_2 \qquad\qquad (6.4.4)$$

Where V_1 is the total resultant force that acts on the canopy, and v_1 is the distance from face-line to the point where V_1 acts. w_1 and w_2 are the distance from faceline to the points where W_1 and W_2 act, respectively. Thus, v_1 is highly dependent on w_1, w_2, W_1, and W_2. It is likely that when the immediate roof is strong and W_2 is large, v_1 is also large. This means that the roof load resultant force acts near or even beyond the gob edge. Conversely, it acts toward the front edge when the immediate roof is weak and caves without overhang. In the ideal conditions, the roof load resultant force acts along the same line as that of the support due to the hydraulic leg loadings. In practice, however, they seldom act along the same line.

6.4.2 Determination of roof loading

Just like pillar design, there have been many methods developed for the determination of roof loading by researchers in various countries where modern longwall mining is practiced (Barczak, 1990; Chen and Peng, 1999; Park et al., 1992; Peng, 1998; Peng et al., 1987; Smart and Aziz, 1986; Trueman et al., 2005; Wilson, 1975). In this section, a few of the more commonly used or more effective ones are covered.

6.4.2.1 Detached roof block method

This method is the simplest one (Fig. 6.4.1) and has been commonly used by support manu-facturers and coal operators. In this method, two factors are used to estimate the height of detached roof block: stratigraphic sequence and bulking factor. Initially, the stratigraphic sequence of boreholes are used to determine the contact plane between the immediate roof that will cave immediately following shield advance and main roof that will overhang. (Note that the immediate roof is not necessarily a single layer, but that it may consist of several different rock types and thicknesses.) The location of the contact plane above the coal seam top is the thickness of the immediate roof, which, in turn, is used to determine whether it will fill up the void created by the complete mining of the coal seam (or mining height, H). The required caving height of the immediate roof that needs to fill up the void created by a certain mining height is governed by the bulking factor, e.g., Equation (3.2.3). If after min-ing, the height of the bulked material equals or exceeds the caving height determined by Equation (3.2.3), it means the caved rock fragments can fill up the voids and provide support to the overlying strata. Consequently, the height of the rock block determined by Equation (3.2.3) will be that which is to be supported by the shield support. The weight of the detached rock block is then:

$$W_f = \frac{H B \gamma_{im} L}{K_o - 1} \qquad\qquad (6.4.5)$$

Where K_o is the original bulking factor of the immediate roof. If $K_o = 1.5–1.1$, then:

$$W_f = (2-10) HB\gamma_{im} L \tag{6.4.6}$$

Therefore, the amount of rock weight on the canopy is equivalent to the rock weight of 2 to 10 times the mining height, depending on the selection of K_o. When using this method, as the support capacity employed by the coal industry increased continuously from the 1970s through the late 1990s and in order to match the increasing load capacity, the required bulking factor must be decreased correspondingly from 1.5 to 1.1 or less which is a considerable range. As stated previously, support capacity determined by using the largest bulking factor of 1.5 in the 1960s to 1970s resulted in an underrated "static" powered support capacity, whereas the use of bulking factor equal to or less than 1.1 produced overrated "static" shield capacity in recent years. If the bulking factor method is adopted for determining the support capacity, the actual bulking factor of the roof strata must be used, and the support capacity so determined may be less than what are being used in practice. Peng (1980) recommended a bulking factor of 1.25.

The increasing roof loading can be attributed to the existence of overhung strata within the caving height determined by the chosen bulking factor, because the overhung strata produce an additional weight on the shield, the magnitude of which depends on the length and thickness of the overhung strata.

If within the required caving height calculated by using a more reasonable bulking factor such as 1.25, there is a stratum or strata that will overhang, Chen and Peng (1999) calculated the overhanging length of each stratum separately and then summed their total weight on the shield (Fig. 6.4.2) using the following equations:

$$L_b = t\sqrt{\frac{T_o}{3q}} \tag{6.4.7}$$

$$L_c = A + B + C + \ell_c \tag{6.4.8}$$

$$F_{rd} = (\text{DF}) \frac{\sum_{i=1}^{k} W_i x_i}{\ell_r} \tag{6.4.9}$$

$$W_i = L_{bi} St_i \gamma_i \tag{6.4.10}$$

$$x_i = \frac{L_i}{2} \tag{6.4.11}$$

Where DF = 1.10–1.25 = design factor, considering the adjacent support's advance after the shearer's cutting

k = number of rock layers within the required caved zone
L_b = length of a fix-ended cantilevered beam
L_{bi} = length of cantilevered beam of the ith rock layer
q = uniform load on the beam per unit length
S = support spacing, 5.75 ft (1.75 m) or 6.56 ft (2 m)
t_i = thickness of the ith rock layer
T_o = in-situ tensile strength of the rock beam which can be estimated by multiplying its lab-determined strength by a reduction factor such as

coal, claystone, or fireclay	0.20
shale or mudstone	0.25–0.30
sandstone, siltstone, limestone, or laminated sandstone	0.30–0.50
massive sandstone	050–0.80

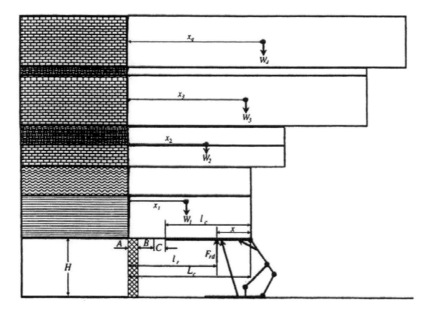

Figure 6.4.2 Beam roof loading model
Source: Chen *et al.* (1999); Chen and Peng (1999)

x_i = distance from the center of gravity of the *i*th rock layer to the first supporting point of coal face
W_i = weight of the *i*th rock layer
γ_I = density of the *i*th rock layer
l_c = shield canopy length
l_r = distance from face line to leg
F_{rd} = resultant force

6.4.2.2 Shield leg pressure measurement method

This method is only applicable for existing longwalls for evaluation of the adequacy of existing shields and improvements needed for the next generation of shields. However, it can also be used for new mines in the same seams or seams with similar geological conditions.

As stated in Section 5.5.2, the final pressure in a shield supporting cycle of an increasing loading type (see Fig. 5.5.3 A) is the maximum load attained in that cycle. Since all shields are equipped with transducers for pressure monitoring, analysis of the leg pressure history for all shields over one or more panels will provide sufficient data for statistical analysis. The average final pressure may be adopted as the desired shield capacity. One may argue that this is a very conservative approach because the final pressure in a supporting cycle usually occurs in a very short period of time when the adjacent shields are released for advance, and that it should better be handled through a yield valve. If this is the case, the time-weighted average resistance or pressure (TWAR or TWAP) may be used.

The aforementioned methods are most applicable for cases when the measured shield leg pressures in a supporting cycle do not or rarely exceed the preset yield pressures. Trueman *et al.* (2005) noted in thick seam mining or shields used with low setting pressures, leg pressures often involved one or more yield cycles. The time interval from one opening of a yield valve to the next is defined as a yield cycle. Based on the yield cycle characteristics, they proposed the following:

1   Adequate capacity and appropriate setting pressure – when leg pressure cycles are compatible to those shown by Figure 5.5.3 A and B, the setting pressure is appropriate, and the yield capacity is adequate.
2   Adequate capacity but too high a setting pressure – after one or more yield cycles, the roof convergence and pressure increment reach equilibrium (Cheng and Peng, 2017).
3   Inadequate shield capacity – when the yield cycle continues without signs of reaching equilibrium between shield leg pressure and roof convergence (see Fig. 5.5.7) (Cheng and Peng, 2017).

In any event, the shield leg pressure data can provide a wealth of information, and they should be used to statistically compare and determine the most likely scenarios of shield loading to be encountered (see Section 5.5, p. 155).

As noted previously, the recorded shield leg pressures vary from cycle to cycle and from shield to shield. Furthermore, the amount of data from the whole face is overwhelming. So in order not to be biased in analyzing the data, statistical methods must be used. Bessinger (1996) stated that the characterization of pressure variations can allow a rational, statistical assessment of the required yield capacity of a new shield design, and he developed a method of estimating the cumulative probability of peak load. He suggested that as a minimum, the 87% cumulative probability roof load value may be chosen. Below this value, an increased frequency of undesirable outcomes had been experienced. In the range of 87 to 93% the results may appear marginal with occasional large-scale roof control problems.

6.4.2.3 DEPOWS (design of powered support selection model)

DEPOWS is a user-friendly menu-driven PC computer model developed by Jiang *et al.* (1989). It consists of four elements, i.e., required vertical load capacity, most suitable support type, minimum horizontal force, and maximum allowable floor pressure. Peng *et al.* (1987) developed a roof-support-floor interaction, displacement-controlled model for determining the required roof load capacity based on the hydraulic leg pressure history monitored over 23 longwall panels in the Appalachian coalfields. Since all of those studies were performed prior to the introduction of electrohydraulic control, manual support settings were employed. They found that face (roof-to-floor) convergence is different due to varying geological conditions and that change in support resistance varies with face convergence. In other words, under the same geological conditions, the final load or the increment of support resistance (i.e., load increment) in a mining cycle is proportional to the setting load. Similarly, under the same setting load, the final load or load increment will be different due to varying roof conditions. Figure 6.4.3 shows the characteristics curve of the model based on this concept, in which:

$$\Delta q = a q_s^2 e^{-cq_s} \qquad (6.4.12)$$

Where Δq is the increment of load density, q_s is the setting load density, and a and c are constants relating to the characteristics of main and immediate roofs, respectively. Factors a and c are determined empirically.

The equations for determining the required load capacity are:

$$\Delta P = 0.4174 aA / c^2 \eta \qquad (6.4.13)$$
$$P_s = 3.2A / c\eta \qquad (6.4.14)$$
$$P_y = P_s + \Delta P \qquad (6.4.15)$$

Where η is the supporting efficiency and is defined as the ratio of the vertical component of load to the total leg load, A is the canopy area, and P_y is the final load in a mining cycle. Note that yielding is not implicit in this analysis.

It must be noted that leg pressure data on which this model was developed were obtained prior to the mid-1980s when support capacity ranged from 280 to 800 tons with an average of less than 500 tons. The recommended support capacities using these empirical equations are considered to be of rational values for "static" capacity and the prevailing support capacities at that time. In addition, the data collected so far on setting vs. final loads for electrohydraulic shields show considerable variation – the new data do not support the model shown in Figure 6.4.3 (Peng et al., 2019).

DEPOW is included here to remind readers that the interaction of setting load and the roofs (immediate and main roofs) is the most important factor in shield load capacity design.

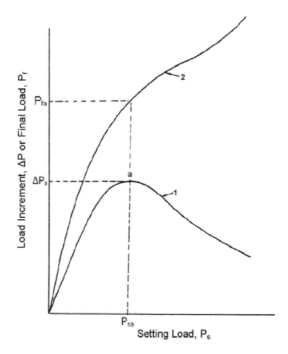

Figure 6.4.3 Characteristics curves of the DEPOW model

Source: Peng et al. (1987)

6.5 Resultant load on the canopy and floor

As the roof load derived in the previous section is applied on the shield canopy, it is taken up by the two shield legs and stabilizing cylinder connecting the canopy and caving shield, and the amount on each of these two components depends on the geometrical relationship of various components. Conventionally, the effect of the stabilizing cylinder and the caving shield is ignored, and roof loading determined in the previous section is used directly as roof loading on the canopy. As the capacity of the stabilizing cylinder has increased (up to 200 tons or 181.8 mt) in recent years, its effect should be considered. In other words, it is the third leg, although much smaller in capacity than the two hydraulic legs. In addition, conventional design does not consider the effect of friction between roof strata and shield canopy.

6.5.1 Resultant load on canopy

6.5.1.1 Magnitude of resultant load

Figure 6.5.1 A (Chen and Peng, 1999) is a schematic drawing that illustrates the state of various forces acting on a two-leg shield support at a longwall face. Taking the canopy as a

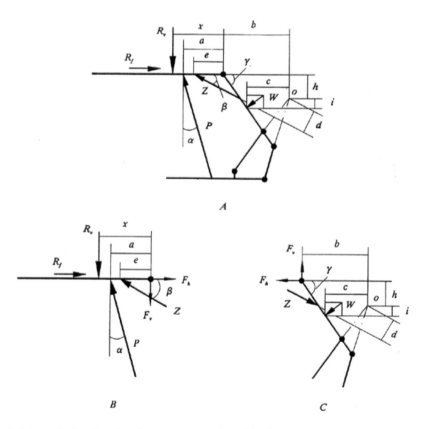

Figure 6.5.1 Analysis of acting force on a two-leg shield

Source: Peng and Chen (1993)

free body (Fig. 6.5.1 B), the equilibrium of the forces acted on the canopy in the vertical and horizontal directions can be expressed, respectively, by:

$$\sum F_y = 0 \ F_v + R_v = P\cos\alpha + Z\sin\beta \tag{6.5.1}$$
$$\sum F_x = 0 \ F_h + R_f = P\sin\alpha + Z\cos\beta \tag{6.5.2}$$

Where F_y is the vertical reaction force of the caving shield-lemniscate assembly in tons, F_h is the horizontal reaction force of the caving shield-lemniscate assembly in tons, R_v is the vertical resultant load or vertical load capacity of the shield support in tons, R_f is the friction force between the canopy and the roof, $R_f = fR_v$ in tons, f is the coefficient of friction between the canopy and the roof, P is support resistance provided by hydraulic legs in tons, Z is resistance of the stabilizing ram in tons, α is angle of inclination of the hydraulic legs from the vertical in degrees, and β is angle of inclination of the stabilizing ram from the horizontal in degrees.

Using the caving shield as a free body (Fig. 6.5.1 C) and taking the moments of various forces about the instantaneous center of the lemniscate links:

$$\sum M_0 = 0 \ F_v b = F_h h + Zd + cW\cos\gamma - i\sin\gamma \tag{6.5.3}$$

Where b is the horizontal distance from the hinge point to the instantaneous center of the lemniscate links in inches, c is the horizontal distance from the line of action of the load on the caving shield to the instantaneous center of the lemniscate links in inches, d is the distance from the extended line of action of the stabilizing ram to the instantaneous center of the lemniscate links in inches, h is the vertical distance from the hinge point to the instantaneous center, o, of the lemniscate links, in inches (if o is below the hinge point, h is positive, otherwise it is negative), and W is the vertical load on the caving shield (assuming that no horizontal force can be transmitted through the loose rock fragments) in tons.

Substituting Equations (6.5.1) and (6.5.2) into Equation (6.5.3)

$$\left(P\cos\alpha + Z\sin\beta - R_v\right)b = \left(P\sin\alpha + Z\cos\beta - fR_v\right)h + Zd \pm W\left(c\cos\gamma - i\sin\gamma\right) \tag{6.5.4}$$

From Equation (6.5.4), the vertical load capacity of the support is:

$$
\begin{aligned}
R_v = {} & \frac{1}{\left(1 - f\dfrac{h}{b}\right)}\left[P\left(\cos\alpha - \frac{h}{b}\sin\alpha\right) + Z\left(\sin\beta - \frac{h}{b}\cos\beta - \frac{d}{b}\right)\right] \\
& -W\left(\frac{i}{b}\sin\gamma - \frac{c}{b}\cos\gamma\right)
\end{aligned}
\tag{6.5.5}
$$

Equation (6.5.5) shows that the magnitude of the vertical resultant load on the canopy is related to leg resistance and angle, stabilizing ram resistance and angle, load on caving shield, and location of the instantaneous center of the lemniscate links (including either below or above the canopy line).

The maximum horizontal friction force that may arise is limited by the coefficient of friction between the roof and the canopy. The cited coefficient of friction value in the literature varies widely, but it has generally been agreed that a coefficient of friction of 0.3 is a reasonable maximum value (Jackson, 1979).

Horizontal friction force acting towards the gob on the canopy of a shield support changes the vertical load capacity of the support or the vertical resultant load. As the coefficient of friction increases, the vertical load capacity of the support will increase. In other words, the support will provide a larger effective resistance to the roof strata with the same P and Z.

The horizontal friction force results from the trend of relative movement between the roof and the canopy. Relative movement may be induced by either the movement of the roof toward the gob during coal mining operations or the trend of the movement of the canopy towards the face after the support has been set up against the roof (Peng et al., 1989). If such a relative movement between the roof and the canopy does not exist, the horizontal friction force vanishes because $f = 0$, and Equation (6.5.5) can be rewritten as:

$$R_v = P\left(\cos\alpha - \frac{h}{b}\sin\alpha\right) + Z\left(\sin\beta - \frac{h}{b}\cos\beta - \frac{d}{b}\right) - W\left(\frac{i}{b}\sin\gamma - \frac{c}{b}\cos\gamma\right) \qquad (6.5.6)$$

If the immediate roof can overhang a certain distance in the gob behind the rear end of the support canopy (e.g., if the roof is medium stable or stable), the load on the caving shield may be ignored, and Equation (6.5.6) becomes:

$$R_v = P\left(\cos\alpha - \frac{h}{b}\sin\alpha\right) + Z\left(\sin\beta - \frac{h}{b}\cos\beta - \frac{d}{b}\right) \qquad (6.5.7)$$

Under this condition, the vertical load capacity of a shield support reaches the maximum for a fixed value of P, Z, and the support structural parameters. From the roof support point of view, only when the second term in the right side of Equation (6.5.6) is greater than zero, then:

$$R_v > P\left(\cos\alpha - \frac{h}{b}\sin\alpha\right) \qquad (6.5.8)$$

When the stabilizing ram is compressed, and Z is positive, then:

$$\sin\beta - \frac{h}{b}\cos\beta - \frac{d}{b} > 0 \qquad (6.5.9)$$

Since the ratio of h and x is usually very small, $(h/b)\cos b$ can be ignored:

$$\beta > \sin^{-1}\left(\frac{d}{b}\right) \qquad (6.5.10)$$

When the stabilizing ram is pulled, and Z is negative, then:

$$-\sin\beta + \frac{h}{b}\sin\beta + \frac{d}{b} > 0 \qquad (6.5.11)$$

$$\beta < \sin^{-1}\left(\frac{d}{b}\right) \qquad (6.5.12)$$

Otherwise:

$$R_v \leq P\left(\cos\alpha - \frac{h}{b}\sin\alpha\right) \qquad (6.5.13)$$

6.5.1.2 Location of the vertical resultant load

The location of the vertical resultant load on the canopy can be found by taking moments about the hinge point (Fig. 6.5.1 B):

$$Rx = aP\cos\alpha + eZ\sin\beta \qquad (6.5.14)$$

$$x = \frac{aP\cos\alpha + eZ\sin\beta}{R_v} \qquad (6.5.15)$$

Where a is the horizontal distance from the canopy-leg joint to the hinge point (or the canopy-caving shield joint) in inches, e is horizontal distance from the line of action of the stabilizing ram to the hinge point in inches, and x is horizontal distance from the line of action of the vertical resultant load to the hinge point in inches.

Substituting Equation (6.5.7) in Equation (6.5.15) and rearranging produces this:

$$Z = \frac{\left[x\left(\cos\alpha - \dfrac{h}{b}\right) - a\cos\alpha\right]}{\left(x\dfrac{h}{b}\cos\beta - \sin\beta + \dfrac{d}{b}\right) + e\sin\beta} P \qquad (6.5.16)$$

For a fixed mining height, the geometrical parameters of the support are constants. Therefore, Equation (6.5.16) can be expressed as:

$$Z = \frac{Ax - B}{Cx + D} \qquad (6.5.17)$$

Where A, B, C, and D are constants with:

$$A = \cos\alpha - \frac{h}{b}\sin\alpha$$

$$B = a\cos\alpha$$

$$C = \frac{h}{b}\cos\beta - \sin\beta + \frac{d}{b} \qquad (6.5.18)$$

$$D = e\sin\beta$$

Under this condition, the stabilizing ram is extended by the pulling force of the canopy such that it may damage the stabilizing ram or the connecting parts of the stabilizing ram, the canopy, and/or the caving shield. Conversely, if R_v acts toward the front edge of the canopy, the stabilizing ram is compressed by a compressive force and acts like a hydraulic leg sharing and supporting the roof with an extra leg. Therefore, two-leg shields are more suitable for unstable roof or medium stable immediate roof where the roof resultant load is located in front of the leg-canopy joint.

In a properly designed and selected powered support, the vertical roof resultant load on the canopy should act along the same line as the vertical resultant resistance of the support due to the hydraulic response to roof loading. If they do not act along the same line, additional force will be needed to counterbalance the moments and added to the support loading.

The ratio of the resultant load to support resistance is usually defined as support efficiency. From Equation (6.5.5) (assuming W can be ignored), support efficiency, η, for a two-leg shield can be expressed as:

$$\eta = \frac{R_v}{P} = \frac{1}{\left(1 - f\dfrac{h}{b}\right)}\left[\cos\alpha - \frac{h}{b}\sin\alpha + \frac{Z}{P}\left(\sin\beta - \frac{h}{b}\cos\beta - \frac{d}{b}\right)\right] \tag{6.5.19}$$

Therefore, support efficiency is related to leg angle, location of the instantaneous center of lemniscate links, the ratio of Z to P, orientation of the stabilizing ram, and the coefficient of friction between the canopy and roof.

6.5.2 Floor pressure under the base plate

6.5.2.1 Location of the vertical resultant force under the base plate

The horizontal friction force acting on the canopy surface toward the gob affects not only the magnitude of the vertical resultant load but also its line of action transmitted from the strata to the support. If such a friction force does not exist, say, no relative movement between the roof and the canopy, then a pure vertical resultant load (or force), R_v acting on the canopy will induce a vertical resultant force, R_f, under the base plate (Fig. 6.5.2 A). Under this condition, $s = n$ with:

$$n = a + k - x \tag{6.5.20}$$

Where s = the horizontal distance from the first contact point between the base plate and the floor to the line of action of the vertical resultant force under the base plate in inches, n = the horizontal distance from the first contact point between the base plate and the floor to the line of action of the resultant load on the canopy in inches, and k = the horizontal distance between the canopy-leg joint and the first contact point between the base plate and the floor in inches.

If the horizontal friction force is present (Fig. 6.5.2 B), the vertical resultant force, R_v, under the base plate will move toward the center line of the base plate. The distance, δ, of the movement depends on the coefficient of friction, $f = \tan\phi$, and the operating height of the support, H, i.e.:

$$s = n + \delta \tag{6.5.21}$$

with:

$$\delta = H \tan\phi \tag{6.5.22}$$

6.5.2.2 Floor pressure calculation

A. RIGID BASE PLATE

Traditionally the determination of floor pressure is always based on the assumption that the base plate is rigid. Depending on the location of the vertical resultant force under the base

(A)   (B)

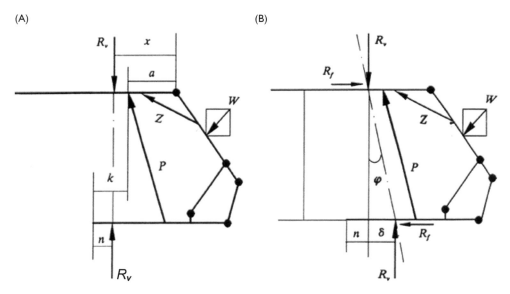

Figure 6.5.2 Location of the vertical resultant force

Source: Peng and Chen (1993)

plate, pressure distribution in the floor can be simplified to a triangular shape, a trapezoidal shape, or a rectangular shape (Fig. 6.5.3) (Jackson, 1979).

If $0 < s < L / 3$ (Fig. 6.5.3 upper), the pressure distribution is triangular. But the length of the pressure envelope, l, is less than the contact length, L, between the base plate and floor. The magnitude of the pressure can be determined by:

$$p_{avg} = \frac{R_v}{lw} \times 2000 \qquad (6.5.23)$$

$$p_{max} = 2 p_{avg} = \frac{2R_v}{lw} \times 2000 \qquad (6.5.24)$$

Where P_{avg} = average floor pressure in psi, l = length of the floor pressure envelope in inches, p_{max} = the maximum floor pressure at the front end of the base plate in psi, and w = width of the base plate in inches.

If $s = L / 3$, then the pressure distribution is still triangular, but $l = L$. Equations (6.5.23) and (6.5.24) can be rewritten as:

$$p_{avg} = \frac{R_v}{Lw} \times 2000 \qquad (6.5.25)$$

$$p_{max} = 2 p_{avg} = \frac{2R_v}{Lw} \times 2000 \qquad (6.5.26)$$

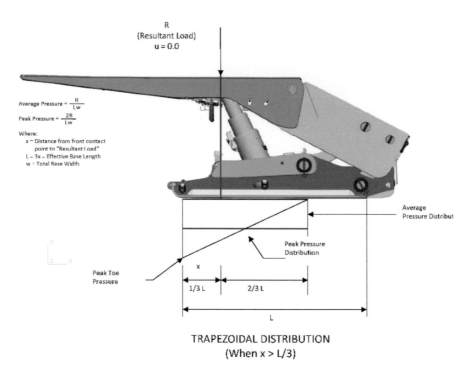

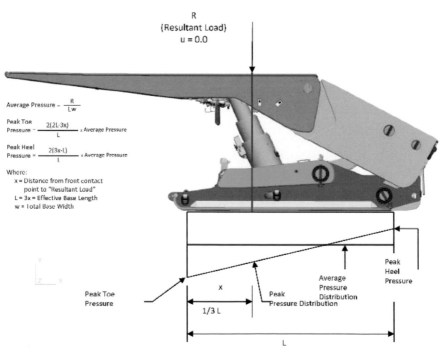

Figure 6.5.3 Floor pressure distribution – rigid base

Source: Courtesy of Caterpillar

If $L/3 < s < L/2$ (Fig. 6.5.3 lower), then the pressure distribution is trapezoidal, and the magnitudes of the pressure are:

$$p_{min} = \frac{2R_v}{Lw}\left(\frac{3s-L}{L}\right) \times 2000 \tag{6.5.27}$$

$$p_{max} = \frac{R_v}{Lw}\left[2 - \frac{6}{L}\left(s - \frac{L}{3}\right)\right] \times 2000 \tag{6.5.28}$$

If $s = L/2$, then the pressure distribution is rectangular and

$$p_{max} = p_{avg} = p_{min} = \frac{R_v}{Lw} \times 2000 \tag{6.5.29}$$

B. ELASTIC BASE PLATE

Finite element analysis was performed to analyze the effect of four factors (i.e., Young's modulus of the floor rock, vertical resultant force, the contact width between the base plate and the floor, and location of the resultant force) on the magnitude of the maximum floor pressure under the base plate of powered supports using the orthogonal factorial experimental design (Dey, 1985; Peng and Chen, 1993).

The distribution of floor pressure is illustrated in Figure 6.5.4 and is completely different from those of the rigid base plate. The maximum floor pressure is always located directly

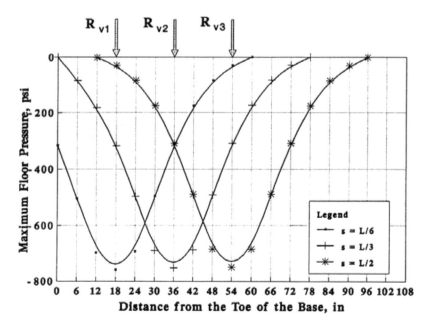

Figure 6.5.4 Floor pressure distribution – elastic base

Source: Peng and Chen (1993)

under the vertical resultant force, and the magnitudes of floor pressure are nearly the same for the same resultant force, floor property, and contact width between base plate and floor.

Based on the finite element computer analysis, a regression equation was developed to determine the maximum floor pressure as follows:

$$p_{max} = 1233 + 213x_1 + 2.72x_2 - 38x_3, \text{psi} \tag{6.5.30}$$

Where x_1 is the Young's modulus of the floor material in 10^6 psi, x_2 is the vertical resultant load in tons, and x_3 is the contact width between the plate and the floor along the faceline direction in inches.

It must be emphasized that the floor strata at longwall faces are elastic, which is far from the rigid body assumed in Jackson's models, and, as for Peng and Chen's model, it is difficult to get representative elastic properties of floor strata. Therefore, the most practical ways are to install the floor lifting device and make the shield base toe longer (see Sections 10.14.3 p. 371).

6.5.3 Determination of canopy and floor pressure distribution by laboratory-controlled tests

Pothini *et al.* (1992) performed laboratory-controlled tests on a full-sized, two-leg shield using the US Bureau of Mines Mine Roof Simulator. In these tests, pressure cells of 6 in. × 6 in. (152.4 mm × 152.4 mm) were placed on a canopy and uniformly spaced under the base plate in various areas to simulate uniform contacts or bias contacts (Fig. 6.5.5). In order to insure contact at specific points, inserts of steel plate, plywood, and rubber pads were

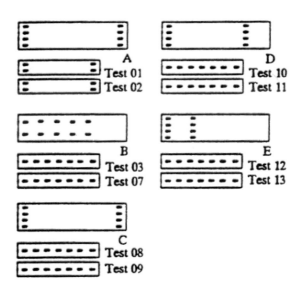

Figure 6.5.5 Full-scale floor pressure testing – contact configurations

Source: Pothini *et al.* (1992)

used between the roof plate and the pressure cells. Steel plates simulate stiff roof contacts, whereas plywood and rubber pads simulate soft roof contacts or debris of various degrees. Under weak roof and compliant floor conditions, as usually is found at mine sites, canopy contact occurs in an area more than 80 in. (2032 mm) from the rear edge, and approximately 50% of the canopy in the tip area has no contact (Fig. 6.5.6). This contact condition produces a pressure/load concentration near the leg connection point. Maximum load on the base plate is located about 10 in. (25.4 cm) from the toe, upon which the ram cylinder of the lifting device acts.

In order to evaluate the interaction of the canopy and base contact conditions, pressure distribution under the base plate was measured using three different canopy contact conditions. Seven pressure cells were evenly spaced under each base plate to simulate full base contact. With contact at the canopy tip, the canopy resultant load was much closer to that of the rear of the canopy at lower leg pressure than the other two contact configurations where there was

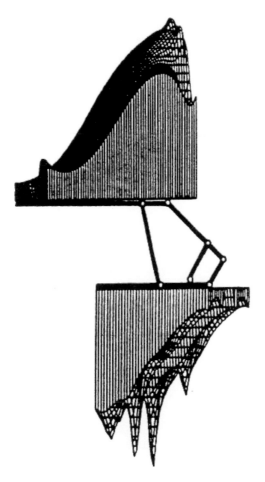

Figure 6.5.6 Measured base pressure distribution for Test 03

Source: Pothini *et al.* (1992)

no contact at the tip of the canopy. This difference in canopy resultant load location shifted the base resultant load slightly toward the rear at low leg pressure. However, as the leg pressure increased, the difference was reduced. When the leg pressure was increased to near or over 5000 psi (34.45 MPa), i.e., more than the setting pressure, the resultant load on the canopy was confined to near the leg-canopy connection point, regardless of the changes in canopy contact configurations. For the base plate, the resultant load was located toward the toe of the base plate as the leg pressure increased regardless of the canopy contact condition.

Load distribution under the base plate for these three canopy contact configurations was similar except a higher toe loading was produced at high pressure when there was canopy tip contact.

Load distribution on the canopy for these three canopy contact configurations was slightly different. The maximum contact loading occurs when the contact line was near the leg connection point, because it required very little force at the rear of the canopy to maintain equilibrium. The most uniform canopy loading occurred when the canopy contacts were evenly distributed on both sides of the leg connection point.

6.5.4 Required active horizontal force

In a weak roof, the unsupported area between the canopy tip and faceline caves easily due to the existence of tensile horizontal stress. When a two-leg shield is used, it can provide an active horizontal force during setting to compensate and convert the tensile horizontal stress to compressive stress, under which rock is stronger and more likely to be stable, because rock is much weaker in tension than in compression. How much active horizontal force is required?

Due to the nature of construction, the shield generates horizontal movement during setting and lowering (Barczak and Oyler, 1991; Peng et al., 1989). Peng and Chen (1991) employed an orthogonal fractional factorial experimental design and finite element analyses and found that the maximum horizontal tensile stress developed in the unsupported area between the canopy tip and faceline is a function of seven factors: Young's modulus of the immediate roof, coal seam, and main roof; immediate roof thickness/mining height; unsupported distance; leg angle; and support capacity. The following regression equation was developed for determining the maximum tensile horizontal stress occurring in the unsupported area:

$$\sigma_h = -247lus\,E_{im} - 218E_c + 28E_m + 14R + 4.49D + 9.45\beta - 0.43P_s \qquad (6.5.31)$$

Where σ_h is the maximum horizontal stress occurring within the unsupported distance between canopy tip and faceline in psi. $E_{im,}$ E_c, and E_m are the Young's modulus of immediate roof, coal, and main roof, respectively in 10^6 psi. R is the ratio of the immediate roof thickness to mining height. D is the unsupported distance between the canopy tip and faceline in inches. b is the leg angle from the horizontal in degrees, and P_s is the shield setting load in tons.

By setting $\sigma_h = 0$, Equation (6.5.31) can be re-rearranged to obtain the minimum shield setting load required to eliminate the horizontal tensile stress in the unsupported area.

$$P_s = \frac{247 + 148E_{im} + 218E_c - 28E_m - 14R - 4.49D - 9.45\beta}{0.43} \qquad (6.5.32)$$

6.6 Determination of floor and roof bearing capacity

The roof load applied on the canopy is immediately transmitted to the floor through the support structures. Thus, the floor is normally considered as the passive reaction frame. In some cases, the occurrence of floor heave will exert positive pressures on the supports.

The principles illustrated here apply to both the roof and the floor. But, since in most cases the floor rocks are much weaker than the roof rocks, the knowledge of the bearing capacity of floor rock is more important.

In the following section, the theories and factors that control the bearing capacity of the floor rock are discussed. The method of determining the bearing capacity has been illustrated elsewhere (Barry and Nair, 1970).

6.6.1 Definition and theories of bearing capacity

The bearing capacity is the maximum load per unit area that the rock can withstand before failure. The mechanisms and modes of bearing capacity failure vary with the rock types (Coates, 1978). In this respect, the floor rocks can be classified into two types (Coates, 1978; Jenkins, 1955): soft and plastic rocks, such as clay shale or fireclay (Peng, 1986), and brittle and elastic rocks, such as sandy shale or sandstone (Coates, 1978).

6.6.1.1 Soft and plastic rocks

Since the material properties of this type of rock are similar to soils, the theories from soil mechanics can be applied. There are many theories on the bearing capacity of soils. The most widely used is Terzaghi's bearing capacity equations, which when applied to the mine floor are as follows:

For a square base:

$$\sigma_b = 1.3\,CN_c + qN_q + 0.4\,\gamma BN_r \tag{6.6.1}$$

For a rectangular base:

$$\sigma_b = \left(1 + 0.3\,\frac{B}{L}\right)CN_c + qN_q + \left(1 - 0.2\,\frac{B}{L}\right)\frac{\gamma B}{2}\,N_r \tag{6.6.2}$$

Where σ_b is the bearing capacity; C is cohesion of the floor rock; γ is the weight per unit volume of the soil; B and L are the width and length of the footing; N_c, N_q, and N_r are bearing capacity factors (Fig. 6.6.2); φ is the angle of internal friction; and q is the uniform loading on both sides of the base. In a two-dimensional model, Equations (6.6.1) and (6.6.2) are applicable for a cross-section parallel to the faceline. The bearing pressures applied by the adjacent supports can be considered as surcharges with the term qN_q added. But when considering the cross-section perpendicular to the face, the term qN_q may be dropped.

The basic concept is that the application of the load P (Fig. 6.6.1 A) pushes the wedge under it (wedge No. 1) downward and causes lateral movements of zones two and three. But the lateral movements are resisted by shear stresses developed along the slope surface and by the rock weights in these zones. As P increases further, it may produce two types of

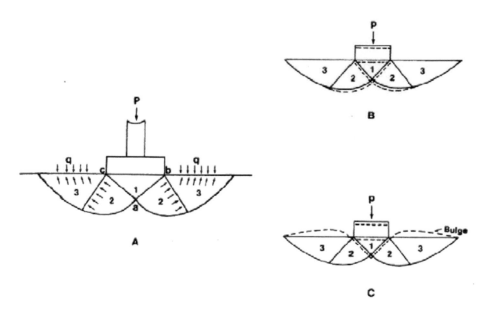

Figure 6.6.1 Mechanics and modes of bearing capacity failure

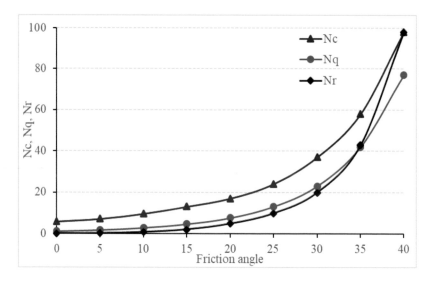

Figure 6.6.2 Terzaghi's bearing capacity factors

bearing capacity failures. The local-shear failure involves considerable vertical settlement before rocks in zones two and three begin to bulge (Fig. 6.6.1 B). Conversely, in general shear failure (Fig. 6.6.1 C), the rocks on both sides heave up rapidly with a slight vertical settlement beneath the base.

Equations (6.6.1) and (6.6.2) are valid for the homogeneous and massive strata. They must be modified for the stratified rocks commonly encountered in underground coal mines. Two

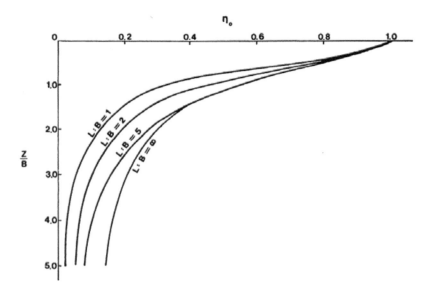

Figure 6.6.3 Nomograph for determining the stress distribution coefficient

extreme cases are discussed here. One is a weak floor stratum overlying a strong one. The shear surfaces will be confined to the weaker one. Consequently, the bearing capacity should be based on the weak stratum, not the strong one.

If the immediate floor is a strong stratum underlain by a weak one, the applied load is spread uniformly over a larger area, reducing the bearing pressure on the weak stratum. In other words, the bearing capacity of the floor is increased. In this case, the bearing capacity is determined by (Bowles, 1965):

$$\sigma_b = \left[\left(1 + 0.3 \frac{B}{L} \right) C N_c + \gamma_1 Z N_q - \gamma_1 Z + \left(1 - 0.2 \frac{B}{L} \right) \frac{\gamma_2 B}{2} N_r \right] \frac{1}{\eta_0} \tag{6.6.3}$$

$$\left(\text{rectangular base} \right)$$

Where γ_1 and Z are the weight per unit volume and thickness of the strong stratum, respectively; N_q, N_c, and N_r are the bearing capacity factor (Fig. 6.6.2); γ_2 is the weight per unit volume of the weak stratum; and η_0 is the stress distribution coefficient (Fig. 6.6.3).

6.6.1.2 Brittle and elastic rocks

For brittle and elastic rocks, Coates and Gyenge (1965) postulated the following equations based on Griffith's strength theory:

$$\sigma_b = 24 T_0 = 3 C_0 \tag{6.6.4}$$

Where T_0 and C_0 are the uniaxial tensile and compressive strengths, respectively.

If a rigid base is used, there will be stress concentrations around the edges of the base. Under this condition, the average bearing pressure at failure is:

$$\sigma_b = K \frac{C_0}{B^n} \qquad (6.6.5)$$

Where K and n are constants, depending on loading conditions and rock properties.

6.6.2 Measured floor bearing capacity

Based on plate bearing and borehole shear tests over seven coal mines in Illinois, where thick soft floor under Herrin #5 and #6 seams is very common, Chugh *et al.* (1989) and Pula *et al.* (1990) have developed regression equations for estimating the ultimate bearing capacity (UBC) and the modulus of deformation at 50%, E_{50}, of the UBC. Their studies employed a small test plate varying in size between 36 in^2 and 1 ft^2 (645 mm^2 and 0.093 m^2). They recommended that the following equations can be used during the pre-mining exploration stage:

$$\sigma_{UBC} = 7.38 - 0.145W \qquad (6.6.6)$$

Or:

$$\sigma_{UBC} = 1405 - 86W - 1.92W_L \qquad (6.6.7)$$
$$E_{50} = 5.3 \times 10^4 \left(W^{-0.468} \right) \qquad (6.6.8)$$

Where σ_{UBC} is the ultimate bearing capacity (UBC) in psi; W is moisture content in percent, and W_L is the average liquid limit for weak floor strata down to a depth of 12 in. (0.3 m) below the coal seam.

6.6.3 Factors controlling the bearing capacity

6.6.3.1 Eccentric loading

If the floor is not uniformly loaded, the reduced bearing capacity can be found by substituting B' for B in Equations (6.6.1) and (6.6.2):

$$B' = B - 2e \qquad (6.6.9)$$

Where e is the distance between the location at which the actual resultant force acts and the location at which the resultant force of uniform loading acts. For uniform loading, $e = 0$, and $B' = B$. If there is eccentricity in two directions, both the width and length should be reduced by the amount shown in Equation (6.6.9).

6.6.3.2 Water

Water is known to reduce considerably the strength of soft and plastic rock. If the rock is saturated with water, γ should be replaced by γ' in Equations (6.6.1) and (6.6.2).

$$\gamma' = \gamma - \gamma_w \qquad (6.6.10)$$

Table 6.6.1 Reduction of bearing strength due to water

Investigator	Floor Rock	Reduction of Strength (%)	Water Condition
Afrouz (1975)	Coal	18	Natural
Afrouz (1975)	Shale	30	Natural
Jenkins (1958)	Shale	25	From dust suppression
Lee (1961)	Shale	35–69	Natural
Platt (1956)	Shale	60–80	From dust suppression

Where γ_w is the weight per unit volume of water. Equation (6.6.10) means that the bearing capacity of the saturated soft rock will be reduced by almost 50%, because γ_w is approximately one-half of γ. Table 6.6.1 lists some measurements on reduction of bearing strength of floor strata due to the presence of water conditions.

6.6.3.3 Base size

Equations (6.6.1) to (6.6.5) indicate that the effects of the width of base plates vary with rock types. For soft and plastic rocks, the bearing capacity increases with the base width or diameter. But for brittle and elastic rocks, it depends on the base types. For flexible bases with uniform pressure distribution, the bearing capacity is independent of the base size, whereas for rigid base, it varies inversely with the base width.

Several investigators (Afrouz, 1975; Barry and Nair, 1970; Jenkins, 1958; Lee, 1961; Platt, 1956) had performed bearing capacity tests on underground coal mine floors and found that:

$$\sigma_b = kA^{-m} \tag{6.6.11}$$

Or:

$$[P_{max} = KB^n] \tag{6.6.12}$$

Where k, K, m, and n are constants. A is the area of the base plate, and σ_b and P_{max} are the maximum pressure and load applied, respectively, when the floor fails. Table 6.6.2 shows the coefficients for failure loads as determined by various investigators.

Table 6.6.2 Bearing capacity coefficients as determined by various investigators

Investigators	Floor Type	K	N	B (in.)
Barry and Nair (1970)	Shale	2.837	1.28	1–12
Lee (1961)	Coal	1.05–4.48	1.0–1.15	1–6
	Sandyshale or Fireclay, Soft clay	0.87–2.95	1.5–2.0	
		0.011–0.15	2.4–3.0	
Jenkins (1958)	Medium hard shale	–	1–2	1–6

6.7 Torsional strength of shield subject to bias loading

Torsional strength of shield is a very critical component of shield design, because the roof and floor at longwall faces are rarely flat and smooth. Rather, they are full of uneven steps, undulations, cavities, etc. Under this condition, contacts between the roof and canopy and between the floor and base plates are normally uneven, resulting in bias loading on the canopy and base plates. Therefore, the resultant load on the canopy may act on or near the edges rather than the center of the canopy along the faceline direction, which in turn causes the support to twist. Since hydraulic legs can only transmit axial forces, the twisting moment induced by the vertical resultant load acting near the edges of the canopy is transmitted through the caving shield and the lemniscate links to the base plate. Torsional forces in various directions are also generated as a result of large steps or uneven contacts between roof and canopy or between base plates and floor or both.

Experiences have shown that cracks develop along the weld joints of the runner plates in the caving shield, and permanent deformation occurs in the pin holes of lemniscate links and various joint pins and/or clevises. These problems are attributable to bending and torsional loadings.

In addition to various assumptions that may not truly represent actual practice, theoretical derivation of torsional stress and strength are very involved (Peng and Chen, 1992), so torsional stress effects can best be and are normally handled by full-scale testing of shields.

6.8 Static structural analysis of shield support

With the latest computer technology, large in memory size and fast in computation speed, formerly prevailing methods of analytical structural analyses (Peng and Chiang, 1984; Barczak and Garson, 1986; Barczak, 1987; Barczak and Schwemmer, 1988) have been replaced by numerical computer analyses. The two major OEMs of shield supports have both employed three-dimensional finite element modeling (DFMEA process for design) to investigate possible failure locations and modes and to weigh the severity and probability of their occurrence. If the analysis shows the stress concentration at certain components is excessive, modifications are then made to improve the design and manufacturing process. The advantage with finite element computer modeling is that parameters are easily changeable and numerous models with different parameters can be run and design improvements can be made prior to fabrication of a full-scale prototype shield for performance testing.

6.9 Testing of full-sized prototype shields and electronic and hydraulic units

6.9.1 Shield structure

The final step in the support design is the manufacturing of a full-sized prototype model for testing. Testing of a full-sized prototype model will provide information about the design deficiencies from which corrections can be made before full production is scheduled. During the period of the 1970s–2000s, underground installations and operations under various field conditions have revealed several design and manufacturing deficiencies. These include failures in the caving shield near the lemniscate link pins and near the canopy hinges (Barczak, 2001a), failure of the canopy due to punch-through of the leg socket, failure of hinge pins due to insufficient sizes, failure of the position-limiting ram between the canopy and the caving shield due to insufficient stroke, incomplete and/or poor-quality welding, poor-quality materials for highly stressed parts, failure of the leg cylinder casing, permanent deformation of the base structure, leg socket failure, failure of canopy structure, and others. A properly designed testing program can uncover weak points for full design changes. Most important of all, there is no design method currently available to consider the "fatigue life" which, in practice, is what a shield is subjected to throughout its service life. A full-scale prototype model testing for 50,000–75,000 supporting cycles of various loading conditions is the best and most assuring way to account for "fatigue life."

There are two types of testing machines. One is a static frame in which the external loadings are generated mainly by applying the hydraulic leg pressures of the support being tested against the test frame. Auxiliary rams are added for external loading, if necessary. The other is the active testing machine in which all external forces are applied. A typical example is the NIOSH Mine Roof Simulator (Fig. 6.9.1) which is a computer-controlled electromechanical hydraulic press. The lower platen is capable of applying 1500 tons (1363.6 mt) vertically and 800 tons (727.3 mt) horizontally. Full load can be applied anywhere over an octagon pattern

Figure 6.9.1 NIOSH mine roof simulator

Source: Garson *et al.* (1982)

on the lower platen. Horizontal and vertical travels are programmable and can be applied simultaneously.

The testing programs vary with the OEMs. They should include the most severe loading conditions expected during the service life of the support. In this respect, any testing program will be less meaningful if the actual loading conditions expected during the service life of the supports are not properly simulated in the testing program.

Lacking the precise information on what actual loading configurations will be during its service life, the most conservative testing program has commonly been used. In this approach, the testing program consists of those loading configurations that produce the most severe deformation at strategic locations. Barczak (1989, 1990) performed numerous tests on an instrumented two-leg shield subjected to various loading configurations using the NIOSH mine roof simulator. He identified several critical load contact configurations, which are defined as those that produce maximum component strain in one or more components, and he recommended those load configurations for prototype testing (Fig. 6.9.2). He also reported that the symmetric base-on-toe and base-on-rear configurations produce larger strain development than the asymmetric configurations of similar base contacts. Overall, the symmetrical base-on-toe configuration is the most critical for a two-leg shield. He suggested that a full canopy and base contact configuration be used as a test standard. Other recommended optional tests evaluate the effects of asymmetric canopy and/or base contacts that produce the maximum out-of-plane stress in the canopy and/or base plate and concentrated load applications that produce leg imbalance. With the exception of the base-on-toe configuration, the effect of the application of horizontal displacement is to decrease base strains and the caving shield-lemniscate link system to increase in strain.

6.9.2 An example of prototype shield testing program

A prototype shield testing program consists of load testing, structural measurements and inspection, and acceptance criteria.

6.9.2.1 Load testing

The latest load testing program of new shield consists of 2 × 30,000 cycles, i.e., run twice the 30,000 cycles program outlined in Table 6.9.1 and Figure 6.9.2. Each run takes about 4–5 weeks. These 30,000 cycles are on major structure components such as canopy, base, caving shield, and lemniscate links. In addition, tests also are performed on other components such as leg retention, leg pocket overload, stabilizing ram (tilt cylinder), working range, articulation, panline hook-up movement, stabilizing ram system shear pin and overload, base-lifting device overload base-lifting/relay bar cycling, and weight and center of gravity (Table 6.9.2).

For each load test of the major components, the test leg pressures will cycle from 50 to 400 + 12% bars. Loading blocks inserted between canopy and upper loading platen of the test machine, and between base plate and lower loading platen of the test machine are used to generate various symmetrical or unsymmetrical bias loadings such as bending and torsion on canopy or base or both. The loading block should be sized such that its maximum pressure will not exceed 28.44–35.55 psi (2.0–2.5 KN cm$^{2-1}$).

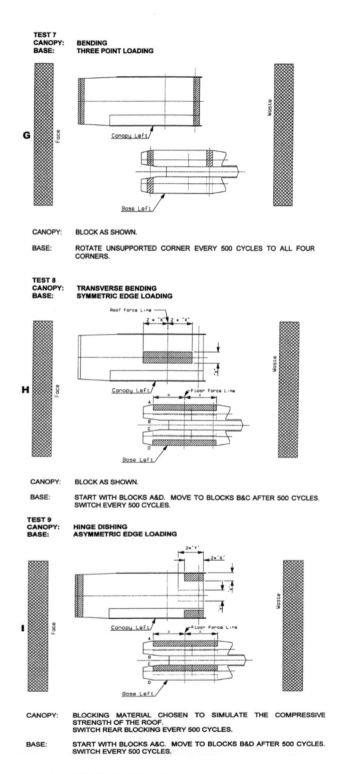

TEST 7
CANOPY: BENDING
BASE: THREE POINT LOADING

CANOPY: BLOCK AS SHOWN.

BASE: ROTATE UNSUPPORTED CORNER EVERY 500 CYCLES TO ALL FOUR CORNERS.

TEST 8
CANOPY: TRANSVERSE BENDING
BASE: SYMMETRIC EDGE LOADING

CANOPY: BLOCK AS SHOWN.

BASE: START WITH BLOCKS A&D. MOVE TO BLOCKS B&C AFTER 500 CYCLES. SWITCH EVERY 500 CYCLES.

TEST 9
CANOPY: HINGE DISHING
BASE: ASYMMETRIC EDGE LOADING

CANOPY: BLOCKING MATERIAL CHOSEN TO SIMULATE THE COMPRESSIVE STRENGTH OF THE ROOF. SWITCH REAR BLOCKING EVERY 500 CYCLES.

BASE: START WITH BLOCKS A&C. MOVE TO BLOCKS B&D AFTER 500 CYCLES. SWITCH EVERY 500 CYCLES.

Figure 6.9.2 An example of OEM shield testing program

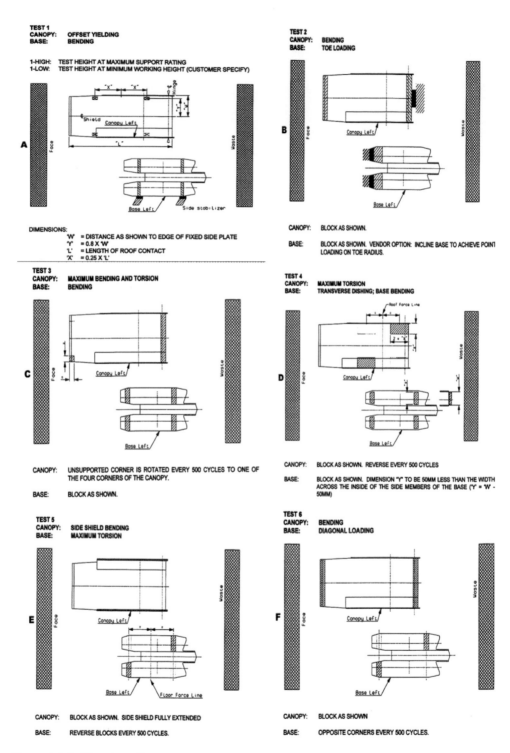

TEST 1
CANOPY: OFFSET YIELDING
BASE: BENDING

1-HIGH: TEST HEIGHT AT MAXIMUM SUPPORT RATING
1-LOW: TEST HEIGHT AT MINIMUM WORKING HEIGHT (CUSTOMER SPECIFY)

A

DIMENSIONS:
'W' = DISTANCE AS SHOWN TO EDGE OF FIXED SIDE PLATE
'Y' = 0.8 X 'W'
'L' = LENGTH OF ROOF CONTACT
'X' = 0.25 X 'L'

TEST 2
CANOPY: BENDING
BASE: TOE LOADING

B

CANOPY: BLOCK AS SHOWN.

BASE: BLOCK AS SHOWN. VENDOR OPTION: INCLINE BASE TO ACHIEVE POINT LOADING ON TOE RADIUS.

TEST 3
CANOPY: MAXIMUM BENDING AND TORSION
BASE: BENDING

C

CANOPY: UNSUPPORTED CORNER IS ROTATED EVERY 500 CYCLES TO ONE OF THE FOUR CORNERS OF THE CANOPY.

BASE: BLOCK AS SHOWN.

TEST 4
CANOPY: MAXIMUM TORSION
BASE: TRANSVERSE DISHING; BASE BENDING

D

CANOPY: BLOCK AS SHOWN. REVERSE EVERY 500 CYCLES

BASE: BLOCK AS SHOWN. DIMENSION "Y" TO BE 50MM LESS THAN THE WIDTH ACROSS THE INSIDE OF THE SIDE MEMBERS OF THE BASE ('Y' = 'W' - 50MM)

TEST 5
CANOPY: SIDE SHIELD BENDING
BASE: MAXIMUM TORSION

E

CANOPY: BLOCK AS SHOWN. SIDE SHIELD FULLY EXTENDED

BASE: REVERSE BLOCKS EVERY 500 CYCLES.

TEST 6
CANOPY: BENDING
BASE: DIAGONAL LOADING

F

CANOPY: BLOCK AS SHOWN

BASE: OPPOSITE CORNERS EVERY 500 CYCLES.

Figure 6.9.2 (Continued)

Table 6.9.1 Load testing program for canopy, base, caving shield, and links

Testing Mode	Number of Cycle
1 Canopy offset yielding and base bending at max height (Fig. 6.9.2 A)	4000
2 Canopy offset yielding and base bending at closed height (Fig. 6.9.2 A)	4000
3 Canopy bending and base toe loading (Fig. 6.9.2 B)	20,000
4 Canopy max bending and torsion and base bending (Fig. 6.9.2 C)	8000
5 Canopy max torsion and base dishing and bending (Fig. 6.9.2 D)	4000
6 Canopy side shield bending and base max torsion (Fig. 6.9.2 E)	4000
7 Canopy bending and base diagonal loading (Fig. 6.9.2 F)	4000
8 Canopy bending and base three-point loading (Fig. 6.9.2 G)	4000
9 Canopy transverse bending and base edge loading (Fig. 6.9.2 H)	4000
10 Canopy hinge dishing and base asymmetric edge loading (Fig. 6.9.2 I)	4000
Total Number of Cycles	**60,000**

Table 6.9.2 Additional tests for prototype shield

Test Number	Test Components	Description	Number of Cycles
11	Leg retention	Canopy and base fixed when one leg retracts at 350 bars	200
12	Leg pocket overload	The sides of pads of canopy and base leg pockets are fixed when both legs are pressurized to 1.5 × yield pressure	1
13	Stabilizing ram	Determine the maximum stabilizing ram's pressure in both directions. The ram will be cyclically tested in each direction at 1.12 × maximum stabilizing yield pressure	2000
14	Range	Measure the shield height when fully extended and closed	
15	Articulation	Determine no interference or damage to the shield when shield is cycled from fully closed to fully extended	
16	Panline hook-up movement	(1) Shield in fully advanced position, apply 11.24 tons (100 KN) load to the relay bar and measure the movement from shield center line. Repeat when pan is advanced. (2) Articulate pan 15° about clevis pin connection when the conveyor is advanced and measure movement from shield center line. Repeat in opposite direction	1 per position
17	Ram system shear pin	When the conveyor is advanced, determine the failure load of shear pin, the weakest component of this system, and its effect on other components	6
18	Base-lift overload	The advancing system is so positioned so that maximum stress occur in parts subject to 50% base-lifting overload. The advancing system will be re-oriented such that various components are stressed to their maximum	On cycle per each position
19	Ram system overload	Set the shield pinned to the panline connection and use a solid pin to induce 1635 tons (1600 KN) force to try to pull it away from the shield	2000
20	Base-lift and relay bar cycling	With the base prevented from lateral movement and base rear from lifting off, and panline connection fully extended, activate the DA ram and base lifting ram to pull the shield forward and lift base, respectively	1000
21	Weight and center of gravity	Measure the weight and center of gravity with and without relay and conveyor attachment by physical test	

6.9.2.2 Structural measurements

The designated dimensions of the prototype shield shall be measured before and after each testing for:

1   Shield cross-section squareness, widths and lengths and deformations of canopy, caving shield, base (split or rigid type).
2   Bores and pin diameters – lemniscate link joints, canopy hinge, stabilizing cylinder, DA ram system, and base-lifting system.

6.9.2.3 Structural inspection

Upon completion of testing, the prototype shield shall be completely disassembled for thorough visual structural inspection for cracks and permanent deformation, including the following:

1   Visual examination of cracks – each crack length is recorded.
2   After stripping of paints, canopy and base leg pockets shall be checked for cracks with dye penetrant or other inspection procedures.

6.9.2.4 Acceptance criteria

Upon completion of testing, the prototype shield will be accepted for full production provided the following:

1   No significant cracks are found in any shield components.
2   No significant cracks in welds, nor major permanent deformations that would affect shield performance are found. Any significant cracks detected in welds would be addressed with improved weld size or design.
3   Limits of permanent deformations for various measurement sections are specified by OEM.
4   Total bore and pin deformation at any measurement plane shall not exceed 0.079 in. (2 mm).
5   Canopy side shields must remain fully operational after load tests.

6.9.3 Electronic and hydraulic components

Environmental testing simulating conditions under which the equipment must perform during its service life (Bessinger, 1996) is required. Many underground conditions exist that are detrimental and yet most likely will be experienced by an electronic control unit. Shield washdown, water immersion, transit drop, and vibration exposure typically cause damage to control units not specifically designed to meet underground service requirements.

The hydrodynamic impulse test simulates shield washdown at the face. In this test, a jet of water is directed to impinge upon the unit at every vulnerable area for up to 5000 cycles without water invasion into the control unit. During this test, all valve components of the system will be cycled as in normal operation with flows and pressures as expected underground for 10,000 cycles without failure.

In a water immersion test, the electrohydraulic control system is immersed in mine water to a depth of 3.28 ft (1 m) for 24 hours. All elements or units must withstand this condition without suffering any water invasion or degradation of performance. Fluorescent may be used to identify leakage paths.

Electronic units shipped by air are subject to several environmental conditions that must be tested to ensure that the units arrive safe and sound. These conditions include extreme changes in temperature, thermal shock, and thermal cycling from −40 to 85°C, low and high atmosphere pressure, acceleration, vibration and free fall drop by workers. Other tests include exposure to high humidity for six to eight hours without condensation, exposure to water mist, and mineral dust (airborne and settled).

In the electrohydraulic system performance evaluation, a bench test of a selected number of electrohydraulic control units is conducted to evaluate software behavior within the operating range of parameters, individually or in group. A bench trial on all new equipment is also recommended.

The correct function of a machine's sequence of operational events and their tolerance to being out-of-sequence should also be evaluated.

After the bench test, a limited scale surface trial is done with the equipment installed on shield supports powered by emulsion at full-system pressure.

Chapter 7

Coal extraction by the shearer

7.1 Introduction

Since its first appearance in 1954, the shearer has undergone continuous changes both in capability and structural design. Today, it is the major cutting machine in longwall mining. There were two types of shearers, single- and double-ended ranging drum. In the earlier models, the drum in the single-drum shearer was mounted on the shearer's body and could not be adjusted for height. As such, it was not suitable for reserves in which there were constant changes in seam thickness and floor undulation. Thus, the single-ended, fixed-drum shearer was used mostly for thin seams of uniform thickness.

A ranging arm with a cutting drum mounted at the very end of it replaced the fixed drum. The ranging arm could be raised up and down by hydraulic control to accommodate the changing seam thickness and floor undulation. But, when the seam exceeded a certain thickness and became thicker than the diameter of the cutting drum, the single-drum shearer could not cut the entire seam height in one cut, and a return cutting trip was necessary to complete a full web cut. Furthermore, if the drum was located on the headgate side, it generally required a niche in the tailgate side. A niche is a precut face end, one web deep and one shearer's length long. With a niche the shearer did not need to cut into the tailgate, and in order to clear itself fully for sumping the next cut, the tailgate needed to be wider than the width of shearer. The reverse was true if the drum was on the tailgate side.

Obviously, there were great operational disadvantages with the single-ended ranging drum shearers. Consequently, the double-ended ranging drum shearer (DERDS) was developed to replace the single-ended ranging drum shearer. It soon became the industry standard and has been used exclusively ever since. The double-ended ranging drum shearer cuts the whole seam height in one trip. The two drums can be positioned to any required height (within the designed range) during cutting and lowered well below the floor level. The arrangement of the drums enables the whole seam to be cut in either direction of travel, thereby ensuring rapid face advance and shortening roof exposure time.

In the 1980s, Joy Technologies developed the 4LS double-ended ranging drum shearer which was thinner, with module construction, for ease of maintenance. It was a multi-motor machine as opposed to previous single-motor ones. Furthermore it was easier to control by using DC haulage, and the 4LS soon became the industry standards during the 1980s and 1990s. Today the 7LS shearers, which is a more powerful and advanced version of the 4LS series, dominates for US longwalls.

Since today's longwalls are highly productive and require high-powered shearers cutting at high speeds, AC haulage that can provide high and variable shearer haulage speeds has replaced the DC haulage.

Prior to the 1970s, all shearers were hauled by a chain anchored at both ends of the long-wall faces. But it was soon replaced by chainless haulage due to safety and efficiency issues. Two types of chainless haulage soon dominated: captivated chain and rack systems. The captivated chain was more popular in the beginning, but now the rack system has become the dominant system due to the ever-increasing installed power of the shearers.

Just like the shield support's capacity, the shearer's capacity also increased continuously prior to 2010 and stabilizes since then (Fig. 1.5.6). Increased power allows the mining technology to apply to thinner and/or dirtier coal seams. For thinner seams, the shearer must cut some roof or floor or both to create a minimal height for comfortable mining operations. Many coal seams contain one or more partings that are normally harder to cut than coal seams themselves. The result of mining in these thin and dirty seams is that clean coal recovery is as low as 30 to 40%.

The current practice is to designate the whole shearer longwall mining system in terms of production capability in tons per hour (tph). For instance, the maximum currently available system is 7000 tph. So for this system, the shearer can cut and produce 7000 tph and the coal conveying system (AFC and stage loader) can carry it away at the same rate.

Today's shearer is highly reliable and allows continuous production up to 4 to 6 million raw tons (that is the size of a longwall panel) of coal without the need for major overhaul.

7.2 Layout of longwall face equipment using the shearer

Figure 7.2.1 shows a typical (left-hand) US longwall face equipment layout employing the shearer as the cutting machine. The shearer is used in all coal seams thicker or mining heights larger than 60–66 in. (1.52–1.68 m). A double-ended ranging drum shearer cuts coal either uni-directionally (uni-di) or bi-directionally (bi-di) and will cut clear at both ends of the panel and into the headgate or tailgate. The shearer rides on the AFC with the gob-side shoes trapped on the haulage system and the face side shoes either on the surface of the face side sigma section of the pans (see inset lower figure of Fig. 7.2.1) or on a ramp plate attached to the bottom of the face side sigma section (see inset upper figures of Fig. 7.2.1). The former arrangement provides a narrower shearer body width, thereby reducing the unsupported distance from canopy tip to faceline. But this arrangement wears the surface of the sigma section faster. The latter arrangement may require a wider shearer body width, and thus be more stable. For today's heavy-duty high-speed cutting shearers, a ramp (toe) plate on the front side of the pan for the shearer's two front shoes to ride on is exclusively used for reason of stability.

After being cut off by the shearer, the coal is loaded by the cutting drum itself onto the armored face conveyor (AFC), which runs the whole face width endlessly. The AFC employs a twin-center chain strand with flight bars at fixed interval to transport coal to the headgate T-junction where it is dumped directly onto a stage loader through a cross frame. A crusher in the inby end of the stage loader will crush oversized lumps and insure a more uniform-sized product for the outby belt conveyor for safe and efficient transportation. The stage loader is usually covered to reduce dust.

The shield supports are lined up behind, and extend the full length of, the AFC providing roof support for the face area.

There are several special features that are unique with US longwall mining technology that allow fast longwall advancement, and thus, contribute to high production and high efficiency (Figs 7.2.2 and 7.2.3).

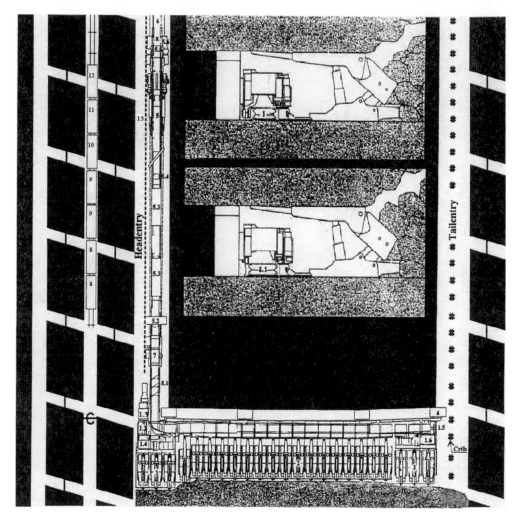

1 Face Conveyor
  1.1 Face Conveyor
  1.2 Head-End Drive Cross-Frame
  1.3 Gearbox
  1.4 Gearbox
  1.5 Tail-End Drive
  1.6 Gearbox3
2 Two-leg or Four-leg Shields with positive base-lifting
3 Face-End Shields
4 Shearer
5 Stage Loader
  5.1 Flex-pans and emergency pull-cord
  5.2 Cable duct

  5.3 Electric controls (optional)
  5.4 Inspection-pan
  5.5 Drive Frame
  5.6 Gearbox
6 Belt Conveyor Tail-Piece
  6.1 Stage Loader Advancing and Belt Tail Anchoring
  Unit
7 Crusher
8 Power Center/Control Panel
9 hydraulic Pump and Emulsion Tank Stations
10 Rock Dust Car
11 Water Pumps
12 Supply/Utility/Parts Cars
13 Monorail

Figure 7.2.1 Typical face equipment layout for shearer longwall system

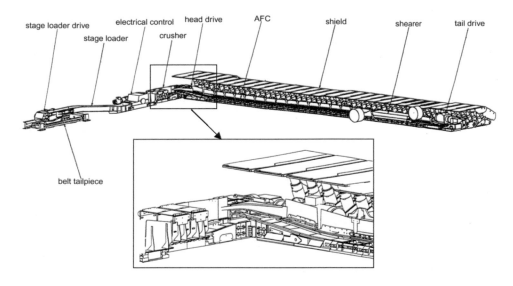

Figure 7.2.2 Face view of right-hand longwall full face equipment layout

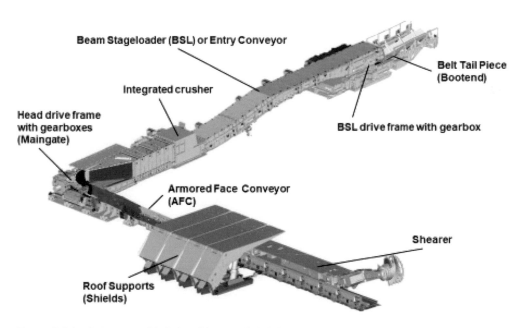

Figure 7.2.3 Gob view of left-hand longwall full face equipment layout

Source: Courtesy of Caterpillar

1   A crawler-mounted belt tail piece for the panel belt conveyor is normally employed so that it can be trammed easily and rapidly in order to keep pace with the fast advancing longwall face. The belt structure 20–30 ft (6.1–9.1 m) immediately outby the tail piece is dismantled in advance.

2   The drive head of the stage loader, where it dumps coal into the panel belt conveyor, rides on the belt tail pieces on a dolly that can travel freely for a distance of 12 to 15 ft (3.7–4.6 m). So as a new cut is made, the gate-end shields advance and push the cross frame and the whole stage loader a web distance ahead.

3   There are four to five flexible short pans, 27.6 in. (700 mm) long, between the cross frame and crusher. This arrangement can absorb the bending and creeping caused by the panline that may be pushed left and right too far out of alignment between the panline and stage loader thereby reducing damage to the coal transfer system.

4   The gate-end shields in the headgate T-junction not only protect the drive head cross frame and drive motors that are mounted on a skid, they are also used to advance the cross frame and drive motors and stage loader forward cut by cut.

5   The AFC tail drive is also mounted on a skid and flush with the rib of the panel coal block without sidestepping into the tailgate. It is also easily advanced by the three gate-end shields cut by cut. This arrangement does not require dismantling the wood cribs (or other types of standing supports) in the tailgate, which are normally installed as a secondary support to cope with the incoming front and side abutment pressures, and insures rapid advance of the longwall.

6   A monorail, 1000 to 2000 ft (305–610 m) long is hung from the roof either on the pillar or panel block side of the belt conveyor in the headgate. Power supply cables and hoses between the power train and stage loader are hung and guided by dollies/ trolleys. These cables and hoses stretch and coil like an accordion easily following the longwall face advance. Figure 7.2.4 shows a basic construction of a monorail system.

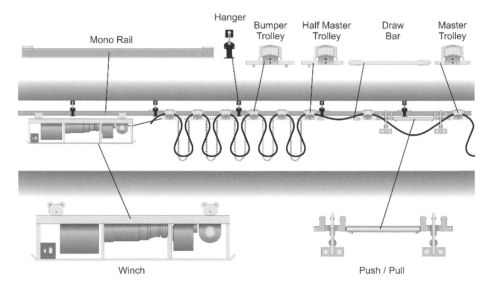

Figure 7.2.4 Typical monorail system

Source: Courtesy of Swanson Industries, Inc.

The electrical cables, hoses, etc. are looped between bumper trolleys. When the monorail is at its full length, those cables and hoses are fully stretched. As the face retreats, the push/pull unit mounted at the inby end pushes them gradually into loops. When the cables/hoses between all bumper trolleys are fully looped, the monorail system is ready to be moved outby, normally when the face has advanced three to five pillar blocks. At this time, the winch mounted at the outby end is used to move and reset the monorail system.

7   The power train, which consists of power centers, hydraulic supply pump system, rock duster, water pumps, tool and supply cars, etc., is normally on a track or skids in the second entry outby the face and can easily be pulled by a small locomotive and moved with the moving longwall face.

7.3 Monorail system

The monorail system for handling the outby cables and hoses was introduced when the face voltage was increased to more than 1000 volts in the late 1980s. The regulations prohibits manual handling of energized high-voltage cables. Prior to the use of monorail, manual labor or a winch was used to dragged along the ground of the belt entry, which was laborious and time-consuming, hindering the fast movement of the face.

Figure 7.2.4 show the basic element of conventional monorail system. The monorail is hung to the roof by hangers at regular interval. On the outby end is a winch, while on the inby end is a push/pull device. In between many bumper trolleys are connected by chain of fixed length in loops (lighter gray). The cable (dark thick line) is looped on one side of the bumper trolley, while on the other side of the rail the hose is looped similar way. The chain loop length is shorter than those of cable and hose such that it ensures the cable and hose will not be over-stretched to damage. A draw bar extends between the half master trolley and master trolley where the push/pull device operates. From master trolley to the stage loader the cable and hose are free as the stage loader advances. The push/pull is used to compress the cable and hose between the half master trolley and master trolley every two shearer's cuts. Figure 7.3.1 shows the more detailed layout of the monorail.

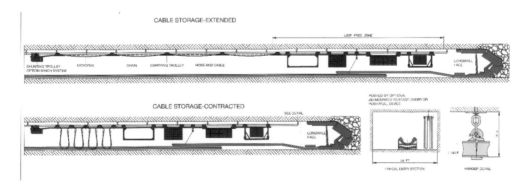

Figure 7.3.1 An example of detailed layout of the monorail

Source: Courtesy of Caterpillar

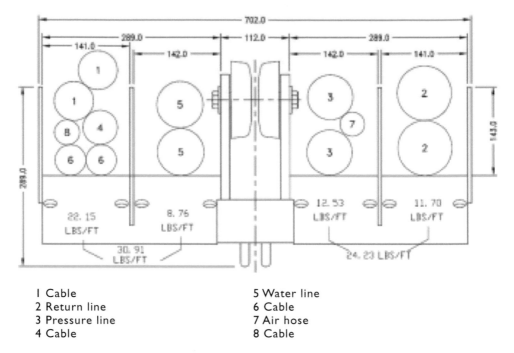

1 Cable
2 Return line
3 Pressure line
4 Cable

5 Water line
6 Cable
7 Air hose
8 Cable

Figure 7.3.2 Two-pocket bumper trolley with typical cables and hoses distribution

Source: Courtesy of Caterpillar

All the trolleys have two pockets (Fig. 7.3.2), one on each side of the rail. Each pocket has a divider to separate different utility lines for safe operations. For example in the left pocket, the first subpocket contains six electric cables (2 × 4/0, 1 × 2/0, 1 × #1, & 2 × #2–5kv). In the right-side pocket, the first subpocket contains two high pressure hydraulic fluid lines and one air hose, while the second subpocket contains two hydraulic fluid return lines.

Depending on the width of belt entry (15.5 or 20 ft = 4.7 or 6.1 m) and size of conveyor belt, space constraint dictates the selection of types of monorail systems such as side by side, over/under or multiple layer. In a side by side system, there two trolleys side by side, one for pipe, and the other for cables and hoses (Fig. 7.3.3 A). In an over/under system, the trolleys for pipe and cables/hoses are over the cables and hoses (Fig. 7.3.3 B). In a multiple layer monorail system, the monorail is a two-tiered trolley for low seams (Fig. 7.3.3 C).

Some longwalls have the hydraulic pumps in a cabinet, plus emulsion tank and filter station hanged on the monorail (Figs 7.3.3 D and 5.3.14). The advantages with this setup are greatly reduced line loss in the pressure lines as well as the return lines, reduced pump's response time, and, during power move, there is no additional time needed for moving the pumps, and less fluid contamination resulting from no line parting (Hutchinson, 2013, 2019).

In the recent development of remote control of longwall automation system, the longwall control center is located in one or two cars hung on the monorail moving with the advance of the face.

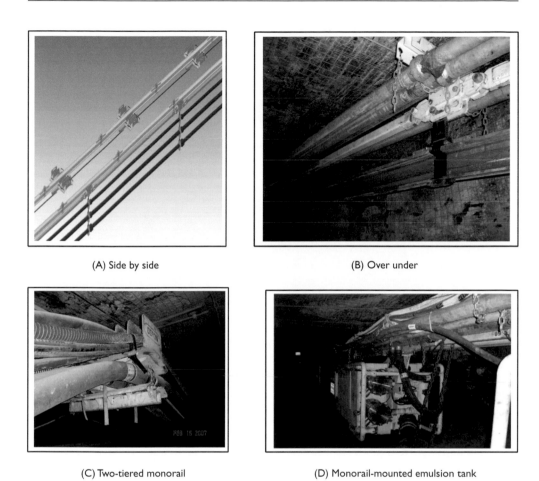

(A) Side by side

(B) Over under

(C) Two-tiered monorail

(D) Monorail-mounted emulsion tank

Figure 7.3.3 Types of monorails

Source: Courtesy of Hutchinson (2015)

7.4 Double-ended ranging drum shearer

7.4.1 Models and major components

There are two major OEMs manufacturing DERDS for US longwalls, the 7LS shearers by Komatsu Mining (Joy) and the Electra range shearer by Caterpillar Global Mining. According to the 2019 US Longwall Census, the 7LS has three models (7LS1, 7LS2, and 7LS5) accounting for 93% of the market share. Caterpillar also has three models (El 1000, EL2000, and EL3000) (Table 7.4.1).

The 7LS series of shearers is a complete line of shearers – 7LS1 through 7LS8 – for seam heights from 5 ft (1.5 m) to 25.9 ft (7.9 m) with total installed power from 1230 to 2935 hp (918–2127 kW). The shearers are 45 to 57 ft (13.7–17.37 m) long, weigh 51.5 to 110 tons (46.8–100 metric tons) with maximum flitting speed up to 68 to 95 fpm (20.7–29.0 m min^{-1}),

Table 7.4.1 Models of shearers used in US longwalls

		7LS1	7LS2	7LS5
JOY	Model			
	Mining Height (m)	1.5-3.0	1.6-3.6	2.0-4.5
	Max Installed power (hp)	1880	1666-2084	2360
CAT	Model	EL1000	EL2000	EL3000
	Mining Height (m)	1.6-3.2	1.8-4.5	2.5-5.5
	Max Installed power (hp)	1880	2390	2805

and are designed to fit the dimensional requirements of the mining height. They possess adequate horsepower for a high production rate; are versatile in accepting different haulage systems and conveyor widths; are reliable, rugged, and easily maintained; and are safe and meet environmental considerations.

The Electra range shearers have three models, EL1000, EL2000, and EL3000. They are designed for seam height of 4.9 to 24.6 ft (1.4–7.5 m) with installed power 1620 to 2800 hp (1174–2229 kW). They are 39 to 47.9 ft (11.9–14.6 m) long and weigh 55 to 110 tons (50 to 100 mt) with maximum haulage speeds from 65.6 to 147.6 fpm (20–45 m/min).

Regardless of models, DERDS consists of the following six major modules (or eight independent components); main body, two haulage sections, one control box or controller case, and two cutting drums with ranging arms (Fig. 7.4.1).

1   Main body (or frame) – the body consists of three or more high-strength steel fabrications bolted together. It is thin and yet sufficiently strong for underground cutting environments. It has no underframe and, thus, provides maximum clearance between the deck plate of the AFC and the lower side of the main body and allows maximum coal flow during a head-to-tail cutting trip. The frame is also so designed that there is no need to access the electrical controller case from the face side.

2   Controller case or control box – This is the center of the main body. It houses the electric control system including vacuum contactors, transformers, the SCR bridge or an AC (or DC) drive, microprocessor, control circuitry, and a data display screen.

3   Haulage sections – for high speed heavy cutting a dual haulage drive system is required to reliably deliver the necessary tractive effort. The two haulage units are bolted and dowelled separately to each side of the controller case or control box. Each haulage unit consists of a haulage motor, a primary traction gear case and a hydraulic system for moving the ranging arm and rotating the cowl. The two AC motors are electrically connected in series. The introduction of variable frequency and variable voltage AC haulage in 1997 gives the operator absolute speed control while providing a greater speed for cutting and flitting. The haulage pull ranges from 63 to 110 tons (57.15–99.79 metric tons).

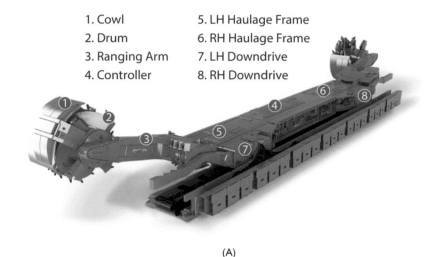

1. Cowl
2. Drum
3. Ranging Arm
4. Controller

5. LH Haulage Frame
6. RH Haulage Frame
7. LH Downdrive
8. RH Downdrive

(A)

Cutting-Drum

Ranging-Arm

Mainframe

Haulage-Unit
&-Downdrive

Hydraulic
Powerpack

Haulage-&-
Transformer-Box

Electrical
Control-Box

Haulage-Unit&-
Downdrive

(B)

Figure 7.4.1 Major components of shearer
Source: Top, Komatsu Mining (Joy); lower, Caterpillar Global Mining

A secondary traction gearcase (or down drive) section, containing the drive sprockets and gob-side trapping shoes, is bolted to the traction case in an arrangement that permits the custom-fitting of the shearer within the AFC and roof support envelope. This arrangement allows the shearer to easily fit any haulage system available in the market.

There are two types of haulage drives: indirect and direct. An indirect drive has a gear mounted out of the transmission gearbox that engages anther gear below it. This lower gear is bolted to the back of the sprocket, so the torque goes through the two gears and drives the sprocket. But this takes up more space and can be more parts cost on replacement (two gears + bottom sprocket). A direct drive has no gear but a "top sprocket" that directly drives the "bottom or rack sprocket" that goes into the rack. This is narrower on the AFC cross-section, and does not impede into the tunnel under the shearer (no gear behind the sprocket). So more tunnel under the shearer for AFC coal conveyance (two sprockets/no gears). The disadvantage is the sprocket has contact at the top from the "top sprocket" and also in the bottom in the rack bar, and wears out more quickly, compared to an indirect drive (gears are driven on top to bottom, then bolted to the lower sprocket) (Fig. 7.4.2).

There are four sliding shoes for guidance of the machine haulage (Fig. 7.4.3), two each on the face and gob sides of the shearer. The two gob-side shoes are trapped on the haulage

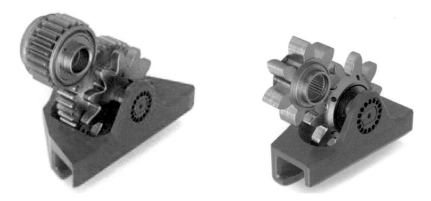

Figure 7.4.2 Indirect (left) and direct (right) haulage drives
Source: Courtesy of Komatsu Mining

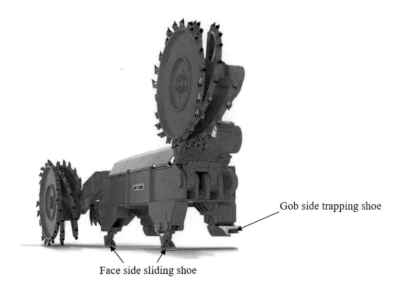

Gob side trapping shoe

Face side sliding shoe

Figure 7.4.3 Full view of shearer
Source: Courtesy of Komatsu Mining

system while the face side shoes are a roller type, moving on a flange (toe) plate attached to the face side of the sigma section of the conveyor pans on the mine floor. This design offers solid support to the shearer but requires a wider machine frame. Another design places the face side skid plates or roller shoes on the top race of the face side sigma section of the conveyor pans, thereby reducing the machine width and the unsupported distance from shield canopy tip to face line.

4 Ranging arm and cutting drum – there are two units of the ranging arms and cutting drums, one at each end of the main body. The ranging arm is thin-profiled to maximize loading ability and to minimize interference problems with cut-out at the tailgate and headgate haulage drive frames, which minimizes obstruction of the panline. The ranging arm contains gearcases and a bore into which the cutter motor is inserted.

The cutting drum is the work horse of the shearer. Each drum has a dedicated motor that ranges from 393 to 1173 hp (289–863 kW), depending on models and OEMs. The drum runs at 37 to 54 rpm with diameter ranges from 48 to 118 in. (1219–2997 mm) and web width from 30 to 42 in. (762–1,067 mm). For more detail about the drum, see Section 7.5.1, Cutting Drums (p. 229)

One special feature of the microprocessor in the shearer is the memory cut system. In this system, the shearer operator manually runs one cut creating an initial face profile, and the shearer then automatically replicates the profile on the subsequent cuts until conditions change. When it drifts out of seam after several automated cuts, maximum five cuts, the operator updates the profile manually, cutting the new cut.

All shearers are equipped with a shearer location device as part of the SISA system (shearer initiation shield advance). Two types of shearer location devices are available: an infrared emitter mounted on the surface of the main body or an odometer, which is a proximity sensor counting the teeth of the haulage sprocket running on the haulage rack. The digitized counter signals are then transmitted to the headgate computer, which, in turn, notifies and issues commands to the shields of interest.

Each drum is equipped with a cowl to load coal onto the chain conveyor, to clean up at the headgate and tailgate drive frames, and also to help control dust. It can rotate a full 360°. Both the ranging arms and cowls are actuated by low speed, high torque hydraulic motors located at the haulage section.

A lumbreaker is also installed at the tail side to break up oversized rock/coal pieces so that it can pass through the tunnel clearance below the shearer.

All shearers are equipped with radio control with which the operator can command specific machine functions staying upwind and minimize the operator's exposure to respirable dust. A special network architecture control system monitors various machine functions to minimize downtime, and the screen display provides full diagnostics to monitor several machine functions, reducing troubleshooting time.

All shearers are equipped with multiple motors. The multi-motor shearer introduced in 1975 with Joy's 1LS consists of an independent motor for each cutting drum, each haulage unit, and other ancillary hydraulic equipment. In a multi-motor shearer, each motor can be replaced or repaired easily and independently. It is applicable to a much wider range of coal seam thicknesses due to its flexible motor size design.

The chainless haulage is exclusively used and powered by AC motors. AC electrical haulage is more efficient, enabling faster acceleration and tramming. The maximum pull force ranges from 63 to 110 tons (57.3 to 100 mt) at a maximum machine speed of up to 120 ft (36.6 m) per minute for cutting and 147 ft (44.8 m) per minute for flitting or light cutting.

The shearer's trailing cable, communication cable, and water hose to the shearer are contained in a cable handler that is brought to the midpoint of the AFC furniture so that it rolls out left or right to either the head or tail end, respectively (Fig. 7.4.4). The cable handler has either one or two compartments.

All modern shearers are equipped with diagnostics and a troubleshooting display. Generally a screen graphic display provides real time information on machine performance,

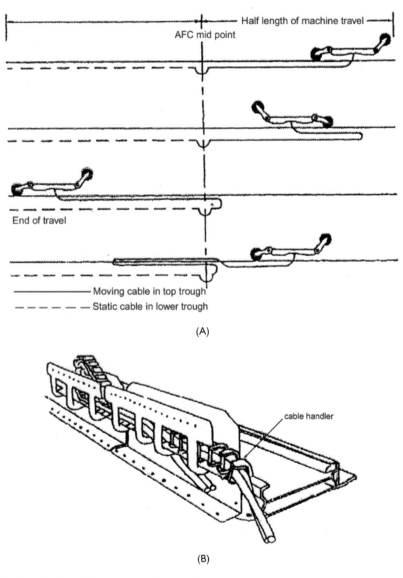

(A)

(B)

Figure 7.4.4 Cable handling system for the shearer

including current/temperature of each motor, phase balance and the status of overload setting for each motor, and machine speed. This information allows the operator or maintenance personnel to monitor the operational characteristics of the machine.

In addition, the shearer is equipped with auxiliary hydraulic pumps and control valves for water spraying devices, cable, chain anchorage and tensioners.

7.4.2 Key factors and dimensions in shearer selection

The following three factors must be considered in selecting the proper shearer model:

1   Cutting power.
2   Haulage power and rack haulage.
3   Key dimensions.
    a   Distance between trapping shoes; gob side vs. face side.
    b   Pan width.
    c   Underframe clearance.
    d   Width of toe plate on face side of pan for shearer's face side shoes.
    e   Shearer body height.

7.4.3 Mining height of the shearer

In selecting the shearer, mining height should first be considered; the diameter of the cutting drum, body height (or thickness), length of the ranging arm, and swing angle must be properly selected. For the double-ended, ranging drum shearer, the maximum mining height cannot exceed twice the diameter of the cutting drum. The mining height can be determined by:

$$H = H_B - \frac{B}{2} + \ell \sin \alpha + \frac{D}{2} \qquad (7.4.1)$$

Where
    H = mining height or seam thickness
    H_B = shearer's body height, i.e. from mine floor to top of shearer's main body = body height (body thickness + height of haulage shoes) + pan height
    B = thickness of main body
    ℓ = length of the ranging arm
    α = angle of the ranging arm from the horizontal line when it is raised to its maximum height
    D = diameter of the cutting drum

7.5 Cutting drums and performance of the shearer

7.5.1 Cutting drums

Coal production using a double-ended ranging drum shearer relies completely on the two drums mounted at each end of the shearer. Since the cutting direction of the shearer is parallel to its travel direction, the drum is designed not only to cut coal but also to load the coal it cuts onto the chain conveyor. The bits mounted on the spiral vanes are used for cutting coal. Since the panline is parallel to the cutting direction, the direction of coal loading from the drum is perpendicular to the cutting direction. The spiral vanes are designed to meet this process for loading coal.

The drum is circular in shape with a diameter normally equivalent to 60 to 80% of the coal seam height, mostly 60 to 62 in. (1524–1575 mm). Its total width or web width is 32 to 45.5 in. (813–1156 mm). Figure 7.5.1 shows a shearer drum in action (A) and an individual drum (B). The drum consists of spiral vanes, bit blocks, bits, and a central barrel where the planetary gear unit and water piping system, if internal sprays are used, are housed. The barrel is circular in shape with a minimum diameter of 31.5 in. (800 mm), depending on the size of the gear train and shell, which can be 1.25, 1.5, or 2 in. (31.8, 38, or 50.8 mm) thick.

The two drums on either end are normally rotated in opposite directions for stability. For instance, if the head drum rotates clockwise (from top to bottom), the tail drum will rotate counterclockwise; conversely, if the head drum rotates counterclockwise (from bottom to top),

Figure 7.5.1 Cutting drum of the shearer: (A) drum in action and (B) overview of drum

the tail drum will rotate clockwise. Drum rotational speed ranges from 37 to 56 rpm. Since cutting force and cutting speed are inversely proportional under constant power, a quiet running shearer needs a bit cutting speed of at least 600 fpm (3 m s^{-1}) but no more than 880 fpm (4.1 m s^{-1}). If the drum speed is too low, its loading efficiency will be low.

7.5.1.1 Starts or vanes

Two or more steel walls, 3–4 in. (76.2–101.6 mm) thick, are welded like a spiral onto the outer surface of the shell of the barrel at a certain angle and evenly spaced around the circumference. These spiraled steel walls are called vanes or starts or scrolls, and they are used to guide the cut-out coal that fills up the space between two adjacent vanes toward the gob side of the drum and then dump it onto the chain conveyor. The vane angle is defined as the angle between the vane and the end surface of the drum (Fig. 7.5.2 A).

The vane angle is rather important in that it defines the vane length that can be installed between the face and gob end of the drum and, consequently, the number of vane starts to cover the whole circumference of the drum. The vane angle is normally 20 to 24° for the whole length. But sometimes two angles are used, 20° at the face end and 24° at the gob end in order to increase loading speed. The vane angle also determines the smoothness and resident time of the coal flow after it has been cut out by the bits.

There are three, four, and six vanes or starts. Most common (up to 85%) are three vanes or starts with one to three bits per circumferential line (Fig. 7.5.2). For low-seam, hard-cutting

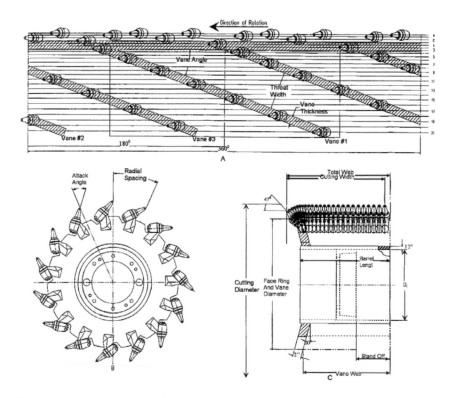

Figure 7.5.2 Detailed schematic drawing of shearer drum showing individual components

conditions, two bits per line may be required. In that case, either three or four vanes can easily accommodate. For very fast cutting and loading, six vanes should be used.

For a three-start drum, each vane starts at the gob side and its length covers a 120–180° interval. In recent years, as drum cutting power increased, the barrel diameter did also. Under constant drum diameter, the vane height is correspondingly reduced to 11 to 14 in. (279–356 mm), depending on the cutting diameter, which in high production longwalls, is not sufficient. For instance, for the most commonly used drum diameter of 62 in. (1575 mm), the vane height is

$$Vane\ height = drum\ diameter - barrel\ diameter - 2 \times (bit\ and\ block\ height)$$
$$= 62 - 31.5 - 2 \times 8.5 = 13.5\ (in.)\ or\ 343\ (mm)$$

In order to increase its height and thus loading volume, a loading skirt, 0.75 in. (19 mm) thick by 3 in. (76 mm) high is mounted on one side above the top surface of the vane (Figs 7.5.1 A and 7.5.3). This remedial measure increases the vane height to 16.5 in. (419 mm). Generally high-strength steel plates are welded on both sides of the top portion to increase its strength and full coverage in some cases at the discharge end (gob side) where wears are the worst (Figs. 7.5.1 A and 7.5.3).

Vane spacing or throat varies with vane angle, vane thickness, and the number of vanes or starts. Normally, it is such that the throat contains a volume loading rate of 55 to 70 ft^3 s^{-1} (1.56–1.98 m^3 s^{-1})

Figure 7.5.3 Close-up view of shearer drum in action showing loading skirts and vane reinforcement at the gob end

7.5.1.2 Bit lacing

The bits are mounted on the bit blocks, which in turn, are welded onto the cutouts of the circumference on the top surface of the vane so that the top of the bit block is flush with the vane's top surface. The number of bits used varies with seam characteristics and cutting parameters and ranges from 35 to 38 for three starts – one bit per line – to 40 to 45 for four starts– two bits per line. The patterns of bit arrangement or bit lacing in a drum are divided into two types depending on location: one pertaining to those mounted on the vanes and the other on the backplate or face side of the drum. The spacing of those in the backplate (or face rings) are much narrower, 0.75 to 1.25 in. (919–31.8 mm), than those on the vanes, 1.5 to 3.0 in. (38.1–76.2 mm). Normally only one bit is mounted on a circumferential line on the vane part while on the backplate, where cutting is much harder due to much higher confinement in all directions, three to six bits per line are used. For harder coals, two bits per line on the vane part may be needed.

The principle in the bit lacing pattern is that cutting should be progressed from the gob side to the face side utilizing the free surface produced by the immediate previous bit cut or cuts and minimizing the energy required for cutting. In Figure 7.5.2 A for instance, bit 21 (on vane number one) cuts first because it is the most outby bit and produces a groove off the free faceline. Then bit 16 (on vane number three) cuts another groove. The spacing between bit 21 and 16 is such that the grooves cut out by both bits will overlap and knock out parts of the ridge-shaped coal between the two grooves. Finally bit 19 (on vane number two) cuts out a groove precisely at this location, i.e., the ridge-shaped coal left between the two grooves made by bits 21 and 16. The grooves cut out by bits 21 and 16 form the first line pattern, while that made by bit 19 is the second line pattern, which is a cutting depth ahead of the first line pattern (Fig. 7.5.4). This cutting process repeats several times until it reaches the face side at the end of the vane (bit line number one in Fig. 7.5.2 A). At the face side, due to confinement at the corner, much heavier cutting is required than along the vanes. Within the vane length there is only one bit in each cutting line spaced at 1 to 3 in. (25.4–76.2 mm) between adjacent lines. Bit line spacing is larger at the gob side reducing gradually toward the face side. The majority are between 1 and 1.5 in. (25.4 and 38.1 mm). At the face side, a backplate is welded at the face end of the barrel at 15 to 30° toward the face side. Four lines of single or multiple bits are mounted on the backplate (face rings) with six bits on A line and three each on B, C, and D lines, forming a cutting pattern of DBACA (repeating three times) (Figs 7.5.2 A and 7.5.4). A special backplate is required for such a dense distribution of the bits. The inclined position of the backplate simplifies the mounting of the bit orientation, resulting in a curved cutout corner, which is easier to cut than a straight 90° corner.

7.5.1.3 Type of bits

Two types of bits are used, conical and radial bits. In the 1970s when longwall mining was introduced, radial bits (Fig. 7.5.5 A) were used exclusively. They were heavier and more expensive. Thus conventional conical bits (Fig. 7.5.5 B), developed and used exclusively in continuous miners, were modified for heavier cutting and gradually replaced the radial bits. The bit tip is sharp and pointed so that it can penetrate and break the cut materials easily with less machine force. It also tends to remain sharp because it has a cylindrical shank that rotates during cutting and wears more uniformly, lasting longer. It can even sharpen itself.

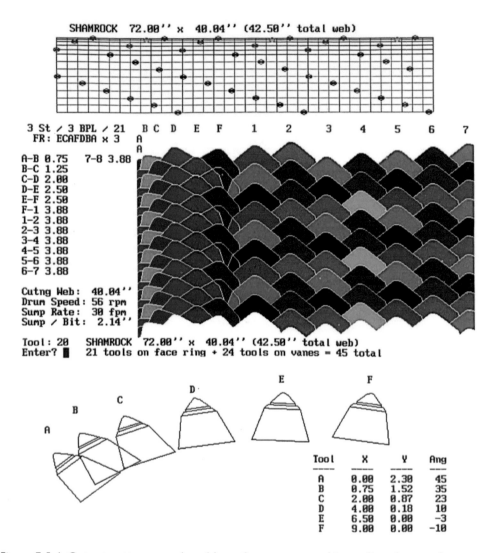

Figure 7.5.4 Cutout patterns produced by a three-start one-bit-per-line shearer drum

But conical bits have a sharp tip that is much smaller than the shank. Therefore, for longwall application, conical bits are much heavier than their counterpart for continuous miners.

Conversely the radial bit can only cut uni-directional. With its shape, the backside of the radial bit tip is always in contact with the cut materials and wears out easily during bit penetration and advance. The radial bit is larger and longer. Its cutting edge is line shaped, therefore, requiring larger machine forces to induce sufficient stress concentration for cutting.

According to McShannon (2006), the size of conical bits and bit blocks have been long increased considerably to withstand the forces of the more powerful machines. This increase has resulted in the drums being over-penetrating and slowing down the shearers because

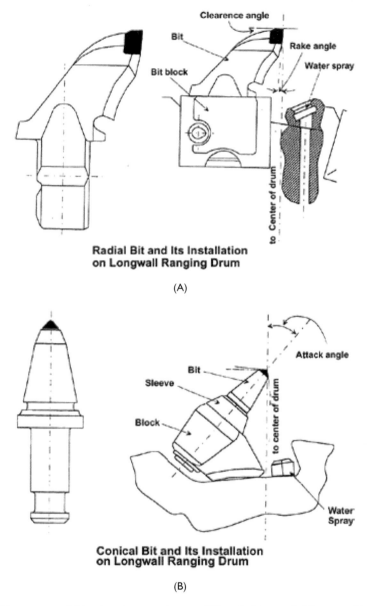

Radial Bit and Its Installation on Longwall Ranging Drum

(A)

Conical Bit and Its Installation on Longwall Ranging Drum

(B)

Figure 7.5.5 Two types of bits for shearer drums: (A) radial bit and (B) conical bit

the large bit blocks act as a lump breaker. The traditional effective reach of the bit cutting systems, including both radial and conical bits, are 4 in. (100 mm). But the new generation of shearers cuts faster and requires a much longer reach. For conical bits, this is extremely difficult, because the bit shank, bit sleeve, and bit block will need to increase proportionally. Conversely, the latest radial bits have a 5.1 in. (130 mm) reach. Due to the required attack angle, the effective reach of the conical bits, when mounted on the vanes of the drum, is shorter than that of radial bits. In addition, radial bits have a slimmer, straight body such that the carbide tip cuts sufficient clearance for the whole body to clear off, rather than the bludgeoning effect of the conical bits. In spite of this conical bits are exclusively used in the US longwalls now.

7.5.2 Mechanism of drum cutting (drum rotation speed vs. shearer haulage speed)

Coal cutting by the shearer is the result of a combined action of drum rotation and horizontal travel of the shearer's body on the panline. Figure 7.5.6 A shows an example of the cutting trace of bits at the face when mounted on a shearer drum operating under the following conditions (Peng and Luo, 1995): drum diameter of 6 ft (1.82 m), drum rotation speed, 60 rpm, and shearer's haulage speed, 30 fpm (0.15 m s^{-1}). It can be seen that the bit's travel locus is a prolate cycloid and for each drum rotation, the bit cut a sickle-shaped slice of coal. The maximum cutting depth, 0.5 ft (0.15 m), occurs at the center horizontal axis of the drum. This is the cutting depth normally referred to as the drum's cutting depth. Obviously, the maximum cutting depth per drum rotation is a function of both the rotational speed and shearer's haulage speed. The slower the drum rotational speed, the larger the cutting depth. Figure 7.5.6 B and C shows examples of how the cutting depth changes with a change in drum rotational

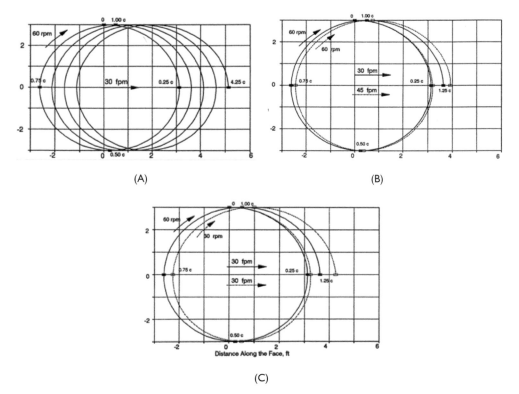

(A)   (B)

(C)

Figure 7.5.6 Cutting traces of a bit for various drum rotational speeds and shearer's haulage speeds

speed and the shearer's haulage speed. The maximum cutting depth at the horizontal level can be determined by

$$d = \frac{12V_S}{nV_R} \tag{7.5.1}$$

Where d = bit cutting depth per drum revolution, in./rev, V_S = shearer haulage speed, ft/min (fpm), n = number of bits per circumferential line, and V_R = drum rotational speed, rpm.

Figure 7.5.7 illustrates the mechanisms of coal cutting by the drum. Under a constant drum rotational speed, when the shearer's haulage speed, V_S, is slow, the bit almost rotates at its original position without actually cutting into the coal face. The "rubbing action" produces considerable fine coal and coal dust. Only when the haulage speed reaches a certain value will the coal be cut loose in lumps.

As the haulage speed increases continuously, the cutting depth of the drum also increases continuously (Fig. 7.5.7 upper). The deeper the cut, the more efficient is the cutting. However, the amount of increase in haulage speed has a limit. When the drum rotational speed is increased from V_{R1} to V_{R2}, its cutting angle decreases from θ_1 to θ_2 (Fig. 7.5.7 lower) while the clearance angle, which is the angle between the coal face and the rear side of the cutting edge of the bit, increases α. Because of this, the bit is overloaded and may be damaged. When the clearance angle decreases, there will be more frictional contact between the coal face and the bit, thereby increasing the bit wear, cutting resistance, and float dust. In order to prevent these phenomena, the cutting angle may be decreased by increasing the clearance angle. As

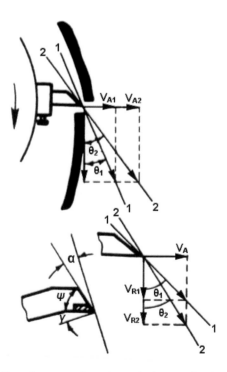

Figure 7.5.7 Bit cutting direction as a function of shearer haulage speed and drum rotational speed

the clearance angle increases, everything being equal, the angle of the cutting tip or lip angle, ψ, will be reduced, which will undoubtedly reduce the bit strength.

From Figure 7.5.6 C, one can further see that when the haulage speed is fixed, reducing the rotational speed of the drum can also increase the cutting angle and cutting depth. Therefore, there should be a proper ratio between the haulage speed and the rotational speed of the drum. If the rotational speed of the drum (or the linear velocity of the cutting bit) far exceeds the haulage speed, it will clearly reduce the cutting depth of the bits, resulting in finer coal being produced. Conversely, if the haulage speed is much larger than the rotational speed, the cutting depth will be too large. For cutting hard coal, the haulage speed should be reduced, while the rotational speed of the drum should be increased. For cutting soft coal, the haulage speed should be increased properly. Today, the computer controlled electric shearer can maintain a proper ratio between the haulage speed and the rotational speed of the drum, depending on the cutting materials encountered.

Figure 7.5.8 A shows the actual cutting depth per drum rotation under various drum rotation speeds and shearer haulage speeds, regardless of drum diameter. Cutting depth per drum

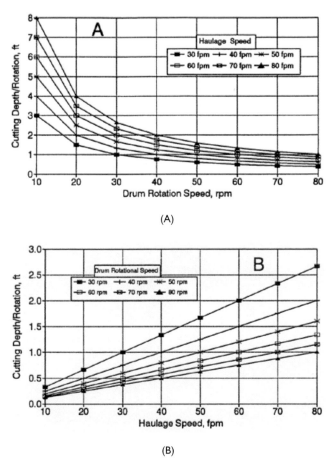

(A)

(B)

Figure 7.5.8 The effect of shearer haulage speed and drum rotational speed on bit cutting depth per drum rotation

rotation decreases exponentially with drum rotation speed under constant shearer haulage speed. The rate of decrease becomes small when the drum rotation speed is greater than 60 rpm. There are practical limits to the number of cuts that a drum can make per rotation. For instance when the haulage speed is 30 fpm (0.15 m s^{-1}), the cutting depth ranges from 1 to 2.6 ft (0.3–0.79 m). Since the effective maximum cutting depth of a bit (including shank) is approximately 4–5 in. (101.6–127 mm), a 1 ft (0.3 m) cutting depth will require two to three bits per line and a 2.6 ft (0.79 m) cutting depth will need six to eight bits per line. The latter is not feasible. The alternative is to increase the drum rotational speed. On the other hand, Figure 7.5.8 B shows that, under constant drum rotation speed, the cutting depth per rotation increases linearly with shearer haulage speed.

7.5.3 Factors for evaluating shearer performance

The performance of shearers is generally evaluated by the following factors: cutting efficiency (or cutting capacity), moment of cutting drum, power consumption, and specific energy consumption. These factors are highly affected by the drum's rotational speed, which reflects the linear velocity of cutting bits, haulage speed, bit cutting depth, and bit wear.

1   Cutting efficiency or cutting capacity is the effective volume of coal per unit time taken down by the shearer, that is:

$$A = \frac{C_v}{t} \tag{7.5.2}$$

because:

$$C_v = sDL_{max} \tag{7.5.3}$$

then:

$$A = sD \frac{L_{max}}{t} = sDV_s \tag{7.5.4}$$

Where
A = cutting efficiency (ft^3 min^{-1} or m^3 min^{-1})
C_v = effective volume of coal taken down in t minutes (ft^3 or m^3)
s = cutting width of the drum (ft or m)
L_{max} = maximum travel distance of the shearer in t minutes (ft or m)
D = drum diameter (ft or m)
V_s = shearer haulage speed (ft min^{-1} or m min^{-1})

From Equation (7.5.4), for a fixed drum width and diameter, the faster the haulage speed, the larger the cutting efficiency. But this is valid only under a fixed rotational speed of the drum. The cutting efficiency for each revolution of the cutting drum is:

$$A_n = \frac{A}{n} = \frac{sDV_s}{N} \tag{7.5.5}$$

Where n is the rotational speed of the drum in rpm. Because:

$$n = \frac{V_R}{\pi D}$$

(7.5.6)

then

$$A_n = s\pi D \frac{V_S}{V_R}$$

(7.5.7)

Where V_R is the linear velocity of the bit (ft min^{-1} or m min^{-1}).

From Equation (7.5.7), the cutting efficiency is affected not only by V_S but also by V_S/V_R. Under a fixed value of V_S, decreasing V_R will increase V_S/V_R and thus increase the cutting efficiency. It has been demonstrated (Brooker, 1979; Chang, 1971; Gregor, 1969; Guillon and Pechalat, 1976) that within a certain limit, decreasing the rotational speed of the drum will greatly increase the cutting efficiency. Within the practical limits of the bit speed, the strength of rock and coal do not vary too much. Conversely, raising the bit speed will increase the bit temperature and increase bit wear, thereby reducing the cutting efficiency. Besides, decreasing V_R implies an increase in cutting depth, which increases the cutting resistance. Experiments (Chiang, 1980) have demonstrated that:

$$P = Fb$$

(7.5.8)

Where
P = cutting resistance of the drum (N)
b = cutting depth (cm)
F = cutting resistance of the coal seam (kN cm^{-1})

The experimental values are shown in Table 7.5.1.

2　The moment of the drum axis reflects directly the magnitude of loading at the drum axis. The moment is the product of the force, P, received by the drum multiplied by the radial distance, r, from the drum axis to the point of action of the force P, or $M = Pr$ (ft-lb or m-kg). The average moment for every point subject to cutting resistance is used to represent the moment of the drum. The larger the moment, the larger the cutting and loading resistances of the drum, which under normal condition, means that more coal is cut loose and loaded.

3　The power consumption of the shearer is the total electrical load. Under normal conditions, it reflects the changes in actual production capacity just as the moment does. A large cutting depth produces more coal and fills more of the space between the spiral vanes of the cutting drum. This will increase the power consumption.

Table 7.5.1 Cutting resistance of coal

Coal	Uniaxial Compressive Strength, psi (MPa)	Cutting Resistance, lbs in.$^{-1}$ (kN cm^{-1})
Soft coal	1422 (9.81)	840 (1.47)
Medium coal	1422–2844 (9.81–19.61)	840–1680 (1.47–2.94)
Hard coal	2844–4266 (19.61–29.41)	1680–2920 (2.94–4.41)

4   Specific energy is the energy required to produce and load a unit weight of coal. Its unit is hp-hr ton^{-1} or kW-hr ton^{-1}. For a fixed drum diameter and cutting width, specific energy is the power required to move one foot of the shearer. Thus, specific energy can accurately reflect the efficiency of the shearer. When the haulage speed is fast and the cutting depth is large, the total power consumption will increase, but the specific energy may decrease; there is an optimal range of specific energy when the haulage speed, cutting depth, and rotational speed of the drum are properly selected. The specific energy for coal ranges from 0.1 to 0.8 kW ton^{-1} but mostly around 0.35–0.5 kW ton^{-1}.

7.5.4 Operational factors controlling shearer performance

The four factors (i.e., cutting efficiency, moment of the drum axis, power consumption, and specific energy) that control the performance of the shearer are highly affected by several major operation-related factors. These include drum rotational speed, haulage speed, cutting depth, seam hardness, drum diameter, and bit conditions (Brooker, 1979; Chang, 1971; Guillon and Pechalat, 1976).

7.5.4.1 Effects of haulage speed and drum rotational speed

Under constant haulage speed, the average cutting and loading moment decreases exponentially with the drum rotational speed. The critical drum speeds above which the moment is more or less independent of the drum speed lie between 80 and 100 rpm, depending on the haulage speed. However, the moment increases with increasing haulage speed, regardless of drum speed.

Similarly, power consumption increases with the haulage speed, and under constant haulage speed, power consumption increases with drum rotational speed.

Under constant haulage speed, the specific energy increases with drum rotational speed. However, the specific energy decreases with the haulage speed. It appears that for a constant drum rotational speed; there is a critical haulage speed above which the specific energy is independent of the haulage speed.

7.5.4.2 Effect of cutting depth

Regardless of the haulage and drum rotational speeds, the average moment increases linearly with the cutting depth. The moment-cutting depth relation is a desirable factor for evaluating the cutting and loading efficiency of the shearer. However, this relationship is affected by the diameter of the drum axis, bit conditions, and seam hardness.

Under constant drum speed, power consumption increases linearly with the cutting depth. But under constant haulage speed, power consumption decreases exponentially with the cutting depth. When the cutting depth exceeds 1.2 to 2.0 in. (30–50 mm), power consumption remains approximately the same. This is due to the fact that under constant haulage speed, the drum speed must be reduced in order to increase the cutting depth. Once the drum speed is reduced, so is the power consumption. When the cutting depth reaches 1.2 to 2.0 in. (30–50 mm), the bit has almost cut all the way into the coal. If the cutting depth is increased further, the bit blocks or spiral vane will make contact with the coal face and may become unstable. In general, the designed cutting depth is approximately 70% of the bit depth. Since the bit depths are generally 2.0 to 3.2 in. (50–81 mm), the cutting depths are 1.4 to 2.2 in. (35–56 mm).

Just like the effect of haulage speed, the specific energy decreases with the cutting depth up to 1.2 to 2.0 in. (3–5 cm) deep, above which it becomes stabilized.

7.5.4.3 Effect of the barrel diameter

Specific energy increases rapidly with coal hardness. The same is true with power consumption and drum moment.

The diameter of the drum axis does not have too much to do with the cutting efficiency, but it is a major factor controlling the loading efficiency. The larger the diameter, the smaller the space between the spiral vanes, and subsequently, the larger the loading moment and resistance. This is especially true for hard coal that breaks in big pieces.

7.5.4.4 Effect of bit conditions

During the production operation, frequent inspection and replacement of worn bits are very important in increasing the cutting efficiency, reducing the power loading on each component, and reducing the specific energy. When the bit is new and sharp, it breaks coal mostly in tension and there is no moment. But when the bit is worn out, it crushes the coal, and there will be an anti-moment against the cut. For a fixed amount of moment, the cutting efficiency of a new bit is higher than the worn bit, because its cutting depth is larger and breaks coal in tension. Rock and coal are much stronger in compression than in shear and in tension. These phenomena are much more evident in hard coals.

In underground operation, the bit condition should be carefully inspected so that it will always break coal in shear and/or in tension. Some bits look still useable, but in effect the carbide tip has been worn out. When the worn surface is used to break the coal, it accelerates bit wear. As the worn area increases, the cutting resistance rapidly increases, thereby greatly reducing the cutting capacity of the bit.

When worn bits are used to cut coal, methane may be ignited by the sparks induced by the friction between the worn bit surface and the coal face. If the seam contains a high volume of methane, this could be extremely dangerous.

7.5.5 Maximum possible cutting speed of the shearer

Modern shearers have a maximum flitting speed of 96 to 147 fpm (0.48–0.75 m s^{-1}). But their maximum cutting speed may be less than that depending on materials being cut and the depth of cut. The other more critical constraining factor is the shield cycle time. If the immediate roof is weak, shields should be advanced following the advance of the shearer as close as possible. Under this principle, the shield cycle time controls the shearer's cutting speed, i.e., if the shield is slow, so is the shearer.

A shield operation cycle consists of three functions in sequence: leg lowering, shield advance, and leg rising or reset. A shield cycle time is the summation of the time allocated for these three functions, which, for example, are these:

Leg lowering	3.5 sec
Shield advance	3.5 sec
Leg rising	3.0 sec
Total	10.0 sec

For this case, only six consecutive shields can be advanced per minute if their cycle time is 10 sec which means 30, 34, and 39 fpm (9, 10.5, and 12 m min⁻¹) of allowable cutting speeds if shields of 5, 5.7, and 6.6 ft (1.52, 1.75, and 2.01 m) widths are used, respectively.

If a faster shearer cutting speed is required, say more than 40 fpm (12.2 m min⁻¹), simultaneous advance of two adjacent or two alternate shields, sometime even three, can be used. In most cases, the two-alternate-shields method is used. In this method, if eight shields, for example, are in a batch, the number one and number three shields will be advanced first, followed by number five and number seven shields, followed by number two and number four shields, and finally by number six and number eight shields.

7.6 Installed power of the shearer

7.6.1 Shearer's power requirement

As discussed in the previous section, the shearer's power consumption, expressed in terms of specific energy in hp-hr ft$^{3\,-1}$ (or hp-hr ton⁻¹) of coal produced, depends on the following:

1   Thickness and hardness of coal including partings.
2   Drum design:

    A   Drum speed.
    B   Drum width.
    C   Drum diameter.
    D   Arrangement and number of spiral vanes and bits.
    E   Bit type and bit wear.

3   AFC chain speed and loading arrangement.

Data in the literature indicate wide ranging values of specific energy required for cutting coal. Ostermann (1966) determined specific energy for German coal from 27.20 to 75.12 hp-hr/ft³ when the cutting web depth was 15–30 in. (381–762 mm), AFC speed at 124 and 177 fpm (0.9m/s), and drum speed at 80 and 102 rpm. Eichbaum and Bendmayr (1974) also determined the specific energy to be 0.01–0.02 hp-hr/ft³ when a double-ended ranging drum shearer with drum diameter 5.3 ft (1.62 m) and drum width 33.5 in. (851 mm) was cutting in a 10 ft (3.05 m) coal seam. COMINEC (1976) estimated a specific energy of 0.016 hp-hr/ft³ for the Lower Kittanning seam and suggested the following formula to determine the required shearer power, P, in horsepower:

$$P = 81.6\, KV_s sH \tag{7.6.1}$$

Where K is the specific energy factor and H is mining height.

From the empirical data gathered over the past 20 years by a US shearer manufacturer, it has been determined that the hp-hr per ton of coal mined in hard cutting conditions is equal to 0.5 or 0.0276 kW-hr per cubic foot of coal. This is referred to as the K factor for power required to cut coal on a longwall face. For softer coal this factor decreases to as much as 0.25 hp-hr per ton or 0.0138 kW-hr ft$^{3-1}$. The value also varies depending upon the haulage speed, but it stays relatively constant when haulage speeds are 10 fpm (0.05 m s^{-1}) or above.

The power required for cutter motors is then determined by:

$$P = QK \tag{7.6.2}$$

Where Q is the mining rate in tons per hour and K is the specific energy consumption factor in hp-hr ton^{-1}.

7.6.2 Example of determination of shearer's power capacity

Design a shearer to comply with the following specifications:

1 A double-ended ranging drum shearer.
2 Machine weight: W = 90,000 lbs.
3 Dual haulage – electric drive.
4 Maximum cutting height: 10 ft.
5 Web thickness: 30 inches.
6 Maximum cutting speed required: 40 fpm on 0° grade.
7 Maximum cutting speed required: 25 fpm on 15° grade.
8 Maximum flitting speed required: 60 fpm on 0° grade.
9 Maximum flitting speed required: 40 fpm on 15° grade.
10 Maximum drum diameter: 72 inches.

Determine the following parameters:

1 Maximum mining rate in tons per minute and tons per hour.
2 Power required for hydraulic pump motor.
3 Power required for two haulage motors.
4 Power required for two cutting motors.
5 Gear ratio for haulage drives.
6 Gear ratios for cutter transmissions.
7 Bit speeds for various drum diameters.

7.6.2.1 Determination of the mining rates, TPM (tons per minutes)

Given: Cutting height: $H = 10.0$ ft
Web depth: $D = 2.5$ ft
Max. cutting speed: $V_s = 40.0$ fpm
Density of coal in solid: $\gamma = 80.0$ lbs • ft^{-3}

$$TPM = \frac{HDV_s\gamma}{2000} = \frac{10 \times 2.5 \times 40 \times 80}{2000} = 40 \text{ tons min}^{-1} \tag{7.6.3}$$

$$TPH = 60TPM = 60 \times 40 = 2,400 \text{ tons hr}^{-1}$$

7.6.2.2 Determination of the power required for hydraulic pump motor

From the fluid mechanics calculations of cylinder sizes and hydraulic motor sizes for the ranging arms and cowl rotation motors, it is determined that a pump rated at 20 gpm at 2000 psi will be adequate. The equation used to calculate the required pump power is:

$$P = \frac{QP}{1714\,E} \tag{7.6.4}$$

where $Q = 20$ gpm, $P = 2000$ psi, and $E = 85\%$ in this case.

$$P = \frac{20 \times 2000}{1714 \times 0.85} = 27.5\,hp$$

Since the hydraulic power required for a shearer is only required intermittently, a 25 hp AC, 950 volt motor will be sufficient.

7.6.2.3 Determination of the power required for the haulage motors

A. MACHINE CUTTING AT 40 FPM

Given:

$W = 90,000$ lbs.
$\mu = 0.25$ (coefficient of friction between the shearer and track).
$F_c = 30,000$ lbs (required cutting force from empirical data).
$F_f = \mu W = 0.25 \times 90,000 = 22,500$ lbs (friction resistance to shearer movement).
$V_s = 40$ fpm.
$F_t = F_f + F_c = 22,500 + 30,000 = 52,500$ lbs.

The equation used to calculate the required haulage motor power is

$$P = \frac{F_t V_s}{33,000\,e} \quad (\text{assume } e = 90\%)$$

$$P = \frac{52,500 \times 40}{33,000 \times 0.90} = 70.7 \text{ hp (total)} \tag{7.6.5}$$

Therefore, each haulage motor should be 35 hp.

B. MACHINE FLITTING AT 60 FPM

In this case $F_t = 22,500$, $V_s = 60$, and $e = 90\%$:

$$P = \frac{22,500 \times 60}{33,000 \times 0.90} = 45\,hp \text{ (total)}$$

Two 35 hp haulage motors will also satisfy this condition.

C. MACHINE CUTTING UP 15° GRADE AT 25 FPM

Given : $V_s = 25\ fpm$ $\qquad\qquad$ $\sin(15°) = 0.259$

$\quad\quad F_c = 30,000\ lbs$ $\qquad\qquad$ $\cos(15°) = 0.966$

$\quad\quad M = 0.25$ $\qquad\qquad$ $W = 90,000\ lbs$

$\quad\quad N = W\ \cos(15°) = 86,940\ lbs$

$\quad\quad F_g = W\ \sin(15°) = 23,310\ lbs$

$\quad\quad F_f = \mu N = 0.25 \times 86,940 = 21,735\ lbs$

$\quad\quad F_t = F_g + F_f + F_c = 23,310 + 21,735 + 30,000 = 75,045\ lbs$

Then

$$P = \frac{F_t V_s}{33,000\ E} = \frac{75,045 \times 25}{33,000 \times 0.90} = 63\ \text{hp (total)}$$

Again, two 35 HP haulage motors will satisfy this condition.

D. MACHINE FLITTING UP 15° GRADE AT 40 FPM

In this case:

$$F_t = F_g + F_f = 23,310 + 21,735 = 45,045\ lbs$$

$$P = \frac{F_t V_s}{33,000\ E} = \frac{45,045 \times 40}{33,000 \times 0.90} = 60.6\ hp\ (total)$$

Similarly, two 35 hp haulage motors will satisfy this condition.

7.6.2.4 Determination of the power required for cutter motors

If the maximum drum size required for this machine is 72 in. with a 30 in. web, then the maximum mining rate for this size drum will be

$$\text{TPH} = \frac{HV_s D\gamma}{2000} \times 60 = \frac{6 \times 2.5 \times 40 \times 80}{2000} \times 60 = 1440\ \text{tons/hr}$$

Using Equation (7.6.2), the power required for the cutting motor powering a ranging arm with a 6 ft diameter drum is:

$$P = 1440 \times 0.50 = 720\ \text{hp}$$

Theoretically, in a 10 ft seam, the other drum would only be required to cut 960 tons per hour (i.e., 2400-1440 tph = 960 tph) and the power required would only be 480 hp. This is not true in actual operation. But if the drum is not cutting either top rock or bottom rock, then the required power is reduced. In actual practice, 600 hp motors on each ranging arm will prove to be sufficient for nearly all applications. The coal must be very hard to obtain a K factor of 0.50. In most applications in the United States, the K factor is probably in the range of 0.25 to 0.35, and less horsepower is then required.

Calculations should be made for both uni-directional and bi-direction cuttings to determine the RMS horsepowers of the cutter motors.

7.6.2.5 Calculating haulage drive gear ratio

The gear ratio of the haulage drive should be such that at the desired maximum haulage speed when cutting coal, the haulage motors should be operating at their rated speed and rated horsepower. If the output sprocket diameter is 14 in. (D) and the motors are rated at 35 hp (P) at 1750 rpm (N), the gear ratio should be:

$$\mathrm{GR} = \frac{0.262\ ND}{V_s} = \frac{0.262 \times 1750 \times 14}{40} = 160:1 \tag{7.6.6}$$

The motor speed at maximum haulage speed should be checked. DC motors rated at 1750 rpm should not be run in excess of 3000 rpm – there is a danger that the motor could fly apart at speeds exceeding 3000 rpm. Motor speed at 60 fpm haulage speed is:

$$N = \frac{GR \times V_s}{0.262\ D} = \frac{160 \times 60}{0.262 \times 14} = 2617\ \mathrm{rpm} \tag{7.6.7}$$

This speed is within the limits of the motor design. The DC haulage motors are controlled by varying the voltage from the SCR drive, thus giving infinite variable haulage speeds from 0 to 60 fpm. A feedback control that monitors the amperage of the cutting motors controls the speed of the haulage motors and prevents overloading of both the cutting and haulage motors.

7.6.2.6 Gear ratios for cutting motors

Selected drum speeds for the ranging arms should be approximately 35, 37, or 54 rpm. The rated speed of cutting motors is approximately 1780 rpm. These motors are generally of a low-slip design for high efficiency.

$$\mathrm{Gear\ Ratio} = \frac{M}{N} \left(\mathrm{where\ M} = \mathrm{motor\ speed\ and\ N} = \mathrm{drum\ speed}\right),$$

So for various drum speeds the gear ratio of the ranging arm should be:

Motor Speed (rpm)	Drum Speed (rpm)	Gear Ratio
1780	37	48.0:1
1780	37	39.5:1
1780	54	33.0:1

7.6.2.7 Bit speeds for 72 in. drum

The equation for calculating the bit speeds for a 72 in. drum is:

$$S = 0.262ND \tag{7.6.8}$$

Drum Speed (rpm)	Bit Speed (rpm)
37	698
45	849
54	1019

7.7 Haulage of the Shearer

Prior to the 1980s, chain haulage was used, but it has been totally replaced by the chainless haulage since then. Today, "haulage" refers to "chainless" haulage, and as such it will be used throughout this book.

7.7.1 Types of haulage

There are many haulage systems offered by various manufacturers under various brand names over the past four decades. However, they can be grouped into two major systems: rack and captivated chain.

When chain haulage was used in the 1970s, the major problems were frequent large vibrations of chain (i.e., too flexible) and chain breakage, endangering miners working at the faces. But if the haulage chain is contained in a guide along the face, it becomes a stiff track and yet has sufficient flexibility to accommodate the deflection of pan snaking. This is the captivated chain system and was the most commonly used system when chainless haulage was first introduced. It offers flexibility and yet is much more stable than chain haulage. As the shearer power increased the system did not have sufficient strength to resist heavy cutting. The more robust rack system was then introduced. As it can better accommodate continuing increases of shearer power, the rack system has gained wide acceptance in recent years. In fact, today it has almost totally replaced the captivated chain system. The rack system is essentially like a railway track, except the crossbars (or ties in the railway track) are convoluted shaped to better fit the teeth of the haulage sprocket. The basic features of these systems are one or two driving sprockets off the shearer's haulage unit engaging on a track, either directly or through idler sprockets, mounted on the spillplate on the gobside of the panline. The racks are half the pan width long, such that one rack is mounted at the center of each pan with a second one straddling every pan connection. There is a horizontal slot in the rack holder that allows the rack to move relative to the pan to minimize pitch variation during pan articulation (See Fig. 7.7.2). A good rack system must have robust components capable of working at high haulage pull. Its track should be sufficiently rigid yet drive smoothly without restricting AFC flexibility. It will self-clean and can be easily adapted to various types of shearers. Above all, its initial and maintenance costs should be low.

According to 2019 US Longwall Census, the distribution of the haulage systems are these:

1   Ultratrac U2000 and its variations 92.5%
    U2000 (27.5%), Super gear rack (12.5%), Jumbotrac 2000 (35%), and 2010 haulage (17.5%).
2   Jumbotrac and its variations 7.5%
    Jumbotrac (5%) and Jumbo gear rack (2.5%).

Special features of the major systems are shown in Table 7.7.1.

7.7.1.1 Ultratrac

The Ultratrac is a one-piece rack that is made of high-strength, forged, hardened steel in involuted tooth profile. They are made in 5, 5.75, and 6.6 ft (1.5, 1.75, and 2 m) sections to fit the corresponding pan widths. The trapping shoe is also articulated with a large wear area and hardened for long life. The sprocket is direct drive, capable of high haulage pull (Fig. 7.7.1).

Table 7.7.1 Major features of various haulage systems

Rack System			
System	Ultratrac 2000	Jumbotrac	SuperGear Track
Pitch (mm)	147/151	147	147/151
Pull force (ton)	112.5	112.5	N/A
Tooth profile	Involute	Involute	Involute
Rack width (mm)	85	90	92
Construction	Forged	Welded	Welded

Figure 7.7.1 Ultratrac 2000

Source: Courtesy of Komatsu Mining

7.7.1.2 Jumbotrac

This is fabricated with side plates and forged involuted tooth profile. They are also made in 5, 5.75, and 6.6 ft (1.5, 1.75, and 2 m) sections for corresponding pan widths (Fig. 7.7.2).

7.7.2 Selection and control of the haulage speed

In Section 7.5.5, the maximum cutting speed of the shearer was shown to be subjected to the constraints of shield cycle time. However if the roof allows, two adjacent or alternate shields can be advanced simultaneously. If this is the case, shield cycle time would not be the limiting factor for the shearer's cutting speed, rather the limiting factor could be the carrying capacity of the AFC.

The maximum flitting haulage speeds for the double-ended, ranging-drum shearer are 95–150 ft min^{-1} (0.48–0.76 m s^{-1}). The tramming speeds for full web cutting are slower, seldom exceeding 55 ft min^{-1} (0.28 m s^{-1}). The higher haulage speeds are for partial web cutting, coal loading, cleaning, and tramming only. There are two rules for selecting the haulage speed for coal cutting: first, the cutting capacity of the shearer and the carrying capacity of the face conveyor must be compatible so that the conveyor will not be overloaded and spill over the line pans; second, the shearer should not be overloaded. The most important factor is the cutting depth of the bit. If it is too deep, the bit blocks and spiral vanes will be pushed into the coal, thereby wasting energy and producing more fine dust.

The hourly production of the shearer can be determined by:

$$A_s = 60SH\gamma V_s \qquad\qquad (7.7.1)$$

Where H = the mining height, s = the cutting width (web) of the drum, V_s = the haulage speed of the shearer, and γ = the weight per unit volume of coal, generally 74.8–87.3 lb ft$^{3-1}$ (1.2–1.4 metric tons m$^{3-1}$).

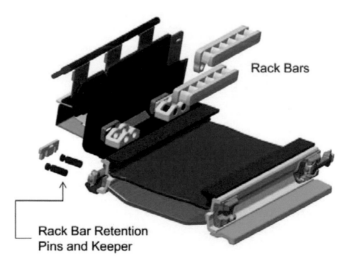

Rack Bars

Rack Bar Retention
Pins and Keeper

Figure 7.7.2 Jumbotrac
Source: Courtesy of Caterpillar

The carrying capacity per hour of the face conveyor, Q, should be compatible to that of the shearer, that is, $Q = A_s$. For example, when $H = 6.56$ ft (2 m), $s = 30$ in. (0.76 m), and $\gamma = 84.2$ lb ft$^{3-1}$ (1.35 tons m$^{3-1}$), if $Q = 2000$ tons hr^{-1} (1814 metric tons hr^{-1}), the maximum haulage speed for the shearer during cutting is:

$$V_s = \frac{Q}{60Hs\gamma} = 47.6 \text{ ft/min (14.5 m/min)}$$

On the other hand, when the rotational speed of the drum, n, remains constant, and if the allowable cutting depth is b, the haulage speed for coal cutting can be obtained as follows:

$$V_s = \frac{Nnb}{500} \ (m/\min) \tag{7.7.2}$$

Where N is the number of bits in each bit-line circumferential cross-section. If $N = 3$, $n = 50$ rpm, and $b = 50$ mm, then

$$V_s = \frac{3 \times 50 \times 50}{500} = 15 \, m/\min$$

In practice, the smaller of the two velocities determined should be adopted, that is, 14.5 m min^{-1}. If the shearer is equipped with a self-adjusting device for the haulage speed, the selected haulage velocity should be set in advance.

7.8 Cutting methods of the shearer

Since the shearer rides on the AFC and uses it as a track for traveling back and forth, and the AFC is laid on the mine floor parallel to the coal face, initiation of coal cutting by the shearer requires the AFC to snake toward the face. By following the snaking toward the face the drum can cut gradually into the coal face until it reaches a full drum width or web. This process is called sumping. The distance of sumping where the pans are snaked runs around 5–10 shield widths. Sumping can be performed at various parts of the face and creates different cutting methods.

Essentially, there are two basic cutting methods, uni-directional (uni-di) and bi-directional (bi-di). The definition of uni-di and bi-di involves both cutting direction and cutting depth. In bi-di, the shearer cuts in both directions, i.e., head-to-tail and tail-to-head and in each direction, a full web-cutting depth is completed. In uni-di, however, it is cutting in one direction in one full web depth in the forward trip, while in the return trip it travels empty, or in one direction, the leading drum cuts the top coal, and in the return trip, it cuts the bottom coal such that in one round trip only one web cut is completed. Thus, bi-di by definition is fairly clear cut. But for uni-di, there are many variations, depending on the sumping location and the manner the leading and trailing drums are performing their cutting/cleaning functions.

For US longwalls, uni-di was most popular accounting for more than 75% of use in the 1970s and early 1980s. But in the late 1980s and early 1990s, as panel width became larger, automation grew, and effective dust control methods were developed, bi-di became much more productive. Consequently, it has become and still is the dominant cutting method. In fact, today, all but a few longwalls employ bi-di.

Normally, when a double-ended ranging drum shearer cuts coal, the leading drum is up to cut the top portion of the coal seam with height equivalent to its drum diameter, while the trailing drum cuts the remaining bottom coal or whatever is left in the bottom portion of the coal seam.

As mentioned in Section 5.4.1 (p. 143), the current two shield control systems, PMC^R-C and RS20s, have a full library of various cutting methods that can be used at the shearer operator's command. It would be impractical to describe each and every one of them. Only a few of the more popular ones are discussed here.

7.8.1 Bi-directional cutting method

In bi-directional cutting, the shearer performs two sumpings and two wedge cuts in a round trip. A complete mining cycle is accomplished both during the head-to-tail and tail-to-head cutting trips. A complete mining cycle includes the extraction of one web depth of whole seam height, followed by the advance of both the conveyor and the shield supports. Under normal conditions, the leading drum cuts the upper 70 to 90% of the seam and the rear drum cuts the remaining bottom and cleans up the floor coal.

Figure 7.8.1 shows the sequence of cutting, especially the face-end operation. In Figure. 7.8.1 A, the numerical numbers indicate the sequence and direction of shearer movement, while Figure 7.8.1 B–D shows shearer's path during face-end operation. In (1), the shearer has just arrived at the tailgate with the tail side drum up and the head side drum down. After having cut out at the tailgate, the shearer is then prepared to begin the face-end operation by raising the head side drum up and lowering the tail side drum down. The shearer follows the conveyor snake toward the head side as shown in (2). At this snake portion, the shearer gradually cuts into the coal face for the next web. After having passed the snake, the shearer has fully sumped a web depth into the coal face for the next web. After this, the conveyor is advanced at the tail end to remove the snake as shown in (3). At this time, the conveyor is positioned straight across the whole face. The shearer then reverses back toward

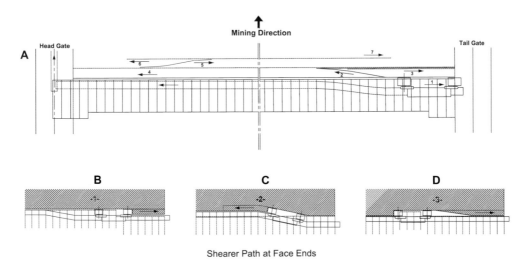

Shearer Path at Face Ends

Figure 7.8.1 Bi-di cutting sequence

the tailgate with the tail drum up and head drum down and cuts off the wedge-shaped coal block. After having cut out at the tailgate, the head drum is raised and the tail drum is lowered. It then begins the next cut, cutting all the way toward the headgate. After the shearer has reached the headgate, it repeats the same face-end operations.

The bi-di method is most suitable for large panel widths, say 700 ft (213 m) or larger. Because face-end operation needs a distance of 74 to 98 ft (22.5–30.0 m) at both ends to perform the task, and within this distance the shearer cuts very slow and takes time, normally 5 to 15 min. But in the remaining portion of the panel width, the shearer can cut at full speed. So the wider the panel, the larger the portion that the shearer can cut at full speed.

In bi-di cutting, the shield supports are advanced immediately following the shearer's cutting. Thus, it is better suitable for weak roof. Both drums are used evenly, but two shearer reverse operations are required at each face end and in each web cut (Peng, 2006, Appendix B).

7.8.2 Uni-directional cutting method

There are many variations in the uni-directional cutting method. A few examples are illustrated in this section.

Figure 7.8.2 shows the complete sequence of the half-face method of shearer operation. For convenience of description, the face width is divided into two equal halves, A and B, with x being the center point. Note that x may be at any other preferred location on the face.

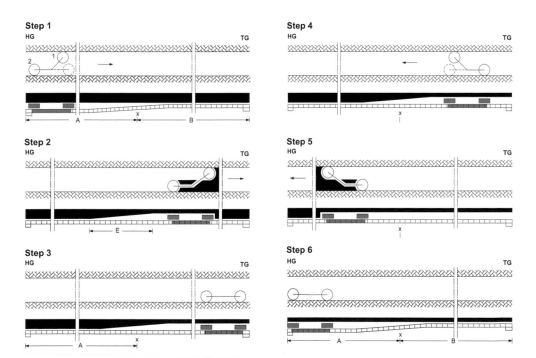

Figure 7.8.2 Half-face cutting sequence
Source: COMINEC (1976)

In step 1, the shearer is ready to travel to and cut the B half where the face conveyor has been fully advanced. The leading (No. 1) drum is raised, and the shearer travels at maximum speed to the midface x without cutting coal. In step 2, at the midface within a distance E, the shearer follows the snaking of the face conveyor, gradually cutting into the coal until it cuts the full web at the end of distance E. The shearer then continues to cut a full web of coal at a slower speed toward the face end of the B half. In step 3, after reaching the face end, the leading (No. 1) drum is lowered and cuts itself free at the corner of the face. Simultaneously, the face conveyor in the A half is advanced, making the face conveyor a perfect straight line across the face. In step 4, the No. 1 drum is then raised, and the shearer is reversed to travel toward the A half. It will begin the trip by cutting the bottom coal between the two drums utilizing the No. 1 drum. In step 5, the shearer travels empty at top speed until it reaches the midface x where it will again sump into a full cut throughout the A half of the face. And, in step 6, when the shearer reaches the end of the face in the A half, the No. 1 drum is raised and the shearer is reversed. At the same time, the face conveyor in the B half is advanced. The shearer is now ready to repeat the cutting cycle again. (Note that the bottom coal between the two drums will have to be cut out by the No. 2 drum at the beginning of the travel toward the B half.) Shield supports can be advanced following the shearer in the B half in (2) and then advance the shields in the A half following the shearer's cutting in (5). Alternatively, all shields are advanced in sequence following the shearer from the face corner of the B half toward the face corner of the A half. The former method offers immediate roof support, while the latter offers delay support in the B half and is only applicable to better roof conditions. However, in the former method, it is more difficult to maintain a straight panline. This is, strictly speaking, a uni-directional cutting method and requires two sumpings in one full web cut and, therefore, takes more time.

Figure 7.8.3 shows another example of the half-face method: in step 1, the shearer is ready to cut from tail to head with the AFC snaking over from about the 30th shield from the tailgate: in step 2, the shearer follows the snaking and sumps full web into the coal and continues to cut toward the headgate; in step 3, the shearer reaches the headgate. The tailside drum is raised up while the headside drum is lowered. The shearer is now ready to clean out the bottom coal left between the two drums. At this time, the AFC is straight across the whole face; in step 4, the shearer first cuts out the bottom coal at the headgate and then runs empty at high speed toward the tailgate until it reaches and cuts the wedge coal at the tailgate; and, in step 5, when the shearer reaches the tailgate, the position of the cutting drums is again reversed to cut the bottom coal, and the cycle repeats.

Figure 7.8.4 is another example of the modified half-face cutting method. In step 1, the sequence begins when the shearer, with the leading (tail) drum up, and the trailing drum (head side) is cutting and cleaning the bottom coal from the headgate toward the tailgate. Shields are advanced following the shearer. In step 2, when the shearer reaches the coal snake it gradually cuts into full face until it reaches the tailgate in step 3. Shields in this area are not advanced at this time. In step 4, the shearer is reversed. With the head side drum up and tailside drum down, it cuts toward the snake in step 5. The bottom coal at the face end between the drums will be cut and cleaned by the tailside drum. Shields in this area are advanced following the shearer's return to the snake. After the snake, both drums are raised and only the top coal (equivalent to the drum's diameter) is cut in step 6. Again strictly speaking, this is a uni-directional cutting because in one round trip it cuts a full web width of coal. Shield advance is delayed until the return trip. But this method only requires one sumping, and a straight face is easier to maintain. The snake could be placed in the headside.

Step 1

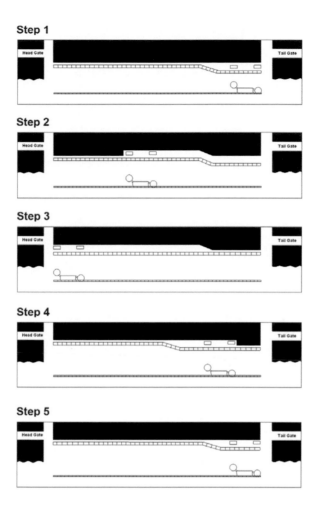

Step 2

Step 3

Step 4

Step 5

Figure 7.8.3 Uni-di cutting sequence

As a matter of fact, this arrangement reduces the shearer operator's exposure to dust. When the snake is on the tailside, as shown, it is used in conjunction with the conventional shearer where the clearance in the pan underneath the shearer body is small and more coal can only be cut on the headside drum. Another shortcoming of this method is that only the headside drum is used to cut nearly all the coal, leaving the other drum idle. To compensate for this problem, the shearer can make a full face cut as it travels from tailgate to headgate and clean bottom coal or run empty from headgate to tailgate until it reaches the coal snake area.

Rutherford (2001, 2005) listed four major cutting methods in Australia and discussed their benefits and disadvantages; bi-di, uni-di, half-web Kaiser cut, and half-web uni-di variant. Figures 7.8.5, 7.8.6, and 7.8.7 show the half-web, uni-di forward snaking, and uni-di backward snaking cutting methods, respectively. Contrary to US practice, the uni-directional half-face method is much more popular than the bi-di cutting in Australia. As shearer power and

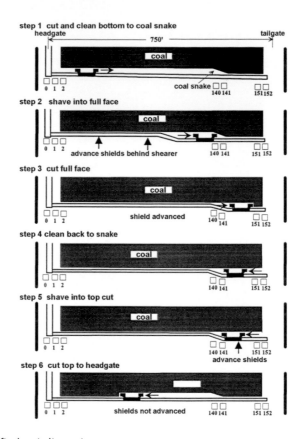

step 1 cut and clean bottom to coal snake

step 2 shave into full face

step 3 cut full face

step 4 clean back to snake

step 5 shave into top cut

step 6 cut top to headgate

Figure 7.8.4 Modified uni-di cutting sequence

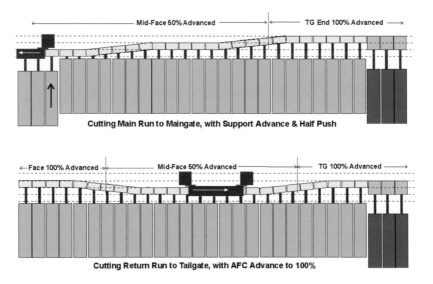

Cutting Main Run to Maingate, with Support Advance & Half Push

Cutting Return Run to Tailgate, with AFC Advance to 100%

Figure 7.8.5 Half-web cutting sequence

Source: Rutherford (2001), tomcat@acenet.com.au

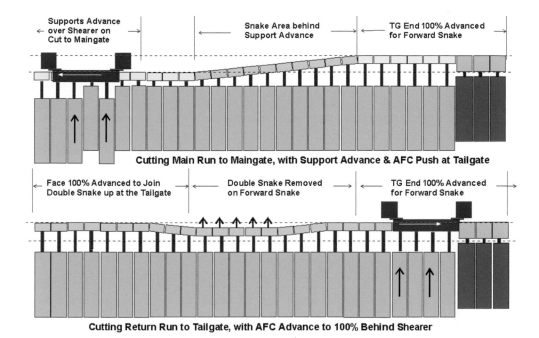

Supports Advance ← over Shearer on Cut to Maingate →

← **Snake Area behind** Support Advance →

← **TG End 100% Advanced** for Forward Snake →

Cutting Main Run to Maingate, with Support Advance & AFC Push at Tailgate

← **Face 100% Advanced to Join** Double Snake up at the Tailgate →

← **Double Snake Removed** on Forward Snake →

← **TG End 100% Advanced** for Forward Snake →

Cutting Return Run to Tailgate, with AFC Advance to 100% Behind Shearer

Figure 7.8.6 Uni-di (forward snake) cutting sequence

Source: Rutherford (2001), tomcat@acenet.com.au

← **Supports Advanced on** Cut to Maingate →

Cutting Main Run to Maingate, with Support Advance

← **Face 100% Advanced** on Backward snake →

← **Snake Area Behind Shearer** →

← **TG Backward Snaked requiring** Support Advance into TG →

Cutting Return Run to Tailgate, with Support Advance over Shearer going into Tailgate and AFC Advance to 100% behind Shearer

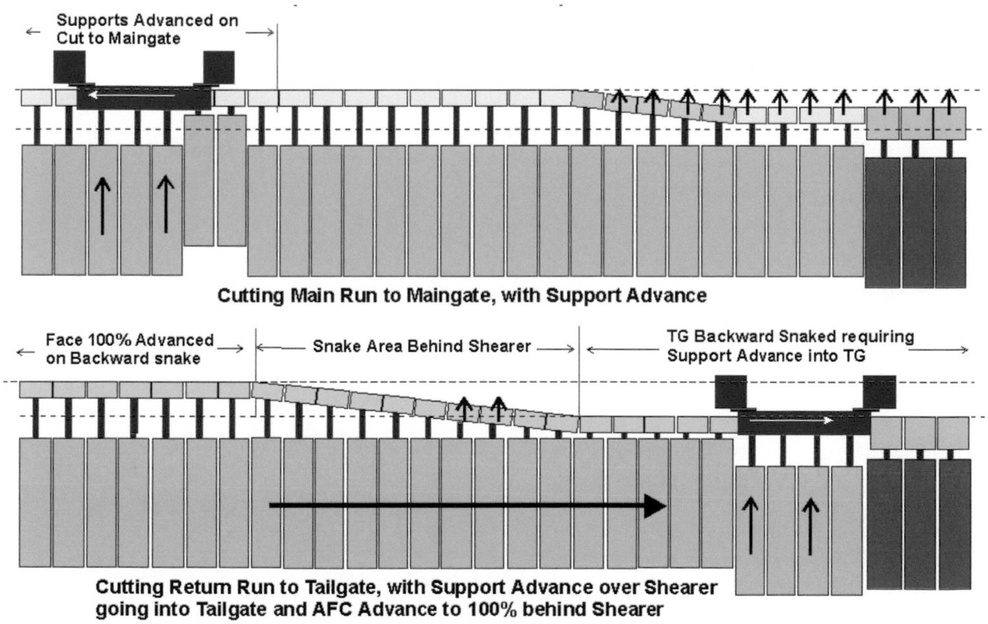

Figure 7.8.7 Uni-di (backward snake) cutting sequence

Source: Rutherford (2001), tomcat@acenet.com.au

haulage speed increase, thicker seams are mined, and environmental concerns increase, the uni-di cutting has become more competitive. The half web system is a uni-di system of cutting in mid-face with bi-di face end operations. This system allows faster shearer cutting sequences because the complicated shuffle associated with bi-di is not required and loading on the shearer is reduced. The half web is achieved by pushing the AFC 50% after the supports have advanced to provide a half-web cut for the return shearer run. This provides a splitting effect on the coal to reduce lumps and can equalize the coal flow in each cutting direction.

7.8.3 Production calculation using bi-di cutting method

An example is used here to illustrate the step-by-step procedure for calculating the production rate when the bi-di cutting method is employed.

7.8.3.1 Mining and geological conditions

Seam	thickness: 78 in.
Seam	depth: 700–900 ft
Average	density of coal:
in-situ (solid)	84 lbs ft$^{3-1}$ (1.345 ton m$^{3-1}$)
broken	65 lbs ft$^{3-1}$ (1 ton m$^{3-1}$) (swelling ratio = 1.3)
Panel	width: 1000 ft rib-to-rib
Panel	length: 15,000 ft

The panel is developed by a three-entry gateroad system.

7.8.3.2 Shearer parameters

Manufacturer/model:	Joy 7LS1A
Drum	diameter (bit tip-to-bit tip): 62 in.
Drum	web: 42 in.
Drum-to-drum	distance: 20 ft
Drum	rotational speed: 56 rpm
Shearer	weight: 56.6 tons
Maximum	haulage speed: 52.5 fpm

7.8.3.3 AFC specifications

Pan	width: 37.4 in. (inside) (1000 mm)
Twin	inboard chain: 42 mm × 152 mm
Chain	speed: 311 fpm (94.8m min^{-1})

7.8.3.4 Assumptions

- Production days are 360 days per year: three 9-hour production shifts/day or 20 shifts per seven-day week (note that Sunday has two production shifts).
- For every shift, the trip time to and from the longwall face takes one hour which allows eight hours of production time.

- The utilization factor for the shearer is 75%.
- Coal preparation plant recovery is 80%.
- Coal is cut into sufficiently small sizes so that it fills up the full space of the pans and that formed by the furniture.

7.8.3.5 Bi-di cutting cycle of operations

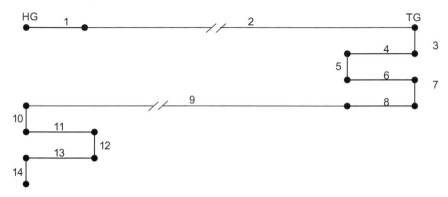

Figure 7.8.8 Bi-di cutting cycle showing task sequencing

Task	Duration (min)
(1) AFC snaking length of approximately 80 ft (24.4 m) is required. Within this distance, the shearer runs from HG to the 80 ft (24.4 m) mark, first cutting the bottom coal between the two drums, and then runs empty from HG to the 80 ft (24.4 m) mark: Cut length of bottom coal (16 in. (406.4 mm) high) = 20 ft (6.1 m) Flitting speed = 52.00 fpm (0.26 m/s)	1.54
(2) Full speed coal cutting from HG to TG: Full seam height coal cutting length: 1000 ft − 80 ft = 920 ft (28.4 m) Cutting speed = 45.00 fpm (0.23 m/s)	20.44
(3) Re-position of cowls and ranging arms to reverse cutting direction from TG to HG.	1.00
(4) Cutting first the bottom coal between the two drums left in the HG to TG run of task 2 (first wedge cut) and then full seam height coal cutting to the 80 ft mark: Cutting length, 1000–920 = 80.00 ft (23.4 m) Cutting speed, 45 fpm × 75% = 34.00 fpm (0.17 m/s)	2.35
(5) Re-position of cowls and ranging arms to reverse cutting direction from HG to TG.	1.00
(6) Cutting first the bottom coal between the two drums left in the first wedge cut in task 4 (second wedge cut) and then full seam height coal cutting to the TG: Cutting length = 80.00 Cutting speed, 45 fpm × 75% = 34 fpm (0.17 m/s)	2.35
(7) Re-position of cowls and ranging arms to reverse cutting direction from TG to HG.	1.00

(Continued)

(Continued)

Task	Duration (min)
(8) AFC snaking length is approximately 80 ft (24.4 m). Within this distance, the shearer cuts the bottom coal between the two drums and then runs empty from TG to the 920 ft (280.4 m) mark: Cut length of bottom coal (16 in. high) = 20 ft (6.1 m) Flitting speed = 52.00 fpm (0.26 m/s)	1.54
(9) Full speed coal cutting from TG to HG: Full seam height coal cutting length 1000 ft − 80 ft = 920 ft (280.4 m) Cutting speed = 45 fpm (0.23 m/s)	20.44
(10) Re-position of cowls and ranging arms to reverse cutting direction from HG to TG.	1.00
(11) Cutting first the bottom coal between the two drums left in the TG to HG run of task 9 (first wedge cut) and then full seam height coal cutting to the 80 ft mark: Cutting length = 80.00 ft (24.4 m) Cutting speed, 45 fpm × 75% = 34.00 fpm (0.17 m/s)	2.35
(12) Re-position of cowls and ranging arms to reverse cutting direction	1.00
(13) Cutting first the bottom coal between the two drums left in the first wedge cut in task 11 (second wedge cut) and then full seam height coal cutting to the HG: Cutting length = 80.00 Cutting speed, 45 fpm × 75% = 34 fpm (0.17 m/s)	2.35
(14) Re-position of cowls and ranging arms for reverse cutting direction and preparing for next cutting cycle	1.00
Total time for one cycle (minutes)	**59.36**

Figure 7.8.9 shows the production rate per cycle.

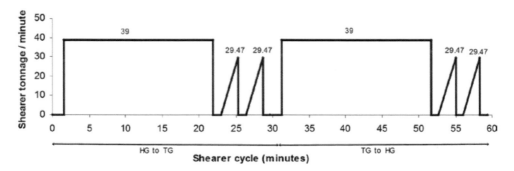

Figure 7.8.9 Production cycle time distribution using bi-di cutting

Shearer tonnage/minute during operation Tasks 2 and 9

$$= 6.5 \text{ ft} \times 3.5 \text{ ft} \times 45 \text{ fpm} \times 84 \text{ lb ft}^{3-1} = 39 \text{ tons}$$

Shearer tonnage/minute during sumping operations (Tasks 4, 6, 11, and 13)

$$= 6.5 \text{ ft} \times 3.5 \text{ ft} \times 34 \text{ fpm} \times 84 \text{ lb ft}^{3-1} = 29.47 \text{ tons}$$

Note that, in reality, during the sumping operation, the shearer starts cutting no coal and ends with cutting 29.47 tons per minute, so a linear variation in cutting capacity of the shearer is assumed from start of sumping till the end of sumping.

7.8.3.6 Production calculations

Total coal production per cutting cycle
= seam height × web width × panel width × 2
= 78 in. × 42 in. × 1,000 ft × 2 = 45,500 ft³ (1288.4 m³)
= 45,500 ft³ × 84 lbs • ft⁻³ (*in situ* density of coal)
= 1,911 tons (short)

Average output per hour	= 1,911 × 60/59.36 × 0.75	= 1,448 tons
Output per shift	= 1,448 × 8	= 11,584 tons
Output per day	= 11,584 × 3	= 34,752 tons
Output per week	= 11,584 × 20	= 231,680 tons
Output per year (raw coal)	= 34,752 × 360	= 12,510,720 tons
Output per year (clean coal)	= 12,510,720 × 0.8	= 10,008,576 tons

7.8.4 Calculation of shearer cutting capacity and AFC carrying capacity

Notations:

H Thickness of seam, ft
D Drum diameter, in.
L Face width, ft
V_s Shearer haulage speed, fpm
V_c AFC chain speed, fpm
N Drum rotation speed, rpm
W AFC width, m
C Clearance under shearer, ft
s Drum cutting width (web), in.
γ Density of coal, tons/ft³
S_f Swell factor (1.3)
a Total vane cross-sectional area, ft²
b Bit cutting depth, in.
A Effective area of load carried by AFC, ft²
n Number of bits in axial cross-section

7.8.4.1 Shearer cutting capacity

Total shearer cutting capacity is:

$$60HsV_s \quad \text{ft}^3 \text{ hr}^{-1} \left(\text{m}^3 \text{ hr}^{-1} \right)$$

Out of this total hourly capacity of the shearer, coal cut by the leading (upper) drum is:

$$60DsV_s \quad \text{ft}^3 \text{ hr}^{-1} \left(\text{m}^3 \text{ hr}^{-1}\right).$$

Coal cut by the rear (lower) drum of the shearer is:

$$60(H-D)sV_s \text{ ft}^3 \text{ hr}^{-1} \left(\text{m}^3 \text{ hr}^{-1}\right).$$

When the shearer moves from tailgate to headgate, coal cut by the rear (lower) drum of the shearer has to pass though the space below the shearer. When the shearer travels from headgate to tailgate, coal cut by the front (upper) drum of the shearer has to pass through the space below the shearer. This clearance space below the shearer puts some restriction on the carrying capacity of the face conveyor.

Assuming the shearer moves at a constant speed, V_s, the capacity of the shearer drum cutting a full web, or the volume of coal that could pass through the shearer drum vane spacing in one hour, is:

$$3.14DNa60 \qquad \text{ft}^3 \text{ hr}^{-1} \left(\text{m}^3 \text{ hr}^{-1}\right).$$

Assuming the shearer drum rotates at a constant speed, N (rpm), the haulage speed of the shearer is:

$$V_s = Nnb/1000 \qquad \text{ft min}^{-1} \left(\text{m min}^{-1}\right).$$

7.8.4.2 Face conveyor carrying capacity

The full carrying capacity of the face conveyor is:

$$60 \, AV_c \qquad \text{ft}^3 \text{ hr}^{-1} \left(\text{m}^3 \text{ hr}^{-1}\right),$$

where $A = A_1 + A_2 + A_3 + A_4$ (Fig. 7.8.10)

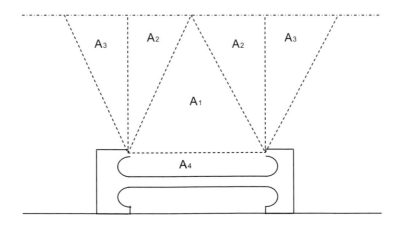

Figure 7.8.10 AFC cross-section showing different traveling speeds

Therefore, the total carrying capacity of face conveyor is:

$$\left(A_1V_C\right)+\left(A_2V_c\right)/2+\left(A_3V_c\right)/5+\left(5A_4V_c\right)/6\mathrm{m}^3 \ \mathrm{min}^{-1}$$

Or:

$$60 \ V_c \left(A_1 + A_2/2 + A_3/5 + 5A_4/6\right) \mathrm{m}^3 \ \mathrm{hr}^{-1}$$

The maximum face conveyor capacity under the AFC is:

$$60WCV_cS_f \qquad \mathrm{ft}^3 \ \mathrm{hr}^{-1}\left(\mathrm{m}^3 \ \mathrm{hr}^{-1}\right).$$

The following conditions must be satisfied:

1   If $2D > H$

    A   Cutting from tailgate to headgate

$$\left(D_sV_sS_f\right)+\left[(H-D)sV_sS_f\right]\leq \text{full haulage carrying capacity}$$

    B   Cutting from headgate to tailgate

$$\left(S_fDsV_s\right)\leq \text{haulage carrying capacity under the shearer}$$

2   If $2D = H$ during both cutting trips

$$2D_sV_sS_f \leq \ \text{full haulage carrying capacity}$$
$$D_sV_sS_f \leq \ \text{haulage carrying capacity under the shearer}$$

7.8.4.3 Conveying area calculation for the conveyor

Conveyor speed $(V_h) = 311$ fpm (91.58 m s^{-1} or 94.8 m min^{-1})
Conveying area under the shearer $= 4.96$ ft^2 (0.4607991 m^2)
Assumption:
Traveling speeds of coal at different cross-sections (see Fig. 7.8.10) are:

$A_1 = 100\%, A_2 = 50\%, A_3 = 20\%$, and $A_4 = 83\%$

$A_1 = 407 \times 694 = 0.282458$ m^2 $\left(3.039 \ \mathrm{ft}^2\right)$

$A_2 = 407 \times 694 \times 0.5 = 0.141229$ m^2 $\left(1.519 \ \mathrm{ft}^2\right)$

$A_3 = 417 \times 694 \ \times 0.2 = 0.057879$ m^2 $\left(0.190 \ \mathrm{ft}^2\right)$

$A_4 = 150 \times 950 \times 0.83 = 0.118275$ m^2 $\left(1.272 \ \mathrm{ft}^2\right)$

$A = A_1 + A_2 + A_3 + A_4 = 6.456635$ ft^2 $\left(0.599841 \ \mathrm{m}^2\right) A_1 = 407 \times 694 = 0.282458$ m^2 (3.039 ft^2)

7.8.4.4 Matching calculations for shearer and conveyor

When the shearer cuts from head to tail, coal cut by the tail drum has to pass under it. The space or clearance under the shearer is limited. The following calculations determine whether this clearance is a limiting factor in production.

A. HEADGATE TO TAILGATE CUTTING TRIP

1   Coal cut by the tail (front) drum that has to pass below the shearer:

$$\text{At 45 fpm} \quad = 62 \text{ in.} \times 42 \text{ in.} \times 45 \text{ fpm}$$

$$= 813.75 \text{ ft}^3 \text{ min}^{-1} \text{ (or) } 23.04 \text{ m}^3 \text{ min}^{-1} \text{ (in-situ volume)}$$

$$= 1057.87 \text{ ft}^3 \text{ min}^{-1} \text{ (or) } 29.95 \text{ m}^3 \text{ min}^{-1} \text{ (broken volume)}$$

$$\text{At 52.5 fpm} \quad = 62 \text{ in.} \times 42 \text{ in.} \times 52.5 \text{ fpm}$$

$$= 949.375 \text{ ft}^3 \text{ min}^{-1} \left(26.883 \text{ m}^3 \text{ min}^{-1} \right) (in\text{-}situ \text{ volume})$$

$$= 1{,}234.17 \text{ ft}^3 \text{ min}^{-1} \left(34.95 \text{ m}^3 \text{ min}^{-1} \right) (\text{broken volume})$$

2   Conveyor carrying capacity under the shearer.

The clearance area under the shearer is 4.96 ft² (0.4607991 m²) for 1 m wide pan and 460 mm clearance under the shearer. Assuming the coal travels at the same speed as the conveyor, the volume per unit time passing under the shearer is:

$$4.96 \text{ ft}^2 \times V_c = 4.96 \text{ ft}^2 \times 311 \text{ fpm}$$

$$= 1542.5 \text{ ft}^3 \text{ min}^{-1} \left(43.68 \text{ m}^3 \text{ min}^{-1} \right)$$

Results:
   When traveling from headgate to tailgate, coal cut by the tail (front) drum of the shearer has to pass below the shearer and should match the carrying capacity of the conveyor under the shearer. Comparing (1) and (2), the carrying capacity of the conveyor under the shearer is much higher than the coal cut by the front drum of the shearer. So, the shearer can be run at its maximum haulage speed.

B.   TAILGATE TO HEADGATE CUTTING TRIP

1   Total coal cut by both drums (front drum + rear drum)

$$\text{At 45 fpm} = 78 \text{ in} \times 42 \text{ in} \times 45 \text{ fpm}$$

$$= 1023.75 \text{ ft}^3 \text{ min}^{-1} \text{ or } 28.98 \text{ m}^3 \text{ min}^{-1} \text{ (in-situ volume)}$$

$$= 1330.87 \text{ ft}^3 \text{ min}^{-1} \text{ or } 37.68 \text{ m}^3 \text{ min}^{-1} \text{ (broken volume)}$$

$$\text{At 52.5 fpm} = 78 \text{ in.} \times 42 \text{ in.} \times 52.5 \text{ fpm}$$

$$= 1194.36 \text{ ft}^3 \text{ min}^{-1} \left(33.82 \text{ m}^3 \text{ min}^{-1} \right) (in\text{-}situ \text{ volume})$$

$$= 1552.6 \text{ ft}^3 \text{ min}^{-1} \left(43.966 \text{ m}^3 \text{ min}^{-1} \right) (\text{broken volume})$$

2   Total conveyor carrying capacity.

$$= 6.45 \text{ ft}^2 \ 311 \text{ fpm}$$

$$= 2006 \text{ ft}^3 \ \text{min}^{-1} \left(56.86 \text{ m}^3 \ \text{min}^{-1}\right)$$

Results:

Comparing (1) and (2), the total carrying capacity of the conveyor is much higher than the total coal cut by the shearer when traveling at its full speed. So the shearer can be run at its full haulage speed.

7.9 Coal loading of the shearer

Coal loading onto the armored face conveyor utilizes the cutting drums and other auxiliary devices. The cutting drum is the critical loading device because it cuts and moves coal simultaneously onto the conveyor via the on-board spiral vanes. The design features of the spiral vanes are very important.

7.9.1 Calculation of drum loading capacity

As soon as the coal is cut loose by the bits, it falls into the vane spacing area and moves toward the gob side of the drum by following the shape of the vanes as the drum rotates. The drum loading capacity can be determined by calculating the volume per revolution. An example calculation is shown here.

Given: drum diameter 60 in. (1524 mm), drum speed 56 rpm, cutting web 39.37 in. (1000 mm), number of vanes or starts 3, vane angle 22°, vane thickness 3 in. (76.2 mm), vane web 36.26 in. (921 mm), vane height 12.25 in. (311 mm), vane spacing (or throat), 17.14 in. (435 mm), vane length 96.70 in. (2456.18 mm), and shearer haulage speed 60 fpm.

Static volume between vanes:

$$96.70 \times 17.14 \times 12.25 = 11.75 \text{ ft}^3 \left(0.333 \text{ m}^3\right)$$

Static volume of the drum is:

$$11.75 \times 3 = 35.25 \text{ ft}^3 \left(0.999 \text{ m}^3\right)$$

Dynamic volume at the operating speed is:

$$35.25 \times 56 = 1974 \text{ ft}^3 \ \text{min}^{-1} \left(55.94 \text{ m}^3 \ \text{min}^{-1}\right)$$

Assuming for a swelling factor of 30% for the loose coal in the vane spacing, the effective dynamic volume, i.e., loading capacity, is:

$$1974/1.3 = 1518.46 \text{ ft}^3 \ \text{min}^{-1} \left(43.03 \text{ m}^3 \ \text{min}^{-1}\right).$$

Shearer's cutting capacity is:

$$60 \times 39.37 \times 60 \times 12 = 984.25 \text{ ft}^3 \text{ min}^{-1} \left(27.892 \text{ m}^3 \text{ min}^{-1}\right)$$

Since the shearer's cutting capacity is only 984.23/1518.46 = 64.8% of the drum loading capacity, the drum will not clog up.

7.9.2 Factors affecting loading efficiency

The following factors control loading efficiency: characteristics of the spiral vanes (or scrolls), that is, vane angle, vane spacing, vane depth, and number of vanes; drum diameter; cutting web; drum rotational speed and direction; haulage speed; cowling; and dip angle of the panline.

7.9.2.1 Cutting drum parameters

After being cut off, the coal comes in contact with one of the vanes which deflect it 90° onto the AFC. The vane angle at the edge with respect to the traveling direction of the shearer and vane tip speed is very important. Since the vane tip speed depends on the vane tip diameter and drum rotational speed, and the volume available in the drum for broken coal depends on vane depth, vane spacing (or throat), vane length (or angle of wrap), and shearer haulage speed, all affect the loading efficiency of drum.

A. VANE ANGLE

This is the angle between the end surface of the drum and the vane, α (Fig. 7.5.2) and it varies with the location of the spiral vane. It is equal at locations measured at the same radial distance from the drum axis. The angle increases toward the axis, so the angle at the bottom of the vane is larger than that at the tip. But, the angle at the tip is usually used to represent the vane angle.

The size of the vane angle has a definitive effect on loading efficiency. The angle must not be too large, nor too small. If the angle is too large, the coal is thrown out farther beyond the conveyor, producing more dust. On the other hand, if the angle is too small, coal cannot flow smoothly onto the conveyor, causing gouging and crowding.

The effect of vane angle on loading efficiency is also related to drum speed. According to research by Brooker (1979), when the drum speeds are between 30 and 40 rpm, the vane angle has no effect on the loading efficiency and can be selected from a wide range between 8 and 30 degrees. The common range is from 22 to 24 degrees.

Angle of wrap is defined as the angle in degrees covered around the circumference at any diameter of the vanes by a vane from its start at the backplate to its finish at the gob end. It can be calculated by:

$$\tan \beta = \frac{360 W_v}{\pi D Z} \tag{7.9.1}$$

Where β = vane angle at diameter D, W_v = vane width, D = drum diameter at any point on the vanes, and Z = angle of wrap. For any given drum diameter, Equation (7.9.1) can be used to calculate the angle of wrap of the vanes for a given vane angle or the vane angle at any point on the vanes for a given angle of wrap. The angle of wrap must be such that there is a sufficient overlap between adjacent vanes to reduce uneven cutting between the face end and gob end of the drum

B. VANE SPACING OR VANE THROAT

This is the spacing between two adjacent vanes along the axial direction. It is closely related to loading efficiency. If the spacing is small, it may be too narrow for smooth coal flow causing crowding and blocking conditions. When the vane depth is shallow, the spacing should be increased to no less than 9 in. (225 mm). The spacing commonly used is 15–21 in. (381–533 mm). If the spacing is constant, it is called the ISO-pitch drum. In an ISO-pitch drum, the vane angles in each spiral vane do not vary. During loading, the loading capacity increases from the face toward the gob end or the conveyor side. But the cutting resistance is just the opposite. In order to maintain a uniform cutting resistance and for convenience of coal loading, the vane spacing and angle can be increased gradually from inside out. This is the variable pitch drum.

C. DRUM DIAMETER

Decreasing drum diameter reduces its loading efficiency. The drum diameter is the cutting diameter, which is the vane diameter, plus twice the bit depth. The vane diameter is the drum barrel diameter, plus twice the vane depth. If the drum barrel diameter is too small, the welding strength between the vane and the drum barrel, and the axial strength of the drum barrel, will not be sufficient. But if it is too big, the vane depth will be shallow; coal will spill easily. The remedial measure is to increase the number of spiral vanes and the drum speed (Eichbaum and Bendmayr, 1974). Although the carrying capacity of each vane is reduced, the unloadings per unit time increases. In general, the vane diameter is about 0.4 to 0.6 times of the cutting diameter.

D. DRUM ROTATIONAL SPEED

The drum rotational speed affects both cutting and loading efficiency. Loading efficiency is low at high drum rotational speed and increases rapidly as rotational speed is reduced. Because when the speed is too high, it will throw the coal out behind the drum, and loading efficiency will be low. Conversely, reducing the drum speed will increase both loading and cutting efficiency. The common drum speeds are 35 to 60 rpm.

E. DIRECTION OF ROTATION

When the drum is cutting coal by rotating from the floor toward the roof, it requires less power than when it is cutting in the reverse direction. But the difference is not large. In selecting the rotational direction of the drum, the stability of the shearer, cutting efficiency, the amount of coal dust produced, whether it will improve the operator's working conditions, and so forth, must be considered

F. FACE GRADIENT

Brooker (1979) reported that when gradients are less than one in eight and one in six in the mining and face line directions, respectively, it will not affect drum loading efficiency.

7.9.2.2 Shearer haulage speed

Brooker (1979) also reported that for any given drum geometry, loading efficiency decreases with increasing haulage speed and that when haulage speed is greater than 19.7 fpm (6 m min⁻¹), loading efficiency is normally high when drum rotational speed is 40 to 60 rpm.

7.9.2.3 Other factors

A. DISTANCE BETWEEN THE DRUM AND THE CONVEYOR

This is the distance from the edge of the spiral vane on the gob side of the drum to the face side sigma plate of the conveyor pan. Some of the coal from the drum drops down in this area. The larger the distance, the more coal will be dropped out.

B. DISCHARGE AREA OF THE COAL

This is the cross-sectional area between the drum and the shearer's body through which the coal passes. If the discharge area is larger, the coal will flow better. But this area is always restricted by the frame of the auxiliary loading devices, ranging arm body, supports for water spraying, etc.

C. CARRYING CAPACITY OF CONVEYOR

If the carrying capacity of the conveyor is low, it cannot increase loading efficiency. The carrying capacity is related to the AFC furniture and the velocity of flight bars. Thus, to increase the loading efficiency, the haulage velocity of the shearer needs to maintain not only a proper ratio to the drum speed but also to the carrying capacity of the face conveyor.

In summary, there are many factors that influence drum loading capacity. It is tied closely to the cutting efficiency of the shearer. If there are conflicting requirements in designing the cutting and loading efficiency, those for the cutting efficiency should be considered first. The same principle applies to the application of the shearer.

7.9.3 Other loading devices

In shearer-operated longwall faces, the spiral vanes in cutting drums are designed to guide and push the majority of broken coal directly into the conveyor. The efficiency of such a coal-loading method can be increased by various auxiliary loading devices.

The cowl is an arc-shaped plate conforming to the circumference of the drum. It is commonly used, because its loading effect is good and it is easy to operate. The distance between the drum and cowl must be appropriate. If the gap is too small, it will restrict the loading capacity due to squeezing and blocking. But if the gap is too large, it will store too much coal and remain there where it is subject to repeated squeezing and crowding. Both will seriously decrease loading efficiency. When the shearer is cutting and loading coal, the cowl is always located behind the drum. Thus before the shearer reverses its traveling direction, the cowl must rotate 180°.

A small amount of floor coal can be scooped up by the face side ramp plate of the conveyor.

Chapter 8

Coal transportation

8.1 Introduction

The armored chain conveyor (AFC), or simply "face conveyor" or "conveyor" as it will be referenced to in this book, was first developed in Germany in the 1940s, and, at that time, it was powered by compressed air when coal was manually loaded. The compressed air was replaced by electric drives in the 1950s. AFC use then spread to other European countries and the United States. As production continued to increase, the power rating of electric drives increased also. In the 1970s, the 2×150 hp (2×112 kW), 1000-volt induction motor with a fluid coupling was used, and, in the 1980s, 402-hp (300-kW), 3300-volt drives with steel or cast iron water filled couplings were the standard. There were considerable problems with the reliability of couplings and power sharing between the head and tail drives. Two-speed motors were introduced to replace the couplings. A typical two-speed motor would be 450/150 hp (336/112 kW) with the AFC being started typically at one third of nominal speed. As the drive for productivity increased in the late 1980s and early 1990s, so did the power requirement for AFC drives.

Experience showed that dual speed motors worked fine up to 704 hp (525 kW). Beyond that, the overloaded AFC would start at a slow speed but trip off at the threshold, and load sharing between head and tail drives was poor and could not be adjusted. In case of a sudden stop, the huge inertia of the two-speed motors would cause serious damage to the transmission elements, sprockets, and chains. To solve these problems, a drive that could soft-start motors and share power between the head and tail drive motors was badly needed. To meet this challenge, DBT introduced the CST (controlled start transmission) in 1994 and Joy developed the TTT (turbo transmission technology) in 1997. Today CST and TTT are the industry standards. In the latest development, Komatsu mining introduced Optidrive that employs VSD (variable speed drive) which is versatile and compact.

Just like with the shield and shearer, the trend of carrying capacity and power rating for AFC has been increasing steadily over the past 40 years (Fig. 1.5.13). It peaked and stabilized at 6500 hp (4710 kW) in 2013. In order to be compatible with this trend, the size of the chain and pan has increased correspondingly. In the 1970s, AFC carrying capacity was mostly in the 1000–2000 tph (tons per hour) range. Today it is 4000–7000 tph; the pans were 15.7–23.6 in. (400–600 mm) in the 1970s vs. 34–53 in. (864–1346 mm) today. The chain was 26 mm in the 1970s running at 220–250 fpm (1.12–1.27 m s^{-1}) vs. 42–52 mm running at 300–451 fpm (1.51–2.29 m s^{-1}) today. And, there were many types of chain strands (single center, twin in-board or twin center, twin outboard, triple) in the 1970s, but today only the twin in-board chain strand (TIB) is used (Figs 1.5.11 and 1.5.12).

8.2 Layout of coal transportation system

In longwall mining, the coal cut down by the shearer is transported by the AFC, which is laid on the floor parallel to the face across the full-face width. The chain assembly (i.e., chain strand and flight bars) in the AFC runs endlessly from the tailgate to the headgate T-junction, where the coal is transferred to the entry belt conveyor through the cross frame, and then onto the mobile stage loader (Fig. 8.2.1). The armored face conveyor has a large carrying capacity. It is structurally very strong, bendable like a snake in vertical and horizontal directions, and low in body height. It serves not only as a track for the shearer but also as the pivot for support advance.

If the conveyor is equipped with a ramp or toe plate on the face side, the conveyor can also help load the coal left on the mine floor.

Although there are several models of AFCs made by different original equipment manufacturers (OEMs), their basic structure and major components are similar in overall configuration.

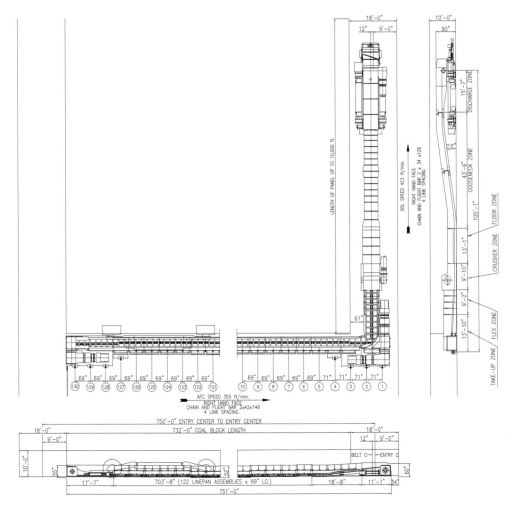

Figure 8.2.1 Layout of AFC on longwall face showing various components

The overall layout on the panel is similar as shown in Figure 8.2.1. An AFC consists of a cross frame, various pans (i.e., line pans, gradient or ramp pans, connecting pans, and adjusting pans), chain links, flight (or scraper) bars, and a tail drive. In addition, furniture such as channels and compartments for handling cables, hoses, and pipes is attached to every pan on the gob side, and, if applicable, a toe or ramp plate on the face side. AFC is advanced or pushed forward by shields through a DA ram/relay bar system from every shield to every pan.

The chain links are connected endlessly, one vertical and one horizontal in alternate sequence, to form a continuous chain between the cross frame at the headgate T-junction and the tail drive at the tailgate T-junction. The flight bars are tied to the chain links at a fixed chain-link interval. During operation, the chain assembly, driven by the motors at the cross frame and tail drives, move on the conveyor deck plate from the tail end toward the head end. As they move, the flight bars, being tied to the chain links, also move and carry the coal toward the head end. The chain links and flight bars (chain assembly) return via the head drive sprocket to the tail end through the underside of the pans, after discharging the coal onto the stage loader.

Depending on the AFC's capacity, two or three drive motors either with Optidrive or CST or TTT are required. If three, the most popular type, two are at the headgate cross frame, one parallel (in-line drive) and one perpendicular (angle drive) to the face line, while the remaining one at the tail drive is parallel to the face line. If only two motors are needed, both can either be at the cross frame or, more commonly, one each would be located at the cross frame and tail drive. If both two are at the cross frame, the tail drive will be a non-driven one.

8.3 Major components of the armored chain conveyor

8.3.1 Head drive and cross frame

The head drive consists of motors, gearboxes, CST or TTTs, and cross frame or simply Optidrive.

Motors with gearbox and CST or TTT are mounted either parallel or perpendicular to the face line on either or both sides of the cross frame, depending on the total power required.

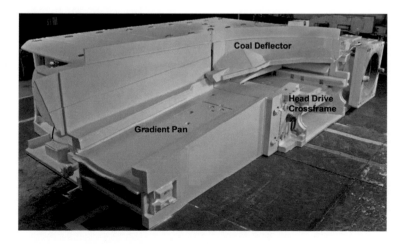

(A)

Figure 8.3.1 Typical head drive layout.

Source: A – Courtesy of Caterpillar; B – courtesy of Komatsu Mining

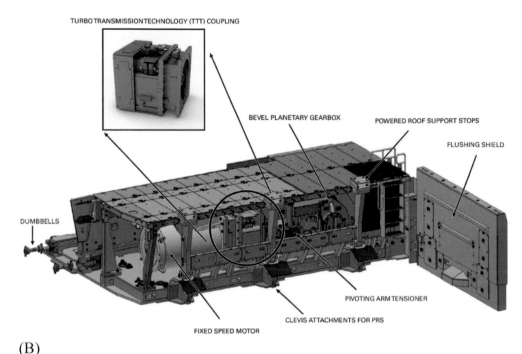

TURBO TRANSMISSION TECHNOLOGY (TTT) COUPLING

BEVEL PLANETARY GEARBOX

POWERED ROOF SUPPORT STOPS

FLUSHING SHIELD

DUMBBELLS

PIVOTING ARM TENSIONER

CLEVIS ATTACHMENTS FOR PRS

FIXED SPEED MOTOR

(B)

Figure 8.3.1 (Continued)

8.3.1.1 Cross frame

In the early years of longwall introduction, coal transfer at the headgate T-junction evolved from the end or T discharge in the 1960s and 1970s through the side discharge and angle frame (Fig. 8.3.2) in 1980 to the cross frame (Fig. 8.3.3) in 1982. All longwalls use the cross frame now.

The cross frame was developed to eliminate the problems traditionally experienced with the transfer of coal from the AFC to the stage loader, in that it reduces carry-back and increases the AFC efficiency. In a cross frame, the top chain assembly of the AFC crosses and stays on top of that of the stage loader. Similarly, the bottom chain assembly of the AFC crosses and stays on top of that of the stage loader. There is no deck plate between the top chain of the AFC and the top chain of the stage loader, thereby allowing the coal to drop directly from the AFC to the stage loader. This arrangement results in a low overall height and less dust in the air.

The most important consideration for AFC is heavy load startup. Insufficient power for a heavy startup load will cause damage to sprockets and the power transmission system. Since the mine electrical system is such that there may be up to a 20% drop in voltage any time, a motor to supply 450% BDT (breakdown torque) is required if this happens. In order to avoid damage to the motors when starting the AFC with a heavy load or sudden blockage of the AFC, the connection between sprocket and motor is not direct, but runs through coupling devices such as CST or TTT, or simply Optidrive that will provide a soft start and protect overloading.

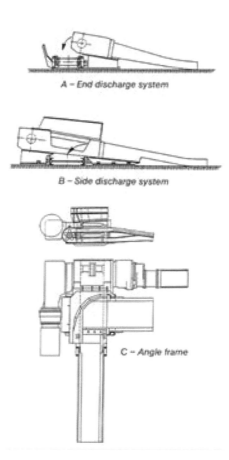

A – End discharge system

B – Side discharge system

C – Angle frame

Figure 8.3.2 Evolution of methods of coal transfer from AFC to stage loader in the 1970s and early 1980s

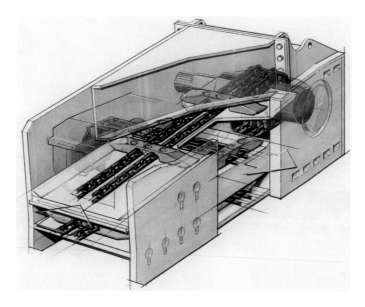

Figure 8.3.3 Cross frame

8.3.1.2 Controlled start transmission

The CST consists of a wet reaction disc pack and the CST gearbox which is a planetary gearbox with the CST clutch (Fig. 8.3.4). The torque transmitted to the output shaft is controlled by varying the clutch pressure. As pressure increases, both the rotating and the stationary reaction plates move together until they contact. During this process the slip is reduced from 100% to 0.2%.

The supply unit contains pumps, filters, sensors and valves integrated in the gearbox system. Fast exchange of data between numerous control units allows for a network of controlled drives for the AFC.

A. SOFT START AND HEAVY LOAD STARTUP

During the motor starts the clutch works at 100% slip allowing a no-load startup of the motor. All drive motors are allowed to run up to full speed sequentially without load. During this period, peak current demand is very low. After all of the motors have reached their nominal speed, the slip is gradually reduced or the clutch pressure is increased continuously to the required breakdown torque, such that torque is gradually applied to the sprocket, allowing a very soft start of the AFC even if it is fully loaded, Only after the last motor has reached full speed, the sprocket begins to run (Fig. 8.3.5). Increase in clutch pressure and motor loading is synchronized all throughout this acceleration period. As a result, the motors start with no load and cause minimum voltage drop; AFC soft-start induces minimal stress on all drive train components; and the synchronized start up of motors maximizes the total available torque. This feature is very important for a soft treatment of the chain and sprockets. Since the PMC-D drive control units are connected to a network, it ensures that all drives

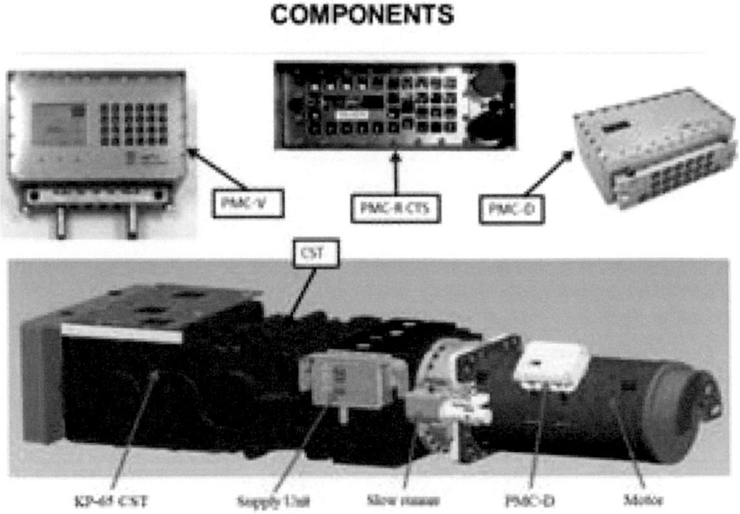

COMPONENTS

Figure 8.3.4 CST control system

Source: Courtesy of Caterpillar

CST Startup Curve

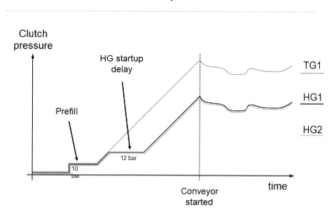

CST Modified Startup Curve

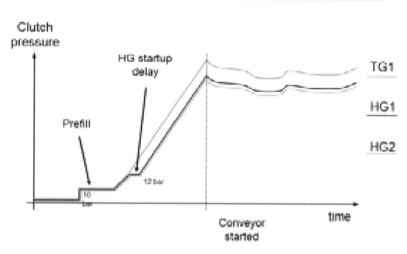

Figure 8.3.5 Controlled start of drive motors in CST drive system: HG motors start delay in B longer than in A

Source: Courtesy of Caterpillar

operate with the same startup characteristics at any time. This unique feature allows a motor to operate with less breakdown torque, because all motors are at the same position on their speed-torque curve at all times. The parameters of the startup procedure such as the starting torque ramp can be changed by the operator if necessary.

B. LOAD SHARING

The power consumption of all drive motors is monitored continuously. If any motor(s) consumes more energy than a preset value, the clutch slip of the motor(s) will be increased to even out the power consumption. This forces the other drive(s) to take over more load and guarantees that all motors are operating with the same power consumption regardless of the load on the AFC. With CST drive, accurate load sharing between drives is realized for the full utilization of installed power and prevention of overheating.

C. OVERLOAD PROTECTION

The input and output speed of each gearbox are constantly measured. In case of a chain stall (blockage) the output speed is decelerated extremely fast to "0" speed. The PMC-D drive control unit immediately senses this deceleration and declutches all gearboxes at the same time (disengaging the masses of inertia of motor and gearbox) before a dangerous peak chain force is reached. The reaction time is approximately 15 milliseconds, which leads to extremely low peak chain forces and an optimum chain protection. With this operating feature, there is practically no wear inside the clutch, because there is no contact between the rotating and stationary clutch plates.

8.3.1.3 Optidrive system and turbo transmission technology

Optidrive is an intelligent variable speed and torque drive. It integrates seamlessly all components of the transmission and provides a complete AFC or stage loader drive system (Fig. 8.3.6).

Variable speed drive (VSD) is a type of adjustable-speed drive to control AC motor speed and torque by varying motor input frequency and voltage.

Optidrive combines the variable speed drive with the motor to achieve better speed and torque control. It has the following capability and advantages:

1 Protection of chain overload and detection of chain break

   In case of sudden increases in shaft torque, it shuts off the motor instantaneously to protect the drive chain and chain from extreme loads. In addition, when it detects sudden decreases in shaft torque, the drive is immediately stopped.

   All motors are independently controlled to eliminate excessive slack chain during startup with high torque at zero speed.

2 Chain speed is automatically adjusted based on AFC loading condition, thereby reducing chain wear by slowing (stopping) the equipment when not required, e.g., during gate end turn around.

   The chain is always fully loaded when it is running, thereby reducing chain travel by up to 42%.

3 Intelligently controlled production rate

   Optidrive integrates with shearer speed control to ensure that when AFC is under used, shearer speed can be increased.

4 Enhanced maintenance and control functionality

   Load sharing is automatically controlled with < 2% variance between motors. Chain mapping and chain position are monitored in real time.

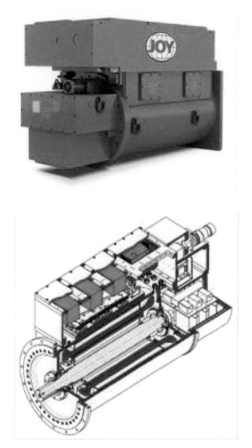

Figure 8.3.6 Optidrive system overview
Source: Courtesy of Komatsu Mining

In summary, Optidrive provides control of motor (independent and synchronized) speed, acceleration and deceleration, resulting in a softer start and stop and faster acceleration and load sharing. The variable chain speeds that match production conditions allow for consistent product flow throughout the mining cycle while reducing equipment wear.

Since Optidrive is an integrated motor and drive system, it saves installation space and requires less maintenance.

The TTT (turbo transmission technology), developed in 1994, is a hydrodynamic water-filled coupling (based on the Voith DTP couplings) that consists of two turbo couplings supported by the electric motor and gearbox bearing. Water is introduced into the coupling through two inlet valves. One is high volume for starting, and the other is low volume for normal running. The water acts as the torque transmission medium inside and cools off the coupling. Water is discharged to the atmosphere during startup and when it is too hot during running. Otherwise, the water is recirculated during normal operation. Since water can be replaced quickly, repeated starts can be done easily during the stop/start sequence. TTT can provide a progressive startup load and results in proactive load sharing and a controlled soft start (Fig. 8.3.7).

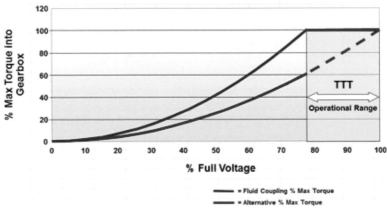

Figure 8.3.7 Turbo transmission technology (TTT)

Source: Courtesy of Komatsu Mining

The TTT range of couplings is made of 562- and 650-sized units allowing up to 22,127 ft lb (30,000 Nm) of torque transmission.

The cross frame together with the drive control system are mounted on a skid plate that is pushed forward with the stage loader by the gateend shields that consist normally of three units.

The drive sprocket assemblies, which are normally of six- and seven-tooth design and oil-filled, are arranged so that they can be removed from the drive unit independently.

8.3.2 Tail drive and tail end

There are two types of tail end: driven and non-driven having no drive unit. All modern high production longwalls are equipped with driven tail end or tail drive. The drive motor is always mounted parallel (in-line type) to the face line on the gob side of the tail drive/pan structure (Fig. 8.3.8). A tensional tail end is normally used regardless of whether it is a driven

(A)

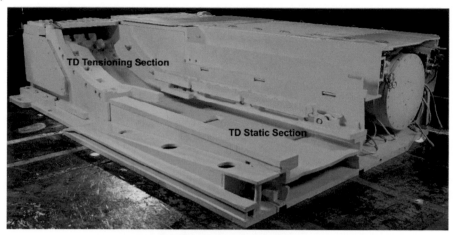

(B)

Figure 8.3.8 Tail drive

Source: A – courtesy of Caterpillar; B – courtesy of Komastsu Mining

or a nondriven end. The adjustable distance usually operated by a hydraulic cylinder ranges from 39.4 to 166.67 in. (1–2 m).

The construction and working principles of the driven return end are similar to those of the head drive in the headgate except the drive frame is flat (Fig. 8.3.8), unlike that of the head drive, which slopes upward. Therefore, the drive frame of the driven return end is short and, in practice, can be connected to the line pan either directly or through an adjusting pan. It is also mounted on a skid plate and pushed forward by the two to three gateend shields.

Figure 8.3.8 B is a low-profile tensionable tail frame and contains two modules. The base frame module is stationary and attached to run of AFC, while the sliding module contains the rotating parts (transmission and sprocket), the position of which is regulated by a hydraulic cylinder to adjust chain tension. It allows the sprocket to be positioned at the lowest possible height, the shearer to cut further past the drive frame maintaining maximum undercut, and greater ranging arm undercut at the end of shearer travel. It also reduces the wear of chain assembly and power consumption.

The power rating of the tail drive preferably should match the size and power of those of the head drives.

When the AFC starts up, the action begins with the tail drive to ensure that the chain is tightened up across the whole face before the head drive starts. Among other things, this process prevents tangling of the AFC chain with the stage loader chain at the cross frame if either or both chains are slacking. If the chain is slack, vibration and jumping of the chain may occur. Most tail drives, therefore, are equipped with a tensioning system in order to maintain a proper tension.

The power rating of the tail drive preferably should match the size and power of those of the head drives.

8.3.3 Automatic tail drive chaining tensioning system

The AFC operational practice is to start the tail drive first in order to pull the bottom chain slack over to the deck top, followed a few seconds later by running the head drives to move the loaded materials on deck top toward the cross frame at the headgate. The reverse is true when stopping the AFC.

Figure 8.3.9 illustrates three possible types of chain tension. The top figure denotes tight or over-tensioned chain in which the upper side of the flight bar runs under the outer guides of the AFC frame. It consumes more power and suffers more wear, while a slack or under-tensioned chain (bottom figure) possesses excessive slack such that the vertical chain links contact one another. A slack chain cannot be operated safely because the loose chain links

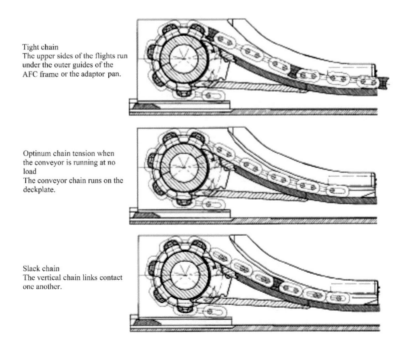

Figure 8.3.9 Three types of chain tension exhibited at tail drive

Source: Courtesy of Caterpillar

could pile up at the tail drive sprocket and damage the system. The middle figure denotes the optimum chain tension in which the chain runs on deck plate when the conveyor is running at no load.

During longwall mining, the shearer often slows down or stops due to various reasons. Consequently, the volume of cut materials on the AFC across the face width is more often than not nonuniform. The conventional method of designing the AFC power capacity is to assume that the chain tension must be adjusted to the maximum possible loading conditions, which means that it is much too high most of the operation time, thereby greatly increasing AFC wear. Therefore, in order to avoid a distinctive slack chain formation, the chain tension is always higher than necessary for normal operating conditions. The automatic chain tensioning system was developed to adapt the chain tension, or the chain power, in optimum fashion to the prevailing AFC loading conditions (Fig. 8.3.9).

Since the head end of AFC is used for transfer cut materials to the stage loader, while the tail end is used to pull slack chain in the bottom race, the automatic chain tensioning system is installed in the tail end or tensionable tail drive.

There are two types of tail drive chain tensioning system, both of which are illustrated in detail in Section 9.2.3 (p. 313).

8.3.4 Pans

Although extremely simple in structure, the pans are the major component of the AFC, mainly because there is a pan for every shield except the gateend shields. These pans are connected one by one by two pan connectors, one on each side (gob and face sides), to form a straight line right behind the face line but in front of the shield line. They serve as the supporting structure and moving roadway for the chain assembly and the loaded coal. They also act as the guide rail for the shearer and are subjected to high resistance and wear. All of these factors require that the pans must be strong, rigid, and highly wear-resistant and, yet, sufficiently flexible to adjust to both horizontal and vertical undulations of the mine floor.

The pans in an armored face conveyor are closed pans and divided into the following types: line or standard pans, ramp or connecting pans, adjusting pans, and split pans.

8.3.4.1 Line or standard pans

Line pans are made in sections, approximately 5 ft (1.5 m), or 5.75 ft (1.75 m), or 6.56 ft (2 m) long, depending on the width of the shield support used. There are two types, cast or rolled. In early years, cast pans were generally less expensive, but produced less precise and smooth profiles than rolled ones. However in recent years cast and rolled pans have the same quality.

Figure 8.3.10 shows a rolled line pan, PF4. The vertical section on each side is constructed from two special side rollings that are welded to a deck plate in the middle, forming the surface for the conveying chain assembly: It looks like a "Σ" or sigma section. Each sigma section is rolled into two parts (i.e., top and bottom) and welded to the deck plate at four spots. The deck plate is 40 to 70 mm thick and the bottom plate is 25 mm thick. The bottom plate is covered with a wear-resistant cover plate. The inner race profile is very smooth providing more contact area, less friction, and more space for coal loading. The bottom

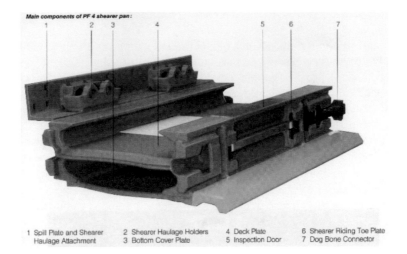

Main components of PF 4 shearer pan:
1   2   3   4   5   6   7

| 1 Spill Plate and Shearer Haulage Attachment | 2 Shearer Haulage Holders | 4 Deck Plate | 6 Shearer Riding Toe Plate |
| | 3 Bottom Cover Plate | 5 Inspection Door | 7 Dog Bone Connector |

Figure 8.3.10 PF4 line pan
Source: Courtesy of Caterpillar

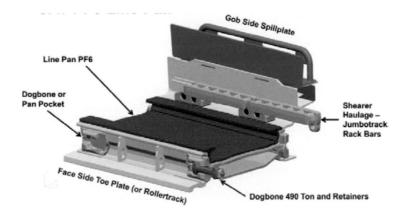

Figure 8.3.11 PF6 line pan
Source: Courtesy of Caterpillar

plate is rounded at the connection between pans to reduce noise. The deck plate is made of wear-hardening material (i.e., wear resistance increases with more coal conveyed) that, for a 40 mm plate, will last for more than 22 million tons of coal conveyed. Line pans are connected by dogbone connectors with a tensile strength of up to 450 tons. The connectors are designed to prevent adjacent pans from separating during conveyor snaking, shield advance, and undulating floor. Thus, at the joints, the pans can be bent. The connector seating is such that the gap between pans is adjustable, the maximum of which is ± 6°. A spill plate with two shearer haulage holders is welded onto the top surface of the gob side sigma section.

Figure 8.3.11 shows the rolled line pan, PF6. A replaceable top trough is welded onto a base plate, 25 mm thick with a side board U profile. When the top race is worn out, the welds can be removed and the top race replaced with a new one. This design is expected to increase the life span of the basic or non-replaceable structure by a factor of three.

A new HD (heavy duty) pan is being developed (Fig. 8.3.12). Its deck plate is 70 mm thick, instead of the traditional 40 mm, to last even longer and for abrasive and corrosive environments and materials.

Figure 8.3.13 is a cast line pan. Dumbbell connectors with tensile strength of 100 to 450 tons are used. For cast line pans, there are regular and composite types. The composite line

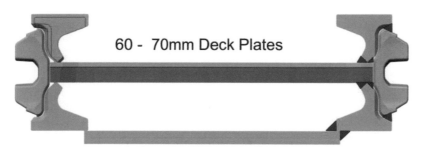

Figure 8.3.12 Newly developed heavy duty pan

Source: Courtesy of Caterpillar

Figure 8.3.13 Line pan with dumbbell connector

Source: Courtesy of Komastu Mining

pan combines pan structure and spill plate cast in one piece. Some sigma sections are cast with the side attachment as an integral piece.

8.3.4.2 Gradient pans

The gradient or ramp pan is used to connect the drive frame to the line pans, with either one or two sections, depending on the manufacturers. The joints are rigid. The ramp pan serves as the transition region from the lower line pan profile to the higher drive frame. Therefore, it is also called a connecting or an intermediate pan. In some cases, a gradient pan is inserted between the ramp pan and the line pan so that the transition slope is milder. A milder slope is desirable for the movement of flight bars.

8.3.4.3 Adjusting pans

Adjusting pans are shorter than the normal section. For example, for the 5 ft (1.5 m) long pans, the half and quarter adjusting pans are 2.5 ft (750 mm) and 1.25 ft (375 mm), respectively. They are used to either extend or shorten the conveyor whenever the face width changes. They are similar in structure to the line pans.

In the 1970s, open bottom pans were used, but the bottom chains would roll up the dust from the floor, and the pan section tended to dig into wet floor. Closed bottom pans have corrected these problems, but once the chain breaks in the bottom section, it is very difficult to locate the point of break for repair (see Section 8.6.3). To cope with this problem, a special type of pan with an inspection door in the deck plate is installed in every sixth to tenth pan to provide access to the return chain in the bottom sections of the pans. (See item 5 in Fig. 8.3.10.)

8.3.5 Furniture and toe plate

8.3.5.1 Furniture

Furniture (Fig. 8.3.14), or spill plate, is used primarily to prevent the coal from spilling over the gob side of the AFC and to enlarge the cross-sectional area of coal loading in order to increase the carrying capacity of the AFC. Various designs have been produced to suit customers' needs. All models have a channel attached to the top on the gob side of the pan for the shearer's electrical cable and water hose that are flexibly hold in a cable handler (see Fig. 7.4.4). In addition, one or more racks below and on the gob side of the channel are available for hydraulic hoses for the shield's hydraulic supply, a water hose for house cleaning, the rock dust pipe, and communication devices. Furthermore, the shearer haulage track is attached to the gob side of the gob-side sigma section as part of the furniture.

8.3.5.2 Toe plate

In some designs, a toe plate or ramp plate is mounted on the face side of the pan (see item 6 in Fig. 8.3.10), which serves as the track for the gob-side shoes or roller of the shearer. It is also used to scoop up the floor coal when the AFC is being advanced.

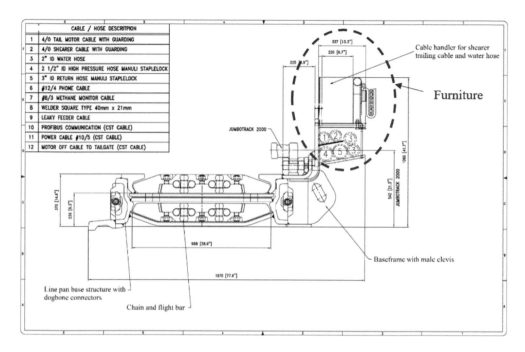

	CABLE / HOSE DESCRIPTION
1	4/0 TAIL MOTOR CABLE WITH GUARDING
2	4/0 SHEARER CABLE WITH GUARDING
3	2" ID WATER HOSE
4	2 1/2" ID HIGH PRESSURE HOSE MANULI STAPLELOCK
5	3" ID RETURN HOSE MANULI STAPLELOCK
6	#12/4 PHONE CABLE
7	#8/3 METHANE MONITOR CABLE
8	WELDER SQUARE TYPE 40mm x 21mm
9	LEAKY FEEDER CABLE
10	PROFIBUS COMMUNICATION (CST CABLE)
11	POWER CABLE #10/5 (CST CABLE)
12	MOTOR OFF CABLE TO TAILGATE (CST CABLE)

Figure 8.3.14 Detailed layout of the panline furniture

Source: Courtesy of Caterpillar

8.3.6 Chain assembly

The chain assembly consists of chain links and flight bars (Fig. 8.3.15). The chain links are made of round-link chains connected in series, one lays horizontally and the next one vertically.

The chain link usually come in an elliptical shape with rounded corners and, therefore, is denoted by its diameter and nominal pitch, such as 48 × 160 mm, i.e., a 48 mm diameter chain with a pitch 160 mm long (a × t in Fig. 8.3.15 C). Recently, as the chain size increases, flat or square vertical chain links (Fig. 8.3.15) are used to reduce the overall AFC height. In order to maintain its strength, the cross-section area of the flat or square vertical chain link is kept the same as the round horizontal chain links. The square vertical links are far less sensitive to chain stretch and allow maximum sprocket tooth size. In order to increase wear resistance, chain links are commonly hardened at the crown contact point. A hardened chain is up to 20% stronger than regular chain. In general, chain strength, T, in tons can be determined by:

$$T = 93d^2 \tag{8.3.1}$$

Where d is the chain link diameter in inches. However, for safety reasons, the maximum allowable chain tension applied is:

$$T_a = 62d^2 \tag{8.3.2}$$

Which gives a safety factor of 1.5.

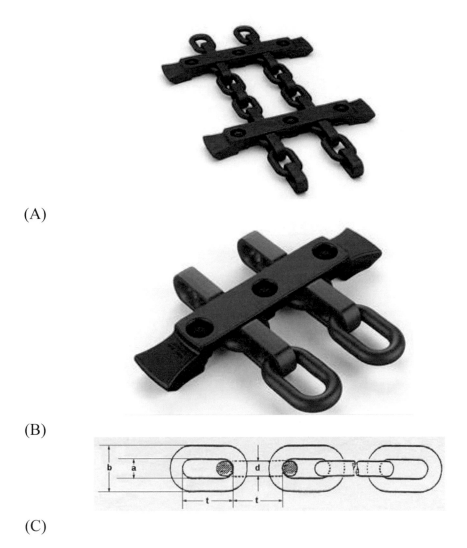

(A)

(B)

(C)

Figure 8.3.15 Chain assembly

Source: Courtesy of Komastu Mining

Equation (8.3.1) shows that, everything being equal, chain strength increases with the square of chain diameter, but chain strength can also be increased with changes in the materials (Bomcke, 1998) or in the manufacturing process (Cooper, 1999).

A short-pitch chain link has also been developed, for examples, a 38 × 126 mm chain rather than the traditional 38 × 137 mm. This allows the use of 38 mm diameter chain in an AFC conventionally designed for a 34 mm diameter chain, thereby prolonging the service life of the conveyor system.

The broadband, low-profile chain consists of a forged, flattened, vertical link and a conventional round-wire horizontal link, and can be manufactured in 34, 42, 50, and 58 mm diameter sizes. Its vertical height is one chain size smaller than the commonly used flat-link models, which allows a mine to increase system power.

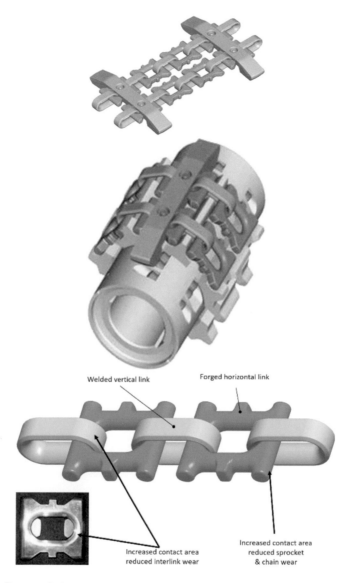

Welded vertical link

Forged horizontal link

Increased contact area
reduced interlink wear

Increased contact area
reduced sprocket
& chain wear

Figure 8.3.16 Power chain

Power chain (Fig. 8.3.16) is a special shape chain designed to reduce contact stress between links and employ the arc-shaped broad cross-section of the flight bars to create a low-profile line pan. As such it reduces wear and prolongs the life of chain and sprocket. Its horizontal link does not contact the deck plate reducing friction for chain movement and increasing deck life. Power chain's interlink contact pressure is 40% and 8% less than that of conventional round link and broadband chain, respectively.

Flight bars are attached to the chain links at regular intervals (Table 8.3.1). They are the major moving parts that carry the coal on the AFC. The amount, size, and hardness of rock to be conveyed determine the type of flight bars and spacing required to guarantee acceptable system availability. Under normal conveying conditions, their spacing varies with chain size

Table 8.3.1 Weight of chain and flights

Pan width [mm]	Chain type	Diameter [mm]	Pitch [mm]	Flight spacing [links]	Flight spacing [mm]	Flight wt [kg]	Flight wt [kg/m]	Chain wt [kg/m]	Total wt [kg/m]	Total lbs ft
732	2	26	92	8	736	23	31.25	27.40	58.65	39.3
732	2	30	108	6	648	30	46.30	27.40	73.70	49.4
600	2	26	92	8	736	34	46.20	27.40	73.60	49.4
743	2	22	86	10	860	20	23.26	19.00	42.26	28.3
832	1	30	108	8	864	29	33.56	36.00	69.56	46.7
932	1	30	108	8	864	34	39.35	36.00	75.35	50.5
932	4	42	146	6	876	41	46.80	68.00	114.80	77.0
932	2	34	126	6	756	40	52.91	45.40	98.31	65.9
932	2	34	126	8	1008	40	39.68	45.40	85.08	57.1
1032	1	30	108	8	864	38	43.98	36.00	79.98	53.6
1032	2	38	146	6	876	45	51.37	54.00	105.37	70.7
1032	2	34	126	8	1008	45	44.64	54.00	98.64	66.2
1132	1	30	108	8	864	42	48.61	36.00	84.61	56.8
1132	1	30	108	10	1080	42	38.89	36.00	74.89	50.2
1132	2	34	126	4	504	45	89.29	45.40	134.69	90.3
1132	2	42	146	6	876	57	65.07	68.00	133.07	89.3
1142	2	48	152	6	912	60	65.79	92.40	158.19	106.1
1142	4	52	175	6	1050	70	66.67	150.00	216.67	145.3
1332	2	34	126	6	756	75	99.21	45.40	144.61	97.0
1332	2	38	126	4	504	65	128.97	60.20	189.17	126.9
1332	2	42	146	6	876	85	128.97	68.00	196.97	132.1
1332	2	30	108	6	648	55	84.88	36.00	120.88	81.1
1332	2	52	175	6	1050	90	85.71	92.40	178.11	119.5
1332	4	56	178	6	1068	120	112.36	156.00	268.36	180.0

and detailed configurations vary with manufacturers. Spacing ranges from five to more than 10 links or 2.5 ft (0.75 m) to 5.0 ft (1.5 m). If there are many large, hard rock fragments, smaller spacing, such as four-link spacing, may be required to avoid locking up and deflecting the flight bars. Flight bars are bolted to the chain links on the horizontal link position at the specified interval either with vertical or horizontal bolts. There are several types of flight bars. The bottom surface of the flight bars are in full or partial contact with the deck plate, and its ends are so shaped that they are in smooth contact with the internal surface of the sigma section. Depending on the design configurations and wear conditions, flight bars generate the majority of noise when the AFC is running. Flight bars, consequently, need to be replaced more frequently compared to other components. The most critical strength areas of the flight bar are the chain pocket and flight tip.

8.4 Selection of armored flexible chain conveyor

Selection of the AFC depends upon carrying capacity, power requirements, and the strength of the chain links. The following factors must be considered: pan width, chain strand type, chain dimension, chain speed, line pan assemblies, auxiliary equipment for the cutting machine, power requirements for the drive units, and arrangement of the drive units, especially the head drive.

In selecting the AFC, variations of seam thickness also must be considered: its capacity must be able to handle cutting the maximum or minimum seam thickness.

Bessinger (2001) listed several factors affecting AFC life.

1   Run-of-mine (ROM) materials to be conveyed. The selected component materials should be resistant to ROM abrasion, and to cause minimal wear on deck plates and sigma sections. A wide selection of abrasion-resistant materials with Brinell hardness 300–500 HB are available.
2   Seam condition and topography. Acidic groundwater or a corrosive environment reduces equipment life. Corroded materials are carried away faster by conveying action. Seam rolls induce different wear areas in pans and flight bars. Hard cutting induces higher forces in the face-to-gob plane and higher wear on the haulage system.
3   Design and manufacturing of the equipment.

8.4.1 Twin in-board chain strand

Figure 1.5.12 shows that, in the 1970s, there were four types of chain strands: single center (SCS), twin in-board (TIB) or double center (DCCS), twin outboard (TOB) or double outboard (DOCS), and triple-chain strand (TCS). TIB has been the preferred chain strand since 1980, and today it is the only type used in US longwalls.

The advantage of locating the chain link at the center of the pan is that it is convenient to bolt the flight bar to the chain links, thereby making it easier to replace a worn flight bar. For the same pan width, a TIB has greater chain strength than a SCS. In a TIB, the two-chain strands must be as close as possible to allow more space for coal handling. The smaller the distance between the two center chains, the greater the oblique position of the flight bar will be once the chains show differential elongations. Wear in a TIB occurs mainly at the deck plate ends.

8.4.2 Determination of AFC capacity

The capacity of an AFC is determined by:

$$Q = 0.03 q_c V_c \left(\text{ton hr}^{-1} \right)$$
(8.4.1)

Where Q is the carrying capacity in tons per hour; V_c is the chain speed which ranges from 250 to 450 ft min^{-1} (1.27–2.29 m s^{-1}); and q_c is the amount of coal loaded per unit length of the conveyor in lbs ft^{-1}.

During operation, coal is loaded uniformly across the pans when fully loaded and pulled along at the same speed as the flight bars; q_c depends on the cross-sectional area of the loaded pans, A_i:

$$q_c = \gamma A_i \left(\text{lbs ft}^{-1} \right)$$
(8.4.2)

Where γ is the weight per unit volume of the loaded coal and ranges from 55 to 95 lbs ft$^{3-1}$, depending on the amount of partings and ash contents, but, on average, $\gamma = 62.5$ lbs ft$^{3-1}$. The cross-sectional area of the loaded pans, A_p depends on the construction and width of the pans, relative chain speed, type of chain strand, height of furniture, and angle of repose of the broken coal. Due to the large cross-sectional area involved in the movement of coal above the pans, the speed of coal flow is not uniform and decreases away from the chain assembly. There are various methods of estimating the coal flow speed within the cross-sectional area (see Fig. 7.8.10).

Underground observations indicate that when the conveyor is lightly loaded, the loaded coal cross-section resembles a modified cone with the sides assuming the dynamic angle of repose, which in most case is less than 30°. The loaded cross-section in this case is:

$$A_i = W_a h_p + 1/4 W_a^2 \tan 30°$$
(8.4.3)

Where W_a is the actual pan width (inner edge of top race to inner edge of top race) and h_p is the height of the top race. When the pan is fully loaded, as in all high-production faces, only the coal lying above the pan width (outer edge of top race to outer edge of top race) is moving with the chain links and flight bars, thus:

$$A_i = W_a h_a$$
(8.4.4)

Where h_a is the actual coal height above the deck plate. Therefore, when fully loaded, the cross-sectional area may depend on the cutting direction. When cutting from head to tail, the actual coal height is restricted by the shearer's body clearance. Conversely, when cutting from tail to head, the actual coal height is restricted by the height of the furniture (or spill plate). Note, however, that Broadfoot and Betz (1997) investigated the AFC coal-loading process and found that different parts of the loaded coal on the AFC travel at different speeds (Fig. 7.8.10, p. 262). In fact, the loaded coal can be divided into four zones, A_1, A_2, A_3, and A_4. A_1 is the triangular area directly above the top races with the base being the distance between the top races. Coals within this zone travel at the same speed as the chain assembly. Coals within A_2 travel at 50% of the chain speed. Coals within A_3 travel at 20% of chain speed. And coals in A_4 (i.e., inside the top race) travel at 83% of chain speed.

Selection of the pan width depends on the carrying capacity of the conveyor, chain speed, assemblies of the auxiliary equipment, and the loading procedure of the shearer. To maintain a smaller prop-free front, the pans should be narrow. The narrower pans also enable the conveyor to be snaked closer to the shearer, thereby reducing the exposure time of the unsupported roof. Fig. 1.5.14 shows the various pan widths used in US longwalls in 2019. It is obvious that pan width is restricted to several models within certain width dimensions. It must be noted, however, that the determination of pan width varies somewhat with manufacturers. Pan width may be measured between the inner or outer edges of the top races or between the inner surfaces of top race grooves.

Pan height is usually 10 to 15 times that of chain size (Westfalia Lunen, 1990), and pan length is 1.5 m or 1.75 m or 2 m if the shield width is 1.5 m or 1.75 m or 2 m, respectively.

Chain speeds up to 450 ft min^{-1} (2.29 m s^{-1}) are commonly used. There are several advantages in employing higher chain speeds. Assuming an equal carrying capacity of the conveyor, the required chain strength and pan width is reduced when higher speeds are used. The height of the loaded section also decreases. This helps the loading and reduces the power consumption of the shearer due to a less constrained flow of broken coal in the conveyor. This is a very important factor for the shearer operating in a bi-directional cutting mode in thin seams where the clearance between the shearer's body and the pans is small. The disadvantage of high chain speed is that it accelerates the wear of the deck plate and chain assembly.

The relative chain speed, with respect to the shearer's haulage speed, rather than the absolute chain speed, is the decisive factor for determining the height of the loaded section and the loading facilities of the shearer. The smallest relative chain speed and the highest loaded cross-section occur during the shearer's trip toward the headgate. In this case, the cutting direction is the same as that of the chain movement, so the loaded cross-section is the largest. Conversely, when the shearer is moving toward the tailgate, the relative speed is highest, and the loaded cross-section is the smallest.

8.4.3 Determination of power requirements and chain strength

The installed motor power must be sufficient to overcome the total frictional resistance due to the weight of the chain assembly and loaded coal running on the conveyor in the top race, the weight of the chain assembly on the mine floor in the lower race, and conveyor snaking. Depending on the degree of sophistication, the determination of the required motor power can be very complex. Two simplified methods commonly employed by AFC manufacturers are discussed here.

8.4.3.1 Method A

A longwall face is equipped with a 3500-ton-per-hour (tph) chain conveyor system using a 39 in. (988 mm) wide pan, and a twin in-board chain strand, 42×146 mm, 1150 ft (351 m) long, running at 311 ft min^{-1} (1.58 m s^{-1}). What is the total horsepower required for the drive motors?

The required total horsepower, P_t is equal to the summation of the power that is required to (1) run empty, P_i; (2) pull back fines in the lower race toward the tail drive, P_r; and (3) pull coal (only) toward the discharge point (i.e., head drive), P_c, or:

$$P_t = P_i + P_r + P_c \qquad (8.4.5)$$

The weight, W_c, of coal (only) on a fully loaded conveyor with a capacity of 3500 tph and chain speed of $V = 311$ fpm is:

$$W_c = \frac{QL}{60V}$$

$$= \frac{3500 \,(\text{tph})}{311 \,(\text{fpm})} \times \frac{1150 \,(\text{ft})}{60 \,(\text{min hr}^{-1})} \times 2000 \,(\text{lbs ton}^{-1})$$

$$= 431,404 \,(\text{lbs}) \tag{8.4.6}$$

Where Q = rated AFC carrying capacity in tph (tons per hour), and L = length of conveyor in feet. The total weight of the chain strand, $W_{st,}$ is:

$$W_{st} = LF_sW_s$$

$$= 1150 \times 4 \times 89.3 = 410,780 \,(\text{lbs}) \tag{8.4.7}$$

Where F_s = a multiplier factor, $F_s = 4$ for a twin in-board chain strand, W_s = weight of chain strand in lbs per foot of length (see Table 8.3.1).
The total weight of the flight bars, $W_{ft,}$ is:

$$W_{ft} = \frac{2LW_f}{\ell_f}$$

$$= \frac{2 \times 1150 \times 125}{2.87} = 100,174 (\text{lbs}) \tag{8.4.8}$$

Where W_f = weight of each flight bar in lbs and l_f = flight bar spacing in feet (see Table 8.3.1).
The horsepower, P, required to pull each type of frictional force is:

$$P = \frac{FV}{550E} \tag{8.4.9}$$

Where F is the force in lbs to be pulled at a speed of V in feet per second and E is the overall power efficiency or:

$$E = E_1 \times E_2 \times E_3 \times E_4$$

$$= 0.96 \times 0.95 \times 0.95 \times 0.85 = 0.7364 \tag{8.4.10}$$

Where E_1 is the efficiency of mechanical drive, E_2 is the efficiency of electric motors, E_3 is the efficiency of hydraulic couplings, and E_4 is the efficiency of the sprocket.
The power required to run an empty conveyor is:

$$P = \frac{\left(W_{st} + W_{ft}\right) f_{mm} V}{550 \times 0.7364} \quad P = \frac{\left(W_{st} + W_{ft}\right) f_{mm} V}{550 \times 0.7364}$$

$$= \frac{\left(410,780 + 100,1784\right) \times 0.3 \times 5.18}{450} = 1,764 (\text{hp}) \tag{8.4.11}$$

Where f_{mm} is the coefficient of friction between chain assembly and conveyor.
The power required to run the coal fine recirculation based on a maximum rate of 2.5% is:

$$P_r = \frac{0.025 W_c f_{cm} V}{550 \times 0.7364}$$

$$= \frac{0.025 \times 431,404 \times 0.4 \times 5.18}{405} = 55 \text{(hp)} \tag{8.4.12}$$

Where f_{cm} is the coefficient of friction between coal and deck plate of conveyor.
The power required to pull coal (only) toward the head end discharge point is:

$$P_c = \frac{W_c f_{cm} V}{405}$$

$$= \frac{431,404 \times 0.4 \times 5.18}{405} = 2207 \text{(hp)} \tag{8.4.13}$$

The total power to run the conveyor at the rated capacity of 1500 tph is then:

$$P_t = P_i + P_r + P_c$$

$$= 1764 + 55 + 2207 = 4026 \, (HP)$$

It must be noted that the total required power determined as illustrated does not consider the electrical voltage supply efficiency.

8.4.3.2 Method B

A longwall face is equipped with a 3500 tph chain conveyor system using a 39 in. (988 mm) wide pan, twin in-board chain strand, 42 × 146 mm, 1150 ft (351 m) long, running at 311 fpm (1.58 m s^{-1}). What is the total horsepower required for the drive motors?

A high-capacity face chain conveyor usually consists of normally three drives, two at the headgate and one at the tailgate. The head end drive supplies the force to pull the top chain and loaded coal to the headgate, whereas the tail end drive supplies the force to pull the bottom chain toward the tail end. (Note that this method of calculation assumes the loaded cross-section is sufficiently large to handle the coal volume.)

A. FOR THE HEAD END DRIVE

The force required to pull the top chain, F_t is:

$$F_t = L \left(W_{sf} + W_{cc} \right) \phi_1 \phi_2 f_t \left(\cos \alpha \pm \sin \alpha \right)$$

$$= 1150 \times \left(89.3 \times 2 + 375.11 \right) \times 1 \times 1 \times 0.32 = 203,765.28 \text{(lbs)} \tag{8.4.14}$$

Where W_{sf} = weight of chain assembly in lbs per foot (note the weight of flight bar is considered to be uniformly distributed per foot of chain strand length); ϕ_1 and ϕ_2 are factors for transverse inclination and conveyor curves, respectively; f_t is the coefficient of friction for the top race (between the chain assembly and deck plate and sigma sections); and α is the conveyor gradient, which is zero in horizontal seams. $W_{cc} = 375.13$ lbs ft^{-1} is the weight of coal per unit length of conveyor running at rated capacity of 3500 tph or:

$$W_{cc} = \frac{3500\,(\text{tph}) \times 2000\,(\text{lbs})}{311\,(\text{fpm}) \times 60\,(\text{min})} = 375.13\,(\text{lbs ft})$$

The required motor horsepower, P_h, is then:

$$P_h = \frac{F_t V}{550 E_h}$$

$$= \frac{203,765.28 \times 5.18}{550 \times 0.8} = 2399\,(\text{hp}) \tag{8.4.15}$$

Where E_h is the head drive efficiency.

B. FOR THE TAIL DRIVE

The force required to pull the bottom chain, F_b, is:

$$F_b = LW_{sf}\phi_1\phi_2 f_b\,(\cos\alpha \pm \sin\alpha)$$

$$= 1150 \times 89.3 \times 2 \times 1 \times 1 \times 0.4 = 82,156\,(\text{lbs}) \tag{8.4.16}$$

Where F_b is the coefficient of friction between the bottom chain and mine floor.
  The required horsepower P_b is then:

$$P_b = \frac{F_b V}{550 E_t}$$

$$= \frac{82,156 \times 5.18}{550 \times 0.8} = 967\,(\text{hp}) \tag{8.4.17}$$

C. TOTAL HORSEPOWER REQUIRED TO RUN THE CONVEYOR IS:

$$P = P_h + P_b = 2399 + 967 = 3366\,(\text{hp})$$

D. MAXIMUM ALLOWABLE HORSEPOWER FOR CHAIN (TABLE 8.3.1) IS:

$$P_{ct} = \frac{2 \times F_{ct} \times V}{550 \times SF \times E_h \times \beta}$$

$$= \frac{2 \times 283.25 \times 2000 \times 5.18}{1.25 \times 0.8 \times 2.4 \times 550} = 4446.2\,(\text{hp}) \tag{8.4.18}$$

Where F_{ct} = chain strength in lbs, SF = safety factor, 1.25, and β = pull-out factor, 2.4.

Table 8.3.2 Selected chain strength

Nominal size [mm]	Chain strength [kN]	Chain Strength [ton]
34	1450	162.98
34	1640	184.34
34	1640	184.34
38	1920	215.81
38	2040	229.30
38	2040	229.30
42	2520	283.25
42	2520	283.25
42	2520	283.25
48	3150	354.06
48	3150	354.06
52	3700	415.88
56	4300	483.32
60	5000	562.00

F. SAFETY FACTOR FOR THE CHAIN IS:

$$SF = \frac{P_{ct}}{P} = \frac{4446.2}{3366} = 1.32 \tag{8.4.19}$$

8.5 Coal transfer system

The modern longwall advances so rapidly that it imposes two requirements on the entry transportation: it must have a high carrying capacity and it must be able to move along as fast and as easy as the face advances.

The flexible belt conveyor is exclusively used as the entry transportation system to satisfy these requirements. Since the belt conveyor is much higher than the armored face conveyor, and it is not convenient to adjust its length constantly, a stage loader is used to transfer coal from the AFC to the entry belt conveyor at the headentry T-junction (Fig. 8.5.1).

8.5.1 Coal transfer from AFC to stage loader

As mentioned in Section 8.3.1, the exclusive application of the cross frame has resolved many disadvantages associated with the earlier transfer systems. This is because, in a cross frame, the coal from the AFC is transferred directly to the stage loader in an integral setup that keeps the entire system low in height, eliminates dust entrainment associated with open transfer, and simplifies system advancement.

8.5.2 Stage loader

The stage loader (beam stage loader, BSL), usually 75–230 ft (22.9–70 m) long, is the connecting device between the AFC on the inby end and panel belt conveyor on the outby end

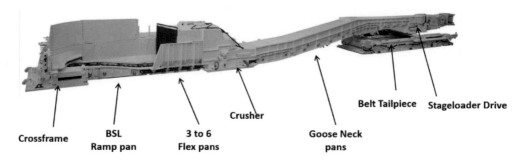

Belt Tailpiece   Stageloader Drive

Crusher

Crossframe   BSL   3 to 6   Goose Neck
            Ramp pan   Flex pans   pans

Figure 8.5.1 Stage loader
Source: Courtesy of Caterpillar

(Fig. 8.5.1). Its carrying capacity is up to 6000 tph with total installed power 966 hp (700 kW) and line pan width up to 1550 mm. Generally, it consists of the following four sections: (1) discharge zone, (2) gooseneck zone, (3) crusher, and (4) take-up zone.

The discharge zone is connected to the panel belt conveyor tail piece by a dolly straddling the belt to allow movement of 12 to 15 ft (3.66–4.57 m) relative to the belt conveyor return end. It includes a drive frame with gearbox, couplings, and a motor; the intermediate pan; the connection pan; and the compensation pan that makes the transition to the line pans. The intermediate pan may be replaced by a telescopic line pan, which allows chain tensioning. Other chain tensioning mechanisms employ a hydraulic tension pan that is an integral part of the drive frame and allows up to 20–39 in. (0.5–1 m) of tensioning length, depending on the panel width.

The gooseneck zone includes a convex pan, line pans, and a concave pan. This arrangement allows the stage loader to rise up at varying degrees (7.5–10°) above the mine floor and discharge onto the belt conveyor.

The crusher may be a continuous flow or an impact head type. Its capacity, up to 800 hp, varies with the characteristics of the conveyed materials. The material to be crushed is conveyed continuously through the crusher by means of the stage loader chain and flights. It is crushed and sized to below a product size, normally about 5.5 to 16 in. (140–406 mm); the product size is usually adjustable. This process produces a more uniform conveyed material and reduces belt spillage. On the inby end of the crusher, four to six flexible, short pans are used to allow bending in both vertical and horizontal directions of the whole stage loader system because the shearer may push the AFC either toward the head or the tail end. This allows the stage loader to operate off square with the AFC without damaging the components of the stage loader.

The take-up zone immediately inby the crusher includes the line pans, the connection pan and the compensation (or adaptor) pan that connects to the return end, an integral part of the cross frame.

The pans and chain assembly used in the stage loader are similar to those employed in the AFC. All pans are closed-bottomed, and the whole stage loader is topped with a cover plate so that no dust will be entrained into the intake air going into the face. Normally, pan width is larger and chain speed is at least 30% faster in the stage loader than those in the AFC. Consequently, the carrying capacity of the stage loader is larger than that of the AFC, ensuring fast and continuing transfer of coal from the AFC to the panel belt conveyor such that the loaded materials on the AFC is removed from the face faster than the face production rate, thereby reducing the likelihood of blockages.

The AFC delivery end and the stage loader receiving end (including the cross frame) are structurally combined into an integral unit.

The stage loader is fully enclosed from outby the crusher to the delivery end, with dust suppression sprays installed along the stage loader, in the crusher and at the delivery end to minimize dust exposure.

The side plates are supporting units and are used to attach brackets for cable and hose installation.

The stage loader as a unit, with its return end mounted on a base plate or skid, is advanced by the hydraulic rams in the gateend shields on the face, advancing over the BSL on an overlap system. The overlap allows approximately four pushes of the AFC before the BSL must be advanced.

A cross-over safety platform to and from the panel coal block side for personnel and cables is located outby the crusher.

In order to minimize the recirculation of coal fines, there should be an adequate height difference at the transfer point between the stage loader and belt conveyor. An adjustable chute assembly may be mounted on the delivery end of the drive frame to ensure optimal delivery.

The panel entry conveyor must always start before the AFC. The AFC must always be stopped before the entry conveyor. If the entry conveyor stops as a result of a fault, the face conveyor must be automatically switched off without delay.

A delay time of around 10 seconds should be allowed for the startup. Be sure to adapt the delay to the conditions in the mine. The time of 10 seconds stated here should be regarded as an approximate value only.

8.5.3 Belt conveyor

As stated in Chapter 1, modern longwall panels are more than 10,000 ft (3049 m) long. The unique feature is that a single belt conveyor flight is used to transport coal out of the panel at the rate of 3500 to 7000 tph (tons per hour). This single, long panel belt system has a main drive and several booster/tripper drives. The location and quantity of boosters depend on many factors including length, lift, and belt PIW. Booster/tripper drives reduce the required belt strength (PIW), thereby reducing the capital cost. Because, for example, a long panel belt conveyor that requires a total of 4000 hp (2900 kW) main drive will require steel belting. If the same system is used with booster drives, a 2000 hp (1449 kW) main drives with two 2 × 500 hp (2 × 362 kW) booster/triple drives will be sufficient. So with the same installed power, the belting required would be much less expensive. With booster/tripper drives, belt tension (or motor torque) must be strictly controlled. Load cells are used at each booster/tripper drive to measure/sense the belt tension at all times. Sometimes a booster/tripper drive needs to put in power (i.e., applying torque to lower tension), and sometimes a booster needs to remove power (i.e., reducing torque to raise belt tension).

For a flexible belt conveyor, the drive unit can be supplied with a single- or double-belt bank storage/belt tension unit. The single bank stores 80 ft (24.4 m) of belting, whereas the double bank can accommodate 160 ft (48.8 m) of belt. Some belt banks can store up to 330 ft (100 m) of belt. A motorized tensioning winch with a load cell is used for tensioning and controlling the amount of belt in circuit.

Figure 8.5.2 shows the working principles of the belt bank in the flexible belt conveyor used in conjunction with the stage loader. The belt bank is located behind the drive head

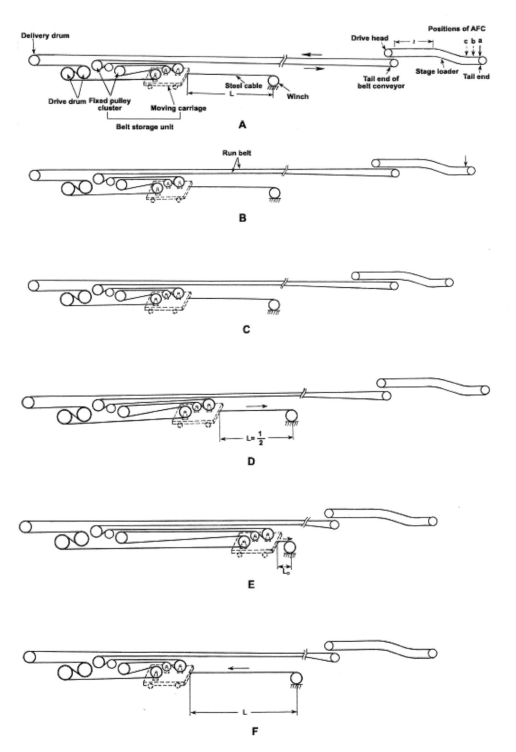

Figure 8.5.2 Working principle of the belt conveyor

of the belt conveyor. After unloading at the drive head, the run belt enters the belt bank through the drive drum. It winds through the deflecting drums mounted at the moving tension car at the rear and the fixed supporting frame at the front. After two deflections, it reaches the tail end of the conveyor and forms a continuous belt conveyor. If the conveyor needs to be shortened, the moving car is pulled by the winch toward the tail end, causing the run belt at the tail end to move toward the head end and storing excess belt in the belt tank. Conversely, when the winch is released, the conveyor will lengthen at the tail end. As the belt is lengthened or shortened, the supporting frame must be added or dismantled accordingly.

Figure 8.5.2 A shows the starting position. Each time the face advances, the connection point between the AFC and stage loader (i.e., cross frame) advances a distance equivalent to a cutting web. At this time the discharge zone or section of the stage loader overlaps the tailpiece of the belt conveyor for 12 to 15 ft (3.66–4.57 m) (Fig. 8.5.2 B). When the discharge section has exhausted its overlapping length (Fig. 8.5.2 C), the tailpiece is trammed forward (outby) for another 12 to 15 ft (3.66 to 4.57 m). After this, the connecting conditions will be similar to that shown in Figure 8.5.2 A. At the same time, the moving tension car is pulled to store the excess belt. Since there are four overlapped turns in the belt bank, only a moving distance of one half is needed for the moving car in order to keep the remaining run belt tight. The haulage cable of the winch is reduced by 1/2 in. (12.7 mm) in length (Fig. 8.5.2 D). As the face continues to advance, the process repeats again and again. When the moving car is pulled to the limiting position, a full roll of belt is stored in the belt bank (Fig. 8.5.2 E). At this time, the winch is released, the belt is unspliced, and the excess roll of belt is removed. The moving tension car is then returned to the original position and the belt is re-spliced, ready for repeating the processes as shown in Fig. 8.5.2 A–E. When the panel is near completion, special provisions must be made because the flexible distance for the conveyor is approximately 165 ft (50 m). Therefore, when the transportation distance in the headentry is less than 200 ft (60 m), the flexible belt conveyor cannot be used for adjusting the distance.

8.5.4 Skip hoisting

There are several longwall mines where skips are used to hoist coal out of the vertical mine shafts. Those mines were designed originally for room and pillar mining in the early 1970s and converted to longwall mining in the late 1970s to the early 1980s. Without exception, the designed mine capacity was far smaller than that of modern longwall mines. Due to its limited capacity and intermittent nature, skip hoisting of coal has often been the bottleneck of production for those longwalls. However, continuing improvements in hoist systems including structure and skip speed have allowed those mines to achieve a comparable production level to those longwall mines equipped with a slope belt system. Improvements included structural and electronic control and automation. Structural improvements include lighter materials for skip construction and more efficient and larger motors. Skip sizes normally range from 27 to 32 tons. Skip hoisting speed is now running at 1700 to 2000 fpm (8.63–10.16 m sec) with minimum acceleration and deceleration periods, resulting in an average speed per trip around 1100 fpm (5.59 m sec). With the implementation of good quality maintenance, the availability of the skip hoisting system is often better than that of an underground belt conveyor system. In fact, skip availability is normally in the high 90% compared to high 80 to low 90% for an underground belt conveyor system.

8.5.5 Underground storage bunker

In order to fully utilize the production potential of modern longwall mining, each and every subsystem must be available and function properly at all times.

After being cut loose by the shearer drums, coal is loaded on the AFC and moved out to the surface through (in sequential order) stage loader, panel belt conveyor, mainline belt conveyor, and slope belt conveyor. On the surface, it normally feeds directly into the preparation plant. That long line of outby belt conveyors, including the preparation plant, is often the source of lost production. In addition, if skip hoisting is used, surge capacity must be available in order to eliminate a transportation bottleneck as stated in Section 8.5.4. In recent years, a remedial measure for reducing production caused by stoppage of the outby transportation subsystems is the use of underground bunkers.

A well-designed bunker stores materials in times of emergency and discharges them back into the transportation system at the appropriate time and rate. A bunker can regulate the flow of coal to avoid the sudden surge of a load and allows the longwall to continue to operate during a stoppage of the outby transportation system. To be effective, a bunker must have sufficient storage capacity so that the longwall can continue to operate for a certain length of time when delays due to the stoppage of the outby system occurs. In this respect, the size of the bunker may vary from one hour to one shift's production or 500 to 7000 tons. Bunkers are generally located near or around the end of the main line belt conveyor at or near the shaft or slope bottom. The size of the opening depends on the size and type of bunkers. For instance, a 5000 to 7500-ton bunker may need an opening in the range of 20 to 22 ft wide by 40 to 75 ft high by 300 to 400 ft long (6.1–6.7 m × 12.2–22.9 m × 91.5–122 m).

There are four types of bunkers (Fama, 2005): moving bed bunker, moving car bunker, bin type bunker, and strata bunker.

8.5.5.1. Moving bed bunker

Figure 8.5.3 shows the schematics of a moving bed bunker.

They are constructed in sections and are normally installed in line with the main line conveyor. The incoming coal for storage is fed at the outby end of the bunker. The bunker is equipped with a chain conveyor on the bottom that can run either backward or forward. Under normal conditions, the chain conveyor is at a standstill, and the materials just flow through (Fig. 8.5.3 A). During storage loading, the chain conveyor is running backward in the inby direction, materials are gradually filling up the bunker (Fig. 8.5.3 B). Sensors are installed inside the bunker to stop in-feeding when the bunker is full. During discharging, the chain conveyor is reversed to run forward in the outby direction (Fig. 8.5.3 C). The capacity of this type of bunker is small, up to 500 tons. Automation, including a controlled discharge rate, is available.

8.5.5.2 Moving car bunker

The moving car bunker uses track-mounted mine cars with a bottom discharge hopper. They are closely coupled together in series to form a train and hauled back and forth by an endless rope (Fig. 8.5.4). The capacity of the moving car bunker varies by up to more than 2000 tons.

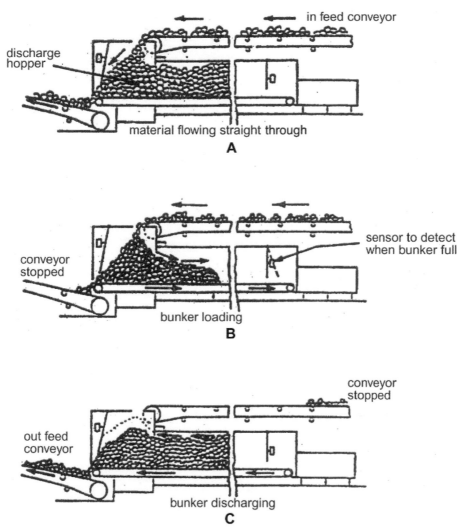

Figure 8.5.3 Moving bed bunker
Source: Modified from Fama, 2005

The bottom of the bunker is a non-driven belt conveyor with fixed pulleys at both ends. The top flight of the conveyor is supported by closely spaced idlers so that the top flight is close to the car bottom to form the base of the bunker. The belt is clamped to the car at the very end of the train so that the belt will move under the action of the loaded cars.

During loading, the bunker is moved inby gradually, whereas during discharging, as the outby conveyor begins to move, it allows the bunker to move outby, such that the bottom of each car is opened in series, allowing the loaded materials to discharge onto the outby conveyor.

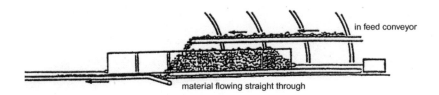

Figure 8.5.4 Moving car bunker
Source: Modified from Fama, 2005

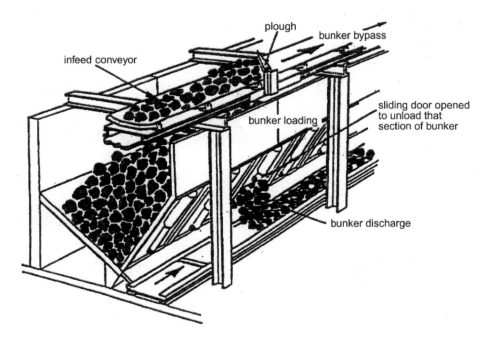

Figure 8.5.5 Bin-type bunker
Source: Modified from Fama, 2005

8.5.5.3 Bin type bunker

Figure 8.5.5 shows a bin type bunker. Unlike the previous two types of bunkers, it is stationary without any moving parts. The bunker is a V-shaped structure installed under the in-feed conveyor. Its capacity varies with the cross-sectional area and length of the hopper and can hold more than 2000 tons.

A plough, which is progressively moving, is used to deflect the materials from the in-feed conveyor onto the bunker section by section until the bunker is full. During discharging, doors are opened hydraulically to allow the materials to fall through and onto the chain conveyor underneath.

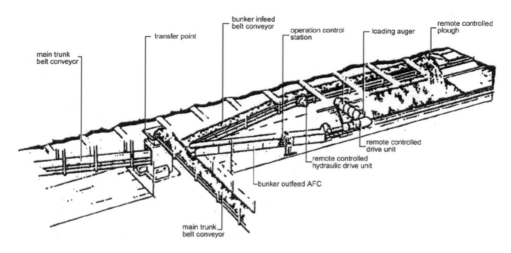

Figure 8.5.6 Strata bunker

Source: Modified from Fama (2005)

8.5.5.4 Strata bunker

An earth bunker utilizes an old entry for coal storage. Coal on the in-feed conveyor, which is mounted high above the entry floor in the roof, is ploughed off remotely onto the entry floor (Fig. 8.5.6). The coal is reclaimed by a loading auger and cowl onto the bunker out-feed chain conveyor and discharged onto the panel belt from the bunker. From there it travels via the transfer point onto the main outby belt conveyor.

Figure 8.5.7 shows a 7500-ton (6818-mt) capacity bunker and ancillary functions facilities located near the slope bottom of a coal mine in the Pittsburgh seam. It is 22 ft (6.7 m) wide by 325 ft (99.1 m) long by 75 ft (22.9 m) high. A reclaim belt slope and tripper-car feed belt extended the bunker length to 500 ft (152.4 m) long. Coal enters the bunker slightly above mine level on a feed belt, where it is dumped and distributed evenly and automatically on the underlying bunker through a "tripper car" running on beam-supported rails. Seventeen feed bins serve as the bottom of the bunker and independently distribute coal onto an underlying reclaim belt, which, in turn, runs up the reclaim slope back to mine level. The height of the bunker is approximately 49 ft (14.9 m) from the bottom of the feed bins to the maximum storage elevation.

8.6 Operation and maintenance of the armored flexible chain conveyor

8.6.1 Operation of AFC

An AFC is designed to operate successfully when it is snaked but is more efficient when arranged in a straight line. Because a "snake" consists of two curves or bends in opposite directions, one compensates for the other with regard to chain length and gives equal chain tension. A single bend or curve, however, is not only detrimental to the conveyor

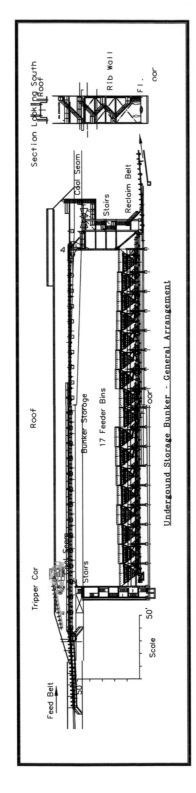

Figure 8.5.7 The side and end profiles of surge bunker in a Pittsburgh seam longwall mine

Source: Robinson *et al.* (2007)

chain and pans, but will rapidly cause the return chain to drop out of the guides, resulting in major stoppage. Therefore, sharp snaking should be avoided as much as possible. The snaked portion should not be less than five to seven units of line pans, depending on pan and web width.

A bend in a conveyor throws all the tension on one chain resulting in unequal elongation and permanent deformation. It is advisable in TIB strand to check the faceline at least once a week. During conveyor advance, resistance is less if the conveyor chain is in motion.

If the coal seam dips and the panel advances on the strike direction, the sequence of the conveyor advance is very important. If it is advanced from the tail end towards the head end, the conveyor tends to slip down. The amount of slippage varies considerably with seams, depending largely on the floor and seam inclination. If it is advanced from the head toward the tail end, the resistance to the advancing ram is larger and may also cause the conveyor to creep toward the tail end. If the conveyor is advanced from both ends toward the center, it may lead to the pans becoming "end-tight," causing considerable resistance to the advancing rams. In severe cases, the center portion of the conveyor may be lifted up.

The simplest yet most effective method for preventing downward sliding or upward creeping of the AFC and shield supports is to adjust the relative direction between the face direction and seam dip.

Bessinger (2001) recommended several tasks to increase AFC life: (1) rotate pans at panel change to avoid the same location being subjected to repeated wear; (2) ensure pan lubrication is adequate, water lubrication of the deck increases life greatly; (3) inspect regularly for deck and side section thickness, check articulation at appropriate locations of the face; and (4) perform panel change maintenance and repair, such as cleaning connector pockets, replacing stretched connectors, building up races, etc.

8.6.2 Maintenance of AFC

Proper chain tension is very important. If the chain has too much slack, it can quickly lead to links being jammed in the strippers, jumping over the sprocket teeth, falling out of the return guides, or even a breakage of the chain. When an AFC is newly installed, it is common for the chain to create slack more rapidly. This is caused by the wear and pitch increases of chain links and the joints between pan closings.

If the two chains in TIB are subjected to unequal tension, the flight bars will eventually assume oblique positions. If this happens, the faceline should be checked for straightness because a slight bend or curve results in chain slack on the inner side of the curve.

An automatic chain tension system must be used and all flight bars inspected. A bent flight bar should either be straightened or replaced.

8.6.3 Adjusting and replacing the chains

8.6.3.1 Causes of broken chains

There are two types of chain breakage:

1   Static failure. The chain will break suddenly if it is subjected to impact loading exceeding the static breaking strength. It usually breaks at the curved portion of the link or at

defective parts, such as an area with defective welding. The ruptured surface is jagged, and the straight portion of the link is lengthened.

2   Fatigue failure. The chain is subjected to repeated loadings, usually less than the static breaking strength, for a long period and breaks in fatigue. The rupture point usually occurs at the junction where the curved portion meets the straight portion. The ruptured surface consists of two parts: the moon-shaped and smooth part is due to fatigue failure, and the jagged part results from static failure.

During operation, the chain is subjected to both static and dynamic loadings. Dynamic loading includes the impact forces produced at starting and sudden stoppage and also at periodic acceleration and deceleration. The forces tend to cause static failure, whereas dynamic loading, although small in magnitude, is usually repeated for a long period and will eventually cause fatigue failure.

Chain wear is another reason for a broken chain. When chains enter the sprockets, relative sliding occurs between them and induces wear at the adjacent links. Under normal operation, there is little wear between the links and the sprockets. But when the links do not match the sprocket teeth (e.g., chain jumping), severe chain wear will occur.

Finally, the friction between the chain and the pan cannot be ignored. It can be severe if pan installation is not proper or severe conveyor snaking occurs.

8.6.3.2 Chain inspection and replacement

Chains have a tendency to lengthen due to wear and shock loadings. Chain elongation is usually used for determining the amount of chain wear and the teeth-matching characteristics of the sprockets. When link elongation exceeds 1.5–2%, it should be replaced.

8.6.3.3 Adjusting the chain

The problem of adjusting chain length occurs when installing the AFC, replacing the chain, changing face width, and tightening the slack chain. During adjustment, the chain is frequently one or two links short. In order not to cut the links and maintain proper intervals between adjacent flight bars, it is advisable to have sets of chains at variable lengths. During installation and adjustment of the conveyor chain, it is not recommended to cut the links randomly, because the chain is very costly. Additionally, it breaks the original long chain, which often causes new problems in subsequent chain adjustment.

8.6.3.4 How to locate a hang-up in the AFC

From time to time, the AFC may get hung-up, and Equation (8.6.1) can be used to determine where it is stuck, assuming the hang-up is located at point x:

$$\left[\frac{S_x}{S_t} = \frac{D_x}{D_t} \right] \tag{8.6.1}$$

Where D_x is the distance from the head drive to point x and D_t is the panel width or ½ of the AFC chain length. The total slack is the sum of slack measured at the head and tail drives. The slack at the head drive can be obtained by starting the head motor(s) only and run the chain to stall and stop the motor(s). The chain will fall back. The amount of fall back is the slack at the head drive. Repeat the same at the tail drive. Note that the method is only as accurate as the measurement taken.

Chapter 9

Automation of longwall components and systems

9.1 Definition

The word "automation" in longwall mining has been used loosely in the coal industry; it means different things to different people. So a definition is in order for this book

Modern longwall mining employs three machines, shearer, AFC, and shields, to produce and move coal. Automation means those three machines function automatically themselves without human intervention. How do these three machines work together to produce and move coal? The shearer, riding on the AFC pans and following the configuration of AFC as traveling track, cut coal of fixed width (web), and shield, with the relay bar connected at the center of its base to the pan in front of it, uses the relay bar to pull itself and pushes the pan forward for a web distance after the shearer has passed by. Miners are employed to operate those three machines to accomplish those functions. The shearer needs two operators, one for the lead and the other for the trailing drum, to steer the drum to cut at the coal-roof rock interface for the lead drum and coal-floor rock interface for the trailing drum. The shieldmen are to advance the shield and push the pans for a web distance following the shearer's cutting. The numerous shields across the face need one or more shieldmen to initiate and complete the shield advance cycle and then push the panline one by one following the shearer. Since every pan is mechanically linked to the shield behind it, AFC does not need an operator. Consequently, "automation" in longwall mining centers on equipping the shearer and shields with "intelligence" to operate by themselves doing what the human operators do: seeing, hearing and feeling. There is only one shearer while there are numerous shields, the motion of which fortunately advances linearly, and all shields are the same. The shearer drums however require the identification of, and follow the ever changing, coal/rock interface, which obviously is much harder to automate.

Therefore "full automation longwall mining" or "intelligent longwall mining" in this book refers to manless mining at the face. In other words, normal coal production operations are conducted without any crew members at the face, nor at any time the production operations require crew's intervention to correct any problem. So it is the highest level of automation and, based on the current technology available, it is not attainable. So "automation" in this book refers to different levels or depth of "full automation." The year the automation technique or device for any of the three machines or their components was introduced was stated if available and periods of improvements toward rising maturity and/or power capacity were recorded wherever possible.

Since longwall mining consists of three major machines, shearer, AFC, and shields, there are two types of "automation": "automation" for each machine and "automation" for the longwall system (i.e., the three machines mutually communicate and operate automatically for coal production). Since each machine has its level of automation developed at certain specific time period, the precise level of automation for the whole system at any instant is difficult to specify. Besides, there remains many unknown and unexpected events to be defined for automation for each machine and longwall system before "full automation" is realized. Consequently, it is important to recognize that the word "automation" is qualitative, not quantitative, in this book.

Ideally, in a fully automated longwall mining operation, the longwall face has three straight lines, faceline, panline, and shield line, in the horizontal plane and in the vertical cross-section, the roof and floor lines are flat and horizontal. Therefore, the system will run automatically back and forth between headgate and tailgate cut after cut without the need for human intervention. Unfortunately, actual roof/rock and floor/coal contacts are rarely flat and uniform, and the three lines (faceline, panline, and shield line) are hard to maintain ideally straight in a wide panel. Consequently how to make those three machines to steer themselves automatically in technological development to cut and move coal follows a long history of learning process. Therefore, the term "automation" can be any milestones leading to the "full automation."

It must be strongly emphasized that underground operational experience has demonstrated in the past four decades that it takes time for a new automation device/technique to reach maturity, i.e., the elapse time between introduction and becoming a reliable one can be very long due to complicated geological and mining conditions underground. Therefore, the introduction of an automation device/technology does not necessarily mean it will work reliably for safe production operations, rather it requires continuing refinements and upgrades, the period of which varies with machine and technology.

9.2 Automation of three face machines

9.2.1 Automation of shields

Automation of shield consists of individual and then multiple shields from a few to all shields in a longwall face.

A shield's cyclic movement is linear and simple, i.e., lower a few inches, advance a web distance, and raise a few inches to set, followed by retracting the relay bar to pull itself forward and then extending the relay bar to push the pan forward for a web distance. However, it has multiple units normally 150–270 for today's panel width and thus requires sequencing. Therefore, the shield is the easiest and most beneficial to automate among the three face machines. This is why shield automation began in early 1980s and is the most matured among the three machines.

Automation of an individual shield is to add a pilot valve and an electromagnetic solenoid valve to the hydraulic valve to control all movement steps in a cycle such that with a light touch of a button, the three separate steps of shield advancing cycle is performed automatically in sequence, rather than a push button for each of the cycle steps when only the manual hydraulic valve is available. This is an electrohydraulic shield that is much faster and safer than only with the hydraulic valve. The shield control unit (SCU, i.e., PMC-R or RS20s) mounted on every shield is in fact a computer. Since there are many units of shields in a face, those SCUs form an internet, enabling multiple shields, up to the whole face, to operate in sequential order accurately by programmed menu.

In the United States, electrohydraulic shields were first installed in 1984 at the Loveridge Mine (now Monongalia County mine) about 18 miles west of Morgantown, WV. It was a landmark event in that it was among the first industries ever to employ a computer in production operations because IBM's personal computer was barely on the market then.

Electrohydraulic shields can be operated individually or in a batch of three to eight units in the beginning for safety reason and later expanded to the full face.

When it was first introduced in 1984, electrohydraulic control shields had many problems, the major ones of which were not resistant to underground environments (moisture, dirt, and dust) and cable faults. It was damaged quickly and required replacement frequently. However, the OEM had worked hard to solve the problems. By around 1990, it had become very reliable and gained industry's full acceptance.

A fully operational floating batch control electrohydraulic system that requires only two to three push buttons and yet enabling a faster shearer cutting, thereby higher and safer production: specifically, (1) shield cycle time reduced from >40 sec (manual) to 6–12 sec, (2) positive setting ensured all shields were properly set and uniformly loaded across the face, and (3) reduction in number of shield men.

Numerous computer programs have been developed to run various combinations of automation sequence, for diagnostics and for corrective measures for error messages such that shield control becomes much more sophisticated and much more precise.

9.2.2 Automation of shearer

Shearer automation is the most difficult part among the three face machines because it has many components doing different interacting tasks that cover and therefore require all aspects of human intelligence: seeing, hearing, and feeling.

An automated shearer will cut the coal and stay in-seam cut after cut without the need for human intervention. In order for the shearer to perform those tasks replacing humans, it has to have the "seeing, hearing, and feeling" ability that human uses to perform the tasks. For the shearer, it means electronic "sensors." Since a shearer has many components that perform different functions, each component needs one or more different types of sensors to accomplish its assigned tasks. For the shearer as a unit, it needs to know its location at the face, and direction and speed of travel (or acceleration and deceleration); for the drums, it needs to know its elevations with respect to the interfaces of roof/coal and floor/coal and shearer's location; for the cowl, it needs to know it is on the head or tail side anywhere at the face. Those features must be able to display in screen all in real time.

The sensors developed for monitoring shearer's location and/or traveling speed and direction are IR, gyroscope or rotary encoder. IR is an infrared emitter and receiver. Alternatively, inclinometers are used to monitor the shearer's pitch, roll, and yaw positions. Inclinometers are also used to monitor the ranging arms for drums' (and bit tip) elevations.

Figure 9.2.1 shows a likely longwall face working condition for an automated shearer. The attitude of the shearer at any instant can be determined by movements in three mutually perpendicular directions; "pitch," "roll," and "yaw," using the terminology by an airplane in the air. However, in longwall automation, "pitch," "roll," and "yaw" denote the inclination of machine chasis in the face advancing and faceline (panel width) directions, and snaking (deviation from straight line) of panline, respectively. A gyroscope can perform all three measurements simultaneously in real time. Otherwise, each of the three movements requires a separate inclinometer to monitor.

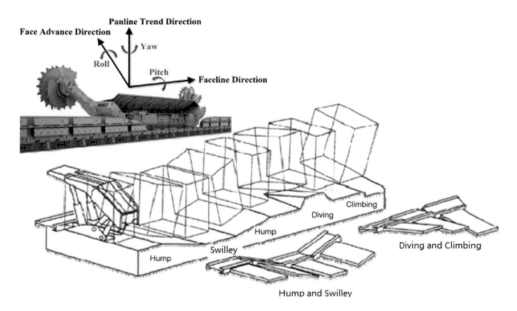

Figure 9.2.1 Definition of shearer attitude (pitch, roll, and yaw) at longwall face

Source: Modified from Caterpillar

As the quality (resolution and reliability) of sensors has been greatly improved over the past few decades, so are improvements in the level of automation in components as well as the shearer as a unit. Consequently, sensor control becomes more precise enabling higher level and maturity of automation today.

Shearer automation began with radio remote control in 1978 (analog) and then in 1984 (digital) (Junker and Lemke, 2018).

The "Memory Cut" technology was introduced in 1986. It has many versions. The most popular one was that the cutting height was kept constant. In this version, the leading drum in the training run was guided by the shearer operator, while the trailing floor drum was fixed by running the shearer with a constant mining height. It produced a more uniform floor horizon, although not necessarily following the true floor-coal interface The data on drum elevation and ranging arm inclination as well as shearer location were recorded. Those data were recalled to run the subsequent shearer's cuts without operator's intervention as long as it stayed in-seam or within the desired horizon. The resulting face cutting profiles were displayed. This level of automation required a ranging arm tilt sensor and shearer location sensor. An on-board central computer was also needed to process the data. The technology then was not reliable and required frequent retraining run. For shearer location, the infrared sensor was developed in the mid-1980s (Fig. 9.2.2 A). It identified the shearer by shield number along the face which was not sufficient for precise "automation" control. The serial link counting the engaged haulage sprockets as the shearer travels across the face was developed for more accurate shearer location detection (Fig. 9.2.2 B). All modern shearers are equipped with both systems.

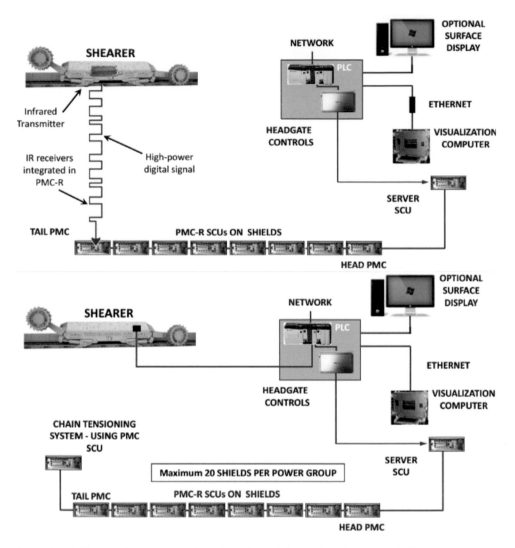

Figure 9.2.2 Shearer location tracking by infrared (upper) and serial link (lower)

Source: Courtesy of Caterpillar

In recent years, the "memory cut" has improved considerably. Roof and floor profiles can be more accurately controlled and mapped. It even allows the shearer operator to change the cutting profile anytime. However, due to geological variations, there are many fine vibrations in the produced roof and floor profiles. Uneven floor (Fig. 9.2.1) not only causes problems in equipment operations, but also makes automation difficult. It follows that floor control, rather than the conventional thinking of roof control should be the automation focus. A smooth floor makes the face equipment sitting and running on it more to act like the designers perceive.

A gamma-ray coal thickness sensor was developed in late 1989 to early 1990s (Bessinger, 1991, 2001; Ingram, 1994; Nelson, 1989) and mounted on the shearer for shearer's cutting horizon control. For this automation system the shearer was equipped with gamma-ray coal thickness sensor for coal/rock interface detection through coal thickness measurement, roll and pitch (inclinometers) sensors for the attitude of shearer body, and tilt meters for ranging arms height. The data were collected at 0.6 m sampling interval. All four sensors and meters were bulky and low in resolution. Unfortunately, the gamma ray coal thickness sensor only worked in shale roof. So the system never gained industry-wide acceptance and never took off.

The next level of shearer automation was the concurrent development of more sophisticated sensors to replace the old ones for more precise control of shearer operations. They include the pitch and roll sensors for sensing the attitude of shearer's body; encoder for the ranging arm for control of drum and cutting bit elevation; and rotary encoder for tracking shearer's location that samples every 60 ms. Other accomplishments include automation of gateend operations and individual components such as ranging arm, cowl, lumpbreaker, haulage speed and water. Software has been developed to control cutting, cowl, haulage, ranging arm, and other components independently. As a result, the shearer can run automatically by itself all day long if it is not out of the seam. However, the shearer still requires one or two operators to operate the training run and constantly evaluate when a new training run is required (Komatsu Mining, 2018).

The latest development is pitch steering in which the floor drum is positioned such that it is steered to a specific pitch angle on the following pan push in order to smooth out the floor steps (Komatsu Mining, 2018).

Pan profiling is another new development. It is designed to command the shearer to follow a defined roadway mapping both along the headgate and tailgate and in-between (i.e., inside the panel across the full face) (Fig. 9.2.3). In this technique, pan profile is mapped as the shearer is traveling. The mapped pan profile is used to project the amount of correction needed ahead of it such as shown in Figure 9.2.3 A when the shearer enters the gateend roadway. Management can also draw the desired profiles along the roadway and across the face based on the 3D panel geological map (Fig. 9.2.3 B), the shearer will determine the

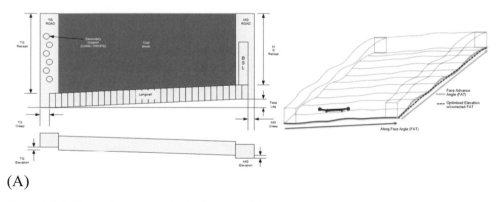

(A)

Figure 9.2.3 Principle and method of pan profiling

Source: Courtesy of Caterpillar

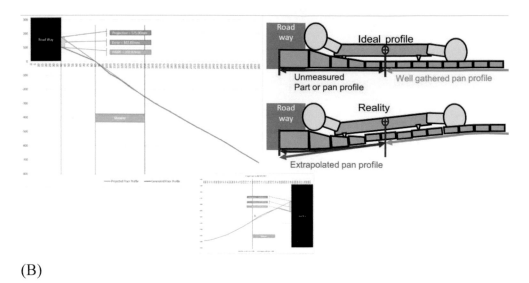

(B)

Figure 9.2.3 (Continued)

correction needed in each cut as the sensors compare the actual profile with the desired profile and adjust itself to follow the desired profiles.

From the late 1990s to 2013, Australia's CSIRO successfully developed systems for face alignment, shearer horizon control, and AFC creep control. Face alignment is the process of maintaining a straight line AFC from head end to tail end, thereby enabling the shearer to cut a straight faceline. The desired face alignment is maintained by accurately measuring the 3D position of the shearer using an Inertial Navigation System (INS) (CSIRO, 2005: Kelly, 2005; Ralston *et al.*, 2015). This information is used to control the movement of shields accurately and consistently. In horizon control, a thermal infrared camera is used to detect the heating of marker bands in the coal seam, using the relative offsets as a reference for correcting the vertical location of the cutting drums. In AFC creep control a sensor based on a 2D laser scanner is installed on the AFC structure at the headgate T-junction to actively measure the closure distance per cutting cycle. These systems, collectively known as LASC (Fig. 9.2.4), have been widely used in the Australian coal industry since 2014, but were only recently introduced into US longwall mining and only two mines have adopted so far in 2019. New systems being deployed in Australia in 2019 by CSIRO include 3D laser scanners for full cross-longwall face measurement, and INS based continuous miner navigation for roadway development.

As the face gets wider, face alignment becomes more important in that face operators cannot easily view across the face to make adjustment to keep the face straight. A crooked panline can distort shearer location from cut to cut, making automation unreliable.

A radio motion monitoring system with accelerometer embedded in the handheld radio has been developed to detect operator's three events of movement: loss of movement, free fall, and sudden impact.

Communication among different subsystems and gateend computer and surface control station has been upgraded to fiber optic as primary and Ethernet as secondary method.

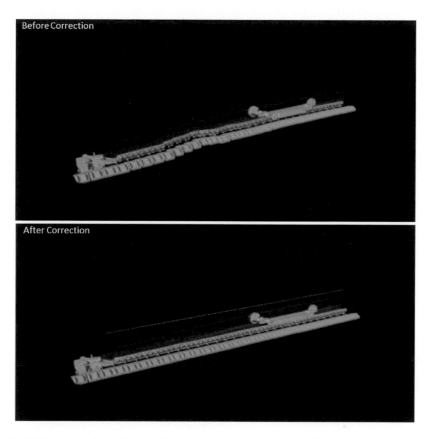

Figure 9.2.4 The curved panline before adjustment (left) and the straightened face after applying LASC corrections (right)

Source: Images courtesy of CSIRO

In summary, with various high quality sensors, the current automated shearer allows the programming of fully automated cutting sequences across the whole face, including gateend turnarounds. The shearer operators can create an initial cutting profile and extraction heights using a graphical planner. The shearer then automatically replicate the profile until conditions change. The operator can then override control of the roof drum to follow the preferred horizon. Using the new horizon data, the remainder of the cutting sequence is fully automated using the pre-defined extraction heights.

9.2.3 Automation of AFC

Since the panline is stationary and attached to, and relies on, the shield to move, it does not need to be automated. Automation of AFC thus concentrates on drive motors to ensure smooth running of chain strand and consists of soft start, load sharing, overload protection, and optimum chain speed control as well as chain tension.

The AFC drive system requires static and dynamic power reserves to startup against high loads and handling of slack chain. In the 1980s, due to smaller chain size and chain strength, the sudden overload of AFC often occurred, which caused chain breakage and/or drive motor burnout. So the soft start technology was developed for AFC drives in 1994 to handle heavy-load startup to guarantee load sharing among the drive motors and avoid slack chain at the head drive frame. It will stop when a sudden overload occurs to relief the pressure and re-clutch quickly. In recent years, the application of variable frequency drive (VFD) will slow down the conveyor speed when encountering overloads.

The chain pretension can be adjusted automatically to the operating conditions by retract-ing or extending the hydraulic cylinder of the tail drive frame. There are two automated chain tensioning systems. In one system the basic electric system consists of two main com-ponents: a hydraulic pressure sensor and a piston displacement sensor in the cylinder. These two components are connected to the electro-hydraulic control unit PMC-R (Fig. 9.2.5 A and B) located on the tail-end drive in the vicinity of the hydraulic system of the chain tensioner and regulates the chain pretension by means of one or more hydraulic tensioner cylinders, depending on the design of the tensionable drive frame. The control unit is a dedicated shield control unit which is connected to headgate computer (Fig. 9.2.5 B).

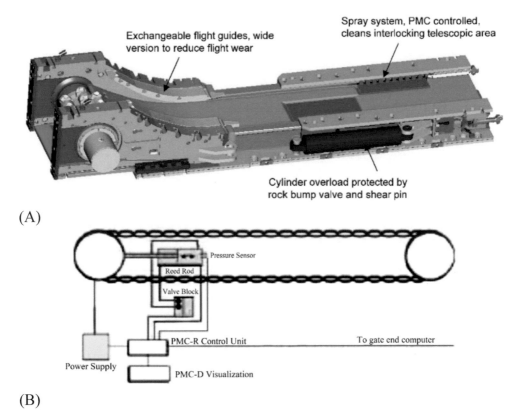

(A)

(B)

Figure 9.2.5 Tail drive automated chain tensioning system

Source: Courtesy of Caterpillar

In the other system, a non-contact load sensor mounted on the sigma section near the tail drive sprocket measures indirectly chain tension and the system automatically adjusting the chain tension (Fig. 9.2.6).

Chain tension control takes place between two key values representing the minimum and maximum chain slacks, either in terms of pressure or piston stroke (displacement) in the

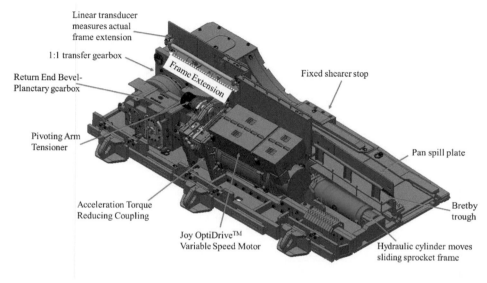

(A)

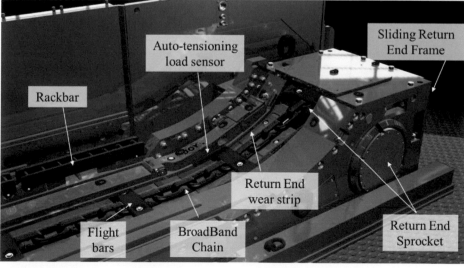

(B)

Figure 9.2.6 Tail drive automated chain tensioning system II

Source: Courtesy of Komatsu Mining

cylinder. Chain slack is maximum when the conveyor is operating with a full load. It is reduced to a minimum when the conveyor is operating under no load. The automatic chain tension control takes place between these two extreme slack chain values. The pressures in the cylinder corresponding to the three chain types of on-site production operation conditions as shown in Figure 8.3.9 (p. 280) are calibrated when the AFC is first installed at the face. These three pressures served as chain tension operation and adjustment criteria. Alternatively, the two extreme values of no load and full load can be calibrated with the cylinder stroke or displacement for displacement control.

During normal production operations, the pressure in the cylinder is monitored to keep it to stay around the optimum slack chain condition (middle figure in Fig. 8.3.9, p. 280).

The automatic chain tensioning system attempts to set the optimum chain tension for all possible loading conditions, resulting in less wear, thus longer service life of chain, less need for maintenance of the chain, chain sprocket and drive components, less friction on all circulating and friction surfaces, and optimum current consumption of the motors.

9.3 Automation of longwall system

According to the chronological development of automation of each of the three machines, "automation" of the US longwall system can be roughly divided into the following three stages with rising maturity of automation longwall system: (1) shearer-initiated-shield advance (SISA); (2) semi-automated longwall face; and (3) remote control of shearer.

9.3.1 Shearer-initiated-shield-advance, mid-1980s–present

The first step in automation of longwall system is to establish communication between automated shearer and automated shields to develop an automation longwall system. In this automation system, as the shearer passes by, the shield behind it will automatically advances one by one, followed by the panline being pushed also one by one automatically. The system requires a shearer location sensor that determines the shearer's location so that the headgate computer know where the shearer is cutting and issues commands to certain shield behind it to begin advancing cycle.

An infrared emitter was developed in mid-1980s (see Fig. 5.4.11) as a shearer location sensor. It is mounted on the main body of the shearer and, as it passes a shield, the infrared radiates toward and covers three shields, one on each side of it that in turn relay the shearer's position and direction of travel to the headgate computer, which then commands the next shield (two to three shields behind the trailing drum) to initiate advancing cycle. This longwall system automation requires the following: (1) a headgate computer, (2) a shearer mounted with an infrared emitter, (3) shields with fully operational electrohydraulic control system, (4) a shearer that can perform "memory cut" with fixed height full face, and (5) the face shift crew may be reduced to four or less. Note that the infrared sensor's radiation covered three shields wide or 4.5 m which was not precise enough to enable consistent function automatically.

It was soon discovered that infrared sensor for identifying shearer location had several shortcomings: in areas where the heavy dust laden air and water spray mists exist, its signal would confuse the receiver leading to missing shearer location. Also since the location is represented by shield number which was 1.5 m and now 1.75 m or 2 m wide, the shearer's location so determined was rough such that after a few web cuts, the location at both head and tail ends may misrepresent. Therefore, a serial link that count the number of the teeth of shearer's driving sprocket, much like the automobile's odometer counter and much more accurate in location identification than the infrared system, was developed and installed in the early 1990s. Today, all shearers are equipped with both systems (infrared and serial) to complement each other. In the 2000s, rotary encoder monitoring the haulage motor gear train was developed and has since been used for shearer location sensor with an accuracy of a few millimeters.

The system worked well for the whole panel width when the shearer employed the uni-direction cutting method. When bi-direction cutting was used, the system worked well within the panel, but more often encountered difficulties in carrying out the face end operation that involves two double-backs and wedge cuts in automation mode.

The problems with face-end operation were resolved in the late 2000s. Since then, the SISA system is fully operational full face wide, including face end turn around, wedge cut and AFC snaking. However, the current automated face end operations require 20 shields of distance to complete and takes more time than what is required for manual operation. For that reason, automated face end operation has not been adopted industry wide as routine practice.

9.3.2 Semi-automated longwall face, 1995–present

This is the system most commonly used in the US longwall mines now. Here "semi-automated" covers various degrees of automation developed up to now. It consists of the following equipment:

1 SISA system.
2 Shearer with ASA (advanced shearer automation) and DCM (dynamic chain control management) or hydraulic chain tensioning systems with or without CSIRO's LASC system (currently only two longwall faces use it). In ASA, the shearer is equipped with pitch and roll sensors, ranging arm inclinometer, serial or encoder for precise shearer location. In DCM, the AFC is equipped with sensors for chain tensioning and uses Optidrive or CST to control load sharing and overload protection on AFC; lumpbreaker and cowl are automatically and independently controlled.

The shearer is operated in memory cut mode: the first cut is operated by the operator. The recorded data (drum heights, pitch and roll sensors at each shear location) are stored in the computer. Those data are recalled to run the second and subsequent cuts. It is much more accurate and reliable than the initial system developed in the 1980s and 1990s. The cut profile can be adjusted anytime by the shearer operator. And in recent years the shearer can also perform pitch steering and pan profiling.

The system still maintains four to eight crew members at the face depending on mining practice and mining and geological conditions: one or two shearer operators, one to three shield men, one electrician-mechanic, one utility man, and a foreman.

9.3.3 Remote control of shearer, 2012–present

It began in 2012 in a coal mine in New Mexico. In order to increase coal miners' safety and production (Zamora and Trackemas, 2013), a shearer operator was moved to a control point outby the headgate T-junction and remotely controlled the shearer's cutting from head to tail, while the other operator controlled the shearer at the face from tail-to-head cutting.

The respirable dust standard was reduced from 2 mg m³ to 1.5 mg m³ effective August 2016. In order to meet the standard, a WV coal mine adopted similar cutting method. In these two remote control systems (Fiscor, 2017):

1   Six to eight video cameras are mounted on the shearer at strategic locations facing different directions such that all essential views of the face in real time are visible for the remote control shearer operator. Zero to four cameras are mounted at AFC head and tail drives, as well as camera mounted in shield across the face. Most are high definition cameras with manual and automated image capture and light control. Images are transmitted through WiFi.
2   A longwall control cache is constructed at T-junction or outby in the mule train where the real time numbers of all monitored parameters for all sensors are available in respective screens. Any time the operator finds deviation, the maintenance guys are notified immediately for checkup. A surface control center with similar function is also installed on the surface.

High definition video cameras are mounted on the shearer, AFC head and tail drives, as well as shields across the face. These cameras are facing different directions and distributed at strategic location on the machines and across the face such that they capture different parts of the face as the shearer travels across the face. The images from each camera are transmitted instantaneously through WiFi to the remote control station located outby the face or on the surface. As the images are merged together on the computer screens, the remote control operator can see clearly the actual face conditions and steer the shearer's cutting drum. The cameras can be operated manually or automatically and with light control. Obviously, the number and distribution of cameras are important factors. Those cameras must also be able to see through thick dust and mist all the time.

With complicated geological and mining conditions, so many unexpected events can occur. For a fully automated face, all of those events must be able to detect in advance and corrective actions implemented timely. It requires sophisticated sensors of proper type and correctly mounted. In the last two decades great strides have been made toward that direction. We are closer but still a long way to full automation!

At this moment, it seems that the remote control of face equipment, in particular the shearer, is the correct step-stone toward full automation as a remedial measure for lack of various reliable precision sensor technologies, i.e., move the face and equipment operators to a safe place away from the face and operate the equipment via video cameras. The problem is to find a video camera that can defy the dirty misty air around the shearer all times and transmit clear pictures to the control center.

The health and safety and production records for the two operational remote control longwall faces have been excellent, except the number of face crew did not reduce much. Perhaps over time it will see the benefits of the manpower reduction once they gain confidence in the system.

The number of face crew remains the same, except one shearer operator is at the remote control center away from the face.

Figure 9.3.1 summarizes the chronology of development of US longwall system automation. The depth of automation of longwall system at any particular year can be found by drawing a vertical line from the year of interest. For example, the 1995 Cumberland Mine system (A line) would have a shield with electrohydraulic and hydraulic water spray, shearer with memory cut and AFC with hydraulic soft start and tail drive tensioner. B line represents state-of-the-art automation longwall system if all available technologies are employed.

9.3.4 Big data, fast communication, and visualization

Modern longwall system is equipped with more than 7000 various types of sensors on every functional component of equipment sampling data in real time. Most of those sensors are actually digital measured and connected to one or more PLCs. Thus the data accumulated are huge at any instant. Numerous software programs have been developed to analyze those data for the individual components and correlation of data among components in individual machine and among machines. The application of modern data transmission technology allows fast communication and enable smooth interlink among the three machines (Figs 9.3.2 and 13. 3.1, p. 474).

Therefore, big data and fast communication technique and dedicated software programs allow real-time presentation of the performance of every component of each individual machine, and any one of the three face machines as well as the whole longwall system (Fig. 9.3.3). Figure 9.3.3 (upper) provides a plan view of the longwall displaying shields, conveyor, and shearer.

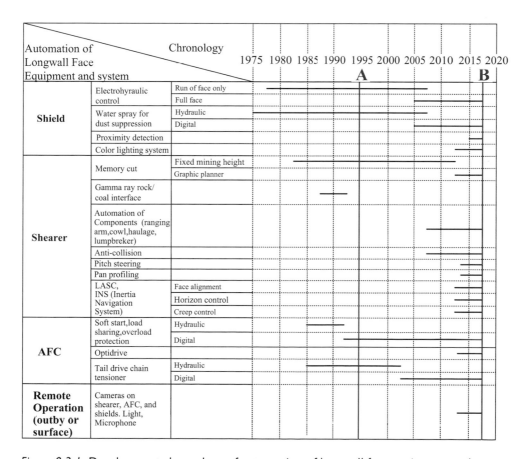

Automation of Longwall Face Equipment and system		Chronology	1975	1980	1985	1990	1995	2000 (A)	2005	2010	2015	2020 (B)
Shield	Electrohyraulic control	Run of face only										
		Full face										
	Water spray for dust suppression	Hydraulic										
		Digital										
	Proximity detection											
	Color lighting system											
Shearer	Memory cut	Fixed mining height										
		Graphic planner										
	Gamma ray rock/coal interface											
	Automation of Components (ranging arm, cowl, haulage, lumpbreker)											
	Anti-collision											
	Pitch steering											
	Pan profiling											
	LASC, INS (Inertia Navigation System)	Face alignment										
		Horizon control										
		Creep control										
AFC	Soft start, load sharing, overload protection	Hydraulic										
		Digital										
	Optidrive											
	Tail drive chain tensioner	Hydraulic										
		Digital										
Remote Operation (outby or surface)	Cameras on shearer, AFC, and shields. Light, Microphone											

Figure 9.3.1 Development chronology of automation of longwall face equipment and system

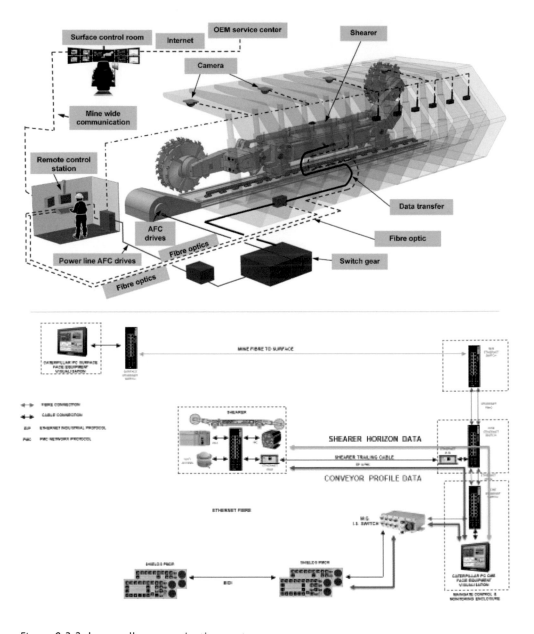

Figure 9.3.2 Longwall communication systems

Source: Upper – courtesy of Langefeld and Paschedag (2018); lower – courtesy of Caterpillar

From this screen the operator sees the current face shape between headgate and tailgate and determines when an alignment is required. The whole screen displays information in descending order, leg pressure, TWAP, and load increment; face height and collision; face position and profile; shearer position; and SCU status. The operator can program the reed rod in DA ram to

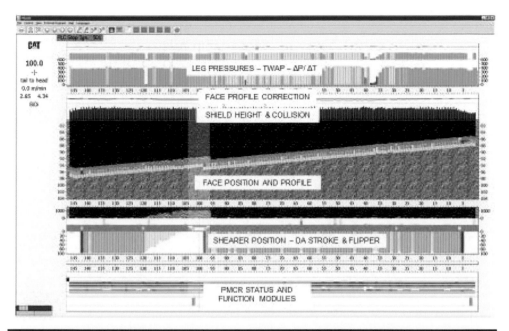

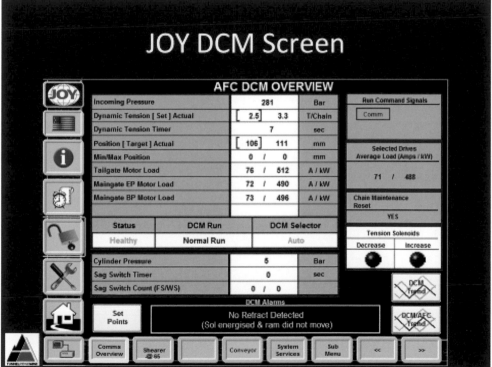

Figure 9.3.3 Visulization screens for individual machines and longwall system

Source: Courtesy of Caterpillar (left); Komastu Mining (right)

adjust panline for whatever correction is needed for the cut. The dynamic chain control management system (DCM) (Fig. 9.3.3 lower) shows among others in descending order incoming pressure, set and actual chain tension, the target and actual piston position of the cylinder and more. The parameters displayed on one or more screen either indivisually or consolidatedly can be customized to reflect the emphasis of individual mine.

3D visualization has also been developed recently (Fig. 9.3.4). The system utilizes SCUs to collect data from each of the three machines and construct a high-fidelity 3D graphic system that allows operators including management to navigate the complete longwall face while

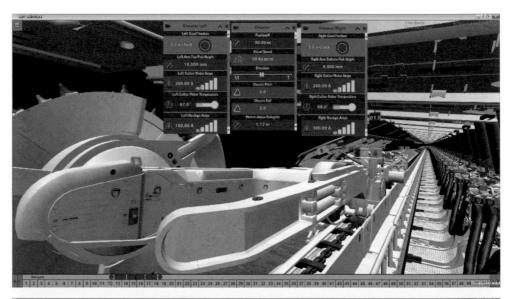

Figure 9.3.4 3D visualization of longwall face, top-shearer and bottom-shield
Source: Courtesy of Komatsu Mining

in production. The operators can select specific data points of any of the three machines and view the longwall system from any angle in real time. It allows the operators to not only see the equipment in operation but also the location of the operator when proximity detection is in effect. The system can be played back.

Visualization screens allow machine operators and mine management to review the real time performance of the components and system and whether they are operating as designed. If not, then the operators can pick it up quickly and inform proper persons for action. Those big data can also be used for diagnostic for operation events and corrective measures issued.

9.3.5 Benefits of automation

In addition to reduction in manpower, automation has the following benefits:

1   Increased coal recovery and quality – since cutting always stays in-seam, run-of-mine coal is much cleaner. In addition, the boundary coal in many coal seams contains higher sulfur content, leaving top coal will enhance product quality.
2   More uniform roof and floor profiles for enhanced roof control (Fig. 9.3.5) – the immediate roof in many coal seams is weak and some top coal must remain to protect it. Automatic horizon control can make this happen easily and consistently. The roof and floor are much smoother. Supports can have better or full contact with the roof and floor, and eliminates severe rolls in both the faceline and mining directions. The mining height is also more uniform from cut to cut. Figure 9.3.5 shows the change in roof profiles during application of shearer automation – degree of automation is increased in two major steps and the period of manual override is constantly reduced. While the roof profile becomes more uniform as the automation application increase.
3   Enhanced health environment – the shearer and shield operators can always stay on the upwind, clean air side and avoid exposure to dusty air and be away from where noise is generated.
4   Extend equipment life by reducing damage to shield and reduction of wear on chain, flight bars and sigma sections.
5   Improve operator safety.

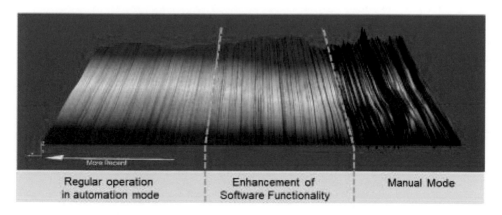

Figure 9.3.5 Automation creates uniform floor profile
Source: Courtesy of Caterpillar

In practice, reduction of face crew was most dramatic when the electrohydraulic control system was widely accepted and applied in the late 1980s during which reduction of shield men often from four to six to one or two occurred in spite of continuing increase in panel width. Thereafter further reduction in manpower with improved level of automation was not apparent.

9.4 The first modern longwall mining system standard

The first semi-automated longwall system began production in 1995 at US Steel Mining Company's Cumberland Mine, Kirby, Pennsylvania. The system consisted of the following: (1) electrohydraulically controlled shields; (2) the shearer with infrared emitter and serial link to enable SISA, and with the memory cut algorithm; (3) AFC with soft start and hydraulic chain tensioner. The mining height was 2.4 m. The system routinely produced 4000–6000 clean tons per shift. For the 300 m wide longwall, shift crew consisted of eight persons: two shearer operators, three shield men, one mechanic, one headgate, and one foreman.

This longwall mining equipment set up and arrangement are designed to allow the face to advance freely without delay caused by any other auxiliary systems (Peng, 2006). Consequently, this system was quickly adopted in US coal industry and later worldwide (Figure. 9.4.1).

Figure 9.4.1 Face equipment setup at the set up room, ready for panel startup mining at Cumberland Mine 1995

9.5 Real-time mining conditions analysis and problem solving

In keeping abreast with the rapid development and application of big data monitoring, transmission and communication between machines, up to 7000 sensors have been installed in the current longwall mining system (Fourie, 2019). Depending on the sampling rate of those sensors, the amount of data obtained at any moment are massive and the usage of this big data has just started. It promises to turn longwall mining from a reactive to a proactive system when the current technology on internet and computer's fast processing power on big data, as well as cloud computing services, are fully utilized. By simultaneously monitoring those 7000 sensors on various machines in the system and system performance, it can not only analyze the performance of individual machines but also the inter-relationship between different data sources in different machines, resulting in great improvement of system performance.

The 7000 sensors are distributed in the whole longwall system including shearer, AFC, shields, stage loader, crusher and belt conveyor. They can roughly be classified into two categories: functional and health and maintenance. Figure 9.5.1 shows examples of locations of some major functional sensors on shearer and each shield.

1   Functional sensors

Shearer: biaxial tilt, inclinometers, gyroscope or encoders, IR emitter, cameras, lights.
Shield and flipper: biaxial tilt (shield only), leg pressures, reed rod, IR receiver.
AFC: biaxial tilt.
Other sensors not shown in Figure 9.5.1 include these:
Water sprays: up to 28 each shield, 130 on shearer, and 20 on AFC.
AFC tail drive tensioner: load, and DA ram pressure and stroke sensors.

2   Health and maintenance sensors

Every motor and gearbox is monitored if applicable for current/voltage, input/output speeds, oil level, oil pressures and temperature, cooling oil level and temperature, outlet/inlet water temperature, vibration, bearing temperature, solenoid states, etc.

Those data are first analyzed, mainly by dedicated software, to evaluate the conditions of individual machines and then compared the results with the results of analyzed data concurrently obtained for other machine(s). Statistical methods are applied to determine if any cause-effect relationship exists between a pair of machine data or among multiple machine data. The individual data and correlated data, if they exist, can be displayed on computer screen in real time on mine site and OEM service centers via the internet. When a set of correlated relationship has calibrated with actual underground events, it can be used for pre-warning of impending events, good or bad.

The advantages of this approach of utilizing the big data is that the traditional way of relying on human's unscientific control of "seeing, hearing, and feeling" is replaced by quantitatively analytics, leading to precision control.

Some examples of the cause-effect relationship are as follows:

1   When the shearer's cutting speed slows down at particular location (represented by shield number) consistently or getting slower for several consecutive cuts, it may signal roof rock invasion downward possibly getting worse cut by cut.
2   When shield leg setting pressure are low, it may indicate encountering weak roof or roof debris is too thick.

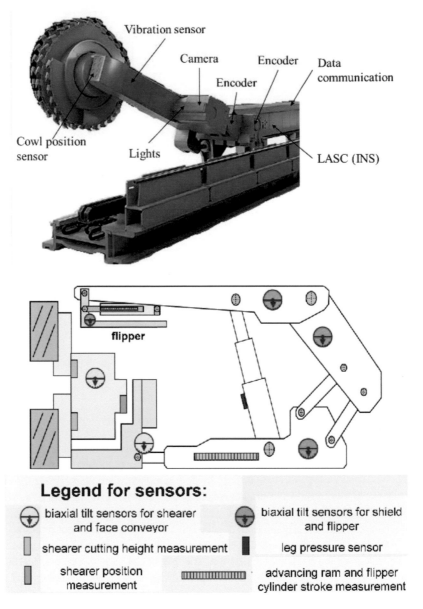

Figure 9.5.1 Example of some major sensors on shearer, shield and AFC

Source: Langefeld and Paschedag (2018)

There are many more cause-effect relationships to be identified when this big data are fully analyzed and relationship established by calibrating with underground events or non-events. Of course, numerous software are waiting to be developed to analyze these big data for improving machine performance and longwall mining operations.

9.6 Automated health and safety technologies since the mid-1990s

The development and introduction of individual face machine automation and later system automation prior to mid-1990s brought about rapid and continuing increase in productivity (see Fig. 1.1.2). However, since then, development of automation, individual or system wide, has been incremental. In the past two decades, the manufacturers have been concentrating on improving the reliability and health and safety features of face equipment, including anti-collision, personal proximity detector, and shield water sprays:

9.6.1 Personal proximity detector

A modern shield is so heavy and fast moving (cycle time less than 8–10 seconds) that it is dangerous to miners working at the face once it moves. Furthermore, shield automation is such that any authorized crew member can initiate shield movement at any place along the face, but, due to the ever-increasing panel width, the ability of the face crew to identify positively the location of other crew becomes less certain, creating further danger to the face crew.

The proximity detector is to determine accurately and reliably if anyone is present at or near the shields of interest and transmit this location information to a control system that will issue commands instantaneously to lock up and halt the movement of the shields of interest.

The system in general consists of more or less three parts:

1   An RFID (radio frequency identification) tag assigned to an individual miner as his/her ID and, when worn, it is used to identify his/her location or presence at a certain location at the face. It transmits RF signals and announced its location by way of signal strength and accelerometer.
2   An RF signal reader or marker or exciter is installed on every shield with a unique shield number ID. It can also receive and emit signals. The detection zone is approximately 10 ft (3 m).
3   A wireless access point or network switch that receives this information and route it to a position software on a server (generally the gateend computer) that in turn transmits this information to the control system on a network.

Both Caterpillar and Komatsu Mining (Joy) have developed a proximity detection system (Fig. 9.6.1) and incorporated into their respective shield control units: PMC-R and RS20s. System accuracy is +/- one support.

9.6.2 Anti-collision technique

Face equipment design in the longwall face allows sufficient clearance between the top of machine body and underside of shield canopy, and the shearer drum cut the panel coal block in front of the shield canopy tip by an unsupported distance of 18–24 in. (0.46–0.61 m) (see Section 10.2 "Longwell Compatibility Test"). So, theoretically, as the shearer moves back and forth between headgate and tailgate, the shearer's body and shield will stay clear of each other. In practice due to the ever-changing mining and geological condition (e.g., changes in roof and floor cutting profiles, undesirable shield advance and canopy

- When a tag is detected in a shield its protection zone extends to an additional shield in each direction
- Any automated sequence that is initiated by the system or another worker will queue for 10 seconds for the tag's alert zone to leave the area before giving up the action

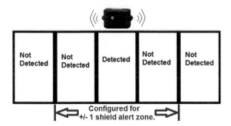

(A)

(B)

(C)

Figure 9.6.1 Principle of personal proximity detectors

Source: A and B – Courtesy of Caterpillar; C – courtesy of Komatsu Mining

attitude either due to setting or ground pressure/roof movement effects), those clearances change constantly in magnitude and very often the clearance disappears and will collide and requires human intervention in advance. Consequently, in the automated longwall shearer face, the shearer must be able to detect potential collisions with shields far in advance and stop or change attitude to be allowed to pass it without collision. This is the anti-collision technique.

In essence, the anti-collision technique involves knowing the location (i.e., exact 3D coordinates) of the tallest part of the shearer's body and cutting drum and shield canopy tip. If they are in colliding course, the shearer will stop before reaching that collision point. Therefore, this information must be available within certain time frame, which is also a function of shearer's speed, ahead of the anticipated collision location. Therefore, the anti-collision techniques involve monitoring and thereby knowing the coordinates of the canopy tip and the tallest part of the shear's body and the leading drum in real time.

The recent model of automated shearers are equipped with sensors that provide information on its instantaneous location and pitch and roll attitudes of the machine body. Consequently, the instantaneous coordinates of the tallest part of the shear's body and the leading drum can be determined. As to the canopy tip position, inclinometers are mounted on various strategic parts (the number and location depends on the developers of techniques). Those data are analyzed in conjunction with the mathematical equations derived from the geometrical relationship of the shield components (Peng and Chiang, 1984) to determine the coordinates of the canopy tip.

Figure 9.6.2 shows a shield anti-collision model with five inclinometers installed, one each in canopy, base plates, caving shield and front and rear lemniscate links, plus convergence monitoring. The heights of the tip, middle, and rear of canopy are measured. Their monitored data are used to model the shield position and hydraulic function status.

Figure 9.6.3 shows the model for collision between shearer and shield. At the set shearer cutting speed, collision detection is determined 10 shields in advance and shearer stops two shields in advance of anticipated shield of collision.

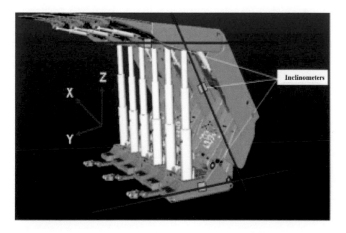

Figure 9.6.2 Shield installed with three inclinometers for unplanned movement detection

Source: Courtesy of Caterpillar

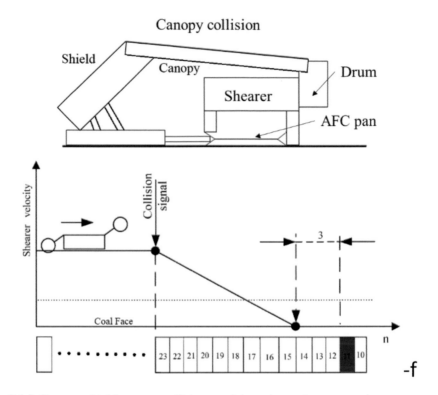

Figure 9.6.3 Shearer-shield canopy collision models and warning protocol

9.6.3 Integrated color lighting

As longwall face becomes longer and equipment automation function increases, the status of each piece of equipment in its automation function sequence is very critical to face crew's safety in carrying out his/her job due to the natural limitation of eye sight. So the implementation of integrated light system with the following color codes is adopted in recent years as an improvement in longwall face safety feature:

1   Yellow light with audible warning indicates AFC start pre-warning and AFC overload warning with load percentage visible on screen.
2   Red light with audible warning denotes warning of automated (blinking) and manual shield movement.
3   Green light on shield means the shield is lockout.
4   Blue light indicates status information available on SCU screen.

9.6.4 Shield sprays for dust suppression

Shield dust is one of the major respirable dust sources.

In weak friable roof, a lot of roof debris, including respirable dust, fall off, mostly in-between adjacent shields as the shield is lowered during the shield advance. A firm roof does too. This is because, as a shield sets, the setting pressure tends to crush asperities on the roof surface. The weaker the roof, the more severe it is. The larger the shield setting pressure and capacity, the more severe it is.

Water sprays on shield for dust suppression started back in early 1980s. It was hydraulically operated, but did not catch on. The new dust standard of 1.5 mg m^3 effective 1 August 2016 renewed interest in water sprays for shield's dust control.

Since shearer's dust sprays already requires close to 300 gpm (1140 liters per minute) of water which tends to flood the floor and weakens the floor rocks, especially for those susceptible to water degradation, water sprays on shields are strictly controlled and used when needed.

Figure 9.6.4 shows a shield equipped with 28 water sprays (note only major ones are shown). They are divided into three major groups: (1) sprays aiming toward the face on the

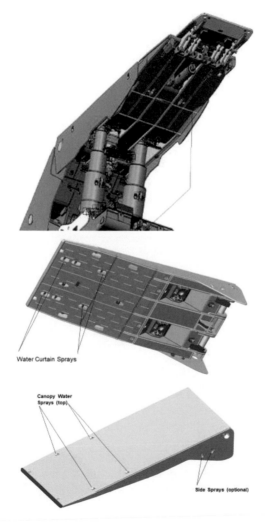

Figure 9.6.4 Major water sprays on shield (yellow dots indicate spray locations)

Source: Courtesy of Caterpillar

front underside of the canopy to prevent shearer's dust from spilling over to the walkway, (2) two rows of sprays, one on each side of canopy top, aiming outward toward the respective side of adjacent shield (they are activated when the respective adjacent shield is being advanced), (3) sprays aiming toward the gob on side shield and lemniscate links, preventing gob dust from spilling over the face area.

Water sprays groups 1 and 3 are activated when the shearer is cutting nearby and turned off when the shearer has passed by some distance. They form a moving umbrella of water cloud from the face side over the shield canopy to caving shield side.

A complementary text on shield water sprays for dust suppression is also included in Section 11.4.2.3C(2), p. 425.

9.7 Discussion

Ever since the introduction of longwall mining in the early 1970s, the shearer operators have been instructed to cut coal in-seam and stay out of roof and floor rocks. This is still the norm in longwall mining. However, the 2019 Coal Age US longwall survey showed that in 21 out of the total 40 US longwalls, the mining height is larger than the coal seam thickness by 6–40 in. (152–102 cm). It either cut into the roof or floor or both with clean coal recovery rate as low as 18%. With today's longwall shearer mining technology, the minimum mining height is about 5.41 ft (165 cm) by machine height while the optimum operation height (crew's comfortable working height) is around 5.8–6.7 ft (178–203 cm). In central and southern Appalachian coalfields where met coal seams are thinner, they can afford larger amount of rejects in coal preparation plant and still make profits due to higher met coal price. Consequently, their cutting heights are 16–40 in. (41–102 cm) larger than the seam thickness. In Northern Appalachian coalfield, the immediate roof, slate, of Pittsburgh seam, 6–20 in. (15–51 cm) thick, tends to fall off immediately after shearer cutting, making the cutting height larger than the seam thickness.

For those seams where mining height is larger than seam thickness, shearer's horizon control, i.e., stay in-seam, is not a critical factor in that it does not need to follow the coal/rock interfaces. Rather the important factor is to maintain uniform flat roof and floor surfaces for smooth operation of shield and AFC pans. With the ever-increasing shearer's cutting power, cutting a shale or weaker roof or floor does not seem to present any problem, except it presents a more challenging job for coal cleaning and surface waste disposal which are much easier to solve. It follows that horizon control for cutting at coal/rock interfaces is not as critical as conventional thinking.

Cutting in the rock consistently causes faster bit wear and more frequent bit change, perhaps requiring at least one change per 1–2 cuts in today's 1200 ft (366 m) wide or wider panels. For current drum bit layout, each bit change needs and causes 20–30 min production delay. Therefore, development of higher wear-resistant bits will greatly help but not eliminating miners to change bits when required.

If horizon control for coal/rock interface is not a critical factor, then horizon control for shearer drum's colliding with shield canopy and/or shield flipper if exists is the most critical factor, if the shearer operators are to be eliminated. Although there are anti-collision technologies reported in the literature (Komatsu, 2018; Kopex Group, 2013), their reliability in production operation is unknown. Therefore, further research is required. The anti-collision technique essentially requires the exact position of the shearer's leading cutting drum and the canopy, and flipper if equipped, of each shield at any instant.

"Horizon control" in longwall shearer automation originally refers mainly to control of roof rock/coal interface (Nelson, 1989; Mowrey, 1991, 1992). It has now been used to refer to various horizons in the coal seam, for instance, different partings at different levels in the coal seam. The latest reference is to floor control as in pitch steering and pan profiling.

9.8 Summary

Automation of longwall machine and system follows the advancement of sensor and communication (or data transmission) technologies and big data analysis. As the sensors are getting more sophisticated and more types of sensors as well as faster communication technology (e.g., 5G) are developed and introduced for mining application, the level and reliability of longwall automation will advance further.

The state-of-the-art semi-automated longwall system consists of electrohydraulic control shields and doubled-ended ranging drum shearer operated in shearer-initiated-shield advance mode with the memory cutting algorithm for the whole face.

Recent adoption and application of remote control of shearer is the correct approach from miner's health and safety as well as coal production points of view considering the complicated mining and geological conditions where many unexpected events can occur suddenly.

The prerequisite for longwall automation is very high equipment reliability, i.e., 100% reliable all of the time if a man-less system is to be achievable.

Various sensors for detecting unexpected underground events as well as their software control algorithm due to complicated mining and geological conditions must be identified, developed and tested successfully in advance before full automation is feasible.

Adoption of automation of longwall system does not guarantee a reduction in face crew. The highest level automation of longwall system today requires 4–10 crew members at various mines. Obviously mine/crew practices and mining and geological conditions still dictate the manpower need at the longwall face.

Chapter 10

Application issues of longwall mining

10.1 Factors to be considered in increasing panel width

As Figure 1.5.3 (p. 13) shows, the panel width increases steadily ranging in 1976 from 200–620 ft (61–189 m) with an average of 420 ft (128 m) to 580–1580 ft (176.9–481.7 m) with an average of 1150 ft (359.8 m) in 2019, an almost threefold increase. Due to a conventional conservative approach, the existing designed capacity of face equipment normally can accommodate the extra requirements for small incremental increases in panel width, thereby eliminating the necessity of considering various factors influenced by increase in panel width. However, if the increase in panel width is substantial, say 300 ft (91.5 m) or more, the existing face equipment and logistics may not be adequate, and many factors must be evaluated for adequacy.

In the early years, widening the panel is to reduce the frequency of face move which takes time and is non-production. As it evolves, wider and longer panels have obvious advantages that outweigh the disadvantages.

Advantages:

1   Improved the ratio of longwall/development coal or reduction of gateroad development per ton of longwall mined coal.
2   Improved productivity by reducing wedge cut time thereby cost/ton.
3   Fewer longwall moves thereby reducing non-production time and improving safety.
4   More complete recovery of reserves by eliminating pillars of reduced entry development.
5   Reduced construction work by eliminating entry development.
6   Wider panels allow more time for gateroad development.

Disadvantages:

1   Adequate ventilation air at tail end is difficult to maintain.
2   Pre-mining degasification needed.
3   Increased gob emissions and floor gas issues.
4   Risk of hidden geologic anomalies, e.g., faults, washouts, reduced seam height etc., increases.

Before implementing the extension of the current panel width, the following issues of equipment and mining conditions should be evaluated (Bryja and Beck, 2005: Trackemas and Peng, 2013):

1   Armored face conveyor (AFC)

    A   Drive motor power capacity and reserve.
    B   AFC maximum loading rate.
    C   Chain strand assembly, chain size and strength, and chain safety factor.
    D   AFC carrying capacity.

2   Shield hydraulic supply system

    A   System pressure losses.
    B   Hydraulic system hosing and arrangement.
    C   Pump capacity and number of pump.

3   Electric supply system

    A   Power center (transformers) capacity.
    B   Electric cable size for shearer, etc.
    C   Panel as well as mine wide power capacity.

4   Ground control

    A   Pillar size and additional loading, if any.
    B   Additional shield loading if any and additional number of shield.
    C   Effect of additional entry stand-up time on pillar and entry stability.
    D   Effect of larger AFC drives on gateend shield pushing force and floor stability as well as entry size.
    E   Effect on the width of setup and recovery rooms.

5   Mine ventilation and methane emission.

    Effect of increasing methane emission and ventilation resistance.

6   Mine and system infrastructure.

    The following factors should be evaluated; tailgate rock dust system, water system for dust suppression, face and power moves, and equipment components lives.

10.2 Longwall compatibility test

Before delivery of equipment by OEM, a longwall face equipment compatibility test is conducted to make sure the system will work as designed for routine longwall mining production operations. This test takes 1–4 weeks and is critical, especially when the three major machines are not from a single OEM but from two or more OEMs. After the test, changes are made as appropriate before delivery to the mine.

The test involves building a mini longwall with the completed yet to be delivered set of equipment, approximately 30 shields long including the head- and tailgate end shields and drives (Fig. 10.2.1).

Just like the pilot shield test, the longwall compatibility test varies but includes both normal and worst case scenerios. The test program involves checking under normal and worst case scenerios will shearer, shield, and AFC be able to operate normally as designed

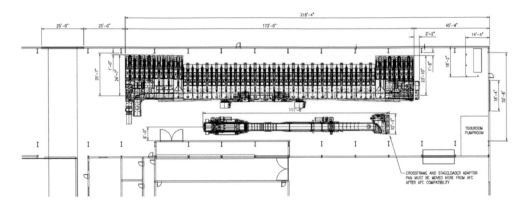

Figure 10.2.1 A mini build longwall for compatibility test

Source: Courtesy of Caterpillar

without incurring any safety and production issues. The following sections detail an example test program.

10.2.1 Shearer

1   When the mini build longwall is set up with AFC straight and flat with shields.

A total of up to 43 dimensional items are to be checked out. Some notable examples are these: floor to top of shearer, shearer and shield clearance at various mining height, canopy tip to face at various mining height, travelway, clearance between cover plate and drum, drum diameter and width, maximum undercut on face run, trapping shoe clearances, height of rack haulage, bottom of shearer to deck top of pan, lumpbreaker (top height, vertical and horizontal clearances to rack, clearance to ranging arm), water clearer split bar to spillplate (horizontal and vertical), shoe centers (gob and face sides), floor to top of handrail and bottom of cable trough.

2   When the mini build longwall is set up with AFC straight and positioned on 6 in. (152.4 mm) hump.

A total of up to seven dimensional items are to be checked out when the shearer is straddling over the 6 in. (152.4 mm) hump, including the following: can the machine operate safely and accommodate full vertical articulation of the AFC? Are there enough clearance between shearer bottom and top of sigma and between rack and drive wheel, tip to face distance when pan is tilted 4–6 degrees toward face and gob, and clearance when two humps are under a single pan.

3   When the mini build longwall is set up with normal snake (run of face).

A total of up to four dimensional items are to be checked out during normal snake and face run: lumpbreaker to spillplate, water clearer splitter bar to lumpbreaker, minimum clearance head and tail side ranging arm engagement with #1 face side pan.

4   When the mini build longwall is set up with AFC straight and positioned on headgate-shearer at headgate.

A total of up to eight dimensional items are to be checked out: gob side ranging arm to face side gradient pan, shearer height, undercut depth and clearance, cowl clearance, tailgate drum to toe plate.

5   When the mini build longwall is set up with worst case with AFC and drive snaked toward face-shearer at headgate.

A total of up to 10 dimensional items are to be checked out, some of which include the following: drum entering and leaving the HG, drum bit to stage loader pan, HG drum to toe plate and toe plate to cowl HG drum during snake, toe plate to cowl.

6   When the mini build longwall is set up with worst case with AFC and drive snake back from face-shearer at headgate.

A total of up to nine dimensional items are to be checked out, some of which include the following: toe plate to drum and cowl, shoe to gob- and face side rack, water clearer splitter bar to spillplate, lumpbreaker to spillplate, traction unit to spillplate, shoe correctly on toe plate, sprocket ride at rack center during snake.

7   When the mini build longwall is set up with shearer with AFC straight and positioned on tail drive-shearer at tailgate.

A total of up to 15 dimensional items are to be checked out, some of which include the following: maximum and minimum undercut, ranging arm gobside to tail drive face side, lumpbreaker height, lumpbreaker to bottom of tail drive, spillplate to lumpbreaker arm and water clearer splitter bar, tail drive face side to cowl gob side, cowl clearance entering the tail drive, toe to HG drum, and HG cowl.

8   When the mini build longwall is set up with worst case with AFC and drive snaked back face-shearer at headgate.

A total of up to seven dimensional items are to be checked out, some of which include the following: drum problem entering and leaving the tailgate area, undercut clearance with tailgate extension and retraction, gob side edge to face side of drive frame, underframce to pan, drum bit to drive frame, trailing drum to toe plate at maximum cut past.

10.2.2 AFC/shearer/shield

A total of up to eight dimensional items are to be checked out, some of which include the following: stop position for shearer welded on at HG and TG. Chain clearance at HG and TG drive frames and sprockets, gateend shield clearance, travelway, canopy tip to face, shield hose connections.

10.3 Life cycle management, expert solutions, and equipment re-use services

10.3.1 Life cycle management

As longwall production increases, production delay of any length becomes very expensive and must be avoided as much as possible. Consistent high production relies highly on equipment functioning without frequent breakdowns, large or small, which in turn requires high-quality maintenance. In addition, today's sophisticated longwall face equipment requires well-trained skilled labor, which many longwall mines are lacking, to keep up with scheduled maintenance. Therefore, in recent years, the OEMs have contracted with coal mines for equipment maintenance work termed as life cycle management of face equipment. The program can improve equipment reliability from low to high eighties without the program to 99.5%+ with the program.

A few examples of life cycle management are these:

1   Maintenance – reduction of equipment delay.

In a life cycle management, the OEM assigns a special employee as life cycle coordinator (LCC) to a longwall mine to perform equipment maintenance work. For example, for an AFC, the LCC is responsible for the following duties. Similar services can be designed for shield and shearer:

A   Underground weekly inspection reports.

1   Monitoring of how the system equipment is running.
2   Keeps track of monthly equipment maintenance due dates.
3   Lists all maintenance, component change outs and weekly issues with equipment.
4   Lists upcoming weekend maintenance jobs.
5   Keeps a running tab of monthly measurable maintenance items.

B   Monthly tasks.

1   Where applicable, equipment vibration test and results.
2   Oil samples are pulled monthly for lab analysis.
3   AFC and stage loader chain measurements are taken once a month.
4   Consignment inventory is tracked monthly and cores are accounted for on any inventory that has been consumed.

C   Incident, failure, component change out reports.

1   A failure report for any failure at the mine.
2   An incident report for any issue causing downtime.
3   A component change out report for normal wear.

2   Maintenance planning.

A   By monitoring chain slack and knowing upcoming maintenance shifts, the mine can pull chain slack on non-production shifts.
B   Keeping track of monthly maintenance due dates allows jobs to be done during maintenance shifts.
C   Monitoring sprocket conditions and previous panel conditions can help you plan sprocket change outs or flips within a month of advance.
D   By monitoring the system, failures can be found beforehand. System running conditions can be changed to get through the week to the next maintenance shift until component can be changed out.

3   End of panel reports.

   A   A detailed end of panel reports summarized all the events reported under A–D at the end of each longwall panel.

   B   Presentation of the report to the rebuild team to help with modifications to prevent recurring failures.

10.3.2 Expert solutions

All longwall face equipment have various level of automation with various types of sensors to monitor various factors for control purpose. The real time monitoring of those sensors including health monitoring produce huge amount of data that are transmitted also in real time to the gateend computer, surface control room, and online to the OEMs service centers 24 hours every day, 7 days per week. Those data are displayed on the computer screen for the status of the longwall operations. They are also analyzed by OEM's professional staffs for various tell-tale trends of the root causes of breakdowns and repetitive alarms, production of shift reports, and prognostic rules. Recommendations and remedial work are proposed and fed back to the mine site immediately. The results are elimination of major delays, optimizing equipment reliability and performance optimization.

For example, Fig. 10.3.1 (top) shows frequent high setting pressures (brown columns) suddenly occur in an otherwise uniform normal setting pressure period, a roof cavity was formed within 12 hours (Fig. 10.3.1 bottom). This cause-effect relationship can be used as a warning of pending roof falls when next time a set of sudden high setting pressures appears.

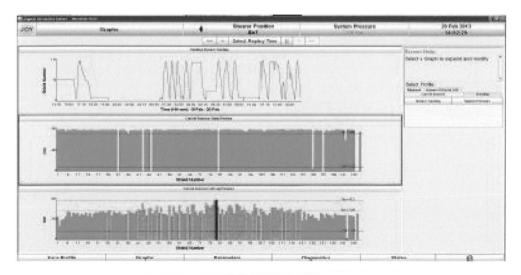

Figure 10.3.1 An unusual event of high setting pressure was detected that led to formation of a roof cavity

Source: Courtesy of Komatsu Mining

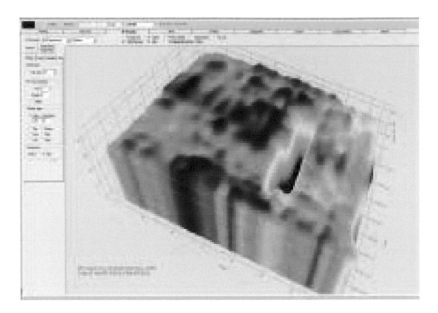

Figure 10.3.1 (Continued)

10.3.3 Shield support re-configuration

During shield's service life of 12–15 years, major geological condition changes may occur such that the original equipment design may not be desirable, for instance, seam height changes and roof strata change that require higher capacity.

Instead of buying a new set of shield for the different geological condition originally designed for, many mines opt to re-build or reconfigure the old shield to fit the new geological conditions. The project consists of the following steps: (1) visual inspection and perform functionality test, (2) clean, strip, and sandblast all structures, (3) fully inspect all structures and welds, (4) fully inspect all leg pockets, (5) inspect all pins and bores, (6) recommendations – what parts for repair, replacement, and/or modifications.

10.3.4 Extended life services

How much life is left for a piece of equipment until the end of service life? Extended life testing is a process used to evaluate the life left in shield support structures. It is a predictive tool that can determine when individual components are likely to reach the end of their serviceable life.

10.4 Basic performance requirements for shields

In high production longwalls, all subsystems must be utilized to their full potential. For shields, the following conditions must be met:

1   Headgate and tailgate.

The headgate and tailgate must be kept straight, parallel, and at the correct width. Failure to do so results in variation in the width of the longwall face.

2   Maintain straight lines at the face.

The coal face, panline, and shield should form three parallel lines at the face area. If the face is not straight, shields will walk into each other and the panline is subject to bending and will wear more quickly. Uneven shearer cutting from unequal panline push by the shields is a major problem, leading to an irregular faceline. All pans must be advanced equally.

As the face continues to widen, maintaining a straight face is getting tougher and tougher without the help of automation guidance. The OEM claims that improved software in pan push control by reed rod sensors making sure every shield's reed rod is extended the same distance can maintain the panline straight. Production operations in the last few years have also proven that LASC system (INS-Inertia Navigation System) (CSIRO, 2005) is effective in keeping a straight faceline.

3   Keep the face area clean.

It is important that the face area be as clean as possible so that every piece of equipment can work to its capacity. Too often debris piles up in front of, on, and/or under the shield base plates, which can obstruct shield advance and make the shield move in different direction. Leaks in hydraulic hoses must be fixed promptly to avoid losing pressure and to keep the floor dry. Shields must be cleaned so that any equipment damage can be identified easily and quickly.

4   Cut the roof and floor smooth.

Every attempt should be made to maintain the cutting horizon to avoid severe steps between webs and from shield to shield both in the roof and floor. This way the shield canopy will make better contact with the roof for better support and will cause less crushing, the source of shield dust upon release for advance. Smooth floor will enable the face machines to utilize to its full potential and facilitate automation.

5   Maintain the setting pressure.

Since the setting pressure is the minimum pressure required to support the roof, it must be maintained at all times. For whatever reason, if the setting pressure drops, it may result in roof falls. If the positive or guaranteed set is available, it should stay on unless roof falls or a bad roof is encountered.

6   Keep the canopy tip to faceline distance minimum.

For weak immediate roofs, it is important to keep the distance from the canopy tip to faceline as short as possible. Too often, due to roof steps, the first contact point between the canopy and roof is not at the tip but some distance behind. When this happens, if the canopy tip to faceline distance is excessive, roof falls may occur. The easiest way to avoid this problem is to reduce the distance between the canopy tip and faceline.

7   Canopy must be maintained flat.

Shield design is such that, when the canopy is flat, there are residual strokes to take up additional loading. If the canopy tilts up or down too much exhausting the cylinder stroke of the stabilizing or tilt cylinder, roof loading during shield service life could easily damage the cylinder and/or steel structure, especially when the positive set function is engaged. The OEM recommends a minimum of 0.8 in. (20 mm) residual stroke (see Fig. 10.14.5).

8   Passing a roof fall cavity.

When a roof fall cavity is encountered, do not set the canopy tip up inside the cavity and against the top of cavity that may cause the stabilizing cylinder fully extended (see Fig. 10.14.5). Rather, set the canopy horizontally and/or, if needed, filling the cavity with a set of cribs on canopy top. This will underpin the roof in the next cut. Normally, the two-leg shield can advance under a roof fall cavity up to 3.5 ft (1 m) safely. For control of a massive roof fall, see next section.

Accumulation of roof debris on the canopy can be avoided by sometime advancing the shield along the roof on contact, or by tipping the canopy tip downward to let the debris slide down, thereby cleaning up the canopy top.

10.5 Control of adverse roof conditions

10.5.1 Control of roof cavities

Roof falls are very much a part of underground coal mining when roof bolting is used as the primary supporting system. The key is, of course, to prevent it from occurring and learn how to deal with it effectively.

The major reason for local roof falls at the longwall face is the lack of an efficient way to support the exposed immediate roof between the faceline and canopy tip and preventing the broken rocks from falling into the working space.

Shield supports can easily negotiate a roof-fall cavity of 3 ft wide by 3 ft high (0.91×0.91 m) without the need for additional artificial supporting materials (Hafera *et al.*, 1989; Peng, 1990). If the cavity is larger than that, then a wood crib is customarily erected on top of the canopy to control the cavity. The crib functions to prevent the cavity from enlarging and acts as a temporary roof for the shield support to work against. The shield support reverts to its usual role after passing the roof fall area.

If a larger roof fall (longer than the canopy length) develops along the direction of the face advance, large blocks of thicker roof strata may pose a serious threat to the shield supports. In this case, an artificial roof may be built on top of the canopy. The artificial roof is made of foam material pumped either from the staging area near the face or from the surface via boreholes (Fig. 10.5.1).

10.5.2 Polyurethane injection

Polyurethane injection is widely used to stabilize adverse roofs at longwall faces and gateroads, either to prevent further deterioration of existing adverse roof or to strengthen the expected adverse roof in advance of mining.

The polyurethane binder system consists of two components, component A (a polymeric isocyanate) and component B (a polyol resin) (Watson and Hussey, 2001). They are mixed and injected under pressure into the rock strata through drill holes using specially developed

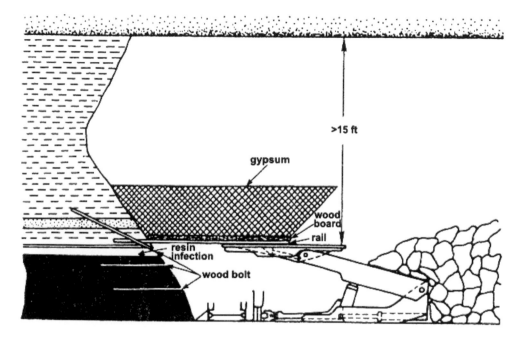

Figure 10.5.1 False roof built to pass through a very large roof fall

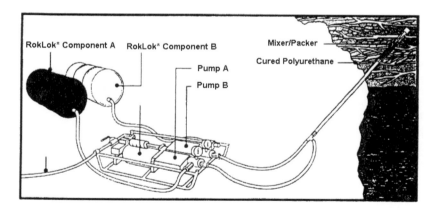

Figure 10.5.2 Polyurethane injection system and example of its application

Source: Watson and Hussey (2001)

equipment (Fig. 10.5.2). The mixed components enter the strata through the mixer/packer unit, fill the voids between the packer and the end of the hole, and proceed to flow into the rocks through fractures and bedding plane separations. The largest cracks that offer the least resistance are filled first, followed by smaller fractures. The result is that the rock strata are restored to their intact condition. The flow direction of the binder's migration within the strata can sometimes be changed, and leaks that appear on the rock's surface can be restricted.

Under laboratory conditions, the mixed components cure in two minutes. At that point rock strata are thought to begin to consolidate. But underground, the mixed components should be allowed to cure for at least two hours, after which the binder system should offer adequate support to enable mining to proceed. In 8 to 10 hours, the binder should have reached its ultimate strength. The final rock binder has a uniaxial compressive strength (UCS) of 10,200 psi (70.34 MPa), uniaxial tensile strength (UTS) 3850 psi (26.5 MPa), and maximum elongation of greater than 17% at ultimate strength.

Successful strata consolidation depends on locating fractured rocks and their areas of extent and fractured density. At longwall faces, injection drill holes oriented at 10° to 20° from horizontal and strategically located at the coal seam/roof interface are recommended. The location, pattern, spacing, depth, and quantity of components for the injection holes require perceptive judgment from a person(s) knowledgeable in both behavioral characteristics of the roof and the injection system.

For best results, the temperature should be equal to or greater than 60° F (16° C). Normal operating pressure measured at the injection pump should be around 400 to 800 psi (2.76–5.52 MPa). Resistance to flow in the pumps, hoses, and mixer/packer accounts for 80–90% of the total pressure. In order to overcome additional hose extension, the recommended operating pressure is 700 to 1100 psi (4.83–7.59 MPa).

10.5.3 Collapse of longwall faces

US longwall history is full of stories involving partial or full-face longwall collapses. There were several cases reported prior to 1990 (Peng and Chiang, 1984, p. 292; Hsiung and Peng, 1985: Hafera et al., 1989) and many more went unreported due to their proprietary nature. The major reason for full- or partial-face collapse and powered support failure was insufficient support capacity. As stated in Section 2.3.1, during the 1980s, there were many supports with rated yield capacity around 400 to 600 tons, insufficient for hard, stable roof that overhung 250 ft (76 m) or more (Hsiung and Peng, 1985). Even if the supports were deemed sufficient by the then state-of-the-art technology, they still could not cope with huge blocks of rock broken along joint sets that over-powered the supports (Peng and Chiang, 1984, p. 292). Furthermore, there were several cases after six to nine years of service life (even though the total number of cycles accumulated was relatively low compared to today's standard), the powered supports were not in good shape, and the roof sheared off along the faceline and crushed the powered supports along the whole face (Hafera et al., 1989).

There are still cases of face collapses in recent years that went unreported in the literature, although much less frequently, in large part due to much greater shield capacity and advanced shield design technology. In these cases, the key problem was poor roof geology, whether massive collapse along preexisting fractures or unconsolidated roof strata or whether shallow cover collapsed from seam top to the surface. Advance knowledge of large fractures and massive poor roof are critical in all cases.

Salvage of a collapsed face always involved the application of explosives. Holes were drilled into the rock blocks that sat on the shield supports, and explosives were loaded and detonated to break it up. Repeated blasting was needed to create gaps to raise the shield until the shields were restored to functional height. It normally took two to four weeks to recover the face.

In Australia, there was a case where the cover was shallow, and the full cover collapsed a short distance from the setup room and buried the whole face. A dragline was used to dig it out.

In modern longwall shields, all legs are installed with a pressure sensor, and the leg pressure is measured and recorded continuously. Those pressure histories should be analyzed as described in Section 5.5 for trend analysis and prediction of incoming roof weightings.

10.6 Two-leg shields vs. four-leg shields and chock shields

Hafera *et al.* (1989) and Peng (1990) performed underground evaluations of five types of shields for two longwall panels over a two-year period (Fig. 10.6.1). The five types of shields were a two-leg 420-ton (382-mt) shield (2/420 or 2/382), a two-leg 450-ton (409-mt) shield (2/450 or 2/409), a two-leg 500-ton (455-mt) shield (2/500 or 2/455), a four-leg 600-ton (545-mt) chock shield (4/600 or 4/545), and a four-leg 610-ton (555-mt)

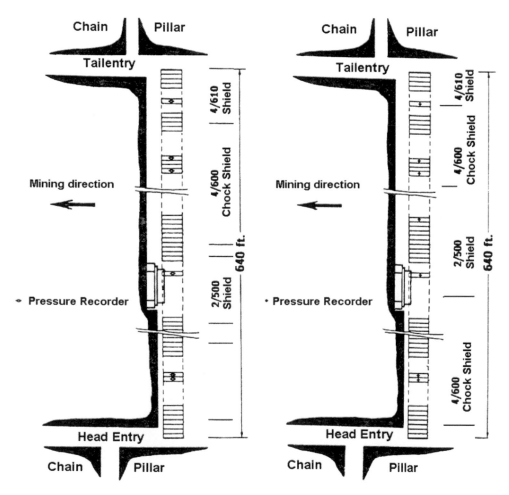

Figure 10.6.1 Layout of five types of shields in two separate longwalls (Hafera *et al.*, 1989; Peng, 1990)

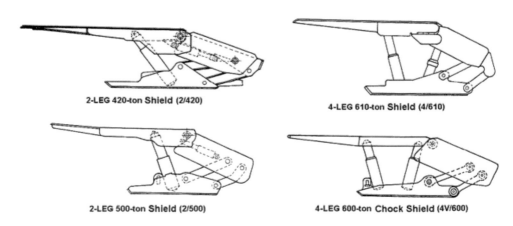

Figure 10.6.2 Four Types of shields (Hafera *et al.*, 1989; Peng, 1990)

shield (6/610 or 6/555) (Fig. 10.6.2). Parameters of evaluation included the following: (1) hydraulic leg pressure for analysis of support load, load increment, supporting efficiency, and active horizontal force; (2) horizontal movement of the canopy during support setting to identify the effect of active horizontal force; (3) roof-to-floor convergence and leg closure; (4) the dimension and location of roof cavities; and (5) canopy contact condition and the distance between canopy tip and faceline. The results are discussed in the next several subsections.

10.6.1 Active horizontal force

There are two types of horizontal forces, passive and active. Passive horizontal force is generated after support setting by overburden strata movement toward the gob, whereas active horizontal force is produced by the relative horizontal movement trend between the roof strata and canopy during the shield-setting period. Active horizontal force will strengthen the immediate roof and prevent roof falls between canopy tip and faceline.

The measured horizontal canopy displacements were 15, 14, 17, 2, and 2 mm, which translated into an active horizontal force of 118, 146, 52, 21, and 30 tons (107, 133, 47, 19 and 27 mt) for the 2/420 (2/382), 2/450 (2/409), 2/500 (2/455), 4/600 (4/545), and 4/610 (6/555) shields/chock shields, respectively. Therefore, the four-leg shields and chock shields have little active horizontal force. Conversely, for the two-leg shields, an active horizontal force is generated during setting, and its magnitude is proportional to its capacity and leg angle from the vertical.

10.6.2 Roof-to-floor convergence and leg closure

Roof-to-floor convergence and leg closure were measured when the shearer was cutting within 150 ft (45.7 m) on both sides of the monitored supports. The amount and rate of roof-to-floor convergence were the highest for 4/600 chock shields, followed by 2/500, and then 2/420. The difference between roof-to-floor convergence and leg closure was about the same between 4/600 chock shield and 2/500 shield but larger than that for the 2/420 shields. A larger difference indicates that there were more bad roofs or debris or both at the contacts between roof and canopy and/or between base plates and floor.

10.6.3 Supporting efficiency

Supporting efficiency is the ratio in percent of the measured load to the rated load capacity. The setting load efficiencies were 78, 58, 51, and 48% for 2/420, 2/500, 4/600, and 4/610 shield/chock shield, respectively. The same trend occurred for final load, load increment, and load density. Therefore, four-leg shields and four-leg chock shields with higher capacity could not control the weak roof efficiently, resulting in poor supporting efficiency.

10.6.4 Roof control

Figure 10.6.3 shows that roof cavity frequency for the 6/600 shields and 4/610 chock shields was much higher than that for 2/500 shields. Roof cavity frequency is defined as the ratio of the number of shields in which a certain size of roof falls had occurred to the total number of shields of the same type.

The first contact between the faceline and canopy surface is closer for the 2/500 shields than 4-leg shields and chock shields as shown in Figure 10.6.4. In terms of controlling weak roof and preventing roof falls between the canopy tip and faceline, two-leg shields were clearly superior to four-leg shields and chock shields.

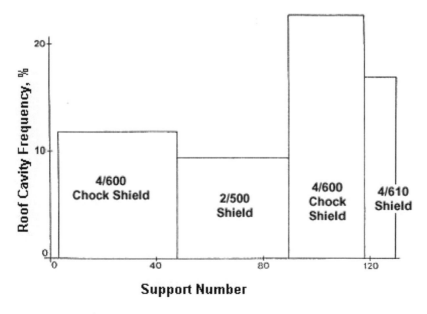

Figure 10.6.3 Distribution of roof fall frequency

Source: Hafera et al. (1989); Peng (1990)

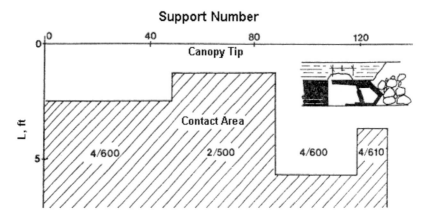

Figure 10.6.4 Distribution of contact condition between canopy and roof
Source: Hafera et al. (1989); Peng (1990)

10.7 How to accurately measure the shield hydraulic leg pressure

In two-leg shields, the hydraulic legs are all two stage, bottom and top pistons. In a two-stage cylinder, the bottom piston will extend first during setting against the roof. When it is fully extended, the top piston will then begin to extend. Similarly, when the shield is released for advance, the bottom piston will retract first. When it is fully retracted, the top piston will begin to retract.

In longwall faces, the extension lengths of the two stages of each leg in a shield usually are different between the two legs. How does this happen and will it affect shield loading? Theoretically, it should not because pressure must be in equilibrium at all times between the two stages. But, according to Barczak and Gearhart (1998), it may not be the same due to the construction of the piston in the cylinder. They used a two-stage leg to illustrate the problem as shown in Figure 10.7.1

Initially for a mining height of 36 in. (914.4 mm), when the leg is set against the roof, the bottom piston is extended to its full stroke, 24 in. (609.6mm) and the top piston extends 12 in. (304.8 mm). If in the second cycle the mining height is 33 in. (838.2mm), then the bottom piston will extend 21 in. (533.4 mm), while the top piston remains at 12 in. (304.8 mm). If in the third cycle the mining height is 39 in. (990.6 mm), the bottom piston will extend fully, while the top piston will extend further to 15 in. (38.1 mm). In the fourth cycle when the mining height returns to the original first cycle height, the bottom and top piston extension will return to that of the first cycle. It can be seen that the top piston extension at any supporting cycle is equal to the initial extension plus the amount of increase in mining height beyond the initial mining height. Unless the bottom piston is fully retracted during support advance, the top piston extension will never be less than the initial extension.

Assuming that, after setting, the bottom piston is fully extended against the mechanical stop so that a portion of the setting pressure is used against the stop. When the roof begins to converge, a portion of the induced roof loading will not be detected by a change in the

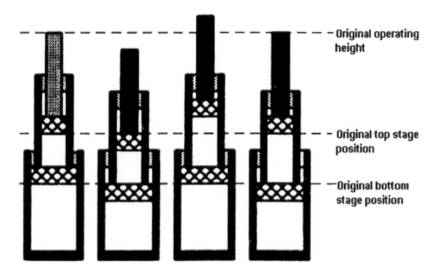

Figure 10.7.1 Stage extensions produced by changes in operating height
Source: Barczak and Gearhart (1998)

hydraulic pressure in the bottom chamber until after that portion of pressure used against the stop is overcome, and the piston begins to compress the fluid. This condition will happen only when the pressure in the top piston, which is at pump pressure at setting, is increased to exceed the setting pressure in the bottom chamber. In the example given for a 2/700 shield, the undetected load was nearly 200 tons.

The results shown in Figure 10.7.1 have serious implications. It means that any time the leg pressure is measured, the sensor must determine if the bottom stage has exhausted its stroke, and if it does, compensate for that. None of the current leg pressure measurement systems have taken this point into consideration.

In practice, the two stages of leg pistons in both legs of a shield are frequently found to be in different extension length. It may be due to uneven roof or simply because for various reasons, mechanical or otherwise, the bottom and top stages do not extend in the sequence as stated earlier.

10.8 Pre-driven recovery room and mining through open entries

10.8.1 Introduction

The continuing effort to increase productivity always calls for the design of wider and longer panels to reduce the number of longwall moves in a budget year. In the 1980s to 1990s, due to smaller bleeder shafts and bleeder fans, cut-through entries from the headgate side to the tailgate side may be required somewhere near the middle of a very long panel. For instance, if the development work for the next panel to be mined runs late, creating cut-through entries at convenient locations to block out the next panel is one of the most frequently used options.

The main problem with those open entries is that the longwall needs to cut through those entries during the retreat mining, which generates the front abutment and raises concern regarding the stability of those open entries.

The longwall recovery plan at the end of panel retreat mining calls for meshing the roof for 10–13 cuts before the stop line. This is a tedious operation and slows down the rate of face advance considerably. In order to reduce this low-production period, the pre-driven open recovery room concept was developed in the late 1980s. Another reason for adopting an open recovery room is that if the roof condition in the designated location is bad, an open recovery room provides an excellent opportunity to pre-support the roof and ensure its stability during the recovery operation.

There are other occasions when an open entry is beneficial. For instance, if a geological fault causes a poor roof condition and is encountered on the panel side during gateroad development, this weak zone may be taken out, during development, by driving an open entry along the fault line, and roof bolting the entry to prevent premature instability during retreat mining. In other instances, mine management may, for whatever reason, decide to extend the length of the previously driven shorter panels, so the bleeder entries in those shorter panels or other previously driven transportation entries can become open entries in the new panels.

10.8.2 Methods of supporting open entries or pre-driven recovery rooms

The primary supports for open (or cut-through) entries and recovery rooms are normally similar to other development entries in the same mines. The secondary or supplementary supports installed to handle abutment pressures during retreat longwall mining can be divided into the following three types (Fig. 10.8.1) (Peng, 2000):

1   Complete backfill of open entries.
2   Supplemental roof and/or rib bolt reinforcement only – no standing support.
3   Rows of standing supports with or without supplemental roof and/or rib bolt reinforcement.

10.8.2.1 Complete backfill of open entries

The most commonly used material for backfill is flyash mixed in various proportions with artificial cements. Depending on the coal and roof rock properties, backfilled materials with a designed uniaxial compressive strength of 400–1000 psi (2.76–6.90 MPa) have been used. In this method, the backfill material is pumped into the entries via surface boreholes. It is the safest of the three methods and must be used when the roof is very weak and cannot maintain any entry roof span when the front abutment pressure arrives. However, since it requires a vast quantity of material to backfill entire entries, it is normally the most expensive. Figure 10.8.1 A shows an example of this backfill system.

10.8.2.2 Supplemental roof and/or rib bolt reinforcement only

If the main and/or the immediate roofs are strong and free of major fractures or geological defects, supplementary roof supports alone (without any standing supports) may be used. The supplemental support system must be well designed to cope with the incoming front abutment pressure, the shields must be in good working condition, and they must have sufficient capacity to support the roof structure as the face cuts into the open entry. The most

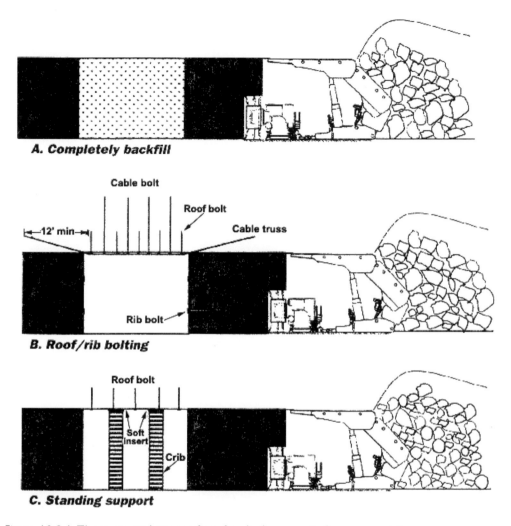

A. Completely backfill

B. Roof/rib bolting

C. Standing support

Figure 10.8.1 Three general types of roof and rib supports for open entries
Source: Peng (2000)

critical period occurs when the face is approaching an open entry with less than 10 to 15 ft (3.1 to 4.6 m) of coal block left. Because this is approximately the period of peak front abutment, and if the coal is soft, sloughing off on both sides of the remaining coal block will suddenly increase the unsupported region, and thereby the roof load. The supplementary support system usually consists of heavy-duty high-strength cable bolts and cable trusses. The cable bolts must be sufficiently long and installed vertically to reach beyond the main roof. For the cable trusses, the inby side of the inclined bolts must be sufficiently long (minimum 12 ft or 3.66 m) and installed at lower angles so that they can prevent premature roof falls in front of the shields as the face approaches that of the open entry. The inby side of the open entry ribs should be supported with rib bolts and wiremesh. Figure 10.8.1 B shows an example of the roof support system. It is also important that shields should have sufficient capacity to maintain the roof level as they approach the open entry, because at this moment the strong roof beam built by the supplemental supports over the open entry is bridged between two abutments: One is the coal pillar on the outby side and the other is the shields (plus whatever is left off the solid coal panel) on the inby side.

10.8.2.3 Standing support with or without supplemental supports

In-between the two extreme methods described above, there are several others that have been used (Fig. 10.8.1 C) (Heasley *et al.*, 2003; Tadolini *et al.*, 2002). Depending on the roof strength and entry width, open entries can be supported by one or more rows of standing supports with or without supplemental roof reinforcement. For stronger roofs devoid of any geological anomalies, one row of high-strength cribs at specified intervals may be sufficient. Conversely, for weaker roofs, three rows (one row each against the inby- and outby-side ribs, and the third row at entry center) of cribs at the specified interval are preferred. Again the inby side cribs are critical and must be designed to act as an abutment to support the constructed roof beam as the face approaches. There are two kinds of concrete-like crib materials available; one is prefabricated and the other is pumped in. The former requires underground vehicular transportation while the latter requires one or more surface boreholes and surface pumping facilities. The crib material should have a uniaxial compressive strength varying from 2000 to 3000 psi (13.7 to 20.7 MPa) and be spaced so that it provides a support density varying from 300 to 1000 psi (2.07 to 6.9 MPa). In addition, it is important that a softer material like wood blocks or a pumpable cement bag be inserted between the top of the standing support and the roof to allow initial elastic convergence of the roof as the abutment pressure sets in so that the cribs are not subjected to undue localized pressures and crack up prematurely.

10.8.3 Support design considerations

1   In the design of open entry support, support resistance and support materials are the two major factors. Support resistance, which more or less determines the type of open entry supports required, must match roof geology, mining, and stress conditions. The support materials must have sufficient strength and be disposable.

2   Intuitively, the second method, supplemental roof and/or rib bolt reinforcement, seems to be the most desirable if all requirements can be met. But the most frequently cited criticism for this type of support is that it is too expensive. All factors considered, if cut-through entries must be used in longwall panels, the fail-proof measure is to drive the roof bolts in an angle approximately 30 degrees to the roofline toward the face and supported the entries with the third method, standing supports with or without supplemental roof/rib bolts.

3   If the first and third methods (complete backfill and standing supports with or without supplemental roof/rib bolts) are employed, consideration should be given to the cuttability and disposal of the backfilled and standing support materials. Harder materials, either by nature or due to compression by abutment pressure, will take longer to cut which may eventually offset the time gained from not meshing. When cutting under high pressure, these materials may also break in large pieces and cause handling problems in the armored face conveyor.

4   Wood cribs alone are too soft to use as a standing support. They also present problems for cutting and handling in the slope belt and separation in the preparation plant.

5 Since the normal width of a recovery room is 22 to 24 ft (6.71–7.32 m) wide, meshing prior to the face reaching the stop line may still be required if the roof is very weak. In such cases, the advantage of using a pre-driven recovery room may not be obvious. Therefore, in order to take full advantage of a pre-driven recovery room, the required open recovery room is in the range of 32 to 36 ft (9.75–10.98 m) wide. In such cases, the third method, rows of standing supports plus supplemental roof bolt reinforcement, must be used.

6 Everything considered, the safest approach for open entry support is the third method, standing support with supplementary support. The standing support system itself or in combination with soft inserts, must be chosen so that it can absorb the initial onslaught of the front abutment pressure without cracking up. In other words, the chosen standing supports should yield somewhat as the front abutment first arrives and then maintain high supporting capacity to the end (see Section 4.7 for various types of standing supports).

10.8.4 Case example

Figure 10.8.2 shows a case example of panel layout in the recovery area where the pre-driven recovery room was employed (Barczak *et al.*, 2007). The panel width was 1250 ft (381.1 m)

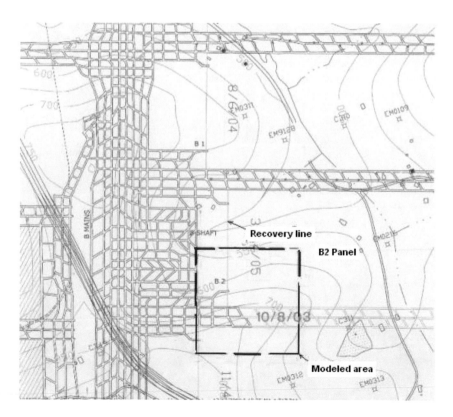

Figure 10.8.2 Panel layout in the recovery room area

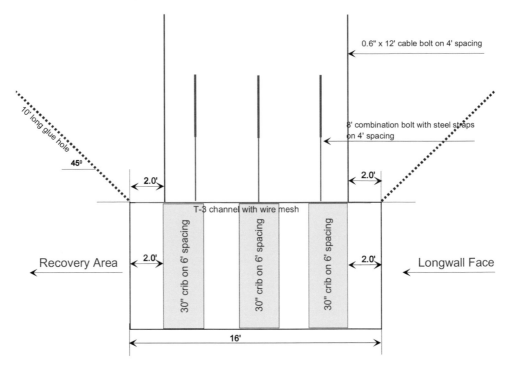

Figure 10.8.3 Roof support for the pre-driven recovery room

and the overburden depth was 500 ft (152.4 m) at the headgate side and 700 ft (213.4 m) at the tailgate side. Both mining and entry heights were 8 ft (2.44 m). The width of the recovery room was 16 ft (4.88 m). After the longwall face cut into the recovery room, another 16 ft (4.88 m) was advanced to the final longwall recovery line.

Figure 10.8.3 shows the support design for the pre-driven recovery room for handling both front abutment pressure and weak immediate roof conditions. The primary supports were combination bolts with steel straps and wire mesh on 4 ft (1.22 m) spacing. Supplementary supports were 12 ft (3.66 m) cable bolts and pumpable cribs. Two cable bolts and three combination bolts were installed in a row on 4 ft (1.22 m) spacing. Cable bolts were anchored 4 ft (1.22 m) on the top. In actual bolt installation, some additional rows of 8 ft (2.44 m) combination bolts without steel straps were installed where the immediate roof was slickensided. Two 10 ft (3.05 m) inclined glue holes were drilled on the entry corners and chemical grout was pumped to glue the immediate roof. Three rows of Heitech pumpable cribs were installed in stagger at the inby, middle, and outby sides of the recovery room. Cribs were 30 in. (762 mm) in diameter on 6 ft (1.83 m) spacing. The capacity of the cribs was about 245 tons (222.73 mt) at peak load with 1 in. (25.4 mm) convergence, and about 100 tons (90.11 mt) at residual load with up to 7 in. (1778 mm) convergence.

The longwall face cut into the recovery room successfully, and the recovery process took about 13 days. Considerable time was saved compared with the previous panel employing the conventional recovery room method.

10.9 Longwall mining under hard-to-cave roof

On occasions, the hard-to-cave strata, mainly thick sandstone or limestone, dives down to near the coal seam top and will overhang for a considerable length before it caves. When and if caving occurs, it generates a wind blast that poses a severe hazard to miners working around the face.

Su *et al.* (2001) described a case in the Pittsburgh seam where a large sandstone channel, 47–70 ft (14.3–21.3 m) thick and within 4–5 ft (1.2–1.5 m) of coal seam top, meandered across three panels, B12, B13, and B14 (Fig. 10.9.1). The uniaxial compressive strength (UCS) and Brazilian tensile strength of sandstone were on average 10,000 psi (68.97 MPa) and 1000 psi (6.9 MPa), respectively. When the sandstone was first mined under in B12 panel, no preparation was done. The face advance rate was slowed to 9.8 ft/shift (2.99 m/shift) and took seven weeks to mine through it, with a total production downtime due to bad top of 30,600 minutes. Surface hydraulic fracturing using nitrogen foam in 5 in. and 6 in.

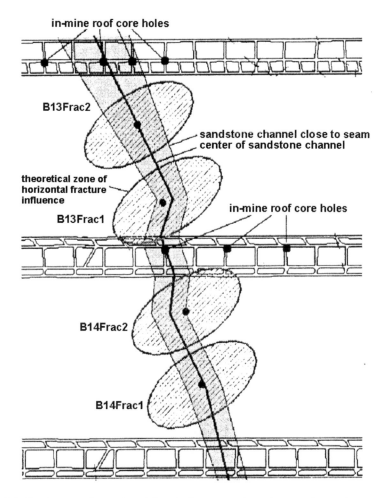

Figure 10.9.1 Sandstone channel and frac holes over three longwall panels

Source: Modified from Su *et al.* (2001)

diameter holes was then implemented in B13 and B14 panels. Due to the shallow depth at which the vertical stress would be the minimum of the three principal stresses, a horizontal (pancake) fracture, rather than a vertical fracture (as often observed for the cases of deep oil reservoirs), was created in the sandstone stratum. Furthermore, since the two horizontal principal stresses were not equal, the horizontal fracture created would be elliptical, not circular, with axes of the ellipse in relative proportion to the principal stress magnitudes. The average advance rates were 16.6 and 17.1 ft/shift (5.06 and 5.21 m/shift) with total downtime of 3655 and 3265 minutes for B13 and B14 panels, respectively. Overall production rates for B13 and B14 panels were 70% better than that for B12 panel.

Mills *et al.* (2000) performed a successful case of forced caving in Australia. The roof of conglomerate was 30–50 m thick with UCS 7,250 ± 1450 psi (50 ± 10 MPa) and tensile strength 580 psi (4 MPa) and overhung over an area 984 ft (300 m) long by 1324 ft (100 m) wide. Once caved, the windblast generated was extremely hazardous to miners at the face. A surface trial of hydraulic fracturing showed that a horizontal fracture could be created parallel to and within 16.4 ft (5 m) of a free surface and that this fracture would grew to a radius of 82 ft (25 m).

Underground observation and surface borehole monitoring also confirmed that, after caving, the shape of the roof top would resemble a flat arch, 39.4–49.2 ft (12–15 m) high at the center. In the first underground hydraulic fracturing operation, an inclined hole was drilled upward from the headgate toward the center of the panel to induce caving. The results indicated that approximately 70,000 tons of rock (55 m of standing gob) had been induced to cave.

10.10 Cutting through a fault

Geological faults of various types are encountered in many coal seams. Depending on the offset displacement between the two opposite sides of the fault and the extent of fractures around it, it may or may not present any problems for mining through it.

There are many potential hazards working with faults, including sudden changes in horizon and grade, thinning of the coal seam, weak and disturbed strata, less stable gateroads, increased gas and water, and potential for equipment damage. Longwall preparation and planning for mining through faults should include the following:

1   Gateroad preparation upon development – additional supports, such as cable bolts and cable trusses, may be needed in an area on either side of the fault.
2   Equipment must be prepared and configured correctly for cutting through the fault zone.
3   Gateroad reinforcement, such as polyurethane injection to strengthen the roof and ribs, should be implemented when the face is within 150–300 ft (45.7–91.5 m) of the fault zone.
4   Longwall navigation plan – all geological structures of the panel as exposed through gateroad development should be mapped.
5   If necessary, the longwall face crew should receive training on the navigation and gateroad preparation plans.

Rowland (2002) developed a fault management plan for thick seam mining. The seam was 18 ft (5.5 m) thick and longwall mining extracted the bottom 14.76 ft (4.5 m), leaving 3.28 ft (1 m) of top coal to assist in controlling the roof. The longwall operated on a uni-di cutting cycle, 33.46 in. (0.85 m) web, and SISA (shearer initiated shield advance). The longwall would mine through major faults of less than full-seam displacement only and move around major faults exceeding full seam displacement. The fault management plan consisted of fault mapping, preparation, manpower reorganization, and operational parameters.

10.10.1 Fault mapping

Fault mapping showed that the longwall intersects a total of 13 vertical faults with displacements or offsets up and down from 4 in. to 12 ft (10 mm–3.7 m) within three crosscut distances. Past experience showed that small wedged blocks formed among the faults would fall off above the shields and cause a total failure of the roof in the headgate. The location and nature of these faults were interpreted from physical measurements in gateroads, surface borehole data, surface 3D seismic, in-seam drilling, and computer modeling of the data.

10.10.2 Preparation

The preparation work consisted of reinforcement of the headgate within the disturbed area. It included polyurethane injection, installation of cable bolts, cable trusses, square sets with link-n-lock cribs on top of them to stabilize roof cavities, micro fine grout (with steel forepole tubes) injected into the face over the top of the cut horizon for various lengths in the faulted ground, and training in the fault management plan.

10.10.3 Manpower reorganization

During the fault drivage period, a geotechnical advisor was assigned on each shift to assist in making all decisions. Items of special attention were elaborated for every member of the shift crew. For instance, see the following:

1   Shearer operators – the cut profile will depend on the planned attitude, not the seam marker used under normal seam condition. Changes in the cutting horizon will be made using minimal increments (maximum 5°). When the shearer is cutting the faulted ground, one shield width should be cut at a time and then debris cleared for shield advance. The shearer, if necessary, must park in a stable roof area.
2   Shield operators – in faulted areas, shields will be advanced with minimum lowering with positive set turned off and by adjacent control, and flippers should remain fully extended. Side shields should be pressurized during shield advance. Shield operators must be aware of the shield's leaning angle and direction, and the gap between adjacent shield canopies should be minimized at all times to maintain shield alignment. AFC will be manually advanced.

10.10.4 Operational parameters

1   Horizon control – the maker bands used in normal seam conditions will not be consistent in the faulted area. So cutting profiles are plotted every 32.8 ft (10 m) to make sure floor slope will not exceed 1:12 – the maximum allowable for shields, and a minimum of 3.28 ft (1 m) roof coal shall be maintained for roof stability. Operational height must be maintained between 13.12 and 14.1 ft (4.0–4.3 m) and the stage loader clearance must also be maintained at all times.

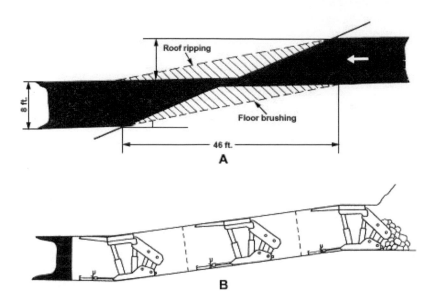

Figure 10.10.1 Passing through a fault by roof ripping and floor brushing

2   AFC alignment and migration, and pan attitude – the face must be maintained as straight as possible at all times. AFC migration should be stabilized prior to any mining operations through the faulted area. The AFC attitude should be checked at least three times per shift (every cut, if the AFC attitude varies considerably) and recorded at least every 10 shields. If the pans are lifted on the face side, the shearer will cut upward, leading to reduced or elimination of drum clearance with the shield canopy.

3   Face angle – in order to minimize the effect of the faulted area, it may be necessary to swing the face through such that the faceline intersects the faults at a designated angle. This will be achieved by controlling the AFC creep toward the headgate. In poor ground, keeping the face moving is the most efficient way to overcome difficulties.

If the offset of the fault is less than the difference between the seam thickness and the closed height of the shields, the coal face can pass through the fault area without roof ripping and floor brushing. If the offset of the fault is greater than that difference, it is necessary to make an artificial slope on the floor by roof ripping and/or floor brushing (Fig. 10.10.1) so that the shearer and shields can pass through successfully. If both the roof and floor are soft, the shearer can cut through them easily. On the other hand, if the roof and floor are very hard, it may be necessary to blast them off.

10.11 Pillar design for gas/oil wells

In eastern and central coalfields, there are many gas wells, both active and abandoned, drilled through coal seams that are being actively mined. There are two ways to deal with those wells. One is to plug them, and the other is to design the panel layout in such a way that gas wells are located in the chain pillar system between the panels or ideally outside the panels.

The former method is very expensive, whereas the latter, when feasible, is more effective. If there are many gas/oil wells in close proximity, pillar sizing becomes a critical issue because, invariably, the location of one or more gas/oil wells would not satisfy the existing pillar plan required by law.

In placing a gas well in the chain pillar system, several factors must be considered in order to assure its integrity after longwall mining. Because longwall mining not only causes surface and subsurface subsidence within and beyond the panel edges, it also induces high abutment pressures around the edges of the panel. The effect of longwall mining on a gas well is determined by considering overburden depth, abutment pressure, pillar size, and the location of the gas well with respect to the panel edge. A properly designed pillar (or pillar system) should protect the gas well located within it from longwall mining effects.

In the state of Pennsylvania (PA DEP, 1957), the adopted pillar plans call for different pillar bearing areas for various depths of mining (Fig. 10.11.1).

The bearing area consists of a solid core pillar plus additional pillars surrounding it. The pillar core starts from 60 ft (18.3 m) square for depths below 150 ft (45.7 m) to a maximum of 100 ft (30.5 m) square for depths above 250 ft (76.2 m). The additional pillars surrounding the core also increase with the depth until it reaches a maximum total pillar dimension of 200 ft (70 m) square with the gas well at the pillar center when the mining depth is 650 ft (198.2 m) or more. In recent years, as long as the total pillar bearing areas meet the size requirement, the location of a gas well in the core pillar is allowed to be off-center as long as the minimum distance from the gas well to panel edge is 50 ft (15.2 m) (Fig. 10.11.2).

Pillar research (Peng et al., 2002) in the past few decades has shown that this supporting plan is over-conservative in many cases. What then is the appropriate pillar size and is the pillar the only factor in gas well protection in longwall mining?

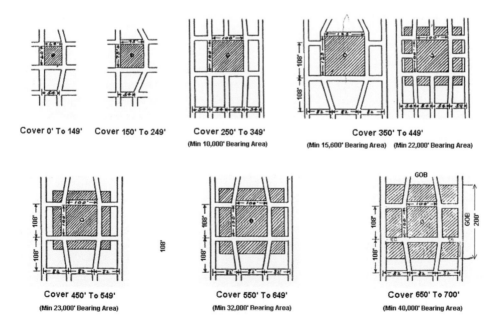

Figure 10.11.1 State of Pennsylvania pillar plan for protection of gas/oil wells

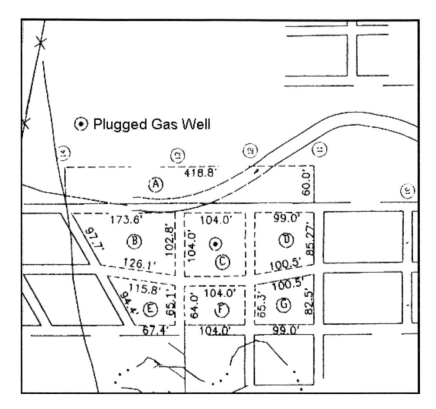

Figure 10.11.2 Alternate pillar plan for gas well protection

Gas wells located within longwall chain pillars and barrier pillars are influenced by longwall mining. Longwall mining causes subsurface subsidence that can induce stresses and deformations in gas well casings. If the induced stresses and deformations are excessive, gas well casings can be damaged or ruptured. Historical data showed that gas well failures are likely to occur in coal seam horizons, overburden strata near coal seams, and floor, and as overburden depth increases, failures tend to occur in coal seam horizons or in weak floor (Zhang *et al.*, 2019).

The stability of gas wells is influenced by pillar stability, the gas well location in respect to gob and pillar rib, floor stability, overburden geology, and the construction of the gas well casings. First of all, stable pillars are a primary requirement for protection of gas wells because gas wells would fail if pillars fail. However, pillar stability is not the only determining factor for gas well stability, the location of the gas wells within the pillars has a great influence on the stability of the gas wells. If a gas well is close to a gob or pillar rib, its casings are likely to be subjected to high stresses induced by large vertical and horizontal subsurface movements. Historical gas well failure cases showed that the gas wells generally failed within about 80 ft from the gob edge. In addition, gas well failure in the floor is associated with the presence of weak claystone in the floor. A weak claystone floor can induce not only vertical stress but also shear stress in gas well casings due to both vertical and horizontal compression of the weak layer. Gas well failure cases in longwall chain pillars suggested that the gas well failure in the floor is a major concern under deep cover if claystone floor is present (Zhang *et al.*, 2019). Moreover, overburden geology, especially weak claystone layers and massive strong sandstone/limestone layers, also has a major influence on horizontal displacement in the overburden over a longwall pillar. Large horizontal displacement up to 5.5 in. was measured above a longwall chain pillar about 120 ft from the gob edge under overburden depth of 642 ft, which is sufficient to cause a gas well casing failure (Su, 2016; Scovazzo, 2018). In addition to the geological and mining factors, the construction of the gas well casings also affects induced stresses in the casings, the tolerance of the casings to the subsurface movements, and thus the gas well stability.

To assess the stability of the gas wells in longwall pillars, it is important to quantify subsurface movements and their effect on gas well casings. Numerical modeling is an effective approach to quantify subsurface deformations and induced stresses in gas well casings by taking into consideration geological and mining factors as well as the parameters of gas well construction.

10.12 Cyclic failure of strong roof strata and seismic events

10.12.1 Cyclic failure of strong roof layers

Experience in longwall mining shows that ground pressure in gateroad systems is low in the first and second panels in a virgin reserve. Ground pressure increases in the third panel, especially the tailgate side, and continues to increase as more panels are developed side by side in sequential order. Depending on geological conditions, ground pressure in the fourth or fifth panel may be so high that roof control problems occur in the gateroad systems. Thereafter, the same events repeat again in more or less the same cyclic manner.

Numerical modeling using the Pittsburgh seam geological column of one to five longwall panels (Peng and Su, 1983) showed that caving height increases with the number of panels and that for the five-panel model, the caving height of the central panel is deeper than those on both sides due to the interaction between panels (Fig. 10.12.1). The 43 ft (13.1 m) immediate roof fails all at once upon mining. Thereafter the failure areas progress upward at a much slower pace until the final stable configuration is reached. The final caving height in the one-panel model stops immediately below the 25 ft (7.62 m) limestone layer. In the two-panel model, it stops short of breaking through the 25 ft (7.62 m) limestone layer; and in the three-panel and four-panel models, it stops just short of breaking through the 20 ft (6.1 m) sandstone layer. In the five-panel model the final caving height at the central panel stops at the middle of the 50 ft (15.2 m) thick limestone layer, with the caving height of the two immediate neighboring panels stopping at the middle of the 20 ft (6.1 m) sandstone layer. The convergence on the gateroads increases with the number of panels mined.

The effectiveness of the bridging and breakthrough of the main roof competent strata depends strongly upon the relative location of these strata with respect to the roof line and their thickness. The closer the strata are to the roof line, the more effective are their bridging and breakthrough. Therefore, the amplitude of cyclic maximum ground pressures on the gateroads depends strongly on the thickness of the main roof competent strata and their relative locations with respect to the roof line. As to the cyclic period, the timing of breaking through the most effective main roof competent strata appears to be the dominant factor. This

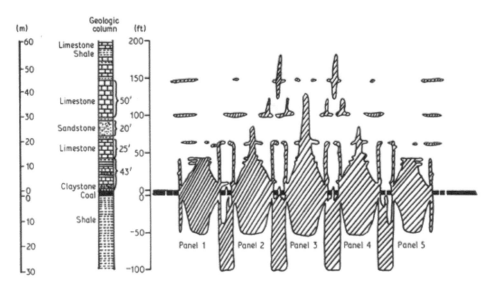

Figure 10.12.1 Areas of roof failures in the five-panel model

Source: Peng and Su (1983)

timing, however, depends not only on the thickness and location of the competent strata but also the panel width, mining height, overburden depth, and strength of the competent layers. The most effective way to control the adverse effects of cyclic ground pressure is to put a barrier pillar of sufficient width at regular intervals, for instance, every four to five panels depending on mining and geological conditions, to isolate the effect of previous panels.

It must be pointed out that weaker rock strata such as sandy or firm shale, if massive in occurrence, may behave similarly to a strong roof stratum. In other words, it is strata stiffness, not merely strata strength, that counts.

10.12.2 Cyclic seismic events in longwall panels

A coal mine in south western Virginia had 12 recorded seismic events in three consecutive panels: four in panel 6R, five in panel 8R, and three in panel 9R (Fig. 10.12.2). When these events occurred, there were no coal bumps at the longwall face, but the face workers observed dusty air at the tailgate end coming from the previous panel gob. Sometimes the air was strong like an air blast. Borehole camera records for the tailgate where seismic events were recorded showed that the tailgate pillars were still stable.

To eliminate/reduce the seismic events, mine management proposed a new panel design to avoid the sudden failure of thick sandstone roof stratum or strata. The key points of the new panel design are these:

1   Reducing the panel width from 1000 ft (305 m) to 700 ft (213 m) rib-to-rib.
2   Keeping the old four gateroad system but changing pillars sizes from 70 × 130 ft (21.3 × 39.6 m), 110 × 280 ft (33.5 × 85.4 m), 30 × 110 ft to 30 × 130 ft (9.1 × 33.5 m to 9.1 × 39.6 m), 150 × 280 ft (45.7 × 85.4 m), 30 × 130 ft (9.1 × 39.6 m), respectively.

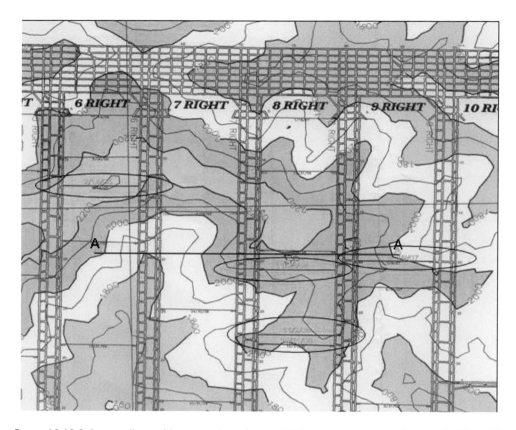

Figure 10.12.2 Longwall panel layout and sandstone thickness contour map for panels 6R to 9R

Mine geology varies over the study area. Typically, below the coal seam, the floor rocks are composed of a few feet of sandy fireclay and very hard sandy shale or sandstone. The typical immediate roof consists of dark silty shale overlaid by two distinctive sandstone strata (lower and upper sandstone). The first sandstone unit is located from 0 to 25 ft (0–7.6 m) above the seam, and ranges in thickness from 0 to 35 ft (0–10.7 m). The second sandstone unit is located approximately between 25 and 38 ft (7.6 and 11.6 m) above the seam, and ranges in thickness from 22 to 52 ft (6.7–15.9 m). The average uniaxial compressive strength of the sandstone strata is 10,532 psi (72.6 MPa).

Table 10.12.1 summarizes the geological conditions above the longwall panels 4R to 10R. The seismic events occurred when the following conditions existed simultaneously:

1   Mining depth was mostly greater than 2000 ft (610 m).
2   The combined thickness of sandstone 1 and 2 was greater than 50 ft (15.2 m).
3   The distance from seam to the bottom of the first sandstone layer was less than 5 ft (1.52 m).
4   The second sandstone was in direct contact with the seam. The most frequent and severe seismic events occurred in panel 8R where the distance was less than 5 ft (1.52 m) all the way.

Table 10.12.1 Geological conditions above the 4R–10R panels

Panel	Mining depth (100 ft)	Sandstone 1 thickness (ft)	Seam top to sandstone 1 (ft)	Seam top to sandstone 2 (ft)	Sandstone 2 thickness (ft)
4R[1]	18–22 (1 pillar block behind the event)	50–60 (1 pillar block behind the event	< 5	0–10	40–50
5R[1]	20–22	40–50	< 5–20	< 10	< 66
6R[1]	22–24	40–50	0–5	0–15	60
7R[1]	18–22	50-> 70	0	< 5	48–65
8R[1]	20–22	60–70	0–5	0-> 25	60
9R[1]	20–22	40–60	< 5	< 10	66
10R[2]	Mostly <16, before first caving < 22	< 40	0–15	10–< 25, mostly < 20	72
8R[2]	> 22 in 7R[2] and 8R[2]	40–60	0–15	>10, mostly > 20	< 72
9R[2]	18–22 in 8R[2]; 16–18 in 9R[2]	40-> 70	0-> 20	>15, mostly 20	84

[1] North end (recovery room side) of the panel.
[2] South end (set-up room side) of the panel.

Surface subsidence measured above panels 8R and 9R was 2 ft (0.61 m) in areas where the combined thickness of sandstone strata was more than 60 ft thick (18.3 m), and 3 ft (0.9 m) in areas where the combined sandstone thickness was 40–60 ft (12.2–18.3 m). For a mining height of 5 ft (1.52 m), this amount of measured subsidence indicates that the sandstone strata even up to more than 70 ft (21 m) thick did cave sometimes after mining.

FLAC[2D] models were used to simulate the geological and geometrical conditions at a borehole in panel 8R where the seismic events/tailgate air blasts were observed. The model considered topographic change where the overburden depth varied from 1900 ft to 2200 ft (579–670 m).

Modeling results confirmed that the overburden depth and the caving height are the key factors for the tailgate instability at the mine. To overcome this problem, a new panel design was proposed by reducing the panel width from 1000 to 700 ft (305 to 213 m) and increasing the size of center abutment pillar from 110 to 150 ft (33.5 to 45.7 m) while reducing the size of side pillars from 70 to 30 ft (21.3 to 9 m) wide.

To check the performance of the proposed new panel design, a 2D model for five consecutive panels at maximum overburden depth of 2200 ft (670 m) and smaller caving height of 4 ft (1.2 m) was performed.

Modeling results show that the new panel design induces a smaller amount of yielding in the sandstone strata. The yielding state of sandstone roof for the new proposed panel design is similar to that of panel 6R and 7R. Therefore, the new proposed panel design of 700 ft (213 m) wide panel and 290 ft (88.4 m) wide yield-abutment-yield gateroad system will provide stable tailgate condition even at high overburden depth (up to 2200 ft or 670 m) and thick sandstone strata (up to 75 ft or 22.8 m). Additional modeling shows that after 7–10 consecutive panels a barrier pillar of 300–500 ft (91–152 m) should be left to isolate these panels from the subsequent new panels.

10.13 Faster or slower longwall advancing rate

10.13.1 Two opposite views

When it comes to the effect of surface subsidence on structural damages and pillar stability, there are two schools of thoughts. The US coal industry believes that a faster and constant face advance rate produces smaller dynamic surface deformation and, thus, is more beneficial to surface structures. The German coal industry believes the exact opposite to be true. Similarly, if two longwalls are approaching each other in opposite directions at faster speeds, will the dynamic-effect associated with each fast moving longwall superimpose such that it requires a larger barrier pillar be left between them? Or, if a smaller barrier pillar is used, shall the two approaching longwalls be slowed down to avoid a dynamic superposition effect?

10.13.2 Fast vs. slow longwall advancing rates

There are several notable differences between US and German coal mining conditions. Among them, US longwalls are mostly in single seams in virgin overburden coalfields compared to Germany's previously mined-out, overlying seams where the overburden had been disturbed many times. US longwalls are also relatively shallow compared to those in Germany. Most important of all, US longwalls advance much faster and do so to reduce surface structural damages, but German longwalls are required to move slowly in order to reduce structural damage. Why the complete opposite concept of protecting surface structures?

Overburden strata in Germany has been disturbed many times and broken up due to the many overlying seams that have been previously mined out (Luo *et al.*, 2001). This type of overburden reacts faster to undermining so that a faster advancing rate will produce a faster subsidence velocity, leading to larger vertical deformation at the center of the subsidence basin and larger tilts with smaller vertical subsidence at the edges of the subsidence basin. As mining proceeds from seam to seam in descending order, the overburden strata become increasingly fractured and loosened. Consequently, the subsidence basin produced by deeper seam mining creates larger tilts and compressive strains. Therefore, the reasons that a faster advance rate could increase subsidence damage potential in German coalfields are these: (1) the dynamic subsidence front over the advancing face induces higher slopes, and tensile and compressive strains, and (2) the faster the ground subsides, the larger are the maximum slope, curvature, and horizontal displacement, and the faster is the change for the dynamic strain from tensile to compressive as well as for the curvature to convert from convex to concave form as the longwall face advances. The shape of the dynamic subsidence profile is heavily dependent on the face advance rate, the degree of disturbance of strata, and the size of a panel as shown in Figure 10.13.1.

With a higher rate of advance of 16.4 or 32.8 ft d^{-1} (5 or 10 m d^{-1}) (curves *b* and *a*), the broken roof strata settle more rapidly behind the face than an intact immediate roof that could cantilever a long distance into the gob. In a fractured rock mass, strata settlement behind the face is accelerated with the increasing rate of advance, and the dynamic subsidence profile becomes steeper (curve *a*). On the other hand, in intact strata, a constant higher face advance rate causes a gentler dynamic subsidence profile (curve *c*).

In the United States, where most longwalls exist in a virgin coalfield, or if under multiple-seam mining, the overburden is not broken up as many times as in Germany, the dynamic subsidence profiles show different characteristics. Figure 10.13.2 shows that the maximum

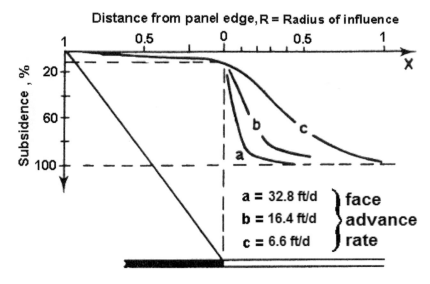

Figure 10.13.1 Dynamic subsidence development curves at different face advance rates, as compared with the normal subsidence development curve.

Source: Modified from Kratzsch (1983)

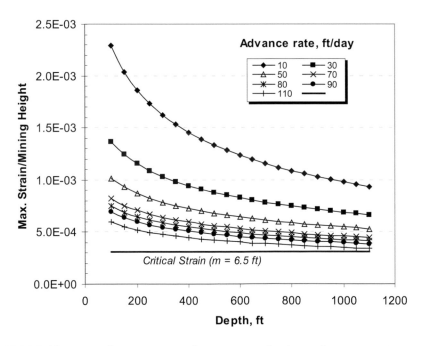

Figure 10.13.2 Maximum dynamic strain decreases as the face advance rate increases for US coalfields

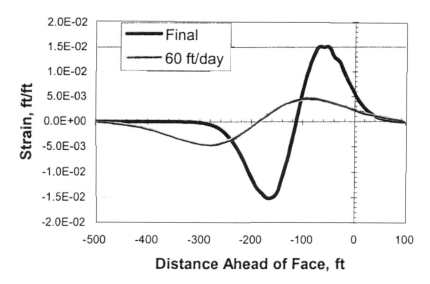

Figure 10.13.3 Distribution of final and normal dynamic strains behind a stopped longwall face

dynamic tensile strain decreases with an increase in the face advance rate and mining depth and with a decrease in mining height. A significant reduction in the maximum dynamic tensile strain can be obtained by increasing the face advance rate when the mining depth is shallow and the initial advance rate is small.

There have been numerous cases in which longwalls were forced to stop due to unforeseen circumstances, causing the surface structures to suffer much worse damage than under normal advance rate. After the face stops, the surface area near the longwall face experiences a residual subsidence phase, during which the surface strain changes from normal dynamic strain at the time of face stoppage to the final strain (Fig. 10.13.3).

Note that the maximum normal dynamic strain is only one-sixth of the final strain. Therefore, a higher face advance rate induces a larger residual subsidence that, in turn, will induce a more intense residual subsidence process (Fig. 10.13.4).

10.13.3 *Barrier pillar stability between two oppositely moving longwalls*

Beck *et al.* (2004) presented a case study in which two longwalls in two separate mines were to be laid out on both sides of the mine boundary, mining in opposite directions. What should be the barrier pillar width between these two longwalls, considering the dynamic effect of the fast-moving longwalls when they were approaching and passing each other? The cover in the area of interest was 550–900 ft (167.7–274.4 m) deep, and the mining height was 8 ft (2.44 m) in both cases. Three-dimensional finite element modeling (without considering the dynamic effect of fast moving longwalls) was performed and a barrier pillar of 231 ft (70.4 m) wide was recommended. The two longwalls, determined to be 900 and 1023 ft (274.4 and 311.9 m) wide, were successfully completed as planned without any incident.

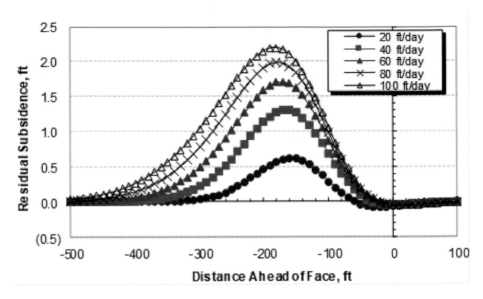

Figure 10.13.4 Residual subsidence distributions around a longwall face for various face advancing rates before its stoppage

10.14 Bottom coal, hard face, soft floor, and AFC creeping – causes, problems, and solutions

10.14.1 Bottom coal

Bottom coal refers to the portion of the coal cut off by the shearer that fails to be loaded onto the AFC for moving out of the face. All longwalls have bottom coal on the mine floor, some a little while others a lot, which is left behind in the gob. What make it worse is that, in recent years, as pan width increases, so does the pan height. It is common for the space between the faceline and gob side edge of AFC pans to be filled up with bottom coal and often to over-flowing. This high bottom coal tends to cause the AFC to climb when the DA ram compresses it too hard. Besides, this bottom coal confines the bottom portion of the coal seam and may make it harder to cut. Furthermore, it covers up any markers that the trailing drum shearer operator uses for horizon control. The worst part about bottom coal is that the shift crew needs to devote one or more trips to clean it up – trips that can be used for production cutting.

Bottom coal is generated due to the following:

1   The floor not being evenly cut and steps being created where bottom coal accumulates.
2   Spillage of coal between drum and the AFC while loading.
3   Face spalling, the severity of which depends on seam height and the front abutment pressure.

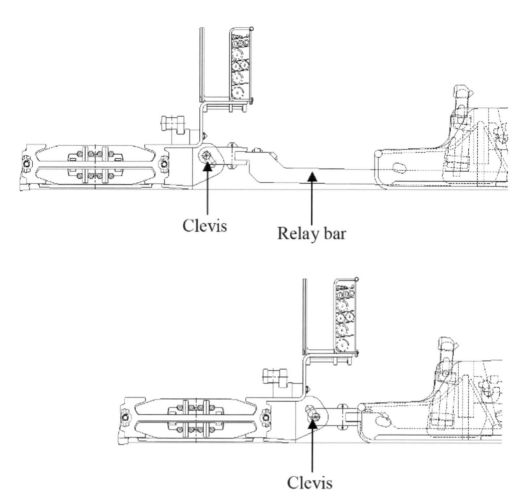

Clevis   Relay bar

Clevis

Figure 10.14.1 AFC pan with slant slot in the clevis for scooping up bottom coal
Source: Courtesy of Caterpillar

In addition to performing timely dedicated clean-up trips, the AFC can be designed to assist in cleaning up the bottom coal during pan push. For example, Figure 10.14.1 shows a pan with a slant slot on its clevis. During pan push, as the relay bar is pushed forward, the connection at the clevis moves up by following the upward slanting profile, thereby raising the gob side pan with reference to the face side. This tends to keep the bottom of the face side pan edge on the floor, preventing bottom coal from passing under the pans. In the other design, the weight distribution of the AFC components is such that the face side is heavier than the gob side, and the location of the clevis/relay bar connection point is such that the

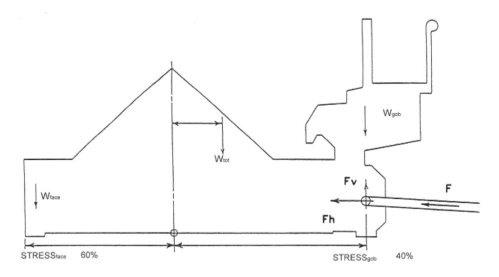

Figure 10.14.2 Weight distribution of AFC components for scooping up bottom coal

relay bar is directed upward toward the face (Fig. 10.14.2) (Hart, 1988). During pan push, the vertical component force of the relay bar is sufficient to raise the gob side pan up with respect to the face side for scooping up the bottom coal.

10.14.2 Hard faces

When the face is straight, the coal is hard and requires more energy to cut. This is referred to as a hard face. Whether a face is hard or soft, it has to do with the stress fields developed in the roof and coal face, the magnitude of which is related to mining height, cleat structure, and the front abutment pressure.

In most coal faces, the high friction contact planes between the coal and the immediate roof rock, and between the coal and the immediate floor rock normally provide lateral confinement and strength to the coal. But this confinement zone is limited to areas close to the contact planes. As seams get thicker and mining height gets higher, the area at mid-height where there is no confinement gets larger. In fact, instead of confinement, there is lateral tension that tends to create vertical cracks more or less parallel to the coal face or ribs, resulting in face or rib sloughage of different degrees. Conversely, when a coal seam is thin and mining height is low, the confinement zone covers the whole seam height. As a result, the face or ribs are intact and straight, resulting in hard cutting.

Most coal seams have cleat structures. Cleats are fractures generated during the coal forming process. They generally occur in two perpendicular sets. The face cleat is more conspicuous, larger, and extends longer, whereas the butt cleat is less conspicuous and shorter in length. Cleats occur in sets and appear in 2–6 in. (5.1–152 mm) interval. If the face cleats are parallel to the coal face or ribs, it is easier for them to fall off, resulting in the face or ribs being soft.

When mining begins at the set-up room, the gob is small and the front abutment is also small, so the face is harder. But as the face advance continues, the gob becomes larger and the front abutment pressure also begins to build up. Right before the first gob caving occurs, the front abutment pressure is largest and, consequently, the face is soft and the cutting is easier.

When roof falls occur in the unsupported area between the face and shield canopy tip or above the canopy, roof beams are broken and stresses in the coal face are released. So, the faces tend to be straight and hard. The effect is much more obvious when the roof fall occurs in the unsupported area. The higher the roof fall, the harder the coal face.

For harder coal faces, a more powerful shearer must be used. If that is not feasible, the most effective way to soften the coal face is to take advantage of the front abutment pressure by properly lowering the shield setting pressure and adjusting the canopy to tip down so that it allows the roof beams to bend down, thereby imposing abutment pressure on the coal face. Several trials may be needed to reach the optimum face condition.

10.14.3 Soft floor

It is well-known that among the various types of powered supports, the two-leg shields produce the highest toe pressures and have the highest probability of making their base toes digging into the floor (Fig. 10.14.3). Therefore, on soft floor, the base toes of a two-leg shield tend to sink into the floor, causing problems during shield advance and delaying production. In dealing with a soft floor problem, the following factors must be considered (Bessinger,

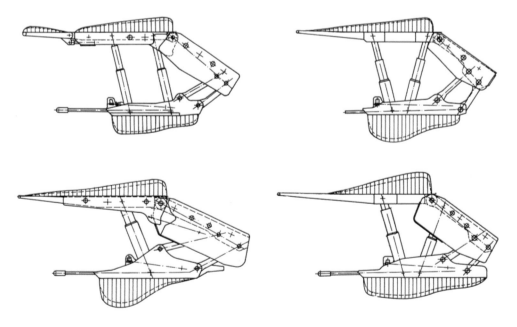

Figure 10.14.3 Comparison of measured pressure distribution on canopy and under the base for various types of shield supports

2006): the thickness of the soft floor, the floating stabilizing cylinder, the base lifting devise, the longer base toe, and the compensation valve.

10.14.3.1 Thickness of the soft floor

A thin layer of soft floor over a thick hard strata generally causes no problems for shield advance. But if the soft floor materials are thick, then the shield design must take it into consideration.

Some floor materials appear to be firm in their natural condition, but become very soft once in contact with water, and in a high production longwall face, there is plenty of water from the dust control system. For example, in the US central coalfields, the floor materials in the Herrin #6 seam, and its equivalent seams in Kentucky, are very thick. Its UCS is 700–1100 psi (4.8–7.58 MPa) when dry but reduced to 300–400 psi (2.07–2.76 MPa) when wet.

10.14.3.2 Floating stabilizing cylinder

The stabilizing cylinder between the shield canopy and caving shield must be double-acting and floating. Its hydraulic supply system must also be independent of that of the shield legs. If the stabilizing cylinder is locked, when the legs are raised, then the base toes will be pulled back toward the gob following a trace resembling an inverted arch creating a tip-toe condition, thereby forcing the toes to dig into the floor. For instance, in the Pittsburgh seam when the yield pressure of the stabilizing cylinder is set at 120 bars, it will crack the floor. But when it is set at 60 bars, it will not crack the floor, because the severity of this tip-toe condition is reduced considerably, and the base/floor contact area at the front end is much larger, thereby reducing the floor pressure.

10.14.3.3 Base lifting device

In soft floor, every shield must be equipped with a base lifting device (see Section 5.3.4). To be effective, a base lifting device must be positive and has the following two features:

A   Minimum range of stroke: 9–11 in. (229–278 mm) above horizon and 6–12 in. (152–305 mm) below horizon.
B   Minimum effective capacity of 30–35 tons (9.1–10.7 mt) and larger than shield weight.

10.14.3.4 Longer base toe

Extending the toe of base plate to closer to the pan (gobside) will not only reduce the peak toe pressure but also move it farther toward the rear (Fig. 10.14.4).

10.14.3.5 Compensation valve

A yield valve or compensation valve of 200 bar, the yield valve end of which is connected to piston side of canopy tip cylinder and on/off switch side is connected directly to the bottom of leg prop, can be used to prevent shield standing on toe when the canopy badly tip up during setting (Fig. 10.14.5). If the canopy tip contacts the roof during leg extension and leg

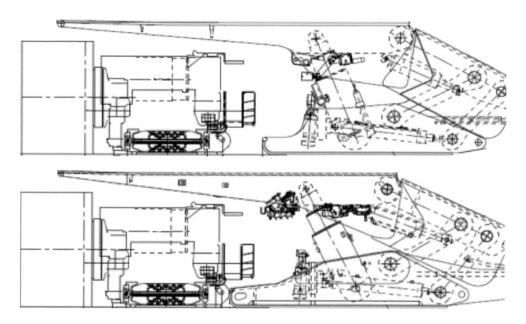

Figure 10.14.4 Comparison of base plate toe length: long (bottom) vs. short (top)

Source: Courtesy of Caterpillar

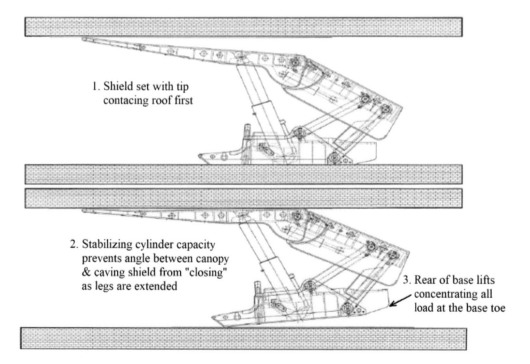

1. Shield set with tip contacing roof first

2. Stabilizing cylinder capacity prevents angle between canopy & caving shield from "closing" as legs are extended

3. Rear of base lifts concentrating all load at the base toe

Figure 10.14.5 Installation of compensation valve to prevent shield toe loading

Source: Courtesy of Caterpillar

cylinder bottom pressure is <100 bar, the valve will yield allowing the canopy cylinder to retract, thereby tipping the canopy down.

10.14.4 AFC creeping

AFC creeping refers to the process in which the tail drive is moving toward and inside the tailgate, or the head drive is moving toward the pillar rib side at the headgate. Creeping occurs more often when mining starts from the setup entry. After the face has advanced some distance along the panel, it usually straightens out. Excess creeping at the headgate side may damage the stage loader, even though there are four to five short flexible pans installed between the cross frame and crusher to accommodate a certain amount of creep.

AFC creeping in flat or near-flat seams is mainly caused by uneven cuts by the shearer in either direction or by repeated long or short cuts in one direction. The common corrective measure is to perform repeated short wedge cuts opposite to the direction of creep. For instance, if the head drive is too close to the pillar rib, wedge cuts are performed at the head end and toward the headgate.

10.15 Stability of gateroad T-junctions subject to high horizontal stresses

10.15.1 Introduction

During the period from the late 1980s to early 2000s, the concept of high horizontal stress played a very important role in coal mine entry stability. In fact most roof falls and entry instability were attributed to the existence and poor panel orientation with respect to high horizontal stress. The most frequently used remedial measure was to change the orientation of entry direction with respect to the major horizontal stress.

Nearly all researchers, when they addressed the effects of high horizontal stress used the concept of stress shadow. They used 2D mine maps with crowded stress trajectories marked at the entry/panel corners to indicate the locations where the horizontal stress concentrates and roof control problems occur (Fig. 10.15.1) (Mark and Mucho, 1994). Since longwall panel layout involves 3D multiple panels with multiple gateroad systems, can the longwall structure be represented by 2D analysis?

Morsy and Peng (2006), employing 3D finite element computer modeling, analyzed stress distributions around the headgate T-junctions. The panel with three-entry system was 700 ft (213 m) deep subject to in-situ high horizontal stresses at four orientations with respect to panel mining direction; 0°, 30°, 60°, and 90°. The entries and crosscuts were 16 ft (4.8 m) wide. The *in-situ* principal stresses are these: σ_3 (minimum principal stress) is vertical and equal to the weight of overburden, σ_1 (maximum principal stress) = 3 σ_3 and is horizontal and σ_2 (intermediate principal stress) = 2 σ_3 is also horizontal. The stresses around the panel and T-junction areas, being different from the *in-situ* stresses, both in orientation and magnitude, are called the induced principal stresses, σ_{1i}, σ_{2i}, and σ_{3i}.

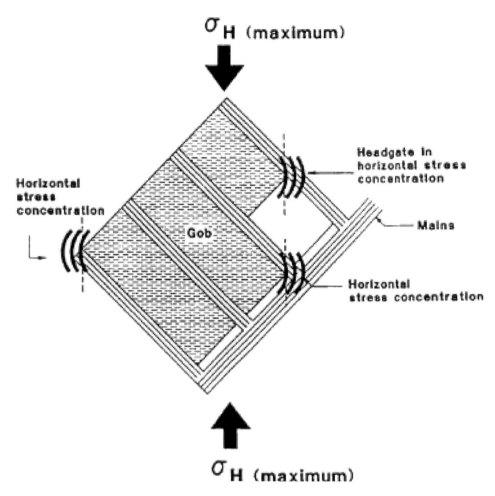

Figure 10.15 1 Horizontal stress concentrations as a result of retreat mining through the stress field

Source: Mark and Mucho (1994)

10.15.2 Changes in the orientations of induced principal stresses

The induced principal stresses have particular orientations around the longwall panel irrespective of panel orientation. Before mining, the *in-situ* minimum principal stress, σ_3, is in vertical direction. After mining, at the face, the induced minimum principal stress, σ_{3i}, relieves and rotates in the vertical plane away from the gob. Above the entries and crosscuts, it is relieved and maintained its vertical orientation. Before mining, the *in-situ* intermediate principal stress, σ_2, acts in the horizontal plane. After mining, the induced intermediate principal stress, σ_{2i}, relieves and rotates in vertical plane towards the gob. Figure 10.15.2 shows the direction of maximum induced principal stress, σ_{1i}, in the immediate roof projected on

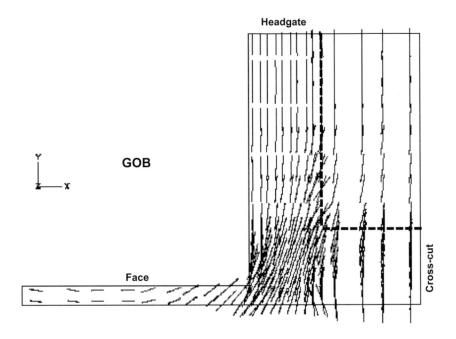

(A) θ = 0°

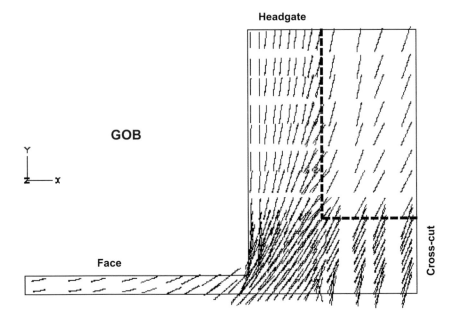

(B) θ = 30°

Figure 10.15.2 Distribution of maximum-induced principal stress in the immediate roof

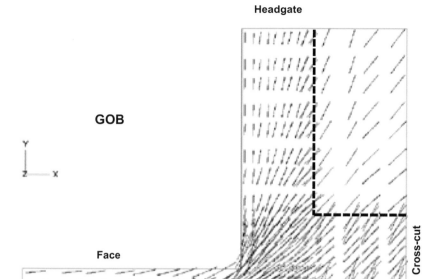

(C) θ = 60°

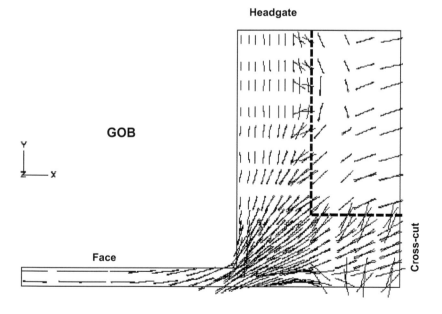

(D) θ = 90°

Figure 10.15.2 (Continued)

the horizontal plan for different orientation for the left-handed panel. It can be seen that, irrespective of panel orientation, the maximum induced principal stresses are approximately oriented parallel to the face and headgate, except at the T-junction where it is oriented away from the face toward the solid ground.

10.15.3 Changes in the magnitude of induced principal stresses

The induced maximum principal stress concentration (IMPSC) in the immediate roof for the left-handed panel is shown in Figure 10.15.3. Zones of IMPSC less than three are stress relief zones which are represented by light fringes, while dark fringes represent high stress concentration zones. The stress relief zone at the face is affected by panel orientation. Zones of significant stress relief are observed at 0° and 30° panel orientations while the relief is very small at 90° panel orientation. The smallest MIPSC at the ribs of the headgate is observed for 0° panel orientation while the largest at 90° panel orientation. The reverse is true for the ribs of crosscuts. The magnitude of MIPSC is inversely proportional to the magnitude of reorientation of the maximum induced principal stress. For example, at the headgate T-junction, the 0° panels cause the maximum induced principal stress to rotate further as compared to the 90° panels; therefore, the IMPSC is smaller for 0° panel. The trends observed for the left-handed panel are applicable to the right-handed panels, except the IMPSC is larger. Figure 10.15.4 shows the IMPSC inside the immediate roof for the left-handed panels at different depths: 0.63, 2.37, and 5.37 ft (0.19, 0.72, and 1.63 m) from the roof line for the 30° panels. Near the roof line it is largest and covers all the ribs of headgate and crosscuts. It decreases both in magnitude and area as it moves higher into the roof. At a depth of $z = 5.37$ ft (1.64 m), stresses concentrate only at the panel edge.

The IMPSC increases with increasing panel orientation angles from 0° to 90° uniformly across the entry width for the right-handed panels. But for the left-hand panels, the 60° and

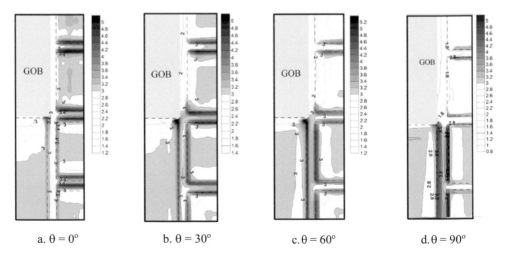

a. θ = 0°   b. θ = 30°   c. θ = 60°   d. θ = 90°

Figure 10.15.3 Maximum principal stress concentrations in the immediate roof for left-handed panel

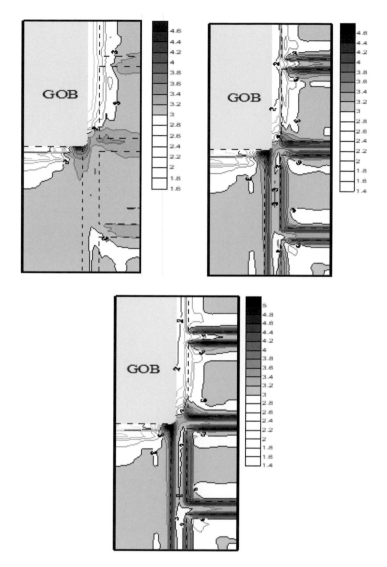

Figure 10.15.4 Maximum principal stress concentration at different depths of the immediate roof for left-handed panel

90° panels are the highest and about the same. Figure 10.15.5 shows the distributions of the induced principal stresses on cross-section Y-Y at a depth of $z = 0.63$ ft in the immediate roof of the headgate and outby the face ($y = 11$ ft). The induced minimum principal stress is distributed almost symmetrical about the headgate center (Fig. 10.15.5 A). Its magnitude is smaller than the in-situ minimum principal stress and its highest values are located at the headgate ribs. The effect of panel orientation on the induced minimum principal stress is only visible at and near the ribs, but this effect is less than 150 psi (1.03 MPa).

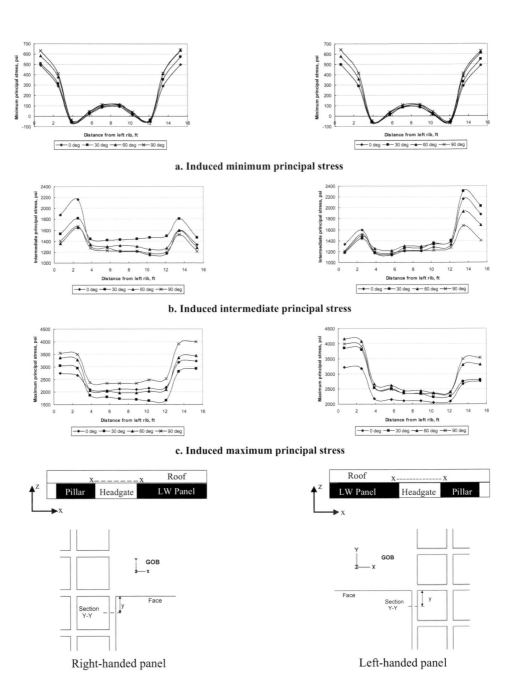

a. Induced minimum principal stress

b. Induced intermediate principal stress

c. Induced maximum principal stress

Right-handed panel

Left-handed panel

Figure 10.15.5 Distribution of induced principal stresses at the immediate roof at distance of 11 ft (3.4 m) outby the face

The induced intermediate principal stress (Fig. 10.15.5 B) is asymmetrical and greatly affected by panel orientation. It is larger, and its magnitude varies with panel orientation, at the left rib for the right-hand panels. But the reverse is true for the left-hand panels. The effect of panel orientation is significant only within 2 ft (0.61 m) from the ribs. Inside and around the center of the headgate, it is relieved (i.e., smaller than 1540 psi) (10.62 MPa) for both the right- and left-hand panels. Note the *in-situ* intermediate principal stress is two times the *in-situ* vertical stress or 1540 psi. The induced intermediate principal stresses are highest for panels oriented at 0° and 30°, and between those two orientations, it is higher in most parts of the entry for the 30° orientation. The effect of panel orientation is more pronounced for the right-hand panels.

Contrary to the induced intermediate principal stress, Figure 10.15.5 C shows that the induced maximum principal stress at the right rib is larger than that at the left rib of the headgate for the right-hand panels. The reverse is true for the left-handed panels. It is relieved inside and around the center of the headgate (i.e., smaller than 2310 psi). Note the *in-situ* maximum principal stress is three times the vertical stress, or about 2310 psi.

As the distance outby the T-junction increases, the induced principal stress distributions remain the same, except for the following:

1   The minimum induced principal stress is relieved more, especially for the right-handed panel.
2   The intermediate and maximum induced principal stresses increase, especially for the right-handed panel.
3   The maximum induced principal stress is more symmetrical and eventually become symmetrical about the headgate center.
4   The difference in stress distribution between the right-handed and left-handed panels becomes smaller.

10.15.4 Headgate stability analysis

As stated in previous sections, all three induced principal stresses differ from the *in-situ* principal stresses and vary from location to location around the longwall face and headgate T-junction areas. Consequently, stability evaluation of these areas must consider all three induced principal stresses. The commonly used Mohr-Coulomb criterion in 2D analysis does not consider the intermediate principal stress. So the Drucker-Prager criterion is used here. According to the Drucker-Prager criterion, roof stability is inversely proportional to the induced maximum principal stress (Morsy and Peng, 2006).

Figure 10.15.6 shows the distributions of yield and stable integration points in the immediate roof of headgate for the left-handed and right-handed panels at different panel orientations. The immediate roof stability is estimated for five sections: 11, 16, 21, 26, and 46 ft (3.4, 4.9, 6.4, 7.9 and 14 m) outby the face.

The stability of headgate is affected by panel orientation for the left-handed panels. Comparing the immediate roof yield %, which is defined as the ratio of number of yielded integration points to the total number of points in the assessed zone, headgate stability is ranked in the following descending order, 0°, 30°, 60°, and 90° (Fig. 10.15.6 A). The same is true for the right-handed panels (Fig. 10.11.6 B), except that within a distance of 21 ft (6.4 m) outby the face, the headgate of 30° right-handed panels is more stable than that for the 0° right-handed panels.

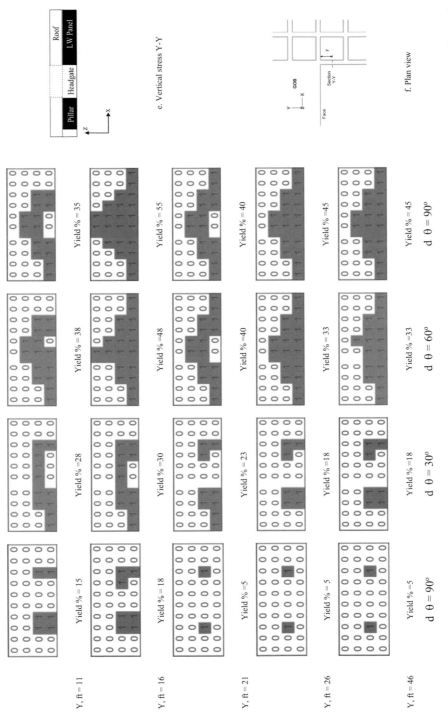

e. Vertical stress Y-Y

f. Plan view

Figure 10.16.6a Yield zones in the immediate roof of the headgate for the left-handed panel (dark shading indicates yielded areas)

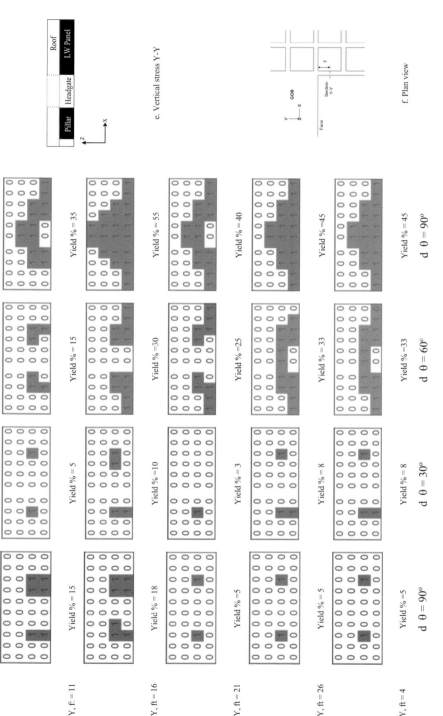

Figure 10.16.6b Yield zones in the immediate roof of the headgate for the right-handed panel (dark shading indicates yielded areas)

The stability of headgates for the 0° and 90° is the same for both the left-handed and right-handed panels. The immediate roof of 30° and 60° right-handed panels is more stable than those for the left-handed panels. No change in the yield % of the headgate for 30°, 60°, and 90° panels beyond a distance of 26 ft (7.9 m) outby the face and 16 ft (4.9 m) outby for 0° panels. There are significant differences in headgate stability between the right and left ribs for the 30-, 60-, and 90-degree panels, within 11 ft (3.4 m) outby the face. For the right-handed panels, however, only 90-degree orientation shows stability difference between the right and left ribs of the headgate.

10.15.5 Summary

As the longwall retreat mining advances, the *in-situ* principal stresses cause local stress re-distribution in and around the gateroad system and longwall face. The induced principal stresses changes both in direction and magnitude. Their applied directions rotate toward the solid ground and away from the excavation, while their change in magnitudes varies with location and among the three principal stresses. Therefore, in order to evaluate the effect of panel orientation with respect to the in-situ horizontal stresses, all three principal stresses must be considered.

Everything being equal, headgate stability varies with location and either being left-handed or right-handed panel. For the left-handed panels, the 0° panel orientation is most stable, followed by 30°. The worst orientations are 60° and 90°, both of which show little difference in yield zone area. For 0° panel orientation, the yield (or unstable) zone is within 16 ft (4.9 m) outby the T-junction, and 21 ft (6.4 m) for the 30° orientation. For the right-handed panels, the best panel orientation is 30°, followed, in sequential order, by 0°, 60° and 90°. For the 30°, orientation, there is very little yield zone all throughout. For the 0° orientation, yield zone occurs within 21 ft (6.4 m) outby the T-junction.

Chapter 11

Ventilation and methane, dust, and noise controls

11.1 Introduction

11.1.1 Historical trends

There are several health, safety, and environmental issues associated with longwall mining, including methane, dust, noise, and surface subsidence. Since methane emission and dust increase with coal production, they are critical factors in high production longwalls. Traditionally, air ventilation is used to dilute methane and dust concentration. But in high production longwalls, additional measures are required in order to reduce methane and dust concentration to the level required by law. Noise is another health issue that is getting more attention in recent years due to the introduction of ever-increasing high power longwall face equipment and much wider panels. All three topics are discussed in this chapter. Surface subsidence will be addressed in Chapter 14.

As production grew, so did the amount of air delivered to the longwall face. MSHA statistics for 1998–1999 (MSHA, 2000) show that the amount of air delivered was 150 to 300% more than required by approved ventilation plans. This trend continues to the present time.

For dust control, wet cutting with water sprays did not work satisfactorily in the 1970s and early 1980s. In order to reduce dust, more than 75% of longwalls then employed uni-directional cutting. As dust control technology improved – including wet cutting with cleaner water and higher water pressure in the 1990s and shield automation developed and implemented, minimum dust standards could be met, and dust control was no longer a commanding factor, even with the implementation of stricter dust standard in 2016.

As panels became wider, the more productive bi-di was adopted as the preferred cutting method in the early and mid-1990s. The current dust control technique was effective when the dust standard was 2.0 mg m$^{3-1}$ as shown by the 2000 MSHA survey (MSHA, 2000). Strict implementation of the same dust control techniques plus digital shield dust control also have successfully met the new dust standard of 1.5 mg m$^{3-1}$ effective August 2016 (National Academies of Sciences, Engineering, Medicine, 2018).

For methane control, major developments lie in the more use of surface directional horizontal boreholes since the mid-2000s. These measures, in addition to being environmentally friendly, are able to degas 50 to 90% of methane content and enable high production operations to proceed even in highly gassy coal seams.

11.1.2 Ventilation, methane, and dust control plans

Each longwall mine is required to submit a ventilation plan and a supplement to the ventilation plan to its respective MSHA district manager. The approved plan must be designed specifically for the individual mine conditions and mining system. It consists of, at least, a description of the method of mining, a longwall ventilation map, the minimum quantity of air to be delivered with the minimum air velocity to specific areas of the longwall panel, and the quantity and location of measurements, all of which vary from mine to mine.

A supplemental plan to maintain compliance with the respirable dust standards and methane monitoring will be submitted as parts of the ventilation plan. It shows how many water sprays and where they are located, type of water sprays, operating pressure, and other information for the shearer and stage loader.

11.2 Longwall specific ventilation

Ventilation is one of the most critical factors when planning for longwall mining. It involves consideration of provisions for adequate gateroad ventilation; face ventilation, including methane and dust control; bleeder system design for control of gases (methane emissions and oxygen) in and around the gob areas; and intake air escapeway and haulage isolation requirements of federal regulations (Fama, 2005).

The basic principle of longwall ventilation is simple: direct a split of intake air from the mains and course it through the headgate gateroads of the panel being mined. When it reaches the last open crosscut, direct the air across the face and down to the tailgate, causing a portion of the air to flow through the gob to the bleeder entry system. Critical tasks include the control of oxygen and methane in the air, the control of respirable dust at the face, the maintenance of the bleeder system, and for high production longwalls it usually requires two intake entries to meet requirements at tailgate T-junction.

11.2.1 Gob and bleeder ventilation system

Air ventilation for US underground coal mines includes face ventilation and the gob and bleeder ventilation systems. They are primarily designed to keep fresh, oxygenated air at the face by moving dust and methane accumulations away from mining activities including the primary airflow paths in the active working sections. For all longwall panels, the face and the gob behind the active face, as well as previously mined-out panels must be ventilated. A certified person must travel along the bleeder and evaluate the effectiveness of the gob and bleeder system at least once a week.

One must first determine the best method of ventilating the proposed longwall district, including each individual panel. This decision centers on the choice of a bleeder system. The most common bleeder system in use in the eastern and southern United States involves a bleeder shaft with a high-pressure fan (usually centrifugal) located behind the worked out areas to the rear of the active working face (Fig. 11.2.1).

Fresh air is coursed from main entries toward the longwall face. When this air reaches the face area, it is directed up the headgate to the longwall face and the tailpiece of the section conveyor belt. A portion of this air is directed outby along the belt to regulators into the main return. The majority of this air is directed down the longwall face toward the tailgate to control dust and dilute methane. A portion of this air flows into the gob, and eventually

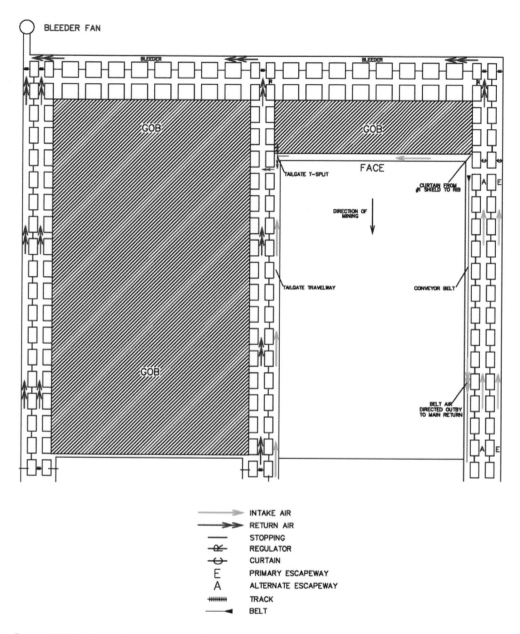

BLEEDER FAN

GOB

GOB

FACE

TAILGATE T—SPLIT

DIRECTION OF MINING

CURTAIN FROM #1 SHIELD TO RIB

TAILGATE TRAVELWAY

CONVEYOR BELT

GOB

BELT AIR DIRECTED OUTBY TO MAIN RETURN

→	INTAKE AIR
⟶	RETURN AIR
—	STOPPING
⟞	REGULATOR
⟝	CURTAIN
E	PRIMARY ESCAPEWAY
A	ALTERNATE ESCAPEWAY
⊞⊞⊞⊞	TRACK
⟶◂	BELT

Figure 11.2.1 Typical longwall ventilation to bleeder system

migrates through the gob to specially designed "bleeder" entries, whose purpose is to direct this air away from active workings to a bleeder air shaft. This air, laden with methane and dust goes out the shaft to the atmosphere. Coal absorbs oxygen, so the bleeder entries also convey air that is somewhat deficient in oxygen to the outside air. The use of a high pressure bleeder fan is important because the fan must overcome the resistance of the gob to the flow of air. Maintenance of the bleeder system is critical to having an effective bleeder system, which maintains the atmosphere above standards established by MSHA and known to provide a minimum level of safety to underground workers. In practice, most mines exceed these minimum standards greatly, as experience dictates that having more air than the minimum specified by the law greatly aids in dust control and control of gases. One advantage of using a bleeder system that takes fresh air from the main entries and directs it toward the bleeder fan is that it is easier to control the air and keep it moving primarily in one direction only except that the air on the conveyor belt is usually directed outby. Otherwise, all air moves from outby to inby.

Another longwall ventilation system makes use of the main return located in the main entries outby the longwall face. Most of the longwall air flows outby, and only a small amount of air makes it to the bleeder entries compared to the previous system that has all air reporting to the bleeder. However, since longwall mining is a type of retreat mining, MSHA requires a bleeder system behind the gob inby the longwall face. The bleeder entries may be tied to the main return or may have a bleeder fan. In either case, the quantity of air flowing through the bleeder entries is much less than the previous bleeder system, as most of the air from the longwall face is directed outby in the tailgate entries. The advantage to this system is that face dust control, particularly near the longwall tailgate tends to be improved because more air reaches the tailgate and does not migrate through the gob as much. However, flow of air laden with gases and dusts is controlled by the air current flowing outby to the main return more so than to the bleeder entry, which is specifically designed to take this contaminated air away from the active workings. Coursing return air toward both the main entries and the bleeder entries means two low-pressure return systems will compete against each other for intake air, and dead spots can develop in the system, where no air flows and gases can accumulate.

Another bleeder system in use today is a "wrap-around" system. Figure 11.2.2 shows a typical wrap-around system for longwall panels with a three-entry development system. This system is commonly used in some underground longwall mines. Fresh air enters the panel in the headgate side through the second and third entry (counting from the left) and travels all the way to the last open crosscut where a curtain in each entry is hung and the air is directed toward the longwall face. Some air is allowed to leak into the gob behind the curtains. When it reaches the first (headgate or belt) entry, it splits; some air will travel outby through the belt entry and exit into the main return. The great majority will travel across the face. The amount of air that exits through the headgate is controlled by a regulator located near the exit point where the headgate intersects the main return. As the intake air travels across the face it will also leak into the gob. When it reaches the tailgate (note from this point on it is the "return air"), it splits again; some will exit the panel through the tailgate while some travels inby through the gob to the bleeder, which is the outermost entry surrounding the panels. The amount of air that exits through the tailgate is controlled by the regulator located near the exit point where the tailgate intersects the main return. Via the bleeder, the return air courses through the gobs of the mined-out panels and will eventually exit into the main return.

11.2.1.1 Headgate ventilation

There are variations of flow pattern; some longwalls seek and receive approval from MSHA to use the headgate (belt or first) entry as the intake airway. In this case, if CO

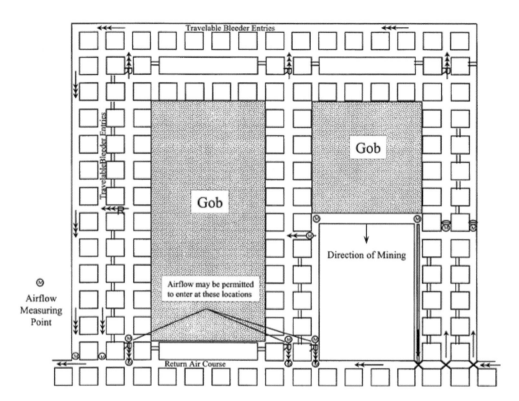

Figure 11.2.2 Longwall wrap-around bleeder ventilation system (circled M = measurement point)

Source: MSHA (2002)

sensors, fire protection, and a monitoring system are installed along the belt conveyor, the belt air can be directed into the face directly as intake air. But if no CO sensors are installed, the air ventilating the belt entry (headgate) must be directed to the return, or to a point outby the face. The belt entry must also be isolated from the longwall intake by permanent stoppings.

A variation of this system is to course intake air through the second entry. When the fresh air reaches the last open crosscut, some air is allowed to return through the third entry. The third entry outby the face normally requires fresh air flow because miners are setting up supplementary supports in this future tailgate. Roof control plans may require that supplementary supports be installed 500–1000 ft (152–305 m) outby the retreating face.

There are also longwalls that send the intake air through all three entries in order to deliver sufficient air to the face, especially in mines with high methane liberation.

Some longwalls install trolley haulage in the belt entry to eliminate a stopping line separating the belt from the powered haulage. Battery powered and diesel track equipment are also popular, because the track haulage system can then be installed in the intake air escapeway.

11.2.1.2 Tailgate ventilation

Tailgate entries may be ventilated by intake or return air. But the tailgate must remain open at all times as an emergency roadway.

1   If the tailgate is used as an intake, it must be isolated from the other entries adjacent to the previously mined-out panel with permanent stoppings. A track may be installed to deliver supplies or for immediate access.
2   A tailgate intake would reduce or eliminate methane and even dust control problems in and around the tailgate side. Air leakage from the tailgate intake would help with diluting methane in the other tailgate entries adjacent to the gob.
3   If the tailgate is used as a return from a split of the intake air coursing across the face, the methane content must be kept below 1% between the tailgate T-junction and the regulator through which the return air exits into the main return.
4   Instead of exiting to the main return, the tailgate return may be re-directed back, if possible, into the gob for better use of air for gob ventilation.

In very gassy mines, however, the methane from the gob area(s) adjacent to the tailgate may migrate into the tailgate return, causing the methane content to consistently exceed 1%. In this case, permanent stoppings must be erected to isolate the tailgate return from the gob of the previous panel and the tailgate return air must be maintained at a higher pressure than the air in the gob.

11.2.1.3 Longwall bleeder systems

Longwall bleeder systems or bleeder entries are special air courses designed to move the methane-air mixture from gob areas away from the active face to the main return air courses. Bleeder entries are connected to gob areas at strategic locations to do the following: (1) control airflow through gobs, (2) induce gob gas drainage, and (3) minimize the hazard of gob gas expansion into fresh air. Bleeder systems are combinations of bleeder entries, entry connections to the gob, and associated ventilation controls. A bleeder system extends from the faceline inby to the intersection of the bleeder split with another air split, excluding the face.

There are many ways to ventilate the face and gateroads, satisfying regulations. But adequate longwall ventilation must also have a well-designed and properly maintained bleeder system. A well-designed and effective bleeder system should not only provide a consistent flow of air through the gob to dilute and remove gob gases, but it should also maintain a continuous positive flow of air away from the face.

Air flow through the gob is primarily controlled by regulators installed at the inby side of the gob (or outby side of the bleeder entries). The regulators must be accessible for adjustments, measurements, and should be located to control as wide an area of the gob as possible.

The maximum methane allowed in the bleeder system is 2%.

11.2.1.4 Bleeder shafts

In lieu of connecting the bleeder entries to the main return entries, bleeder shafts are drilled approximately every four to five panels, located either on or immediately off the bleeder entries, and a ventilation fan (normally the centrifugal type) is installed with a capacity ranging from 50,000 to 300,000 cfm (cubic feet per minute or 23.6–141.7 m^3 s^{-1}) at a pressure of up to 40 in. water gage (0.01 MPa). These shafts, 5–12 ft (1.52–3.66 m) in diameter, are normally blind-hole drilled because it can be done more quickly. See Figure 11.2.1 for an example of a bleeder system using a bleeder shaft.

11.2.1.5 Flow through bleeder fan systems

Figure 11.2.3 shows a typical flow-through bleeder fan system for longwall panels with a three-entry development system. This ventilation plan is suitable for all mines.

The air-flow patterns are the same as those in the wrap-around system, except the return air in the bleeder is exhausted into the atmosphere at the rear end (gob side) of the panel or through a bleeder shaft with the aid of a bleeder fan. This system is most suitable for very large panels or panels with tight gobs that require larger air pressure to pull the air through the gobs and bleeders.

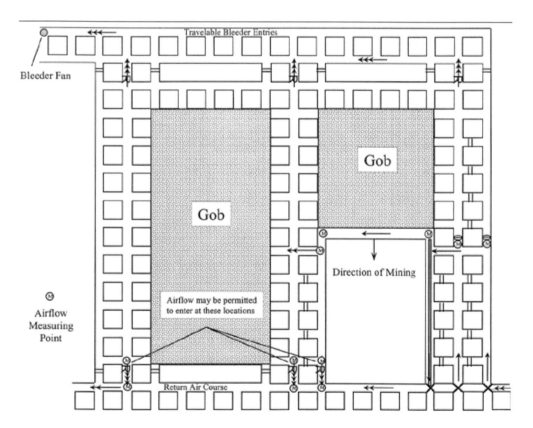

Figure 11.2.3 Longwall flow-through bleeder ventilation system
Source: MSHA (2002)

11.2.2 Bleederless longwall ventilation systems

In western coalfields, some coal seams are prone to spontaneous combustion and contain hydrogen sulfide (H_2S), which is poisonous and explosive. In order to reduce the risk of a gob fire or the production of combustible gases, a bleederless longwall ventilation system was introduced in early 2000s and approved for the gob behind the active longwall face and abandoned longwall panels.

Figures 11.2.4 and 11.2.5 show that gob seals of either composite block or wood block, which are installed in every crosscut, and stoppings of either pro-panel or collapsible metal design are installed every three crosscuts in the second entry on the headgate side. When the panel is completed, explosion-proof seals will be constructed at the mouth of each panel.

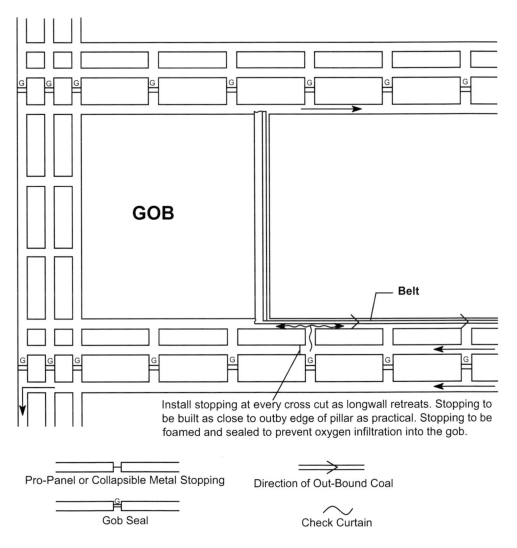

GOB

Belt

Install stopping at every cross cut as longwall retreats. Stopping to be built as close to outby edge of pillar as practical. Stopping to be foamed and sealed to prevent oxygen infiltration into the gob.

Pro-Panel or Collapsible Metal Stopping

Direction of Out-Bound Coal

Gob Seal

Check Curtain

Figure 11.2.4 Typical bleederless longwall ventilation before connection to setup entries

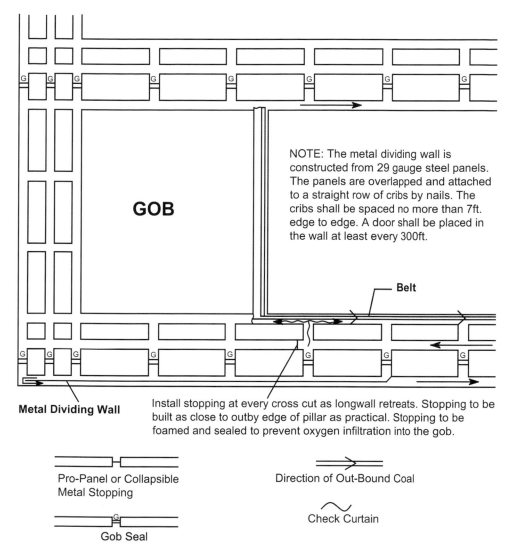

NOTE: The metal dividing wall is constructed from 29 gauge steel panels. The panels are overlapped and attached to a straight row of cribs by nails. The cribs shall be spaced no more than 7ft. edge to edge. A door shall be placed in the wall at least every 300ft.

GOB

— **Belt**

Metal Dividing Wall

Install stopping at every cross cut as longwall retreats. Stopping to be built as close to outby edge of pillar as practical. Stopping to be foamed and sealed to prevent oxygen infiltration into the gob.

Pro-Panel or Collapsible Metal Stopping

Direction of Out-Bound Coal

Gob Seal

Check Curtain

Figure 11.2.5 Typical bleederless longwall ventilation after connection to setup entries

As part of the bleederless system, an atmospheric monitoring system consisting of underground CO sensors and surface monitor/alarm instrument station will be installed to monitor CO concentration in the longwall return air course.

11.2.3 Ventilation measurements and requirements

11.2.3.1 Ventilation requirements

The minimum air quantity needed to enter the longwall face as measured at the intersection of the last open crosscut and belt entry is 30,000 cfm or ft³ min⁻¹ (850 m³ min⁻¹). Some

MSHA districts also require that air velocity be no less than 400 fpm (2 m s⁻¹) and 200 fpm (1 m s⁻¹) at the tenth shield from the headgate and at the tailgate or just outby the last shield in the tailgate, respectively. Measurement of tailgate air velocity indicates how much air at the headgate makes it down to the face without leaking into the gob. The air quantity measured at the tailgate must be less than that on the face so that air reversal does not occur on the face. Nearly all longwalls exceed the minimum air quantity to control methane and respirable dust

11.2.3.2 Ventilation measurements

Regulations require that once a week a fireboss evaluates the gob and bleeder system by taking measurements at strategic locations (as specified in Fig. 11.2.2) and keeps a permanent record. Measurements at each point include the methane and oxygen content of the air. The fireboss must also test and determine the air flow quantity and direction at the air inlet to the gob and at a return split.

According to MSHA (2000), the actual air quantity measured at the intersection of the last open crosscut and belt entry ranged from 30,000 to 207,000 cfm (14–98 m³ s⁻¹) with 63% of them ranging from 30,000 to 70,000 cfm (14–33.1 m³ s⁻¹). There was no correlation between panel width or panel length and the amount of air delivered to the longwall face. Almost half of the longwalls delivered more than 150% of the total air quantity required.

Air velocity as measured at the tenth shield from the headgate ranged from 265 to 650 fpm (1.35–3.3 m s⁻¹). More than 40% of longwalls had air velocities at the headgate exceeding 150% of their approved plan with one exceeding 1000 fpm (5.08 m s⁻¹). Tailgate air velocity ranged from 201 to 900 fpm (1.04–4.57 m s⁻¹) with one longwall exceeding 1000 fpm (5.08 m s⁻¹).

The trends of these measured statistics continue to present time, if not more.

11.3 Methane control

11.3.1 Introduction

Methane is a hazardous gas that has the potential to explode. Although it does not promote combustion, methane will be ignited when its concentration is between 5% and 16% (9.5% is the most dangerous), and the air temperature is from 1200 to 1382 °F (650–750 °C). Some coal seams and rock strata contain large amounts of methane, and under high pressure, the coal and gas will burst out suddenly and simultaneously. This is called gas outburst.

11.3.2 Sources of methane

During the coalification process, the coal-forming materials, which consist mainly of plants, undergoes a series of physical and chemical reactions and produces a large volume of methane (CH_4). The coal then undergoes changes in chemical composition and structure under high pressure and temperature. This process of metamorphism also produces methane. In general the amount of methane produced is proportional to the ranking of metamorphism.

Most of the methane produced during coalification and metamorphism escapes to the atmosphere through fissures in the strata. A small part stays in the fissures in the surrounding strata, and still another small part remains in the coal. This is the methane which is commonly referred to. The methane stays in the coal or the fissures in the surrounding strata either in a free or adsorbed state. The free methane moves freely in the coal or the fissures and fractures in the strata, whereas the gas particles in the adsorbed methane tightly adhere to the surface of the interior fissures or the interior of the coal molecules. Under natural conditions, 1 ton of coal may contain 1235 to 1588 ft^3 (35–45 m^3) of methane, with approximately 90% in the adsorbed state. Under certain conditions, the free and adsorbed states are in equilibrium. As the pressure, temperature, and mining conditions change, the equilibrium will be destroyed. When the pressure is increased or the temperature is decreased, some parts of the free methane will become adsorbed. Conversely, some of the adsorbed methane will be released to become free methane.

During mining operations, the coal seams and the surrounding strata are subjected to continuous fracturing, which increases the passageways for methane and destroys the equilibrium between the free and adsorbed methane that exists under natural conditions. As a result, some of the adsorbed methane will be freed. Thus under normal conditions and as longwall mining progresses, the methane is released directly and continuously from the coal at the face as well as from mined broken coal being carried to the surface. Because pressure at the face is lower than the coal block and surrounding strata, this causes the migration of gas from the interior of the coal block and surrounding strata. This is the basic form of methane emission. Only methane in the free gas state can flow into mine workings. The flow is best considered as a two-step process: first, it diffuses through the micropore structure of the coal, and, second, it flows along interconnected fissures in the coal bed. Methane moves by diffusion from the desorption site through a solid coal lump until it intercepts a fracture in the coal (Curl, 1978).

The sources of methane emission at the longwall face are as follows (Trackemas and Peng, 2013): (1) gas released from the coal broken by the shearer, (2) gas emitted from the broken coal on the face conveyor (AFC), (3) gas emitted from the coal transported on the belt conveyor, and (4) background gas emitted from the coal face and from the adjoining ribs in the intake gateroads.

Methane emission into a mine occurs at a steady state. But when geological anomalies such as faults are encountered, sudden gush of methane release may occur.

11.3.3 Factors controlling methane emission

The amount of methane emission varies from seam to seam and from mine to mine. There are three major factors that control the amount of methane emission.

11.3.3.1 Methane content of the seam and surrounding strata

Gas content is the amount of gas contained in a ton of coal including both adsorbed and free gases. It is the most important factor controlling the amount of methane to be released after mining. In addition to being proportional to the ranking of the metamorphism, methane content in the coal seam and the surrounding strata also depends on the seam depth and geological conditions.

The most accurate method for determining gas content in coal is best measured directly (Diamond, 1993).

Thakur (2006) classified the gassiness of coal seams into three categories in terms of seam depth and gas content as shown in Table 11.3.1.

Table 11.3.1 Category of gassiness of coal seams and degasification method

Category	Depth, ft	Gas content, ft³/t	Required degasification method	
			Pre-mining	Post mining
Mildly gassy	< 100	< 100	None	2 gob wells/panel
Moderately gassy	600–1500	100–300	In-mine horizontal drilling	5–6 gob wells/panel
Highly gassy	1500–3000	300–700	Vertical frac well In-mine drilling	20–40 gob wells/panel

1 ft = 0.3 m: 100 ft³ t⁻¹ = 2.58 m³ mt⁻¹

Source: Modified from Thakur (2006)

11.3.3.2 Atmospheric pressure

Whenever the atmospheric pressure changes, it will also cause the air pressure underground to change. Therefore, as the surface atmospheric pressure decreases, it will increase methane emissions underground. The surface atmospheric pressure change will not significantly affect the amount of methane normally emitted from the exposed coal faces and ribs. But it will greatly affect the amount of methane emission from the gob or the areas where methane accumulates.

11.3.3.3 Air volume and pressure of mine ventilation

Mine ventilation is the primary method for diluting the methane emission at the face. A large amount of air is circulated throughout the mine workings to dilute methane concentration and carry the methane to the surface via bleeder entries and ventilation shafts.

During a normal production period, methane concentration is diluted to below the lowest legal limits (1% at the face and return and 2% in the bleeders) allowed by mainly adjusting the volume of the ventilated air. The required volume of air in a working face can be determined by:

$$Q_{air} = \frac{Q_{gas}}{1440C} K \tag{11.3.1}$$

Where Q_{air} is the required fresh air volume in ft³ min⁻¹ (m³ min⁻¹). Q_{gas} = the absolute amount of gas emission per unit time in the whole mine in ft³ min⁻¹ (m³ min⁻¹). $C = 1\%$ is the maximum allowable methane concentration in the return air. $K = 1.5$ is the non-uniform coefficient of gas emission.

As far as air pressure is concerned, if mine ventilation is the exhaust type, the amount of methane emission will increase with increasing (negative) air pressure. Conversely, if the ventilation is the blowing or forced-air type, it decreases with increasing (positive) air pressure. In practice, one of the objectives for changing the air pressure is to regulate the air volume. But a change in air pressure will also be accompanied by changes in the leakage conditions in the gob. For instance, in exhaust-type ventilation, when the air volume is increased, air leakage from the gob is also increased due to the increase in negative air pressure. As a result, a fixed amount of air with a high concentration of methane will flow from the gob into the normal ventilation flow, causing a sudden increase of methane content in the return air. Only after a certain period of time will the amount of methane emission from the gob, and subsequently the methane concentration in the return air, return to normal conditions. Therefore, when the production process requires a change of air volume in the panel, especially increasing the air volume, the methane concentration in the return air must be carefully monitored.

11.3.4 Methane drainage

According to Thakur (2014), a well-designed mine ventilation system can handle specific methane emission, which is the total amount of gas released from the mine divided by the total amount of coal mined, up to 1000 ft^3 ton^{-1} (25.8 m^3 mt^{-1}). Above that threshold, methane drainage is needed. Mines with specific methane emission higher than 4000 ft^3 ton^{-1} (103 m^3 mt^{-1}) can be safely mined by a well-planned methane drainage system and a well-designed ventilation system.

The gas emission space which is defined as the mining disturbed areas from which gas flows into the ventilated mine workings, may extend to 270 ft (82.3 m) below the coal seam being mined and approximately 1000 ft (305 m) above it (Thakur, 2006).

Various methane drainage methods have been developed to capture the released methane so the mine ventilation air does not need to handle all of it. To be fully effective in terms of timing and space, methane drainage methods can be divided into two types: pre-mining and post-mining methods. The pre-mining methods consist of (Thakur, 2006): (1) horizontal in-seam boreholes, (2) surface stimulated vertical boreholes (frac wells), (3) surface direction-ally drilled boreholes, and (4) dual well drilling system (Zupanick, 2005). The post-mining method is mainly the vertical gob well method.

In practice, a combination of these methods is used to degasify coal seams as much as possible before and during mining to reduce the amount of gas emissions from the gob into the ventilation system. The optimum methane drainage system for a mine depends on the amount of methane produced, emission pattern, and geology of coal seam and surrounding strata. It may need to be evaluated and adjusted on a continuing basis to ensure optimum capture of methane over time.

11.3.4.1 Pre-mining methane drainage methods

A. HORIZONTAL IN-SEAM BOREHOLES

(1) General   In-seam horizontal holes are very popular. It can be either short holes or long and ultra-long holes. Figure 11.3.1 shows the short holes drainage system layout. They are normally less than 1000 ft (305 m) long drilled from the headgate of the panel to be mined during entry development (holes A). The holes are drilled perpendicular to the rib on 200 ft (61 m) centers to within about 150 ft (46 m) of the tailgate. These holes are designed to remove a substantial amount of gas from the middle and the completion end of the longwall panel, since the area drilled first is the last to be mined. In addition, holes are drilled from the tailgate to within 50 ft (15 m) of the headgate on the other side of the panel (holes B and C in Fig. 11.3.1). For best results, horizontal holes with a 3–6 in. diameter should be drilled

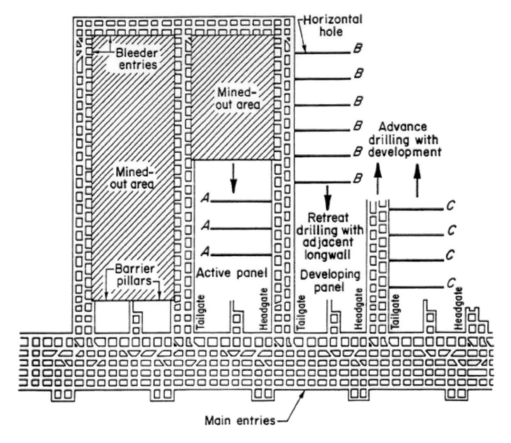

Figure 11.3.1 Short in-seam horizontal holes for methane drainage
Source: Diamond (1993)

perpendicular to face cleats and spaced at 120–500 ft (35–150 m) apart depending on coal permeability, gas content, time to mining, and drilling economics, at least six months in advance of mining to allow sufficient time for methane to drain off (Diamond, 1993). Aul and Ray (1991) reported that short holes can remove 30% *in-situ* gas in two months and 80% *in-situ* gas in 10 months in highly gassy mines.

Ultra-long holes up to 5650 ft (1723 m) are drilled from the ribs of gateroad entries at various angles at different points. Borehole patterns vary with seam gas content and permeability. They can be drilled in the gateroad in advance of continuous miner development and turned to run near and parallel to both the headgate and tailgate of the panel (Fig. 11.3.2). These long holes will isolate the gateroad development section and eliminate gas flow from the panel coal block, preventing entry development delays due to gas inflow. They can also be drilled perpendicular to and from the mains to the projected gateroads and panels far ahead of panel development (Fig. 11.3.3) (REI Drilling, 2006). The key point in Figure 11.3.3 is the combined effect of vertical frac wells and in-seam boreholes. The horizontal holes are directionally drilled around the half-lengths of hydraulic fractures such that the horizontal holes reduce reservoir pressure and increases the production of the frac wells.

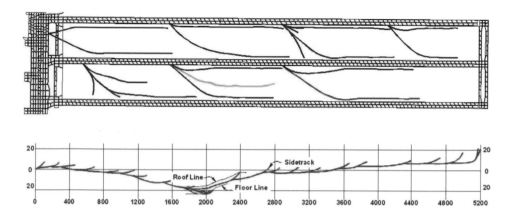

Figure 11.3.2 Ultra-long holes drilled from gateroads in advance of gateroad development
Source: Target Drilling (2006)

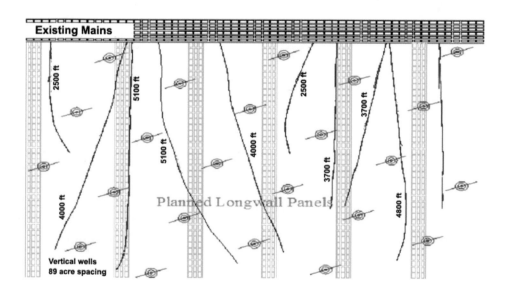

Figure 11.3.3 Ultra-long holes drilled perpendicular to and from mains to projected gateroads and panels ahead of panel development (REI Drilling, 2006). Double circles are surface vertical frac wells with projected fracture direction

Ultra-long hole drilling can map major geological structures. The cross-sectional profile in Figure 11.3.2 shows the seam profile and major geological features such as faults (Target Drilling, 2006).

Multiple boreholes can also be drilled from one location as shown in Figure 11.3.4 (Target Drilling, 2006). If there are seams overlying or underlying the seams being mined, boreholes can also be drilled to the overlying or underlying seam to pre-drain the methane in these seams and reduce the amount of gas flowing into the mine workings or gobs of the seam (Fig. 11.3.5) (REI Drilling, 2006). The overlying holes are placed above the lowest source seam so that these holes remain intact after under-mining, for gob gas recovery. The borehole in the underlying seam produces gas when longwall mining start as it is likely filled with

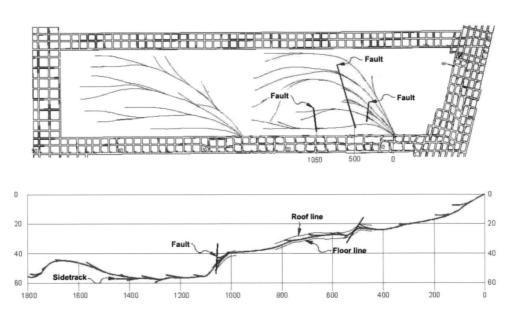

Figure 11.3.4 Multiple long holes drilled ahead of gateroad development and panel mining
Source: Target Drilling (2006)

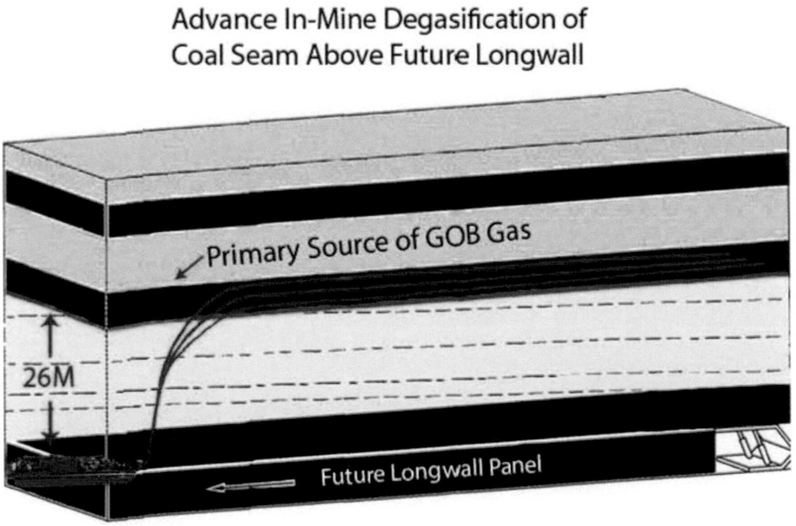

Advance In-Mine Degasification of Coal Seam Above Future Longwall

Figure 11.3.5 Ultra-long holes drilled to reach overlying and underlying seam and gob gas
Source: REI Drilling (2006)

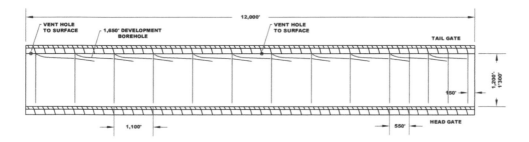

Figure 13.3.6 Typical in-mine long hole drilling pattern

Source: Thakur (2017)

water. This hole drilling system is designed to reduce the gas content of the adjacent seams, and then use them for gob gas recovery.

Thakur (2006) reported that in highly permeable coal seams, nearly 50% of methane can be removed by this technique.

Figure 13.3.6 shows a typical drill pattern of long holes layout (Thakur, 2017). Close to 50% of all *in-situ* gas can be drained in 6–18 months and for low diffusivity coal it will take up to 1000 days to drain 65% of the *in-situ* gas.

(2) Drilling program   In an in-seam horizontal methane drainage hole, the first 20–40 ft (6.1–12.2 m) will be drilled to a diameter about twice the diameter of the designed hole diameter so that an adequate sized steel and/or plastic standpipe may be grouted in (Fig. 11.3.7).

A methane diffusion zone, if needed, must be on the downwind side of the drill and extends no more than 250 ft (76.2 m). Ventilation air leaving the diffusion zone will contain methane less than 1%.

Methane examination should be conducted around the stuffing box every 20 minutes.

(3) Borehole piping system   Every standpipe has a valve for shut in (Figs 11.3.7 and 11.3.8). The methane enters the piping system through the standpipe and then a flexible hose or polyethylene pipe leads to a water/gas separator tank, if needed. As the methane gas flows through the separator tank, water will settle out from the gas.

Gas is transported through the pipeline from the horizontal borehole feeder piping to the vertical piping system (Figs 11.3.9 and 11.3.10) through which the gas is delivered to the surface installation. All gas transmission pipelines will be located in return airways.

B. SURFACE STIMULATED VERTICAL BOREHOLES (FRAC WELLS)

When coal seam permeability is low, 1–10 milidarcies, conventional boreholes are not effective for methane drainage without artificially stimulated fractures. For this reason, hydraulic fracturing, a well-established technology for the petroleum industry, is used to enhance coalbed methane recovery. It involves pumping a high pressure fluid to fracture the coal seam. As the coal fractures under pressure, a proppant of sand or other materials may also be injected

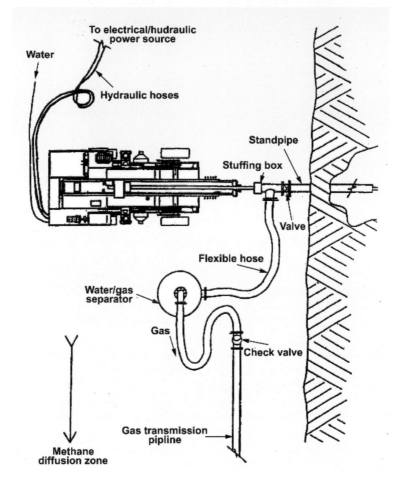

Figure 11.3.7 Typical drilling setup

Source: Fama (2005)

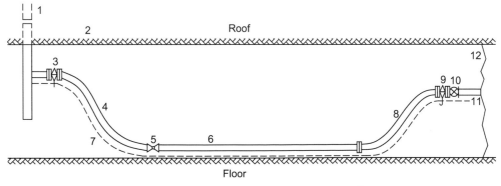

1 Vertical Borehole
2 Mine Roof
3 Separator and Trap 4"
4 4" Stainless Flex Line
5 4" Butterfly Valve
6 4" DuPont Aldyl "A" Pipe

7 Nitrogen LIne $\frac{1}{4}$"
8 3" Stainless Flex Line
9 3" Separator and Trap
10 2" Automatic Shutoff Valve
11 Stand Pipe in Rib
12 Mine Rib

Figure 11.3.8 Methane drainage underground piping installation

Source: Fama (2005)

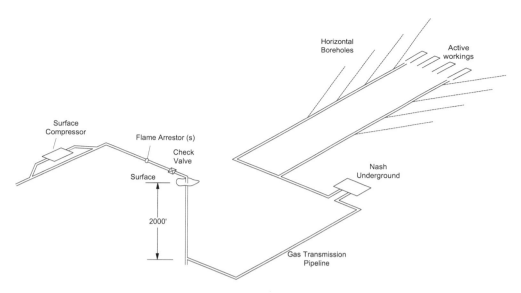

Figure 11.3.9 Methane drainage surface piping system
Source: Fama (2005)

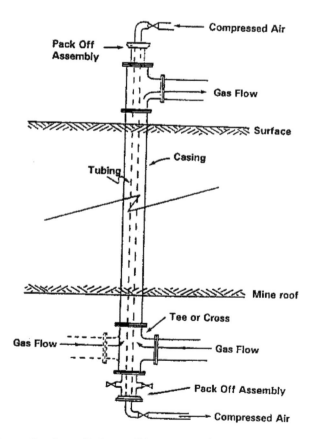

Figure 11.3.10 Generalized ventilation well/compressed air piping
Source: Fama (2005)

to keep the fractures open so methane can flow freely along the fractures to the borehole. Due to its complex natural cleat and fracture system, coal seam stimulation with hydraulic fracturing requires special attention so that it does not damage the coal seams for subsequent mining. To avoid preventing the desorption of methane for emission, the preferred chemicals used for drilling are gel or polymer mud or heavy salt fluid with potassium chloride as an additive. To minimize damage to coal seams, the preferred fracturing fluids are water, cross-linked gel, and nitrogen foam. The coal seams after hydraulic fracturing must be dewatered before methane begins to flow.

Hydraulic fracturing usually produces a single, planar, vertical fracture extending in two wings (180° apart) on opposite side of the borehole and 200–500 ft (61–152 m) long on each side. It provides a highly conductive flow paths for water and gas to flow into the borehole. Multiple coal seams can be stimulated in a borehole.

In production coal seams currently being mined in the United States many are shallow such that the fracture created by hydraulic fracturing is horizontal (pancake fracture), rather than vertical as in petroleum industry (Su et al., 2001).

Vertical boreholes drilled into virgin coal seams frequently produce large quantity of water and a small amount of gas in the first several months of operation. As more water is removed, it decreases the coal seam pressure and allows methane to desorb from the coal matrix and flows through the cleat system to the borehole. The water is separated from the gas and then treated and disposed of.

Frac wells are typically spaced on a grid between 1300 and 1800 ft (396–549 m). A large amount of time is required to drain gas to the levels required for mining. For frac wells gas drainage may not be uniform, and it may need several years to reach the same level of methane drainage as that achieved by a pattern of in-seam horizontal holes.

Thakur (2006) reported that under ideal conditions, frac wells can remove 60 to 70% of the methane in the coal seams.

C. SURFACE DIRECTIONALLY DRILLED BOREHOLES

For methane drainage, surface directionally drilled boreholes combine the best attributes of underground horizontal in-seam boreholes and surface vertical boreholes. It can begin with vertical downhole drilling until it reaches the targeted coal seam and then turn horizontally in the coal seam. Alternately, it can start at various angles and then turn to horizontal after it reaches the targeted coal seam. Figure 11.3.11 shows an example (Target Drilling, 2006).

D. DUAL WELL DRILLING SYSTEM

In surface horizontal hole-drilling technology, the hydrostatic load of a column of straight water drilling fluid, in most cases, is greater than the internal pressure of the reservoir such that it damages the potential of methane gas production. The pressure may force drilling fluid and cuttings into the coal cleat structures and plug the methane flow paths. To correct this problem, Zupanick (2005) described a low pressure or an "underbalanced" dual well drilling system that had been practiced in the gassy Pocahontas #3 seam in southern West Virginia. In this system, a vertical "cavity" well, 8.75 in. (222.25 mm) in diameter, is pre-drilled down through the coal seam of interest (Fig. 11.3.12 A). The hole is cased with three concentric casing strings between which two concrete layers are poured to strengthen the borehole lining. The casings extend all the way down to 50 ft (15 m) above the coal seam of interest,

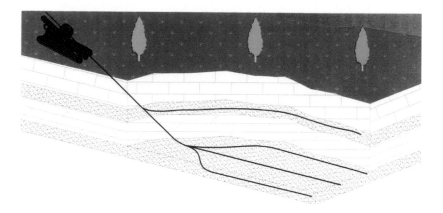

Figure 11.3.11 Surface slant holes for methane drainage
Source: Target Drilling (2006)

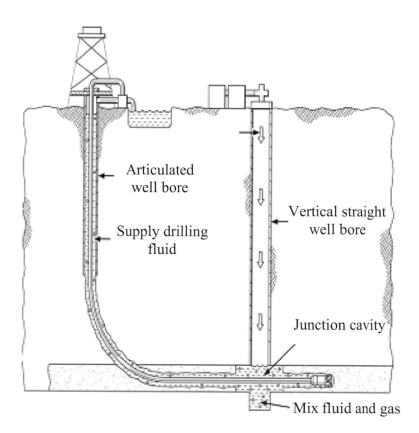

(A)

Figure 11.3.12 Underbalanced surface drilling: (A) duel well system and (B) solids control
in production phase
Source: Zupanick (2005)

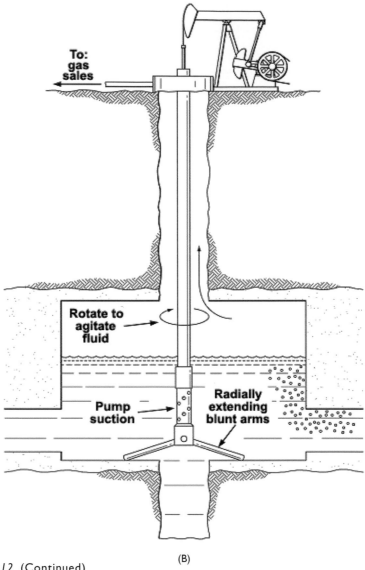

(B)

Figure 11.3.12 (Continued)

leaving an open hole above and below the targeted coal seam. The wellbore around the coal seam is then enlarged to around 6 ft (1.8 m) in diameter by an under-reaming tool. The coordinates of this enlarged wellbore are then carefully surveyed and recorded for interception by the second surface borehole. The second borehole is located about 300 ft (91 m) from the first hole and drilled vertically down to within 200 ft (61 m) of the seam of interest. From that point directional drilling techniques are applied to articulate the wellbore from vertical to horizontal so that the horizontal section will be located in the coal seam before it intersect the first well at the enlarged section. The path of the horizontal wellbore is then carefully directed toward the surveyed coordinates of the enlarged section of the first wellbore. Once the articulated second well has intercepted the first well, underbalanced horizontal drilling begins. Compressed air is then introduced to the cavity well. The air and fluid mixture will then return to the fluid junction, thereby reducing the density of the vertical fluid column, creating and maintaining an "underbalanced" drilling condition.

A large horizontal drainage network is then developed with horizontal holes 4000–5000 ft (1220–1524 m) long. The pinnate well configuration, which resembles the opposing shaft of a feather or the growth structure of a tree leaf, consists of uniformly spaced lateral holes, 4.25 in. (108 mm) in diameter, drilled left and right of the central wellbore, is used to tap the methane reserves (Fig. 11.3.13). In each case, the total horizontal length to the end of each

Figure 11.3.13 Multiple pinnate patterns
Source: Zupanick (2005)

lateral is approximately equal to that of the central lateral. Four central wellbores, one for each quadrant, will cover the 360° drainage area. With this system, each vertical borehole can drill 60,000 ft (18,293 m) lateral holes and cover 1200 acres (4.86 km^2) of surface area. The use of horizontal laterals results in minimal surface disturbance and associated costs, uniform drainage of the reservoir, and production during the dewatering phase.

The system provides efficient downhole water separation, allowing water to be easily pumped off. Water is removed through a pipe inserted down the center of the vertical borehole while gas is extracted through the surrounding annulus (Fig. 11.3.12 B).

Precision directional drilling relies on a real-time oriented gamma ray (OGR) tool and electromagnetic (EM) measurement-while-drilling (MWD) technology to keep its horizontal laterals in the seam. The EM MWD technology allows the drillers to receive directional, pressure, gamma ray, and resistively measurements while drilling an underbalanced hole. Low-frequency EM waves transmit the down-the-hole-measured data in real time to a surface antenna.

Before mine through, water infusion is recommended to seal off any small, residual gas flow. The infused water volume should slightly exceed the coal cleat void space of the drainage area. Specially developed gel materials may be used to fill up the borehole. With this dual well system, 80 to 90% of the gas could be recovered before mining.

Target Drilling applying its patented duel coalbed methane well techniques (Fig. 11.3.14) has successfully completed coal bed laterals in northern Appalachian coalfield exceeding 5789 ft (1765 m) (Kravits and DuBois, 2014). In Figure 11.3.14 the sequence of constructing the duel wells system is this: drill production well, drill and case vertical portion of the well, drill, ream and case access curve portion, drill connector lateral, drill first coal lateral, then drill second and third coal laterals. Figure 11.3.15 is the comprehensive and in-mine

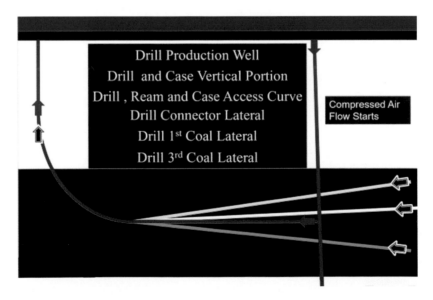

Figure 11.3.14 Target Drilling's patented dual coalbed methane well technique

Source: Kravits and Yearsley (2017)

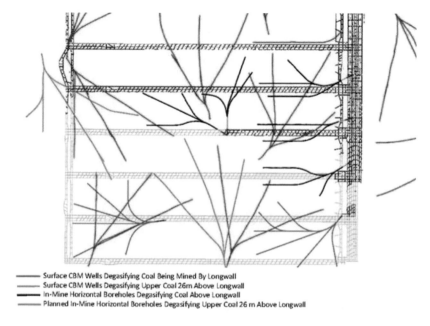

------ Surface CBM Wells Degasifying Coal Being Mined By Longwall
------ Surface CBM Wells Degasifying Upper Coal 26m Above Longwall
------ In-Mine Horizontal Boreholes Degasifying Coal Above Longwall
------ Planned In-Mine Horizontal Boreholes Degasifying Upper Coal 26 m Above Longwall

Figure 11.3.15 Comprehensive and in-mine degasification

Source: Kravits and Yearsley (2017)

degasification plan developed by Target Drilling for a Pittsburgh seam mine (Kravits and Yearsley, 2017) using its patented duel coalbed methane well techniques.

The application of surface directional horizontal boreholes techniques in Australia and Chinese coal fields have met with failures due to low permeability of coal seams. Perhaps the horizontal boreholes coupled with fracturing along the borehole is required to increase its perrmeability.

11.3.4.2 Post-mining methane drainage method

In the United States, the most common post-mining method for methane drainage in longwall mining is by surface vertical gob wells. The strategy is based on the fact that methane is trapped in the rider coal and rock strata above the coal seam of interest and is released upon caving of the immediate roof behind the shields in the gob. Vertical gob holes have proven to be a very effective means for removing methane from the gobs.

A. LOCATION

Before retreat mining begins, 3 to 30 surface boreholes, depending on the panel length and gas content, are drilled along a line about 350 ft (107 m) from the tailgate of the panel. The first borehole is usually located approximately 150–500 ft (46–152 m) from the panel setup room or near the sloping edge of the surface subsidence basin (Fig. 11.3.16). Holes should be drilled far in advance of retreat mining.

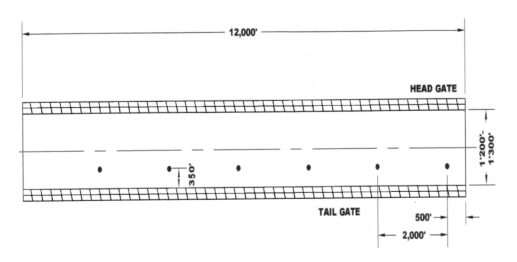

Figure 11.3.16 Layout of gob holes

Source: Modified from Thakur (2017)

B. DESIGN

Methane boreholes consist of an 8–12 in. (203.2–304.8 mm) diameter hole cased with a 6–8 in. (152.4–203.4 mm) casing. The casing consists of ±200 ft (61 m) of slotted steel pipe starting from ±15 ft (4.57 m) of the bottom of the borehole (note that the slotted length allows the gas to flow into it depending on the gas emission space.) and then with solid casing to the surface. The solid casing will be grouted in the borehole and a seal located at the bottom of the casing to prevent water invasion (Fig. 11.3.17 A).

C. PERFORMANCE

Exhaust fans, 10 to 75 hp (7.45–55.92 kW) delivering 1500 to 1750 cfm (0.71–0.83 m³ sec-1) will be used for the gob holes. Methane flow begins when the longwall face reaches to a few meters within the borehole. At this time the fan is energized, and the hole begin its active period. The initial methane flow rate is high and then decreases steadily. Typically, the fan will operate for three to five months. The methane production can be expressed by the following equation:

$$Q = \alpha t^n \tag{11.3.2}$$

Where Q is total methane flow in m³, t is time in days, α and n are constants characteristic of the gob and coal seam in the methane emission space, respectively. To ensure consistent production, a vacuum pump is often used on the surface (Fig. 11.3.17 B).

According to Thakur (2006), the capture ratios of vertical gob holes vary from 30 to 80% depending on the number and size of gob holes per panel and the size of vacuum pumps

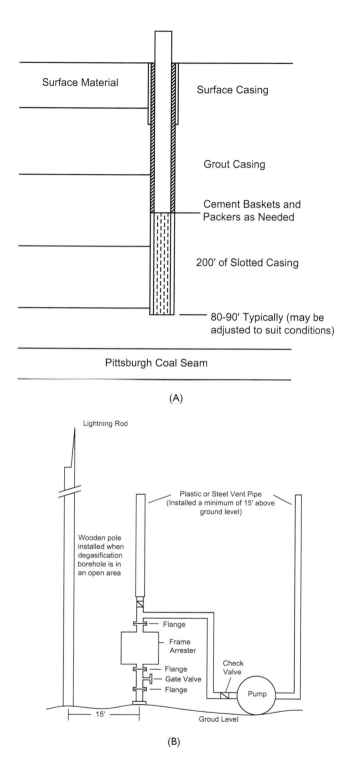

Surface Material

Surface Casing

Grout Casing

Cement Baskets and
Packers as Needed

200' of Slotted Casing

80-90' Typically (may be
adjusted to suit conditions)

Pittsburgh Coal Seam

(A)

Lightning Rod

Plastic or Steel Vent Pipe
(Installed a minimum of 15' above
ground level)

Wooden pole
installed when
degasification
borehole is in
an open area

Flange

Frame
Arrester

Check
Valve

Flange
Gate Valve
Flange

Pump

15'

Groud Level

(B)

Figure 11.3.17 Gob hole: (A) typical degas hole and (B) typical installation of degas hole
with or without pump

D. SAFETY PROCEDURES

Gob holes are periodically checked to monitor the quantity and concentration of methane flowing from the boreholes. If the methane concentration falls below 25%, the motor stops, and the gob hole go into a free flow status. At this time, methane concentration at the damper is of a high percentage but of low quantity. All gob holes are equipped with a damper. It prevents air from flowing down the borehole and yet allows gas to be exhausted whenever the fan motor cannot be operated.

Flame arresters are installed on each active gob hole utilizing exhaust blowers. Lighting protection is also incorporated into each installation.

11.4 Dust control

11.4.1 Introduction

11.4.1.1 Definition, characteristics, and health hazards of respirable coal mine dust

The Federal Coal Mine Health and Safety Act of 1969 established mandatory health standards for coal miners

> to provide to the greatest extent possible, that the working conditions in each underground coal mine are sufficiently free of respirable dust concentrations in the mine atmosphere to permit each miner the opportunity to work underground during the period of his entire adult working life without incurring any disability from pneumoconiosis or any other occupation-related disease during or at the end of such period.

That goal is to be achieved by establishing a combination of three programs: mandatory RCMD (respirable coal mine dust) exposure standards, medical examinations and compensation (including research and development).

Medical surveillance of coal miners since the enactment of the 1969 Health and Safety Act indicates that coal miners are subjected to risk of lung disease depending on the duration and intensity of RCMD exposure (National Academies of Sciences, Engineering, Medicine, 2018). The most commonly recognized respiratory disease in coal mining is coal workers' pneumoconiosis (CWP). Exposure to RCMD damages and inflames lung cells resulted in injury leading to chronic lung disease, with disease progression due to the result of sustained inflammatory effect of the retained dust particles in the interior of lung. Exposure to elevated respirable silica causes greater lung injury (Castronova, 2000).

RCMD contains 40–95% coal, mostly toward the lower range (NIOSH, 1995). In addition, it may contain crystalline silica and silicates, disease particles, rock dust products (mainly calcium carbonate), metals, and other organic compounds. Particle characteristics analysis of RCMD samples by Johann-Essex et al. (2017) showed considerable variability in particle mineralogy, size, and shape in mine environment among regions of coalfields and even among coal seams.

According to WHO (World Health Organization, 1999), "dusts are solid particles, ranging in size from below 1μm up to at least 100 μm, which may be or become airborne, depending on their origin, physical characteristics and ambient conditions". Initially, RCMD was defined as those particles less than 5 μm in diameter based on miners' autopsy data and

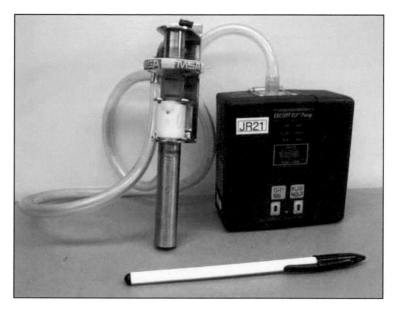

(A) Gravimetric sampling pump, cyclone, and filter cassette

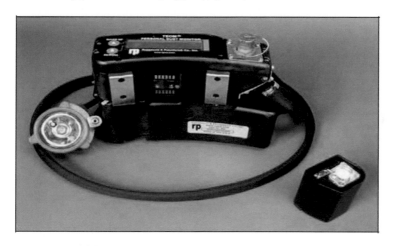

(B) PDM (personal dust monitor) with TEOM removed

Figure 11.4.1 MHSA approved dust samplers

Source: Rider and Colinet (2010)

later when technology became available, particle size analysis showed that the aerodynamic mass size distribution of coal mine dust was almost entirely larger than 3 µm (Burkhart *et al.*, 1987). Generally, the aerodynamic diameter of particles in coal mines are in the range 1–30 µm with a mean particle size (in mass) of approximately 10 µm (National Academies of Sciences, Engineering, Medicine, 2018). The 2014 dust rule states "any respirable dust in the mine atmosphere is considered respirable coal mine dust (RCMD) to which miners are exposed and when measured, is counted for determining compliance with the respirable dust standard." Therefore, for regulatory purpose, RCMD is defined as "that collected with a sampling device approved by the Secretary of Labor and the Secretary of Health and Human Services in accordance with (30 CFR) part 74(30 CFR)." The sampling device currently approved by MSHA (Mine Safety and Health Administration, US Department of Labor) is a personal dust monitor (PDM or CPDM), which is a personal wearable respirable dust sampler and gravimetric analysis instrument designed for use in underground coal mines (Fig. 11.4.1).

11.4.1.2 Respirable coal mine dust standard

According to the 1969 Health and Safety Act, effective 30 June 1970, the average RCMD concentration in the active sections of underground coal mines must be 3.0 mg m$^{3-1}$ or less, which was further reduced to 2.0 mg m$^{3-1}$ on 30 December 1972. That Act also stipulates that the RCMD standard is further reduced if the quartz content exceeds 5%. In other words, the quartz content must be less than 0.1 mg m$^{3-1}$. The reduced respirable dust limit when the amount of silica is greater than 5% is

$$\frac{10}{S_iO_2(\%)} \text{ mg m}^{3-1} \tag{11.4.1}$$

For example, if the silica content in the samples is 7%, the reduced standard will be $10/7 = 1.43$ mg m$^{3-1}$.

MSHA promulgated a new dust rule in 2014 to be effective 1 August 2016 that lowered the dust standard to 1.5 mg m$^{3-1}$. Further, the standard in the intake airways of the mine and in the mine atmosphere is 0.5 mg m$^{3-1}$. The limit of quartz content remains the same, i.e., 0.1 mg m$^{3-1}$ (100 micrograms per cubic meter or μg m$^{3-1}$).

11.4.1.3 Major dust sources in longwall faces

Major dust sources of airborne dust particles in coal mines are these:

1   Particles in the intake air entering the mine for mine ventilation purpose.
2   Application of rock dust products.
3   Dust generated from mining operations at the longwall face area.

The first two sources are not directly generated by coal mining process. It is the third source that is most critical to coal miners working at the face for coal production.

According to Kissell *et al.* (2003), the four major dust sources in a longwall face when the shearer is running are, in sequential order, shearer 53%, shields 23%, stage loader-crusher 15%, and intake air 9%. Undoubtedly, the cutting drum in the shearer is the number one source because that is where coal is cut. So drum design and operating condition are very important. A shields dust source, in addition to weak roof, is due to the fact that when a high capacity shield is set against the roof, it may crush the roof at local contact points, and when it is released for advance, the dust falls off between and in front of the shields. In the case of the stage loader, the dust source is mainly from the crusher when it crushes coal lumps, especially in recent years, roof or floor rock or both are cut to make the minimum mining height for the shearer longwall faces.

11.4.2 Methods of dust control

11.4.2.1 Intake air

Intake air may contain RCMD as high as 0.42 mg m$^{3-1}$ at the last open crosscut (Rider and Colinet, 2007), which is 28% of the new dust limit of 1.5 mg m$^{3-1}$ and is quite significant. Since the intake air travels from outside through mains and panel gateroads before it reaches the longwall face, it will carry the dust encountered along the way. In this respect, there are two major contributing factors: (1) air speed is too high as to entrain the dust into the air and carry with it and (2) supporting activities outby the last open crosscut generates too much dust. Therefore, in order to reduce the intake air RCMD, a proper air velocity shall be maintained. In addition, spraying water in the gateroads to keep the dust sufficiently wet, or reduction of outby supporting activities (i.e., construction, supply haulage, and roadway clean-up activities) during coal production period or when miners are working inby will help. Surfactants or chemical water additives such as soaps and detergents may be used to increase wetting effect of dust particles.

Many longwall mines use the belt entry for intake. In order to reduce RCMD, the belt structure and components must be well kept to minimize spillage and the belt surface (both conveying and non-conveying sides) must also be kept clean with wipers and water sprays at the belt end and transfer points.

When the intake air reaches the last open crosscut and before it enters the longwall face, it passes through the stage loader which contains a major dust source, a crusher used to crush large sized coal/rock before being dumped on the belt. Therefore, in order to eliminate dust generated by the crusher and coal conveying, the stage loader must be completely enclosed and seals are well maintained to prevent leakage. Water sprays at least three in a bar are installed at locations before, on, and after the crusher as well as before dumping on the belt at a minimum water pressure of 70 psi (0.48 MPa). A scrubber (fan-powered or high-pressure water powered) may also be mounted on top of the crusher to prevent the fugitive dust from escaping from the crusher areas (Fig. 11.4.2).

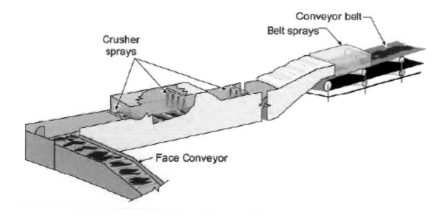

Figure 11.4.2 Enclosed stage loader and location of water sprays
Source: Rider and Colinet (2007)

11.4.2.2 Rock dust

Coal dust is known to cause dust explosion in underground coal mines. Rock dusting on exposed coal surfaces is a proven method for preventing dust explosion. Rock dusting regulations (30 CFR 75, 402–403) state that all underground areas of a coal mine, except those areas in which the dust is too wet or too high in incombustible content to propagate an explosion, shall be rock dusted to within 40 ft (12.2 m) of all working faces (30CFR 75.402–403) and all freshly mined surfaces shall not be exposed without rock dusting for more than a shift.

To offset the RCMD generated by dry rock dusting, some mines have turned to wet dusting or not rock dusting while miners are inby working.

The mandated measurement of RCMD concentrations as practiced include rock dust particles entrained in the mine air. However, there have been no data quantifying the contribution of rock dusting particles to the measured RCMD, and the mandated measurements of RCMD so far show that compliance with rock dusting regulations does not seem to present a problem in meeting the RCMD standard (National Academies of Sciences, Engineering, Medicine, 2018).

11.4.2.3 Longwall face

Based on the dust sources, the method of dust control can be divided into the following categories and applied either individually or various combinations of them in strategic locations and on all face machines:

1   Reduction of dust production involves techniques that reduce dust generation at the source, for example, coal cutting.
2   Dilution and move of dust by sufficient air volume and velocity.
3   Dust suppression uses techniques to prevent dust from entering into the airflow or that convert the airborne dust to settled dust.
4   Dust redirecting employs techniques to direct the dust-laden air to flow away from the operators.
5   Dust collecting by employing dust-collecting devices. The dust-laden air flow is filtered before it reaches the operators.

Among these five categories of dust control, adequate ventilation, dust suppression and dust redirecting are widely used in the United States and have proven to be very effective in meeting the dust control standard.

A. OPTIMUM SHEARER CUTTING PARAMETERS

Although it is inevitable that dust is produced during the coal cutting process, the best approach is to reduce dust at the source. One way to achieve this is to correctly select the operational parameters of the shearer.

Properly increasing the haulage speed of the shearer will increase the size of coal lumps and decrease dust generation. Similarly, properly reducing the drum speed will reduce the airborne dust blown away by the rotation. However, to be beneficial, a lower drum speed must be used in conjunction with a proper haulage speed.

Bit lacing in the cutting drum also has considerable effect on dust generation. Generally speaking, as the bit spacing gets closer the more fine coal is produced, and due to crowding and squeezing, the fine coal is more likely to receive secondary breakage resulting in more dust generation. Comparison between one and two bits per line shows that with one bit per line dust is reduced 20%. However, as the shearer cutting continues to increase to enhance production, two bits per line has become more popular in recent years

The drum vane angle is directly related to drum loading efficiency (see Section 7.9.2). If the vane angle is too big, the dust is easily blown away, resulting in more airborne dust. If it is too small, the coal will be subjected to repeated squeezing and breaking, ending up with more dust. Previous studies (Niewiadomski *et al.*, 1982) have shown that optimizing the vane angle may result in reducing the liberated respirable dust by as much as 50%.

B. AIR FLOW AND AIR VELOCITY

MSHA (2019) recommends the following equation to determine the air quantity per ton of material mined:

$$R = CQ / ST \tag{11.4.2}$$

where R = airflow-to-tons ratio, cfm/ton; C = actual DO (designated occupation) dust concentration, mg m$^{3-1}$; Q = face airflow, cfm; S = applicable dust standard, 1.5 mg m$^{3-1}$; and T = production, tons.

If the face air is less than 600 fpm (3 m s^{-1}), increasing air velocity and air volume will reduce dust (Organiscak and Colinet, 1999) at the face and in the areas downwind of the shearer. When air velocity was increased from 250 fpm (1.3 m s^{-1}) to 450 fpm (2.29 m s^{-1}), the dust level was reduced across the face for all conditions. A higher air velocity means an increased air flow and will hold the dust cloud against the face for a greater distance with a lower peak concentration. (Rider *et al.*, 2001).

To assure sufficient air volume and velocity at the face area, air leakages must be blocked off. The most likely place for air leakage is the gap between the chain pillar rib and the first gate-end shield at the headentry T-junction (Fig. 11.4.3 A) (Niewiadomski *et al.*, 1982). When the intake fresh air reaches the T-junction, a portion of the air flow passes this gap and

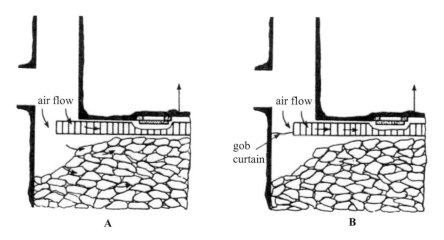

A   B

Figure 11.4.3 Air leakage and its prevention at headgate T-junction area
Source: Niewiadomski *et al.* (1982)

enters the gob. To prevent such an air loss, a curtain should be hung to block off the gap and moved with the advancing shield (Fig. 11.4.3 B). The adoption of a gob curtain at the gap increases the air velocity significantly.

When the shearer cuts out at the headgate, the high velocity intake air flows through the cutting drum and carries with it some airborne dust to the shearer operators (Fig. 11.4.4 B) (Niewiadomski et al., 1982). Although this occurs only in a short period of time, it increases the shearer operators' dust exposure. The remedial measure is to hang a curtain between the stage loader and panel rib at approximately 4–6 ft (1.2–1.8 m) outby the faceline (Fig. 11.4.4 A). The curtain guides the fresh air intake to bypass the cutting drum. The curtain will need to be moved with the face.

Since all shields of recent models are equipped with a full side-shield, full engagement of the side shield closes the gap between shields, reduces air leakage to the gob, and increases air velocity across the face.

C. WATER SPRAYS

Water sprays are the most common and effective method for dust control. They can be used for both dust suppression and re-directing air flow. Accordingly, they are installed to suppress dust at all major dust sources, including cutting, loading, transporting, and support advancing. The basic principle of dust suppression by water sprays is that the high-velocity water droplets impact and wet the surfaces of the airborne dust particles, thereby capturing them to become the settled dust (Jayaraman and Grigal, 1977). The ideal conditions for capturing airborne dust require that water ejected from a nozzle forms a large-area water cloud that is composed of optimum-sized droplets. According to Courtney and Cheng (1977), high-velocity droplets approximately 200 μm in diameter are best for capturing dust particles, whereas droplets approximately 500 μm in diameter are best for impact. To assure sufficient water pressure and quantity at the nozzle outlet, there must be an effective water supply system. This includes a larger pump capacity, less pressure loss in the hoses, and improved nozzle structure and water filtration system.

Kissell et al. (2003) stated that dust generation by shearer is reduced by increasing the quantity of water to the drum, currently up to 300 gpm (1140 L min^{-1}); that the greater the number of sprays, the more effective for dust control (at least one spray for every bit on the drum); that pick-point spray should be used; that all water piping components must be properly sized; and that spray water must be cleaned with 50 μm filters or less.

(1) Water sprays on the shearers   According to the location of the nozzles, water sprays on the shearer can be divided into internal and external types. In the internal or wethead type, the nozzles and water piping are installed inside the cutting drum. Other nozzles, installed on the cowls, ranging arms, shearer body, etc., belong to the external type.

There are various ways of installing the internal water sprays, but the most effective one is back-flushing, pick-point spray (Fig. 11.4.5). An MSHA survey (2000) indicated that water pressure on the drums for all US longwalls ranged from 41 to more than 150 psi (0.283 to >1.034 MPa), mostly 71 to 100 psi (0.49–0.69 MPa) with an average of 91 psi (0.628 MPa) and that some longwalls also employs a wetting agent.

The bits must be kept sharp. A sharp bit breaks larger coal fragments while a dull bit tends to crush the coal and produce smaller particles.

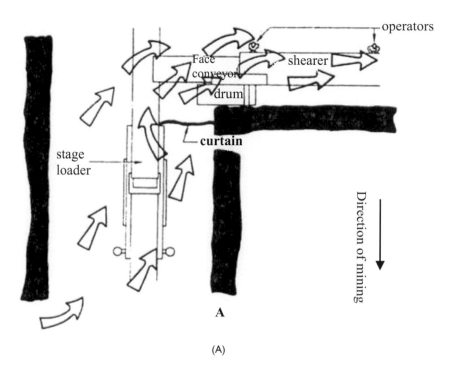

(A)

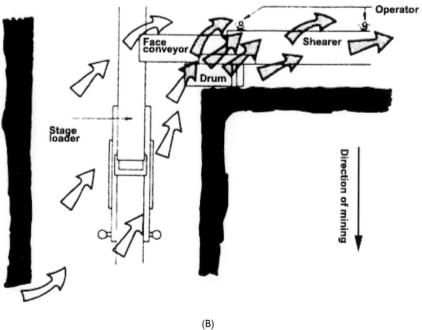

(B)

Figure 11.4.4 (A) A curtain shields the shearer drum from the fresh intake air; (B) fresh intake air blows dust from shearer drum over the operators at the headgate T-junction

Source: Niewiadomski *et al.* (1982)

(A)

(B)

Figure 11.4.5 Back-flushed bits: (A) conical bit; (B) radial bit

In addition to suppressing the dust, the internal water sprays cool the bits, thereby reducing bit wear and avoiding ignitions.

External sprays installed at the ranging arms and on the shearer body can be so arranged that they are very effective in redirecting the dust-laden air flow. According to studies (Foster Miller, 1982; Kissell *et al.*, 1981), when the double-ended ranging drum shearer equipped with a conventional water spray system is cutting upwind from tailgate to headgate (Fig. 11.4.6), a plume of dust-laden air is formed in front of the leading drum, which is responsible for the majority of cutting and loading. The water sprays installed at the ranging arm and in the front part of the shearer body push the dusty air upwind as far as 15 ft (4.6 m) beyond the leading edge of the leading drum. There the dusty air meets with the opposing incoming fresh air. As a result, a turbulent flow forms and spreads into the shearer operators' walkway. The remedial measure is to employ the shearer clearer system (Fig. 11.4.7 A). In this system, a splitter arm is mounted on the gob side at the end of the shearer body on the leading drum side (Fig. 11.4.7 B) extending at least to the edge of the cutting drum. Water sprays are mounted on the bar pointing to the drum. The number of water sprays varies depending on the dust make.

The splitter arm normally extends 18 in. (0.46 m) beyond the outer cutting edge of the drum. A belting is also hung from the splitter arm down to the line pan. Thus the splitter arm with belting divides the incoming fresh air into two splits. One split meets with the dust-laden air, which is then pushed to the face side by the water sprays mounted on the splitter arm. The other split, which is clean, flows into the walkway. The water mists from the water sprays at the splitter arm and the shearer clearer form a water curtain that keeps the dusty air to the face side. Additional high-capacity, low-pressure sprays are mounted evenly along the length of the splitter arm on the gob side of the belting and directed downward to block any fugitive dust from escaping beyond the splitter arm into the walkway (Fig. 11.4.8). Furthermore the splitter arms should be installed level with the shearer body and parallel to floor to prevent the dust cloud from migrating over or under the splitter arm and into the walkway, especially in higher seams (Rider and Colinet, 2010).

Five or more water sprays are mounted on the face side in the middle of the shearer body for maintaining the moving speed of the dust-laden air and keeping it on the face side of the

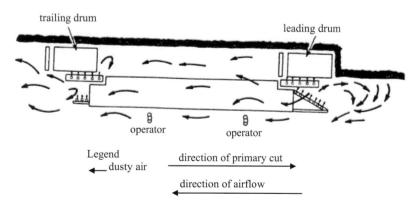

Figure 11.4.6 Dust plume is pushed upstream by a conventional water spray system
Source: Kissell *et al.* (1981)

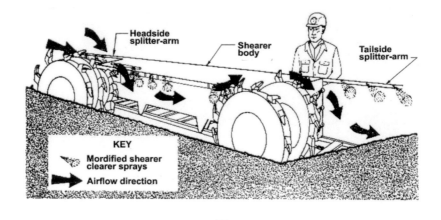

(A)

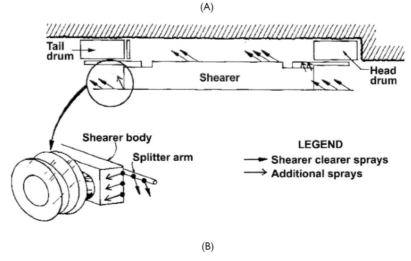

(B)

Figure 11.4.7 The shearer clearer system

Source: Cecala and Jankowski (1986)

Figure 11.4.8 Venturi sprays mounted on headgate splitter arm and high-capacity low-pressure flat fan sprays directing downward mounted on the gob side of the belting

Source: Rider and Colinet (2010)

shearer. Three more water sprays are mounted on the tail end of the shearer body. They are oriented such that the dust-laden air is pushed over and around the trailing drum. An MSHA survey (2000) indicated that the water pressure for water clearer sprays for all US longwalls ranged from 41 to 200 psi (0.283–1.379 MPa), mostly 61 to 140 psi (0.421–0.966 MPa), with an average of 114 psi (0.786 MPa).

Other water sprays are added for higher efficient dust control, such as crescent sprays mounted on each ranging arm and directed inward toward the cutting drum and lumpbreaker sprays at the end of lump breaker directing downward toward the AFC pans (Rider and Colinet, 2010).

A series of underground tests showed that the shearer-clearer spray system reduced operator exposure from shearer-generated dust by about 50% when cutting against face ventilation and by at least 30% when cutting with ventilation (Ruggieri *et al.*, 1983; Jayaraman *et al.*, 1985).

It must be emphasized that the water sprays and their mounting structure must be well kept all the time in order to achieve its deigned dust control purpose.

There are six types of water spray nozzles used for dust control: hollow cone, atomizing, full cone, flat cone, solid stream, and venturi nozzles (Fig. 11.4.9). Effective dust control very much depends on choosing the correct water spray nozzle. Table 11.4.1 summarizes the type of spray nozzle and its characteristics and application (MSHA, 2019; Rider and Colinet, 2010).

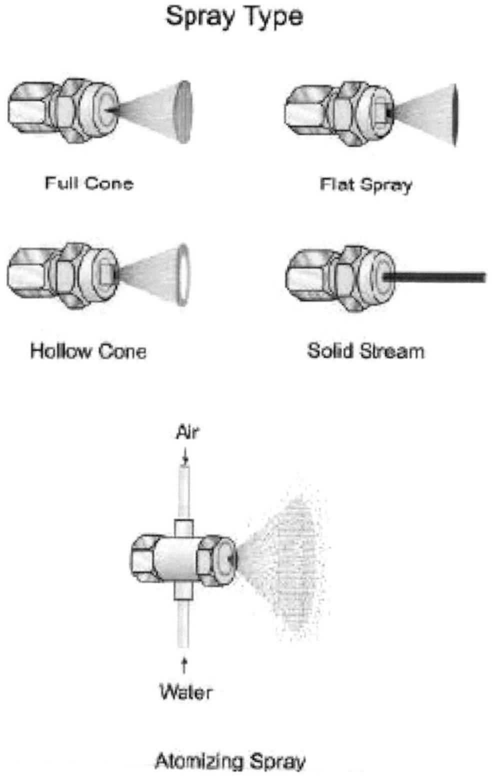

Spray Type

Full Cone   Flat Spray

Hollow Cone   Solid Stream

Air

Water

Atomizing Spray

Figure 11.4.9 Five type of water sprays, not shown is Venturi type
Source: Rider and Colinet (2010)

Table 11.4.1 Summary of spray cone features and application

Dust control method	Water spray type	Spray characteristics	Application
Suppression, air moving	Hollow cone	A circular ring pattern, small droplets in wide angle	Water clearer system, crusher, belt transfer points
Suppression	Atomizing	Full cone spray pattern, very small droplets. Required compressed air	Scrubbing
Wetting	Full cone	Square or circular pattern of larger uniform droplets at high velocity	Belt wiper (nonconveying side), before belt transfer points
Wetting	Flat cone	Tapered edge rectangular or even pattern. Large size droplets at high velocity	Belt wiper (conveying side), before belt transfer points
Wetting	Solid stream	A flat cone with circular spray pattern	Pick point flushing
Air moving	Venturi	Droplets vary, because any nozzle can be placed in venture shroud	Splitter arm

Figure 11.4.10 Raised deflector plate can enhance the effectiveness of the directional spray system

Source: Rider and Colinet (2010)

In terms of size of water droplets produced, atomizing cone is the smallest, followed in sequential order by hollow cone, flat cone, and full cone.

Higher pressure and low flow produce smaller droplets. Small droplets have higher initial velocity but diminish quickly. Larger droplets however retain velocity longer and travel further. When the droplet diameter is larger than that of the dust particles, a dust particle tends to pass around the water droplet. Conversely, the droplet and dust particle tend to collide and settle.

The shearer deflector plate (Fig. 11.4.10) is designed to protect shearer operators from debris flying off the face. In a raised position, it not only makes the roof/coal debris landed on it to slide down toward the face side, preventing it from accumulating on the top of the machine body, but also helps confine the contaminated air close to the face. The deflector plate should be raised as high as face conditions allow to provide maximum protection.

(2) Water sprays on shields   As mentioned earlier, during support advance when the canopy is being lowered, the dust from roof crushing will fall into the face area. The amount of dust will depend on the severity of roof crushing and the caving line of the immediate roof. Water sprays installed on the canopy of every shield with water sprays directing toward the roof, coal face, side, and rear of canopy are very effective for dust control during support advance. Water sprays on shield can be divided into four categories all using solid cone nozzles (Fig. 11.4.11 and Fig. 9.6.4):

1   Water curtain spray: a total of four, two in tandem on each side on the underside in the front end of the canopy. The sprays aim toward the face to settle the dust coming from the shearer drums as the coal is being mined. This curtain will be located in reference to the shearer position and will follow the shearer as it travels across the face.

Figure 11.4.11 Shields' traveling water curtain
Source: Rider and Colinet (2010)

2   Side plate sprays: a total of four, two on each side of the canopy in front of the legs position.
3   Lemniscate link sprays: a total of four, one on each lemniscate link slightly aiming toward the gob.
4   Roof sprays: a total of four, two on each side on the top side of the canopy aiming toward the roof of the adjacent shield which is being advanced.

Thus there are a total of 16 water sprays on a shield. Water sprays are controlled by the SCU (shield control unit). They are programmed to turn on when the shearer is approaching and turn off when the shearer has passed for certain distance and the shield itself has completed movement for certain time. Note that when a shield is advancing, the roof sprays on the left and right side shields adjacent (only the two sprays closest to) to the shield to be advanced will turn on to suppress the roof dust and turn off when the shield advance has completed.

(3) Passive dust control   The powered air-purified respirator, commonly known as the Racel or air-stream helmet, is another way to decrease dust exposure of longwall face workers, especially shearer operators (Fig. 11.4.12). These respirators are the size of a motorcycle helmet and include a full-face shield. Their size makes them difficult to use in lower seams. An MSHA survey (2000) indicated they were used by 54% of the longwalls, primarily in seams higher than 60 in. (1.52 m).

11.4.3 Dust control practice in US longwalls

MSHA survey (Ondrey *et al.*, 1995; MSHA, 2000) on the types of controls used to reduce dust generated at the crusher/stageloader, shearer, and shield supports indicated that dust sources included the intake/belt, headgate area, shearer, and roof support movement, and that longwall

Figure 11.4.12 Shearer operators wearing air-stream helmet

Table 11.4.2 US longwalls dust control practice*

Cutting cycle	Bi-direction
Ventilation control	Air velocity = 400 – > 600 fpm
	Air volume (cfm) = up to 10 × production (tons/shift)
	Gob & cut out curtains
Shearer drum	One water spray per bit @ up to 150 psi
	Foam through drum
Shearer body	Shearer clearer @ 200 psi
Crusher/stage loader	Enclosed
	Enclosure water sprays @ 80 psi
	Dust scrubber @ 5000 cfm
Shields	Shield mounted automatic water sprays
Automation	Shearer remote control
	Electrohydraulic control shields
Barriers	Shearer deflector plate or
	Belting on top of shearer body
	Face curtain
	Water curtain
Surfactants	Soaps, detergents dissolved in water

*Although this survey was performed in 2000, it is believed that this practice or higher standards are being maintained until present time.

sections had maintained dust levels below the applicable dust standard. The dust control methods included the following: high volume ventilation, headgate dust collectors, high-pressure water or foam through the shearer drum, enclosures and barriers, automated roof supports, and shearer remote control. More detail about these parameters is shown in Table 11.4.2.

An example of dust control plan for the longwall working section is cited here as follows: Joy 7LS remote control shearer is used. The DO is a tailgate side shearer operator. Belt entry is used as intake air way.

1   The minimum quantity of air reaching the intake end of the face shall be 50,000 cfm (23.6 m^3 s^{-1}).
2   The minimum velocity of air maintained across the face shall be:

300 fpm (1.52 m s^{-1}) at shield #8 from the headgate.
200 fpm (1.02 m s^{-1}) at shield #8 from the tailgate.

3   Table 11.4.3 shows the water spray system implemented for dust control.
4   During cutting of coal, at least 90% of all water sprays will be maintained operable. In spray blocks with three or fewer sprays, all sprays shall be operable. At the beginning of a cut 100% of sprays will be operable and all missing/damaged carbide bits will be replaced.
5   The operating parameters of the sprays on the stage loader scrubber shall be set within the range specified.
6   Scrubber examination will be in accordance with 30CFR 75.362(a)(2) and 30CFR75.371(j), on shift examination using an anemometer and dimensions. Maintenance and repairs will be in accordance with OEM's recommendations.
7   The designated occupation sampler shall be the person located as described in 30CFR 70.208(b)(7).

Table 11.4.3 Water spray system implemented for dust control

Equipment	Number of spray	Type	Minimum pressure	Minimum flow rate
ACS drum	39	Cone	80 psi hdgt	0.75 gpm
C&A drum	33	cone	80 psi tlgt	0.75 gpm
Shearer body	16	Venturi, fan or cone	80 psi	0.95 gpm
Stageloader crusher	Minimum 4	cone	25–45 psi	5–7 gpm total
Stage loader scrubber	Minimum 16	cone	25–45 psi	0.1–0.75 gpm
Headgate ranging arm	8	Fan and/or cone	80 psi	0.75 gpm
Tailgate ranging arm	8	Fan and/or cone	80 psi	0.75 gpm
Lumpbreaker	4	Fan and/or cone	50 psi	0.75 gpm
Spray bar	Minimum 4	Fan and/or cone	80 psi	0.75 gpm

1 psi = 0.0069 MPa: 1 gpm = 3.8 L min^{-1}

11.4.4 Measurements and modeling of air velocity and dust distribution

11.4.4.1 Measurements of dust concentration and air velocity distributions

The significant effect of a water clearer system can be clearly observed underground. Figures 11.4.13 and 11.4.14 show comparisons of measured respirable dust distributions between head-to-tail and tail-to-head cuttings along the AFC and walkway, respectively (Chiang *et al.*, 1996; Peng *et al.*, 1996).

Overall, dust concentration generated during head-to-tail cutting is higher than that from tail-to-head cutting, especially along the AFC downwind of the shearer. The difference in dust concentration between both cutting directions along the walkway is not significant. During head-to-tail cutting, a higher dust concentration caused by the shield movement occurs about 50 ft (15.2 m) upwind of the shearer. But this high dust concentration zone is carried to the face side by the lateral flow of the external water spray system. Due to the effect of the external water spray system, a lower dust concentration (about 0.7–1.5 mg m$^{3-1}$) in the walkway always occurs within 30 ft (9.1 m) on both sides of the shearer during both cutting directions.

Figures 11.4.15 and 11.4.16 show the measured air velocity fields around the shearer during head-to-tail and tail-to-head cuttings, respectively. The air velocity distribution around the shearer is very complicated due primarily to the change in ventilation cross-section and, second, to the effect of the external water spray system. The air velocity, both longitudinal (faceline direction) and transverse (mining direction) components, during tail-to-head cutting is greater than that during head-to-tail cutting. The transverse component of air velocity is less than 200 fpm (1.02 m s^{-1}).

In summary, under normal face conditions, cutting direction is not a critical factor affecting the respirable dust level exposure to the shearer operators and shield movers. The shield movers in the electrohydraulic control shield face can always stay on the upwind, clean air side regardless of cutting direction.

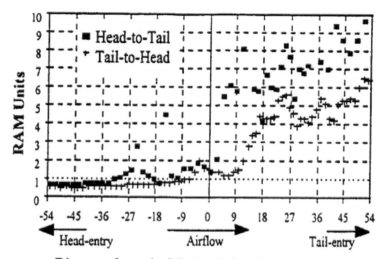

Figure 11.4.13 Comparison of dust distribution along AFC between different cutting directions

Source: Chiang et al. (1996); Peng et al. (1996)

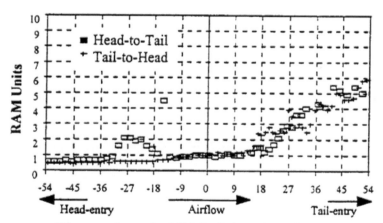

Figure 11.4.14 Comparison of dust distribution along walkway between different cutting directions

Source: Chiang et al. (1996); Peng et al. (1996)

11.4.4.2 Three-dimensional/two-phase model for air velocity and dust concentration distribution

Based on the measured air velocity and dust concentration distributions stated in the previous section, a two-dimensional and a three-dimensional two-phase (air and dust) model was developed by Peng *et al.* (1996) and Chiang *et al.* (1996), respectively.

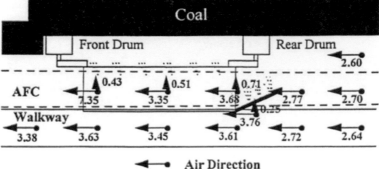

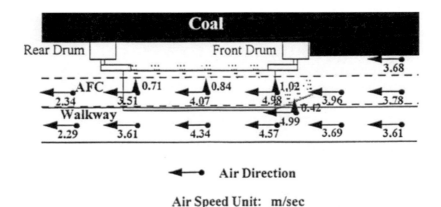

Figure 11.4.15 Air velocity near the shearer during head-to-tail cutting

Source: Chiang *et al.* (1996); Peng *et al.* (1996)

Figure 11.4.16 Air velocity near the shearer during tail-to-head cutting

Source: Chiang *et al.* (1996); Peng *et al.* (1996)

Figure 11.4.17 shows the air velocity distribution at the breathing level predicted by the model. The mean velocity at the inlet is 3.62 m s^{-1} during head-to-tail cutting. After shield advance, the ventilation cross-section reduces and the velocity increases. Conversely, the velocities where shields have not been advanced reduce. During tail-to-head cutting, the inlet velocity is also 3.62 m s^{-1} (11.87 ft s^{-1}). But the velocity around the shearer is higher.

Figure 11.4.18 shows the predicted dust distribution at the breathing level. The basic dust concentration in the intake air is 0.5 mg m^{-3}. Support advance produces dust less than 2 mg m^{-3} at the breathing level in the walkway. Around the cutting drum, dust concentration is greater than 20 mg m^{-3}, but it diffuses quickly toward the walkway at the downwind side. Although the dust level varies from place to place, it does not exceed 2 mg m^{-3}. Similar statements can be made for tail-to-head cutting.

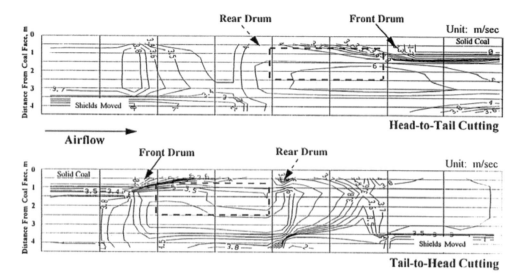

Figure 11.4.17 Distribution of air velocity around shearer plane view at breathing level
Source: Chiang *et al.* (1996); Peng *et al.* (1996)

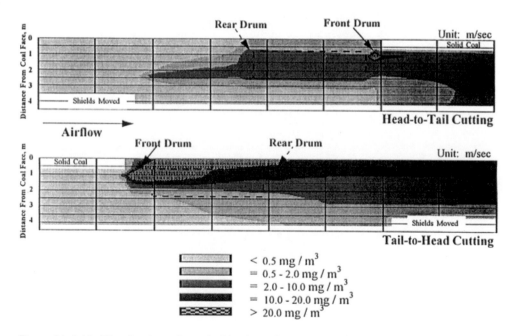

Figure 11.4.18 Distribution of respirable dust plane view at breathing level
Source: Chiang *et al.* (1996); Peng *et al.* (1996)

11.5 Noise control

Noise is unwanted sound. It is associated with equipment operations in coal mining. For longwall mining, many large piece of equipment are concentrated in a confined space of the face areas, and they are running continuously. Furthermore, as the power of all types of longwall face equipment and panel width continue to increase, so will the noise level that is generated from those pieces of equipment.

11.5.1 Definitions and background

Sounds travel in waves in the air. As these waves travel, their strength decreases. In fact, their amplitude or strength reduces by one half when the distance they travel is doubled. As the waves impinge upon the ear drum, the human ear can hear, without damage, pressure (strength) levels that are about 1,000,000 times stronger than the lowest pressure level it can detect. Accordingly noise measuring instruments require a very large range and are conveniently scaled in decibels (dB).

Sound strength, or sound pressure level (SPL), is rated in terms of decibels, the scale of which is logarithmic to allow rating over the entire sound pressure range with two- or three-digit numbers. The reference 0 dB is the level at which the average ear first detects any noise. At 80 dB, which is 10,000 times the sound pressure at 0 dB, the sound pressure is comfortable. Between 80 and 90 dB there may be intolerant sound pressures to some. Above 90 dB the noise is extremely annoying.

Extended exposure to noise can cause permanent damage to human ears known as noise-induced hearing loss (NIHL). A NIOSH survey (Franks, 1996) indicated that 70–90% of miners have a NIHL sufficient to be classified as a hearing disability. MSHA survey (Seiler et al., 1994) also found that among the 60,000+ noise survey data, more than 25% exceeded the permissible exposure level (PEL), which is defined as an eight-hour time-weighted average (TWA_8) of 90 dBA and nearly 77% exceeded the action level of 85 TWA_8 for enrollment in a hearing conservation program.

11.5.2 Noise regulations

The Federal Coal Mine Health and Safety Act of 1969 as amended in 1977 established requirements for protecting coal miners from excessive noise. The latest regulations allow a PEL of 90 dBA TWA and an action level for workers exposed to a dosage over 50% of the PEL (which is 85 dBA TWA). It also disallows hearing protection devices (HPDs) as a compliance measure, therefore emphasizing the use of engineering and/or administrative controls in reducing noise to workers.

11.5.3 Noise sources and noise control at longwall faces

According to Bauer et al. (2005), the noise levels of similar equipment in different mines are relatively consistent. The major noise sources were stage loader, 86–100 dBA, hydraulic pump car 78–99 dBA, shearer (head-to-tail) 88–99 dBA, shearer (tail-to-head) 86–96 dBA, AFC (full) 78–80 dBA, and section belt 83–96 dBA. They also found that workers' dosages of exposure were similar from mine to mine but varied from shift to shift, specifically (Table 11.5.1).

Table 11.5.1 Measured PEL for longwall face crew

Workers monitored	Range of PEL dose, %
Head shearer operator	125–179
Tail shearer operator	124–240
Shield man	49–138
Mechanics	37–99
Foreman	102–123

Source: Bauer et al. (2005)

Figure 11.5.1 Stiffen the sound radiating structures of shearer drum to reduce noise: (A) gussets, (B) thickening plates, and (C) ribs on the face ring

Source: Camargo et al. (2016)

The stage loader operator's dosage was the largest, followed by the shearer and hydraulic pump car operators. Among the workers, the shearer operators and headgate/stage loader operator consistently received the largest dosage above the PEL.

Engineering noise control for underground longwall face equipment that employs technologies to reduce noise generation at the source or isolates noises from workers is still in its development stage. One example is shown in Figure 5.3.14 (p. 143) for reduction of noise from a hydraulic pump station. Workers exposed to excessive levels of noise are encouraged to wear hearing protection devices.

NIOSH numerical model analyses (Camargo et al., 2016) showed that the major noise source during shearer's coal cutting is bit vibration and that the bit force components are 4500 lbs, 1500 lbs, and 1500 lbs in the axial, bending and transverse directions, respectively. The force amplitude is relatively constant up to around 100 Hz. From this point on, the amplitude of the force decreases at a rate inversely proportional to the square of frequency. Within the cutting drum, the noise radiates from three parts: center cylindrical body (15%), inner vane segment (34%), and outer vane segment (51%). Therefore, the noise reduction method is to stiffen the outer vane segment and bit block by mounting thin plate to it as shown in Figure 11.5.1. Underground operation test showed that it can reduce noise by 3 dB.

Table 11.5.2 Measured noise levels for different sections of stage loader

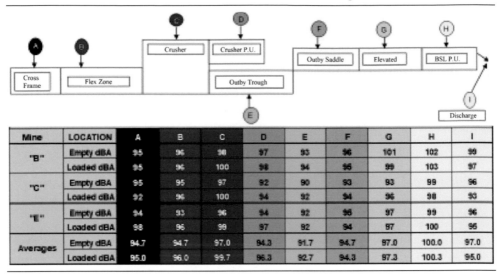

Mine	LOCATION	A	B	C	D	E	F	G	H	I
"B"	Empty dBA	95	96	98	97	93	96	101	102	99
	Loaded dBA	95	96	100	98	94	96	99	103	97
"C"	Empty dBA	95	95	97	92	90	93	83	99	96
	Loaded dBA	92	96	100	94	92	94	96	98	93
"E"	Empty dBA	94	93	96	94	92	96	97	99	96
	Loaded dBA	98	96	98	97	92	94	97	100	96
Averages	Empty dBA	94.7	94.7	97.0	94.3	91.7	94.7	97.0	100.0	97.0
	Loaded dBA	95.0	96.0	99.7	96.3	92.7	94.3	97.3	100.3	95.0

Source: Modified from Caterpillar

Table 11.5.2 lists the noise levels measured in three longwall mines. The noise levels for different sections of the stage loader exceeded 90 dB. The highest noise level occurred at the drive section exceeding 100 dB. NIOSH also investigated the source of stage loader and found that the major noise source is the drive motor at the out end where coal is dumped onto the belt (Camargo *et al.*, 2013).

Chapter 12

Longwall face move

12.1 Introduction

Longwall production generally stops when one panel has been completed and the equipment is being moved to the next panel for setup. In the 1970s, when all major equipment had to be disassembled for a face move, it was not unusual for a face move to take six to eight weeks. As new technology was developed to move the equipment in a whole unit, especially for shields, and miners gained experience, move time reduced steadily. Today, in spite of the much wider panels and heavier equipment, a complete longwall move can be done in two to three weeks, and it will take about 6 to 10 days if only shields are involved in the move, i.e., if the stageloader, AFC and shearer are set up in advance at the new face. Just like all other subsystems of longwall mining, many improvements have been made. Miners and management alike are very good at moving the face. In fact it has become routine to many in the industry. Many longwall mines have gone through 60–120 longwall face moves in the past 40+ years.

In today's high production longwalls, a face move is a very expensive project that should be limited both in time duration and the number of moves per year. Depending on the production level and capital invested, it costs US$ 500 to US$ 2,000 per minute when longwall production is stopped.

Production interruption can be avoided if there is an extra set of longwall equipment that can be set up and ready in the new panel before the old panel is mined out. In addition to keeping the production interruption to a minimum, this arrangement provides ample time to check out the equipment for repair or overhaul. It also gives sufficient time to assure high standard in setting up the face equipment at the new panel and for recovering the face equipment at the old panel. Moreover, it eliminates the close coordination in management and organization required in a face move.

But having an extra set of face equipment requires a large capital investment. This may not be justified economically, because in the United States, most mines have only one operating longwall. In this case, the equipment used for the old panel will also be used for the new panel, requiring moving all equipment from the old panel to the next one. This is a complete longwall face move.

In recent years, many mines have bought one or more extra sets of stageloader, AFC and shearers. The extra set of stageloader, AFC and shearer are set up in advance at the set-up room of the new panel so that when the time comes for the face move, only shields need to be moved to the new face. This shortens the face move time considerably. The time saved for moving the stage loader, shearer, and AFC is used for producing coal. In today's high production longwalls, this is definitely a worthy investment. Furthermore, the first shearer can be sent back to OEM to be rebuilt. This way, the rebuilt shearer will be ready for advanced setup in the new panel in the next face move.

The longwall face move is a big job, because it involves safely moving a variety of mining equipment up to 1600 ft (488 m) in length and can weigh a combined total of 10,000 tons (9090 mt), from the old to the new panel. The major factors in reducing longwall move time, which ranges from six days to three weeks as stated earlier, are good planning and coordination, precise execution and good teamwork, proper tools and equipment, and, above all, consistent hard work.

It is important to note that the longwall move has evolved into a sophisticated art. There are many small details that differ from mine to mine. In spite of this, a longwall face move can be roughly divided into the following four procedural steps: pre-move preparation, move preparation, move and installation, and completion.

12.2 Pre-move preparation

12.2.1 Goals of the longwall face move

It is very important to set goals, no matter how experienced the crew. A longwall move is successful if the following four goals are achieved.

First and foremost, the longwall move should be completed without injuries or violations. A longwall move has high injury potential, so safety cannot be over-emphasized. There are 20 to 30 job categories, depending on mine practices, in a longwall move. Every job should have a JSA (job safety analysis) that should be thoroughly explained to the individuals assigned to it, even if they have gone through it many times before.

Second, the longwall move should be accomplished in so many shifts as realistically set by management. This can only be done with a good plan, training, awareness, and teamwork, good communications, and proper tools and equipment.

Third, the move plan is not final. It will be updated with new ideas and new procedures through suggestions from everyone involved. It is important that everyone takes notes during the move and have discussion after the move. New ideas and new procedures that will contribute to a more safe and productive longwall move will be recognized and incorporated into the future plan.

Fourth, the longwall face is to be disassembled, not torn down. This means all items, when appropriate, should be moved with "re-use" in mind for the new setup room in the new panel. Hose ends should be taped, hoses and cables coiled up and bound neatly, bolts should be placed back in the holes from which they came out, etc. The extra time it takes to disassemble the face will easily be made up when the face is placed back together. This means that all work areas should be kept as clean as possible.

12.2.2 Pre-planning

Well-conceived and detailed pre-planning are the key to a safe, speedy, and economical longwall face move. Preplanning begins as soon as the last longwall move is completed. Any improvements that can be made for the next move should be suggested and thoroughly discussed. Actual planning begins, at the latest, one month before the scheduled face move. All key personnel involved in the move should meet to discuss the pre-move preparation and the actual move procedures, including any anticipated problems and their preventive measures.

The critical path analysis is an effective method for scheduling a longwall move (Peng and Chiang, 1984; Patrick, 1992). This procedure organizes all activities in a longwall move into a unit. The critical path flow chart denotes the sequence of activities, their estimated time of completion, and the man-power requirement. It also identifies the sequence of controlling activities at various periods. Each worker is made aware of his/her role for the speedy and safe completion of the longwall move. With this plan, the supervisors can clearly oversee the face move and make adjustments as required.

The L. H. Gantt graph is another method for scheduling the time and man-power distribution in a sequence of activities. Although simple in nature, it is not as complete nor as clear as the critical path flow chart.

12.2.3 Manpower organization and training

A longwall move is a highly skilled operation. But it differs considerably from normal production at the face. Detailed operational rules, including move procedures, task distribution, JSA (job safety analysis), etc., must be drafted, based on the panel condition, for crew training. This is especially true regarding the hydraulic, electrical, and mechanical components of the longwall face equipment.

During the longwall move period, labor organization should change correspondingly. If there is plenty of skilled labor, it is most advisable to adopt three nine-hour shifts per day and seven days per week. The two consecutive shifts must overlap at least one hour at the working face.

As far as management personnel are concerned, especially those engaged in the first longwall move, they must study the related literature and visit other mines to observe and/or discuss longwall moves.

12.2.4 Preparation of supplies, tools, and equipment

The most critical supplies in a longwall move are used for roof control purpose. These include wood posts, crib blocks/specialty standing supports, geogrids, and roof bolting materials, including regular roof bolts and specialty bolts such as cable bolts. If rail transportation is used for hauling shields, rails and ties are needed. If the emulsion tank is to be cleaned up during the longwall move, there must be plenty of emulsified fluid for refilling. In addition, according to the wear conditions found in advanced inspections of the face equipment, replacement parts must be ordered immediately.

Special tools are necessary for a longwall face move. During pre-planning, a person must be assigned to make sure that the tools are on hand. If any parts require any of these special tools for installation and dismantling, no other tool should be permitted so parts are not damaged and the quality of the equipment installation is not compromised.

Battery-operated scoops and diesel haulers (Fig. 12.2.1) have become an integral part of longwall moves. Their hauling capacity ranges from 30 to 60+ tons, and they are used to pick up, load, and move large heavy equipment in whole units, e.g., shield, pans, chain assembly, head and tail drives, etc. Therefore, they must be in good working order.

Equipment transportation may be accomplished by using all tracks or all rubber-tired vehicles or a combination of both.

(A)

(B)

Figure 12.2.1 Shield movers: (A) the SH650 battery powered shield hauler has a lift capacity of 50 tons (45 mtons). (B) a shield mover is carrying a shield out of the tailgate T-junction

Source: A – courtesy of Caterpillar

Most activities in a longwall face move involve dismantling, transportation, and installing equipment. Shields have the largest number of units which are not only heavy and bulky but also in large number. The most frequent operations in longwall face move are turning, lifting, and pulling. Due to restricted space and roof control, special equipment is used for this purpose. The equipment must satisfy certain requirements, such as reduction in labor, energy, and exposure to potential hazards (Adam *et al.*, 1984).

To facilitate turning equipment, especially shields, various specially designed moving equipment is available. For example, the scoop, shield hauler, Petito Mule, and F Bar.

12.2.5 Transportation route and communication system

The termination point of the panel being mined should be located at the middle of an intersection between the headgate and the crosscut in the head end and between the tailgate and the crosscut in the tail end. Only with this arrangement will there be ample room for recovery operations. Resin bolts, cable bolts (if applicable), geogrids, and/or steel straps must be used to reinforce the crosscuts and entries from the mouth of the panel to one break inby the termination point

Every route that leads to the new panel face must be thoroughly studied in terms of entry size, slope, and curve; roof and floor conditions; entry maintenance conditions and repair requirement; conditions of the existing rails; the length of the route; and so on. The final route will be selected as a result of thorough technical and economical analyses.

There are three ways to pull out the face equipment from the recovery room: from tailgate to headgate, from headgate to tailgate, and from mid-face to both headgate and tailgates. They must be selected based on face conditions, such as face inclination, face width, transportation method, mine conditions, roof conditions, and practice at the headgate and tailgate. The two former methods of recovering face equipment, that is, from tailgate to headgate or from headgate to tailgate, are uni-directional, whereas the third method is bi-directional. Undoubtedly, the bi-directional method (Fig. 12.2.2) reduces move time considerably. It also helps avoid high pressure around the panel center and minimize potential roof falls during shield recovery. But it requires two shifts of crews with two sets of moving equipment for simultaneous operations. It demands high standards in management, communication, and planning. In addition, if the roof at the mid-face collapses after support withdrawal, the ventilation system must be adjusted accordingly. Among the three methods, the first one, from tailgate to headgate for obvious reasons, is more popular. But there are also many longwalls employing the bi-directional method to shorten move time. Using this method, a complete longwall move for a 1000 ft (305 m) wide face can be completed in six to seven days ordinarily. The extra set of moving equipment and manpower required to implement this method is well worth it.

After the transportation route has been selected, the existing tracks, if any, must be built up, that is, add extra ties, blocks, and level. New tracks, if necessary, must be added for empty transfer cars, supply cars, and transportation vehicles. Moreover, storage areas for tools and supplies should be prepared for both the recovery and setup rooms. All required supplies must be checked to make sure they are ready and sufficient. Generally, a storage area should be located near the recovery and the setup rooms in order to reduce transportation and retrieval time.

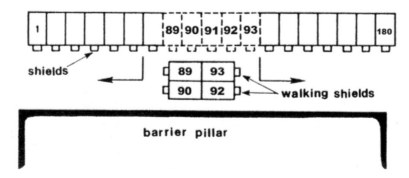

Figure 12.2.2 An example of bi-directional recovery method

The preparatory works at the setup room must be completed during this time. This includes making the new face straight and the floor leveled. When all pre-move preparations have been completed, an inspector should be assigned to check out the work. This can best be done by moving an actual or simulated transportation system along the selected routes and recording all the problems encountered as a basis for modification.

Figures 12.2.3 and 12.2.4 show two examples of the general layout around and outby the recovery room and storage areas designated for equipment at the completion of the pre-move preparations. This arrangement varies from mine to mine depending on crew practices. Ample storage space is an absolute necessity, and spaces must be designated for each type of equipment/tool to avoid confusion.

A control center must be established on the surface. A communication system linking the control center, recovery room, setup room, transfer points, shaft bottom, storage areas, and repair shops must also be established. The control center monitors all the move activities. It maintains the up-to-the-minute information on the progress of the face move for any modifications, if necessary.

Signals must be erected at strategic points such as the transfer points and the lifting points along the face in order to avoid any mishaps.

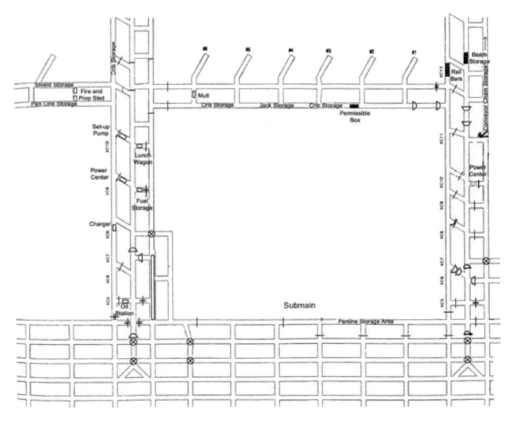

Figure 12.2.3 An example layout of recovery room area employing multiple cross-panel entries and chutes, and storage areas

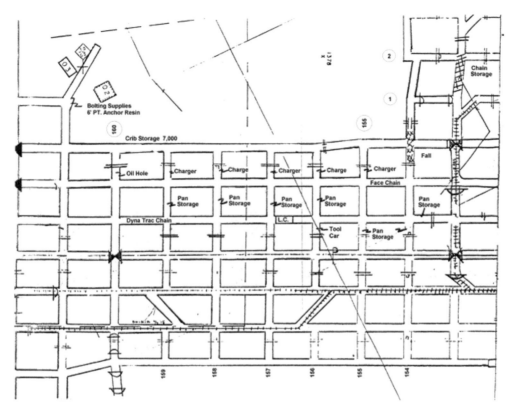

Figure 12.2.4 An example layout of recovery room area employing single chute and storage areas

Finally, it would be most advisable that during the preparation meeting, the group go through the move procedures step-by-step, and check out every item that needs to be prepared, making sure they are ready in advance.

12.3 Move preparation

Move preparation consists of the erection of a recovery room, selection of a type of face move, and inspection and repair of face equipment.

12.3.1 Erection of recovery room

12.3.1.1 Recovery chutes

Normally, at the designated location of recovery room, there is no entry development. Before and when the retreating face reaches that location, the front abutment may be too much to maintain entry stability if an open entry is predeveloped before longwall face arrival. Therefore, near the face stopline on the outby side, short entries should be developed from either the tailgate or headgate, or a special cross-panel entry (entries) should be constructed some distance outby the designated location of the recovery room, or a combination of any of these methods should be developed. Figures 12.2.3 and 12.2.4 show two extreme types of recovery chutes. The former, for soft floor conditions, has two cross-panel entries and six recovery chutes that are evenly distributed over the panel width, while the latter, for firm floor conditions, has only one recovery chute at the headgate side.

Recovery chutes are used to pull the equipment out from the recovery room. They provide more rooms and traveling routes for equipment retrieval, thereby avoiding congestion and, in case of soft floor, help avoid tearing the floor down, which will hinder equipment removal. Recovery chutes must be well-supported in order to cope with the incoming and standing front abutment pressures.

12.3.1.2 Meshing cycle

Meshing is to cover the immediate roof with synthetic polyester grids or geogrids or mine grids in the final stretch of panel mining all the way to the stop line.

Preparation for a longwall move begins approximately 9 to 13 cutting passes (webs) or 30 to 45 ft (9.1–13.7 m) from the pre-determined termination point, depending on the shield length and recovery room width (Fig. 12.3.1).

The principle is that when shields reach their final destination location, the rear end of the mesh will be well behind the shields on the mine floor where rock fragments will help anchor it, and, most importantly, the mesh is so strong that it maintains its integrity and prevents rock fragments from filtering into the recovery room when mining stops. Furthermore, during shield removal, it will keep the recovery room clean of roof falls. In the past, wire meshes were exclusively used. However, in recent years synthetic geogrids such as Minnex (Steffenino, 2019) are exclusively used, replacing the wire meshes. Because the wire mesh breaks frequently under heavy roof pressure, and rock debris fell through the cavities, especially when the shield was pulled out of line but before cribs were installed in its place. Also, wire meshes come in small panels that need to be seamed together during installation. In addition, wire mesh is stiff and heavy with less maneuverability.

The geogrids are very strong. For instance the tensile strength of Minex in both directions ranges from 12,000 to 50,000 lbs (53.4 to 222 KN). They are made of sections 8.2 or 16.4 ft long by 341 ft wide (2.5 or 5 × 104 m wide) and sewn together to fit the width of the longwall panel. Figure 12.3.2 shows the plan view of Minex made for a 1560 ft (475.6m)

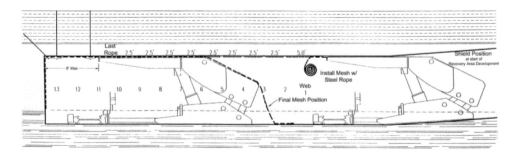

Figure 12.3.1 Meshing in preparation for construction of recovery room

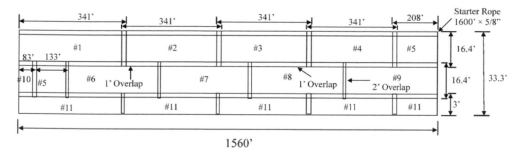

Figure 12.3.2 Geogrid designed for a 1560-ft (475.6-m)-wide panel. top: map drawing showing mesh components and bottom: geogrid spread out in the field showing
detailed construction

Source: Courtesy of Tensar International Inc.

wide longwall panel recovery. Note the starter end of the geogrid is rolled with a starter rope,
5/8 in. (16 mm) or larger in diameter.

The geogrids are rolled up and loaded on a truck (Fig. 12.3.3) and shipped to the mine site
where it is unloaded and loaded onto flat cars shipping through the slope or shaft to the tailgate
side of the panel recovery room. The geogrid rolls are then dragged by the shearer to stretch
from the tailgate end to the headgate end. With the shield canopy tip down, the starter rope
end is laid over the top side of the canopy tip leaving the remaining portion of the geogrid
hung under the canopy tip (Fig. 12.3.4 A). Each end of the starter rope is firmly anchored at
the headgate and tailgate, respectively, such that the geogrid is fixed at the starter end and will
not move with shield advance. As the shearer completes a web cut, the geogrid is rolled up on
top of the shield as it advances (Fig. 12.3.4 B and C) until it reaches the stop line and extends
down to cover parts or the whole coal face (Fig. 12.3.4 D) depending on mine practice.

Clearly, geogrids for meshing cycle is far superior to wire mesh in that it is stronger and
yet much easier to handle during the whole process of preparing and constructing a recovery
room. The whole meshing cycle can be done in one shift.

Figure 12.3.3 A geogrid roll (left) that is rolled up on a truck ready (right) to be transported to mine site

Source: Courtesy of Tensar International Inc.

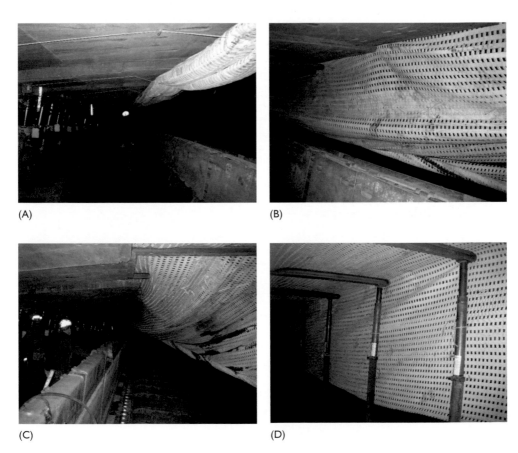

(A)

(B)

(C)

(D)

Figure 12.3.4 Sequential order of meshing cycle: top left – start; top right and lower left –
intermediary steps, and lower right – final step ready for roof bolting

Source: Courtesy of Tensar International Inc

Figure 12.3.5 Artist's view of recovery room using Minegrid
Source: Huesker Minegrid ® longwall recovery screen

Figure 12.3.5 is an artist's view of using geogrid for after completion of meshing cycle in preparation for installation of recovery room.

12.3.1.3 Width and height of recovery room

When the face has advanced to the second or third from the last cut depending on the required recovery room width, shields are not advanced any more. At this time the advancing ram for each shield may or may not be disconnected from the conveyor pan. The conveyor is advanced by inserting a ram jack extension or wood bar of approximately one or more web lengths between the conveyor and the advancing ram. The last cut is made from the tailgate to the headgate. During the last cut, the coal face must be cut smooth and straight (Fig. 12.3.6). After the last cut, recovery room width, i.e., the distance from the faceline to the front toe of the shield base plate, should be at least 16–20 ft (4.88–6.1 m) or at least approximately equal to the total length of the collapsed shield. This constitutes the recovery room or entry. Note that the mesh must continue to cover the unsupported area between the canopy tip and faceline, and, in many cases, will drop down and cover all or only the top portion of the face (Fig. 12.3.4 D).

Depending on the equipment design, especially shields, there may be a minimum operating height in which the shield retrieval operation is optimal. This minimum height may differ from the longwall mining height. If this occurs, it is very important that, during the meshing period, the mining height gradually be adjusted to this minimum moving height.

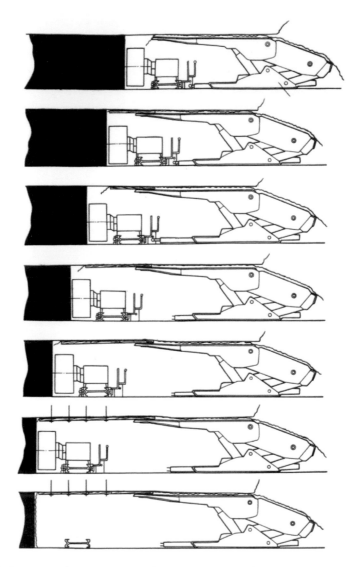

Figure 12.3.6 Sequence for preparation for construction of recovery room

Source: Modified from Conroy and Curth (1981)

When the immediate roof is stable, the roof control method may be simpler during preparation for construction of recovery room. But mine grids (or geogrids) is still used to cover the roof and face as stated before. This is to prevent the sloughed coal debris from the face and/or rock fragments in the roof and the gob from entering into the recovery room prior to, during, and after shield withdrawal.

12.3.1.4 Bolting cycle

After the shearer has reached its final stop line, it will make another trip to clean up the floor. Two to three baby-boom or slimline roof bolters riding on the panline (Fig. 12.3.7) are then

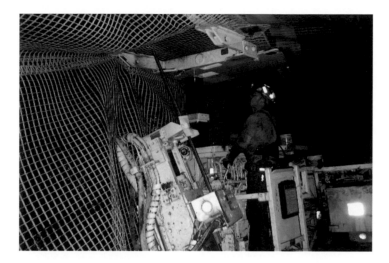

Figure 12.3.7 A slimeline roof bolter installing an angle bolt over the mine grids in the recovery room

Source: Courtesy of JH Fletcher & Co.

brought in to roof-bolt the area between the canopy tip and faceline. To speed up the roof bolting process, each roof bolter will cover different sections of the face simultaneously, depending on mine practice. The roof bolting system (bolt type, bolt dimension, and pattern) employed varies with roof conditions.

12.3.2 Pre-driven recovery room

As stated in the previous section, all conventional longwall recovery plans call for meshing the roof for 10 to 13 cuts before the stop line. This is a tedious operation and slows down the rate of face advance considerably. In order to reduce this period of low production, the pre-driven open recovery room concept was developed in the late 1980s. If the roof condition in the designated recovery room location is bad, an open recovery room provides an excellent opportunity to pre-support the roof and ensures its stability during the recovery operation. The pre-driven recovery room must be at least 20–24 ft (6.1–7.3 m) wide and requires special supports to cope with the incoming and, after mining stops, standing front abutment pressure. There had been several costly failures in the past, mainly due to improper support designs. Recent designs had conclusively demonstrated that the pre-driven recovery room concept is a viable approach. In spite of this, the pre-driven recovery room concept is still rarely used by the industry.

For a more detailed discussion about this topic including its support design, see Section 10.8.

12.3.3 Equipment inspection and repair

Before the longwall face reaches the termination point, a thorough and careful inspection of all face equipment must be made and a detailed record maintained. Based on this inspection, a maintenance and repair plan should be laid out.

12.4 Move

12.4.1 Move procedures

When the equipment is being moved, it is necessary to adjust ventilation for the crew to working inby. The detailed moving procedure varies from mine to mine and from time to time. The move is dependent on spare equipment that is available for advance setup in the setup room of the new panel. In the following description, it is assumed that a complete longwall face move has occurred.

When the face equipment is withdrawn uni-directionally, the sequence of the major equipment withdrawal in the shearer face proceeds as follows:

1 Remove and transport entry belt-conveyor and its accessories.
2 Move the hydraulic supply system to the new panel and install a setup pump for the old panel, or vice versa.
3 Remove the monorail system.
4 Remove and transport the stage loader and its accessories.
5 Remove and transport the face conveyor and the shearer in the following sequence: chains and flight bars, head drive and crusher, cable handler, shearer, communication and safety systems, cables and water lines, conveyor pans, and tail drive. Many mines choose to remove the three tailgate gate-end shields and tail drive from the tailgate by shield movers.
6 Remove and transport the face lighting system.
7 Clear the recovery room.
8 Remove and transport shields. This is the major task of a longwall move, because there are so many units and their total weight is at least two-thirds of the equipment to be moved. The methods of removing shields differ from each mine as discussed in the next section.

To facilitate the recovery, items to be sent to the setup room, to storage, or to the surface should be clearly identified with different paints.

12.4.2 Removal of head and tail drives, stage loader, and belt tailpiece

As soon as the last coal is conveyed out of the belt tailpiece and the belt is no longer needed, it is retracted at the winding station. The belt tailpiece is disconnected from the stage loader and trammed at least one crosscut outby. At this time, cribs are built to support the delivery end of the stage loader.

As soon as the shearer has completed the clean-up run, The AFC is run and stopped when a chain connector is near the tail drive. Separate the chain connector and lay the chain back. Cut tail drive loose from cables, gob plate, cowl cover, etc., and separate the tail drive from the pans and pull it off. The tail drive is ready to be moved to storage in the tailgate side.

At the same time, the head drive is disconnected from the drive motors and the stageloader is scooped away for temporary storage. The cables and hoses in the head drive will be pulled out of the trough and dogbones are removed and bagged. The assembly is disconnected from the sprocket. The relay bars are disconnected and transported out for storage.

The stage loader is separated into six composite sections and moved out for storage separately: between drive frame and flex pans, between flex pans and crusher, between crusher and cross-over, between cross-over and first goose neck connection, and between the first goose neck connection and the delivery section, including the stageloader drive.

12.4.3 Removal of monorail

The monorail is a beam that runs from an outby area, comparable to the location of the power center, to the stageloader. Cables and hoses are hung from the monorail on dollies. All of the dollies and their loads are moved in one piece by the monorail mover from the old face to the new face. The monorail mover consists of modified mine cars connected in a series, one for 10 dollies. Dollies are winched through the cars with the cables and hoses remaining in stretched form between dollies. A splice bar is used to connect the beam ends so that the system can negotiate curves during transportation.

12.4.4 Removal and transport of the shearer

As soon as the shearer has completed the clean-up run, it will park at the first line pan location from the head end. Cut or remove the chain and remove the head drive and ramp pan. The shearer is now ready to be removed.

Track is built up to the first line pan where the shearer is parked. The shearer dolly (Fig. 12.4.1) is brought up to the shearer so the shearer can simply tram itself into it and be towed away to the designated storage place where the drums and ranging arms are disassembled before being transported outside for rebuilt.

12.4.5 Removal of chain assembly

The chain assembly can be removed from either end or from both ends in whole or in cut sections, depending on the AFC length.

As soon as the shearer is moved away, special chain cars are brought up to the face. A wire rope is attached to the chain, over the car, and connected to a motor. The motor pulls the chain into the car and drops it in. The chain is normally removed by section, e.g., in multiples of 164 pieces. After the chain connector reaches the car, the rope will be removed and the sequence will be repeated again until the chain is completely removed. A scoop may be used to assist in loading.

12.4.6 Removal of pans

As soon as sufficient chain has been recovered, the panline is dismantled in two- or three-pan sections. The dogbone connectors, retainers, and haulage pins are put in a bag and secured to the panline. The pan sections are removed and loaded onto the shield dollies or scoops at both ends and hauled away for storage (Fig. 12.4.2).

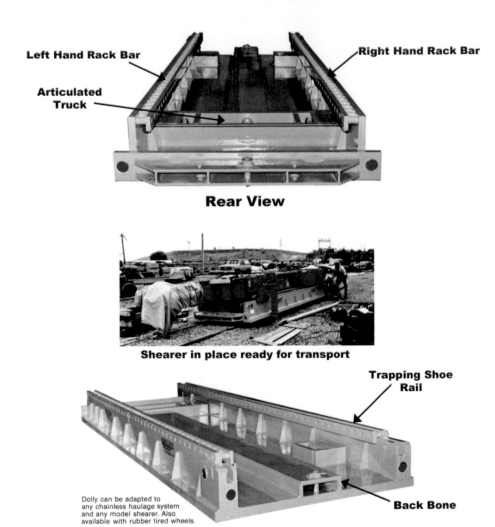

Left Hand Rack Bar

Right Hand Rack Bar

Articulated Truck

Rear View

Shearer in place ready for transport

Trapping Shoe Rail

Dolly can be adapted to any chainless haulage system and any model shearer. Also available with rubber tired wheels.

Back Bone

Front View

Figure 12.4.1 Shearer dolly

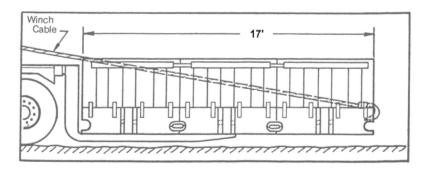

Winch Cable

17'

Figure 12.4.2 Using scoop to transport AFC pans

12.4.7 Removal and transport of shields

12.4.7.1 Preparation for removal

Shield removal is the key part of a longwall face move, because there are so many units. The key to a successful start-up on the new face is related closely to the quality of the job done in preparing for shield removal. The electronhydraulic control system is extremely sensitive to any type of contamination, dust, moisture, debris, etc. Therefore, all hoses, cables, communication system, and components must be carefully tagged, bagged, sealed, and secured. Shields are hosed clean before the removal process begins.

12.4.7.2 Sequence of removal

When removing shields, the major considerations are to control the roof and to prevent the caved fragments in the roof and gob from entering the face area. Before shield removal, the roof is covered completely with mine grids or geogrids that are designed to be sufficiently strong to prevent rock debris from falling between and behind the shields. But after a shield is removed, standing supports (e.g., cribs, propsetters, ACS props, etc.) must be erected in its place so that the mine grids or geogrids stay intact.

The sequence of support removal varies with the method chosen. If the tailgate to headgate method, which is the more popular one, is chosen, shields are removed from the tail end toward the head end where they are loaded for transportation to the new panel. In this method, the three gate-end shields at the tailgate side are pulled out from the tailgate, the fourth shield (or the first face shield) from the tail end is first pulled out of line and turned to face the headgate until it becomes parallel to the faceline. The fifth (or the second face shield) from the tail end is also pulled and turned to line up with the fourth shield side by side. These two shields are called the walking shields or simply walkers (Fig. 12.4.3). They serve as the gobline supports for the space between the canopy tip and faceline as shields are removed one by one from the tail end to head end. One or more standing supports are installed in its place as each shield is pulled out of line. A special setup pump can be used to power the walking shields. The walking shields may draw their power from the hydraulic power supply via the inter-shield, high pressure hose line. The third face shield (the sixth shield from the tailgate) is then pulled and turned and hauled out of the face (Fig. 12.4.3). After the shield has been pulled out of place, secondary or standing supports are erected in its place to keep the roof from caving and the geogrid from breaking up. Those secondary or standing supports must be high and strong enough to maintain the minimum height so as not to affect the removal of the immediate next shield. The secondary or standing supports can be one or more wood cribs of various types, prop setters, 100-ton ACS props, etc., depending on roof conditions. Note when the three tailgate gate-end shields are pulled out from the tailgate, similar secondary or standing supports are erected in their place.

Following the complete removal of the fourth shield, the two walking shields are moved forward a shield width. This process is repeated for each shield removal until all shields are removed.

The number of walking shields employed for establishing the gobline depends very much on the roof condition. Two walkers are most common, but some mines employ one while others use three walking shields.

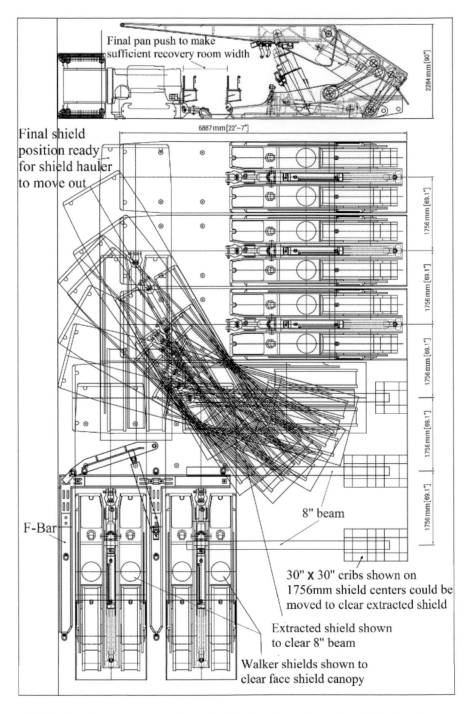

Figure 12.4.3 Step-by-step recovering of a shield and erection of a crib in its place
Source: Courtesy of Caterpillar Inc.

12.4.7.3 Method of removal

A. PETITO MULE

The "mule" is a skid tractor and has been designed to carry part of the load, while skidding most of it. In other words, it puts the weight of the load helps in the work yet allows a tremendous skid factor advantage. There are two models, cat-mounted or rubber-tired. The boom's lifting capacity increases as it is raised to a different angle. Therefore, the boom has minimal lifting capacity when it is leveled, reaching its maximum when raised to the highest position (Fig. 12.4.4).

Figure 12.4.5 shows the steps of shield removal using the mule. Step A – the mule is positioned adjacent to the shield to be removed. The boom is swung 45° toward the shield. Step B – with hook line retracted to the end, the boom is then extended toward the shield for hook up. The hook should be anchored at the front toe such that only the rear end of the shield rests on the floor. Stabilizing jacks are lowered and the shield is then swung out of line toward the face. The boom may be extended further in order to ensure clearance of the next shield. Step C – the boom is retracted and the shield is turned to transport position. At this time, the shield is pulled to the nearest recovery chute where a shield transporter or shield hauler will pick it up and haul it away to a shield dolly or flat car on track to the new panel. In many instances, the mule drags the shield all the way out to the headgate transfer point, provided the floor is solid and would not be easily damaged by being repeatedly run over by the mule.

Figure 12.4.4 Petito mule for shield removal

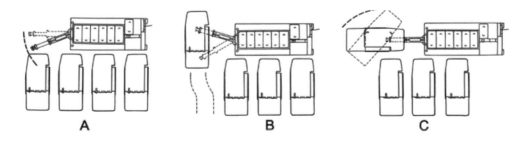

Figure 12.4.5 Method of shield removal using Petito mule

B. F-BAR

The F-bar, as its name implies, is shaped like the letter "F" with two horizontal and one vertical bars. The two horizontal bars are anchored to the base plates of the two walking shields with the lower bar between the shields and the upper bar on the face side of the shield next to the face. A double-acting ram with a mechanical arm is mounted inside the vertical bar. There is a mesh guard along the vertical bar to protect operators from broken chain (Fig. 12.4.6).

The double-acting ram draws its power from the same power source as the walking shields. Two scoops are used to remove the shields from the face. To pull a shield, a cable of proper length is attached to the holes in the base plates of the shields and then a rope is attached to the other end of it. The rope is run through the sheave block attachment on the face side of the F-bar to the scoop. Variations of this process may occur. Several sheave block attachments

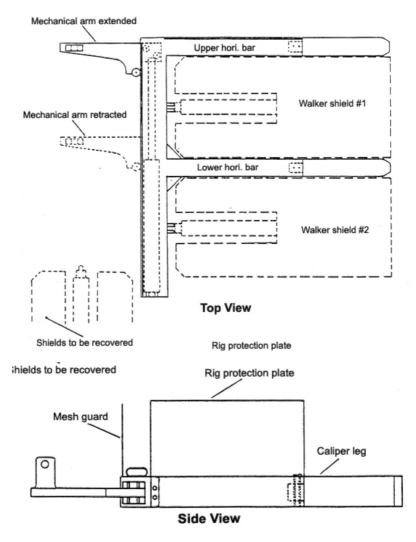

Figure 12.4.6 F-bar for shield removal

are located on the F-bar to ensure that the best location is available for pulling the shield safely and efficiently. After the shield is properly positioned in the scoop, it is taken to the shield dolly via the track for transportation to the new face.

C. MRS (MOBILE ROOF SUPPORT)

The MRS was developed in 1988 for pillar extraction in room and pillar mining (Fig. 12.4.7). This support unit is a crawler-mounted, four-leg hydraulic unit and functions very much like a four-leg chock shield with 600 and 800 ton capacity (Wilson, 1991).

Shield removal begins by bringing two MRSs to mid face, one from the head end and the other from the tail end, if possible (Richards, 1998). This helps avoid having to reroute the cable for one MRS if both are brought in from the head end. Both MRSs are brought to the same location just beyond the first four shields selected for removal. Note that the two MRSs are in back-to-back position (Fig. 12.4.8 A), i.e., one facing the head end while the other faces the tail end. Remove shield number one and build a crib as near to the gob as possible. At the same time, move the MRS units forward but slightly into the cavity left by removal of shield number one (Fig. 12.4.8 B). Shield number two is removed, a crib is built as near the gob as possible, and move the MRSs further in to fill the cavity. This process is repeated until seven shields are removed and both MRSs are fully alongside the gobline (Fig. 12.4.8 C). At this time, seven cribs are built on the face side of the MRSs and shield removal is begun in both directions. Two cribs are built for each shield removed until the first fall occurs. Then two cribs are built for every fifth shield removed. This is the minimum standard, conditions may require more supports.

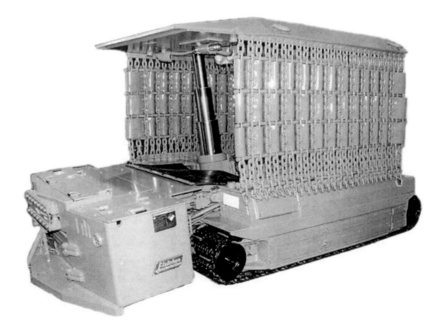

Figure 12.4.7 Mobile roof support (MRS)
Source: Courtesy of J.H. Fletcher & Co.

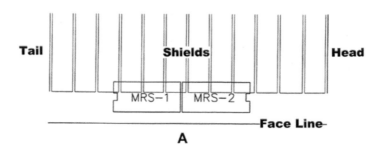

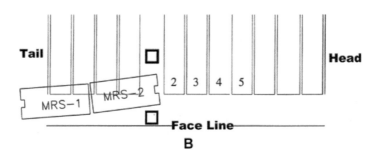

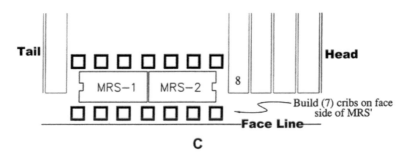

Figure 12.4.8 Method of shield removal using MRS
Source: Richards (1998)

12.5 Installation of the new face at the set-up room

12.5.1 Preparation of installation

12.5.1.1 Face preparation

Maintaining a straight faceline is the key to the quality of installation. Therefore, it is essential that before installation, a detailed survey is made to insure the straightness of the faceline. Those areas that do not meet the requirements must be corrected immediately. If the face is prone to slough, rib supports may be needed.

A smooth and level floor is also essential to equipment installation. If there are undulating areas, they must be leveled off or backfilled. Before installation, rock debris and broken coal on the floor must also be removed.

12.5.1.2 Equipment preparation

Arranging the sequence of transportation for the equipment and various parts of a piece of equipment is an important function and must be performed carefully. It directly affects the speed and efficiency of the installation. Also before moving into the new face, all equipment must be strictly inspected. Each piece of equipment and its parts must be assigned a sequential number in accordance with the method of introducing them into the new face (i.e., from head to tail or vice versa) and be transported in that order. This way, there will be no need to rearrange the freight cars in underground transportation.

The direction in which equipment is introduced into the new face is also important. For example, whether a shield should be moved into the setup room with the front end or the rear end facing the front depends on the equipment that is used to unload and turn it into the final position at the unloading point. The principle is that the amount of shield turning should be minimized to save installation time. Another example would be the arrangement of conveyor line pans unless they can be used in both directions. When moving into the new face, the male ends of the line pans must always face the head end drive unit. This is especially true for thin-seam longwall faces. In addition, all look-alike equipment must be coded with sequential numbers based on the sequence of introducing them into the new face. These include the drive units and the ramp pans for the head end and the tail end.

12.5.2 Sequence of installation

Equipment can be introduced either from the tail end or the head end. Selection between the two depends very much on different pieces of equipment and the manner they are removed from the recovery room. Introducing equipment from the tail end is more popular. The sequence is as follows:

1   Install tail drive and chain assemblies.
2   Install line pans from tail to head.
3   Install shearer.
4   Install head drive.
5   Install stage loader, belt tailpiece, and monorail.
6   Install shields.
7   Complete installation.

To reduce interference between pieces of equipment and to speed up the installation, it is advisable for some equipment, especially the shearer, to be introduced from the tailgate if conditions permit. If the shearer is introduced from the headgate side and the shearer is installed last, the head drive will have to be installed after the shearer is installed. However, as far as the installation of the face conveyor is concerned, the ideal sequence is to install the head drive first and then the conveyor pans from the head end toward the tail end, and finally the chain units are installed. Shields may be installed from the tail to the head end or vice versa.

Similarly, in order not to damage the floor from scoop's repeat trips, an inby parallel entry with chutes connected to the setup room is normally available for shield installation.

After the tailgate gateend shields have been installed, the shearer is introduced from the tail end, followed by the tail drive. If the conveyor tail end is a nondriven type, the shearer can be loaded directly onto the conveyor.

Chapter 13

Longwall power distribution and system control

13.1 Introduction

With the increase of horsepower required for longwall face equipment in the past four decades, power supply has become an important factor. A 1000 kVA power center was commonly found for the longwall sections in the 1980s. Today, it has increased to more than 11,000 kVA. Similarly, 1000 volt face equipment was exclusively used in the 1980s. Today, most longwalls employ either 2400 or 4160 volts to deliver higher power to the power centers and, subsequently, to the face equipment (see Fig. 1.5.8). This chapter will address high-voltage power distribution for modern longwall sections.

As automation continues to grow and improve, smooth interaction between subsystems and subsystem automation becomes the most critical element in the whole system. This chapter will address the logic of individual subsystems and give examples of how they interact with other subsystems to achieve overall system automation.

13.2 Electrical power distribution

13.2.1 Electrical power distribution circuit

Electrical power is delivered from utility companies to the mine site, where it is then delivered underground to the longwall section and the individual face equipment (Fig. 13.2.1). In order to reduce voltage drop during power transmission, high voltage power is used. In fact, the higher the voltage, the smaller the cable required and the less voltage drop.

Utility companies normally deliver power to mine sites at 24 to 138 kV. At the mine site, a substation using transformers reduces the voltage to 12 to 14 kV, which is then delivered to the power centers at the longwall sections. Transformers at the power center reduce the voltage to 4160 V or less for the face equipment. Since this power transmission process involves high voltage, various safeguard and protection equipment are required to prevent electrical hazards. These hazards include fires and explosions, burns from an electric arc, shock, and electrocution. There are five main protection systems: ground fault, ground wire monitoring, undervoltage, overload, and short circuit (instantaneous).

The mine electrical distribution system consists of the following four parts (MSHA, 1999): power source, conductors, load, and control devices. The definition, construction, and selection of the devices for each of the four parts are illustrated in the following sections.

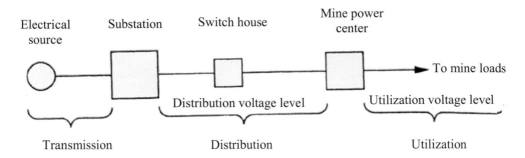

Figure 13.2.1 Basic power system
Source: Morley (1990)

13.2.1.1 Power source

The power source in an electrical circuit can include transformers, rectifiers, batteries, and generators. It supplies electrical voltage (or pressure) to "push" the current through the system (or electrical circuit). Transformers are the power source in substations and power centers. Electrical power is the rate at which the voltage and current get work done as expressed in the formula here.

$$\text{Power}(\text{VA}) = \text{Voltage}(\text{volts})\text{Current}(\text{amps})$$

Or:

$$P = V \times I \qquad (13.2.1)$$

A transformer is a device having two tightly coupled coils or windings on a common iron core. The windings are arranged so that the input voltage to the primary winding will be reduced or increased to a certain voltage as output from the secondary winding, in proportion to the ratio of the number of turns of primary to secondary winding. For example, a transformer has a turn ratio of 15:1. If the input voltage to the primary winding is 7200 V, the output voltage of the secondary winding is:

$$\frac{N_2}{N_1} = \frac{V_2}{V_1} \text{ or } \frac{1}{15} = \frac{V_2}{7200} \qquad (13.2.2)$$

$$V_2 = 480(\text{volts})$$

Where N_1 and N_2 are the number of turns for the primary and secondary windings, respectively, and V_1 and V_2 are the voltage for the primary and secondary windings, respectively.

A safe estimation to be utilized when estimating the needed transformer capacity is about one kVA per horsepower.

13.2.1.2 Conductor

The conductors include feeder cable, trailing cable, trolley wire, and mine track. Conductors provide a path for the current to flow. They carry the electricity from the substation where the power is received from the utility company line, through the power center, to the face equipment. Feeder cables are stationary and run from the substation through shafts, entries, or surface boreholes to section power center. Trailing cables are portable cables that connect equipment to a power center.

Cables are made of conductors, shielding, insulation, and the jacket. Line currents are carried by conductors. The conductors are usually composed of many fine copper wires wound into strands, and varying numbers of strands are wound into the conductors. The conductors come in several sizes (Fig. 13.2.2) with different current-carrying ability or ampacities

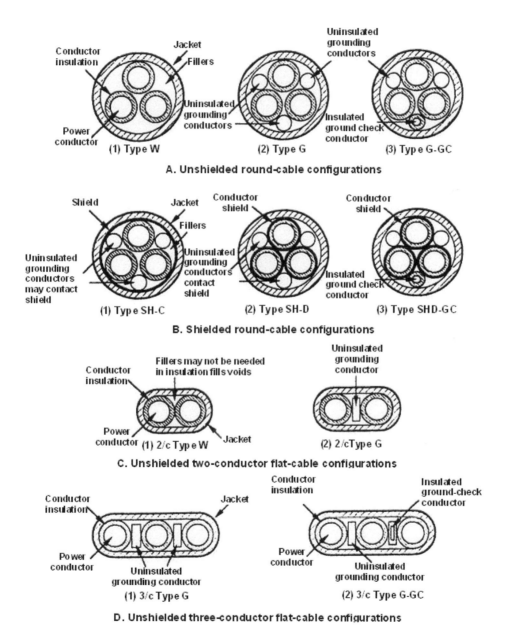

Figure 13.2.2 Several commonly used cables in coal mining
Source: Morley (1990)

(Morley, 1990). The conductors are surrounded by insulation, which is designed to separate power and grounding circuits and to protect mine personnel from electric shock. The thickness of a specific insulating material determines the voltage rating of the cable. The cable jacket is the cover material that protects the inner components and holds the assembly in the designed configuration. Cable shielding confines the electrical field to the inside of the cable insulation. Cable selection mainly involves the determination of cable length, voltage rating, and cable ampacity. The current and voltage rating are two major factors in determining the correct size of trailing cables.

A. CURRENT

The ampacity of various types and sizes of cables are typically defined by the cable provider, following the Insulated Cable Engineers Association (ICEA) ampacities. For example, the ampacity chart (Table 13.2.1) is for AmercCable's Tiger®Brand 5000 Volt Type SHD-GC three conductor cable. This is the cable commonly used for longwall and continuous miner circuits. Each cable manufacturer can provide more specific ampacities. It is important to always consult each cable's manufacturer prior to sizing cable.

B. VOLTAGE

Adequate voltage must be provided at machine terminals for proper starting and operation. The desired voltage tolerance on all rotating machines is ±10% for normal load conditions.

C. REQUIRED CABLE SIZE – SAMPLE CALCULATION

Determine the proper cable size and machine voltage for a shearer operating in the following conditions: panel width = 1000 ft (305 m), power center located 1000 ft (305 m) total cable length outby, and a source voltage = 4160 V. To properly size the cable, it is best to consult nameplate data to determine full load current of the device. For example, a Komatsu 7LS1D Shearer data name plate indicates a full load current of 280 Amps.

1   Determine the minimum cable size for the aforementioned 7LS1D Shearer.

By consulting Table 13.2.1, a 4/0 shearer cable would adequately carry 321 amps thus it is sufficiently sized.

2   Determine the voltage at the shearer.

The voltage at the shearer is calculated by the following formula:

$$V_{load} = V_{source} - V_{drop}$$

(13.2.3)

Table 13.2.1 Underground cable ampacities

Type SHD-GC 3/C • CPE Jacket • 5000 Volts

36–515-	Power Conductors			Grounding Conductors		Jacket thickness mils	Nominal outside dimensions in.	Approx weight lbs per 1000 ft.	Ampacity/ 40°C ambient temp
	Size AWG	No of wires per conductor	Insulation thickness mils	Size AWG	No of wires per conductor				
006	6	133	110	10	49	185	1.56	1560	93
004	4	259	110	8	133	185	1.68	1920	122
002	2	259	110	6	133	205	1.87	2500	159
001	1	259	110	5	133	205	1.95	2860	184
010	1/0	266	110	4	259	220	2.08	3390	211
020	2/0	323	110	3	259	220	2.20	3830	244
030	3/0	418	110	2	259	235	2.36	4418	279
040	4/0	532	110	1	259	235	2.50	5300	321
250	250	627	120	1/0	266	250	2.69	6450	355
350	350	888	120	2/0	323	265	2.95	7880	435
500	500	1221	120	4/0	532	280	3.31	10,440	536

Source: 90°C conductor temperature and 40°C ambient air, per ICEA S-75–381, Table H-1

Chart Compliments of Global Mine Service Inc and Nexans AmerCable Incorporated

Table 13.2.2 Cable resistance

size (solid)	Area CM*	Resistance per 1000 ft (ohms) @ 20 C	Diameter (inches)	Maximum current** (amperes)
0000	211,600	0.049	0.46	380
000	167,810	0.0618	0.40965	328
00	133,080	0.078	0.3648	283
0	105,530	0.0983	0.32485	245
1	83,694	0.124	0.2893	211
2	66,373	0.1563	0.25763	181
3	52,634	0.197	0.22942	158
4	41,742	0.2485	0.20431	135
5	33,102	0.3133	0.18194	118
6	26,250	0.3951	0.16202	101
7	20,816	0.4982	0.14428	89
8	16,509	0.6282	0.12849	73
9	13,094	0.7921	0.11443	64
10	10,381	0.9989	0.10189	55

Source: http://hyperphysics.phy-astr.gsu.edu

Where V_{load} = voltage @ Shearer, V_{source} = voltage @ power train and V_{drop} = voltage lost across the 2000 ft cable when the shearer is at the tailgate.

$$V_{drop} = \left[(\text{ohms/ft}) \times \text{ft}\right] \times \text{FLA} \,(\text{full load ampacity}) \tag{13.2.4}$$

In order to determine the voltage at the shearer, voltage drop across the cable must first be determined. Table 13.2.2 shows the cable resistance values.

$$V_{drop} = \left[(\text{ohms/ft}) \times \text{ft}\right] \times \text{FLA}$$
$$V_{drop} = \left[(.049 \text{ ohms/1000ft}) \times 2000\text{ft}\right] \times 280 \text{ Amps}$$
$$V_{drop} = 27.44 \text{ volts}$$

Now that the V_{drop} has been established, voltage at the shearer (V_{Load}) when the shearer is at the tailgate can be calculated by use of the formula here:

$$V_{load} = V_{source} - V_{drop}$$
$$V_{load} = 4160 \text{ volts} - 27.44 \text{ volts}$$
$$V_{load} = 4132.56 \text{ volts}$$

These calculations for voltage at the shearer are used for normal operations at full load. However, during startup, even if all motors do not start at the same time, the voltage at the shearer can be lower. This is because a motor takes up to five times, sometimes more, the full load current during startup and the power factor is lower. This phenomenon is often referred to as inrush.

4   Determine the available torque of the AFC

It is assumed that the breakdown torque, $T_{BD,}$ of this motor under normal voltage condition is 350% of the full load torque. The available breakdown torque is proportional to the square of the applied voltage.

A   2300 volt tailgate motor

The motor is rated at 750 hp (543.5 kW) and breakdown current is 720 amperes. Assuming a 2000 ft (610 m) long 4/0 AWG cable is used and the power center output voltage is at motor nameplate, the voltage (V_{motor}) and the breakdown torque (T_{BD}) at the tailgate motor are:

$$V_{motor} = V_{source} - V_{drop}$$
$$= 2300 \text{ volts} - [(0.049 \text{ ohms}/1000 \text{ ft}) \times 2000 \text{ ft}] \times 720 \text{ amps}$$
$$= 2229 \text{ volts}$$
$$= 97\% \text{ of nameplate}$$
$$T_{BD} = (0.97)^2 \times 350\%$$
$$= 329\% \text{ of full-load torque}$$

B   4,160 volt tailgate motor

The motor is rated at 750 hp (543.5 kW) and the breakdown current is 400 amperes. Assuming a 2000 ft (610 m) long 2/0 AWG cable is used and the power center output voltage is at motor nameplate, the voltage (V_{motor}) and the breakdown torque, $T_{BD,}$ at the tailgate motor are:

$$V_{motor} = V_{source} - V_{drop}$$
$$= 4160 \text{ volts} - (.0779 \text{ ohms}/1000 \text{ ft}) \times 2000 \text{ ft} \times 400 \text{ amps}$$
$$= 4097 \text{ volts}$$
$$= 98.5 \% \text{ of nameplate}$$
$$T_{BD} = (0.95)^2 \times 350 \%$$
$$= 339.5 \% \text{ of full-load torque.}$$

13.2.1.3 Load

Loads include electric motors, lights, and resistors. They use the current to accomplish work.

Motors convert electrical energy to mechanical force through rotation or torque that performs work. The speed-torque relationship of a motor is used to find the most suitable drive for a given machine as shown in Figure 13.2.3 (Morley, 1990), where locked-motor torque (or starting or breakaway torque) is the torque developed by a motor as soon as the power is applied at zero rpm, breakdown torque (or maximum torque) is the maximum torque achieved by a motor with the rated power input, full-load torque is the torque required to provide the rated output at the rated speed with rated power applied, and pullout torque is the maximum torque produced by a motor without stalling.

It is critical that the motors generate sufficient effective torque to perform certain jobs. As an example, Figure 13.2.4 shows the voltage-torque-force relationship for a shearer drum.

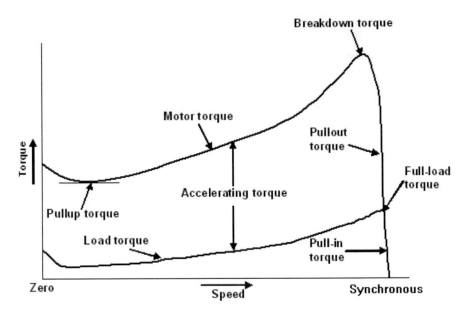

Figure 13.2.3 General speed-torque motor characteristics

Source: Morley (1990)

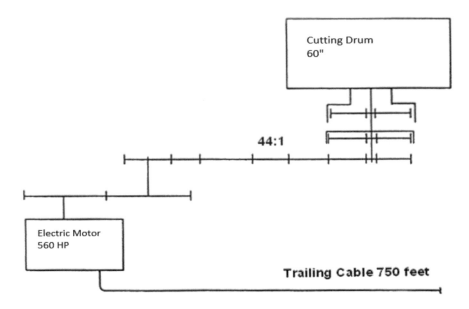

Figure 13.2.4 Motor voltage-torque and drum force relationship

Assuming the 60 in. (1524 mm) diameter drum is equipped with a 560 hp (406 KW) motor, running at 1800 rpms, nameplate voltage of 2300 volts, and a gear ratio of 44:1. The force generated and loss of force at bit tip profile at 100% and 92% voltages can be calculated as follows:

A   100% voltage = 2300 volts

Full load torque from motor shaft = 5252 × motor hp/motor rpm

$$= 5252 \times 560 \text{ hp}/1{,}800 \text{ rpm}$$

$$= 1634 \text{ ft} - \text{lbs}$$

Drum shaft torque at 44:1 = Input torque × gear ratio

$$= 1634 \text{ ft-lbs} \times 44$$

$$= 71{,}896 \text{ ft} - \text{lbs}$$

Force at bit tip profile = torque/distance

$$= 71{,}896 \text{ ft-lbs} / 2.5 \text{ ft}$$

$$= 28{,}758 \text{ lbs}$$

B   92% voltage = 2116 volts

Full load torque from motor shaft $= \left(v\%^2 \right) \times$ full load torque

$$= \left(0.92^2 \right) \times 1634 \text{ ft} - \text{lbs}$$

$$= 1383 \text{ ft} - \text{lbs}$$

Drum shaft torque at 44:1 = input torque × gear ratio

$$= 1383 \text{ ft-lbs} \times 44$$

$$= 60{,}852 \text{ ft} - \text{lbs}$$

Force at bit tip profile = torque/distance

$$= 60{,}852 \text{ ft} - \text{lbs}/2.5 \text{ ft}$$

$$= 24{,}341 \text{ lbs}$$

C   Loss of force at bit tip profile

% Bit Tip Loss of Force $= \left(F_{100\%} - F_{92\%} \right)/F_{100\%}$

$$= \left(28{,}758 \text{ lbs} - 24{,}341 \text{ lbs} \right) / 28{,}758 \text{ lbs}$$

$$= 15.4$$

In other words, an 8% voltage loss reduces the torque by more than 15.4%.

13.2.1.4 Control devices

Control devices include circuit breakers, fuses, switches, relays, starters, contactors, and disconnects. They permit or stop the current, either manually or automatically, for service or protection purpose.

A. CIRCUIT BREAKER

A circuit breaker is an automatically operated electrical switch, which is designed to protect an electrical circuit from damage caused by overload or short-circuit. Circuit breakers are implemented with a solenoid (electromagnet) which increases its strength as current increases. The circuit breaker's contacts are held closed by a latch. As the current in the solenoid increases, the magnetic force releases the latch and allows the contacts to open by spring action (short circuit). Another method of sensing the current in the circuit is a bi-metallic strip, which heats and bends with increased current, releasing the latch (overload). Circuit breakers incorporate both techniques.

Circuit breakers providing short circuit protection for trailing cables are set not to exceed the maximum allowable instantaneous current specified in 30 CFR 75.601–1 *Short Circuit Protection: ratings and settings of circuit breakers*.

B. FUSES

Fuses are the simplest devices for interruption of an electrical circuit. They protect the circuit with which they are installed in series by melting a fusible link. They act not only as the sensing device but also the interrupting device.

C. SWITCH, DISCONNECT, AND RELAY

A switch is a manual device for closing, opening, or changing connections of electrical circuit. It includes the disconnect and the load interrupter.

A relay is an electrical switch that opens and closes under control of another electrical circuit. In the original form, the switch is operated by an electromagnet to open or close one or many sets of contacts. Because a relay is able to control an output circuit of higher power than the input circuit, it can be considered, in a broad sense, to be a form of electrical amplifier. It is also used to separate electrical circuits. For instance, an intrinsically safe circuit from regular circuits.

13.2.2 Substation and power center

13.2.2.1 Substation

A substation refers to an electrical substation at the mine site. It is designed to receive the electric power furnished by a utility company by way of high-voltage transmission lines. It is also designed to reduce high voltage to a lower voltage for transmission to underground. A substation contains the following pieces of equipment and accessories: meter, voltage stabilizer, circuit breakers, lightning arresters, power lines, disconnects, transformer installation, fence and gate, ground field, capacitor banks, and danger signs.

13.2.2.2 Power center

The power center located outby the longwall section is used for transformation, rectification, switching, and protection in an electrical power system. Its functions include regulations on the use and protection of power circuits, guarding of energized components, protection of outgoing trailing cables, and prevention of shock and fire hazards.

The most common and important function of a power center is the transformation and distribution of power. Normally, the electrical circuit from the surface substation to the underground power center employs high voltage, 4160 to 13,200 volts, to transmit power. At the power center, this high voltage is reduced to lesser voltage for face equipment use. The transformation is done by a transformer bank, which produces a lot of heat, so transformer insulation is designed to withstand heat under normal load.

13.2.3 Typical longwall power distribution

In this section, two examples show layouts of common power distribution schemes for longwall mines with a single longwall face.

Figure 13.2.5 shows a power distribution layout for panels up to 1500 ft (457 m) wide by 15,000 ft (4573 m) long (Allekotte, 2006). The utility company delivers the power between 30,000 and 138,000 volts to the mine site substation where a 20 MVA transformer converts it down to 12,470 volts. The output of the transformer is divided into two feeds. One is for surface installation/facilities, such as preparation plant, warehouse, ventilation fan, bathhouse, etc. The second feed goes underground and splits into two circuits, one for development sections, and the other for the longwall section.

For the longwall circuit, there are two power centers at the power train, instead of one, due to size constraints. The hydraulic power center, 5 MVA from 12,470 volts to 4160 volts and less, provides power to hydraulic pumps for the shields. The longwall power center has two transformers, one 8 MVA from 12,470 volts to 4160 volts and one 200 kVA from 4160 volts to 220/110 volts. The 8 MVA transformer provides power to operate the shearer, AFC drives, crusher, stage loader, etc. The 200 kVA transformer is for small consumers such as small pumps, chargers, lights, etc.

In the second example (Fig. 13.2.6), the incoming voltage from the utility company is 69,000 volts. At the substation, it branches into three circuits: underground, fan, and surface (Fig. 13.2.6 A). The surface circuit provides power for the shaft elevator, slope belt drive, and hoist. The fan circuit provides power to the fan. In the underground circuit, the transformers are rated at 10 MVA and convert the voltage from 69,000 to 12,470 volts. The output of the transformers is also branched into two circuits, longwall and continuous miner (CM) units, both of which are dropped underground through a borehole of 1000 ft (305 m) deep (Fig. 13.2.6 B). At the borehole bottom, each circuit has a circuit breaker.

The longwall circuit (Fig. 13.2.6 B) is routed to the active longwall panel, where two separate power centers at the headgate side, each 2500 KVA, convert 12,470/7200 volts to 480 volts. One power center is located at the entrance for the head drive of the panel belt conveyor, while the other power center is placed at some distance inside the panel for the booster drive of the panel belt conveyor. At the longwall section power train, there are also two power centers, one with 7000 kVA converts 12,470/7200 volts to 4160 volts for the shearer, AFC drives, etc., while the other with 4000 kVA converts 12,470/7200 volts to 995 and 480 volts for other equipment. There is a need to power up communication and tracking modules where power centers do not exist. This is accomplished by installing a small 1.5 KVA 12,470 volt to 120 volt single phase transformer in the high voltage circuit.

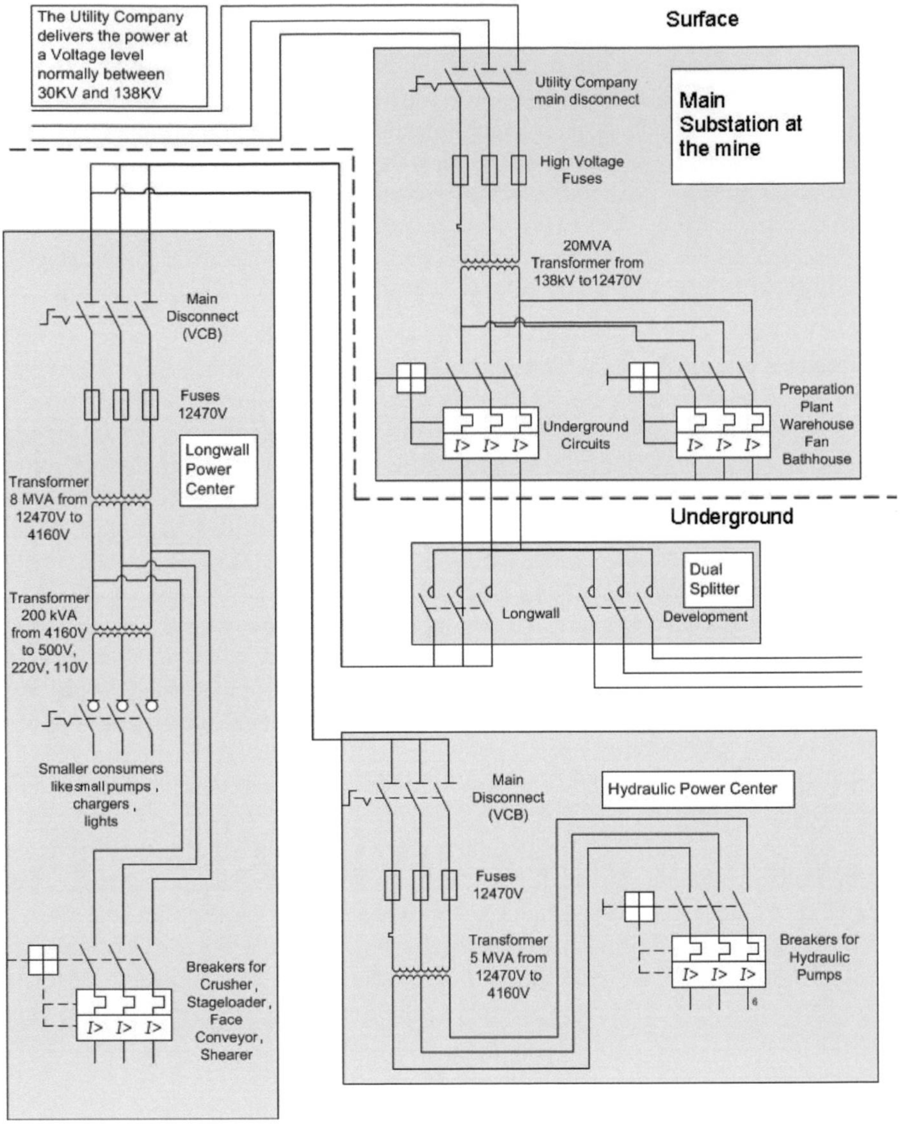

Figure 13.2.5 Power distribution for a longwall
Source: Allekotte (2006)

The development section circuit (Fig. 13.2.6 B) provides power for (1) a 1500 kVA power center used for rail and rubber-tired transportation. (2) two 2500 KVA power centers supplying power to VFD belt starters, and (3) a 1000 kVA power center at the tailgate side panel entrance that is used for charging and recovery. In each CM or gateroad development section, a dual circuit breaker is installed at the section entrance for separate

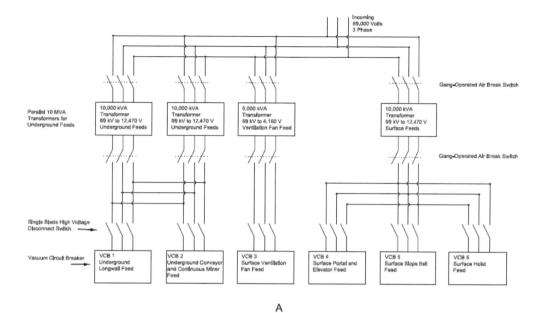

A

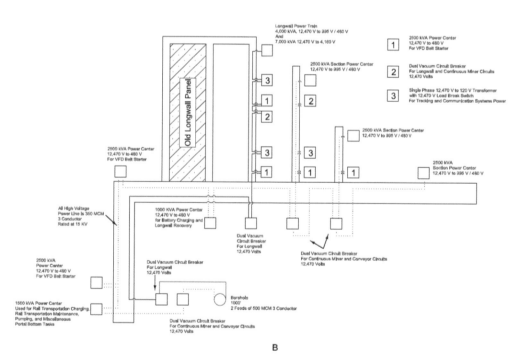

B

Figure 13.2.6 A typical power distribution layout for a mine with a single unit longwall: (A) surface substation; (B) underground power layout

Source: Courtesy of Murray Energy Corporation

control of the section belt conveyor and CM unit. At the section entrance, a 2500 kVA power center converts 12,470/7200 volts to 480 volts for the belt drive. At the active section, a dual circuit breaker provides flexibility for the power supply. The section power center is 2500 kVA and converts 12,740/7200 volts to 995 and 480 volts. For the mains development section, there is one power center that converts 12,470/7200 volts to 995 and 480 volts.

13.2.4 Longwall face lighting system

For a longwall section, MSHA regulations require that the illumination level at the coal face, exposed roof, and travelway on the longwall, five feet outby the headgate shields, and five feet in every direction from the main controller, must be two foot-candles on any four square feet (two by two feet or 0.61×0.61 m) region.

A system of cables and connectors deliver the intrinsically safe power from the section power center to the individual fixtures. Just like all other longwall face equipment, the longwall lighting system has evolved from standalone systems to become components integrated into the longwall mining system. Based on the historical evolution, there are three lighting systems (Huczko, 2006):

1   Standalone system – this is the traditional, self-contained system. All system components, including main power cables, power supplies, system power distribution cabling, and lights are dedicated to the lighting system.
2   Combined system – the main power cabling and the explosion-proof enclosure for lighting and shield control units (SCUs) are combined, resulting in a single main power cable on the longwall face. Each system relies on an independent power distribution cable for delivering power from the explosion-proof enclosure to individual subsystem components, i.e., light fixtures and SCUs.
3   Integrated system – in this system, the light fixtures derive intrinsically safe power directly from the SCUs, resulting in only one cable running from the section power center to the longwall face, one set of explosion proof power supplies, and one intrinsically safe cable from shield to shield. This system requires that the system components for both the lighting subsystem and SCU be highly compatible.

The integrated system is the most common one, while only a few longwalls employ the traditional standalone system.

The current light production technology is fluorescent tube light due to its high efficiency to convert electricity to light. This technology is expected to be replaced by solid state lighting in the near future.

13.3 Longwall system control

The longwall mining system consists of the following subsystems: the shearer, shields, AFC, stage loader, panel belt conveyor, and main belt conveyor. Each subsystem consists of one or more system components. For a longwall system to operate continuously, all subsystems and their components must be integrated or automated in order to reach the full potential of a continuous mining system. Today, most subsystems are either partially or fully automated. In order to be automated, subsystems and subsystem components must be able to communicate freely with each other. Communication consists mainly of data recognition and/or confirmation for actions to be taken.

Examples of data information available for system and subsystem integration include the following:

1   The shearer – subsystem components are mainly motors for ranging arms (drum cutting), traction, and hydraulics. These motors are monitored for temperature, current, and materials eroded. In addition to machine health diagnostics, this data is used to control the shearer. For instance, when the cutting force increases suddenly, the shearer will slow down automatically.

    To integrate the shearer with other subsystems such as shields, the shearer must provide its location and moving direction so that the MCU (master control unit installed at the headgate) can determine what shields can move.

2   Shields – major subsystem components include sensors for leg pressures and DA ram displacement. Measured leg pressures are used to ensure all shields are properly set against the roof. In this respect, the hydraulic pump station is another component that must ensure a timely supply of fluid at a sufficient quantity and pressure. The measured DA ram displacement is used to ensure that the shield is advanced fully to a pre-determined stroke so that both the panline and shield are straight and parallel to the faceline. The shield will receive data information, such as the shearer location, from the shearer in order to determine if it is time to advance.

3   AFC – major subsystem components include the Optidrive or CST or TTT drive and chain tension system. When starting the AFC, the Optidrive or CST or TTT drive ensures that all motors are soft-started in sequence; the tail drive pulls the bottom chain first, followed by the head drive pulling the top chain to ensure no slack in the chain, especially in the bottom strand at the headgate. For integration of transportation subsystems, the most outby belt conveyor section has to start first, followed by the next-to-most outby section, and so on until it comes to the stage loader, followed finally by the AFC.

In order to implement the aforementioned subsystem components and subsystem integration, communication links between subsystem components and subsystems must be established. Communication involves access to and/or providing monitored data to dedicated computers in subsystem components or subsystems.

Figure 13.3.1 shows a simplified communication system for a longwall. It runs on an ethernet network, very much like the commonly used e-mail network. Each subsystem has its own computers and can operate on its own even if the network is down. If the network is down, communication between subsystems is interrupted.

On the surface, the host system includes the server, network controller, and processor as well as monitors. There are also PCs, work stations, and laptops connected to the network.

Underground, the network extends to the longwall section, belt drive, CO monitoring, pumps, etc. In the longwall section, the shearer is commonly hardwired, and its data is transmitted through high tension couplers to the head gate control unit. The shearer can also communicate with the shields through a wireless system such as infrared transmitters/receivers. Another modernized option for shearer communication is through fiber optic connections. Most modern longwalls utilize the hard wire method vs. the fiber optic method. As face automation becomes predominant faster data transfer is required. Fiber optic–based

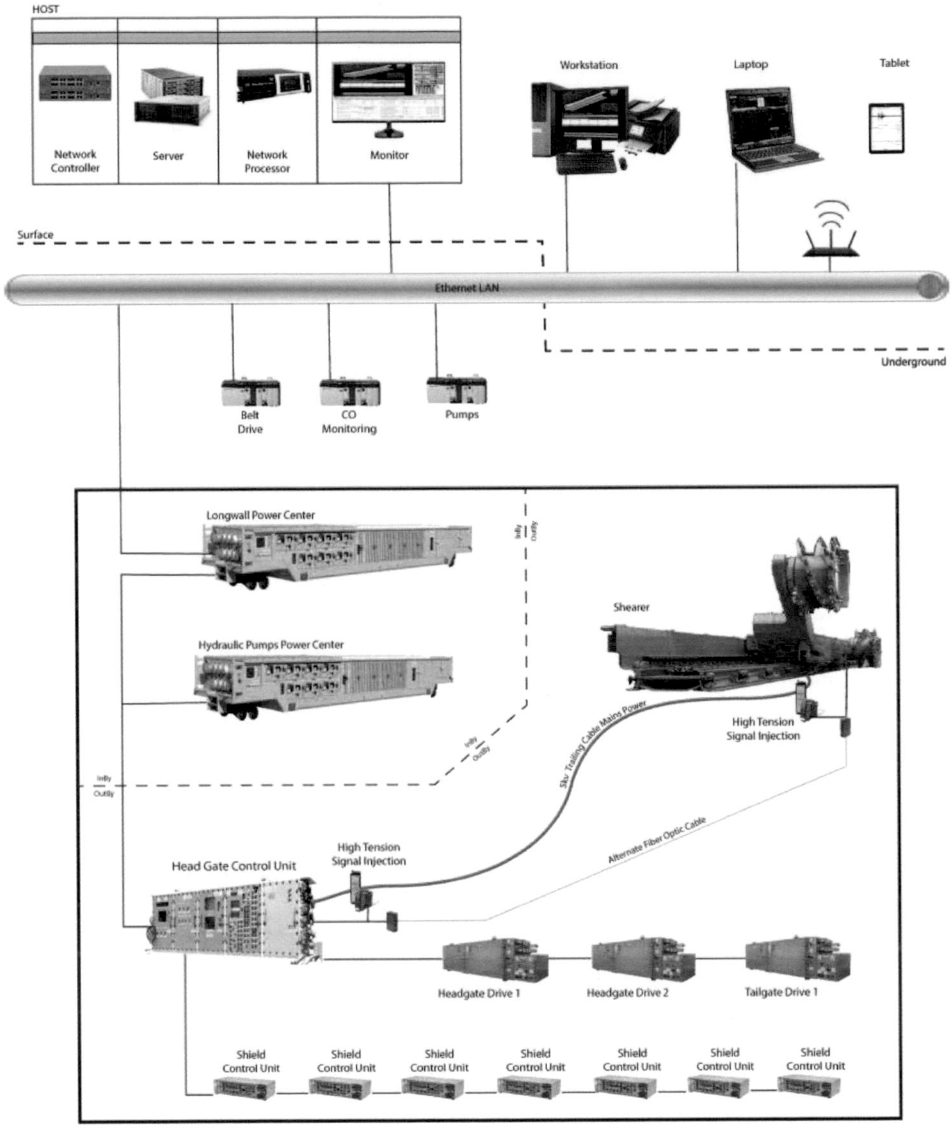

Figure 13.3.1 Simplified communication system for a longwall

Source: Courtesy of Murray Energy Corp.

communication is a better option for transferring mass blocks of data at a fast rate. However, considering the delicate nature of fiber optic cables it is not ideal for most mining environments, although fiber optic cables are being used in parts of overall communication links. The preferred method of shearer communication is the hardware method so long as the communication system can tolerate the slower nature of hardwired networks.

The AFC head drives, either Optidrive or CST or TTT, and all shields are linked to the headgate master control unit (MCU). Data information for all computers for systems and subsystem components are available at the server at the host site on the surface. The system is also designed so that the subsystem components can communicate directly on a need-to-have basis. For instance, each SCU can access data from all other SCUs without the need to go to the server to get the data.

To illustrate how communication works between subsystems and between subsystem components, let's look at the shield control system. Before a shield is allowed to advance, the SCU in that shield must check if all of the following conditions are met. If any one of these conditions fails, it cannot move (Allekotte, 2006):

1   Is it in automatic mode? To confirm this, the shield needs data from its own computer if it is set in automatic mode.
2   Does it have connection to the other shields and the master control unit? To confirm this, the shield needs to check if cables connected to all other shields and MCU are intact.
3   Does it have sufficient hydraulic pressure to set the shield after moving? To confirm this, the shield needs pressure data from the pump station system and from its own pressure sensors.
4   Is the hydraulic pressure in the return line lower than a given value? To confirm this, the shield needs data from the pump station system.
5   Are the emergency stops on? This includes two emergency stops in each SCU, one for adjacent shields and itself and the other is for the whole face.
6   Are all the cables to the valves intact without any short-circuit or broken wire?
7   Is the voltage from the power supply sufficient to switch the valves?
8   Is the shearer outside a restricted area of the shield? To confirm this, the shield needs to know where the shearer is.
9   Do the adjacent shields on both sides have adequate pressure? To confirm this, the shield needs access to data in adjacent shields.
10   Is the distance of the AFC greater than a preset value? To confirm this, the shield needs access to data from Reed Rod sensors for that shield.
11   Is the pre-warning circuit working? This is the beeping sound that gives warning for pending shield movement.

An important example of system communication is when the shearer operator pushes the "start" button, assuming the system is in automatic mode, the following sequence of communication among various subsystems is implemented to ensure maximum system efficiency and safety:

1   Is the panel belt system running? This is to check and ensure if the outby transportation subsystems are running. If "yes," then go to step 2.
2   Start the crusher. If the crusher is running without blockage, then go to step 3.
3   Warning "beeping" is turned on, followed by starting the stageloader.
4   Is the stage loader running? This is to check and ensure coal from the AFC will not be blocked at the cross frame. If "yes, then go to step 5.
5   Warning "beeping" is turned on before the AFC starts, and then go to step 6.
6   The AFC motors start without load.
7   Load is put on motors either by gradually providing pressure to the clutch (CST), or filling water to fluid coupling (TTT). At the same time, the chain tensioning system at the tailgate starts to control and prevent for a slack chain, especially at the head drive.

8   During operation, the following communication occurs continuously:

A   Is the load sharing among motors equal? During operation, the amperage of each motor is monitored constantly. If the amperage of any motor is higher or lower than the others, the system will adjust until all motors are carrying the same load. Under low load conditions. Some systems require, an artificial load sharing towards the tailgate to avoid slack chain at the headgate. Due to difference in wear condition in different sections of the chain assembly, its weight is different and, consequently, requires different motor power to operate. This is why the loads of all motors vary constantly. In addition, different motor speeds, even a few rpm change, induce different chain tension. Therefore, motor speed must be monitored constantly and adjusted to ensure uniform chain tension.

B   Is the total amount of coal on the AFC too much? If "yes," then a message is sent to the shearer to slow down.

C   Is the cutter motor torque too high? If "yes," then the shearer will reduce speed until the cutter motor torque is in an acceptable range.

D   The shearer can perform the following automatic cutting systems:

a   Memory cut – the shearer operators operate the initial cut and the shearer memorizes the cutting profile. In the second and subsequent cuts, the shearer will run itself, cutting the same profile.

b   Constant mining height cut – the shearer is programmed to cut a constant height with fixed horizon of leading and trailing drums.

c   Programmed cutting profiles.

Chapter 14

Surface subsidence

14.1 Introduction

Since longwall mining involves total extraction of a large block of coal, it disturbs the whole overburden from the seam level all the way to the surface, as described in Section 3.2 (Fig. 3.2.1). Unlike other issues, such as respirable dust and noise that affect miners underground, surface subsidence involves the general public. Therefore subsidence is not only a technical but also a public relations issue. When the public gets involved, surface subsidence provokes a lot of emotion and the mine company usually winds up in court, with an eventual legal solution to problems, while the technical side of the story is of secondary importance.

Surface subsidence changes surface topography. As a result it also affects surface water and groundwater flows in some areas.

Subsidence regulations began in the late 1950s when the state of Pennsylvania enacted the pillar supporting plan to protect surface structures (PA DEP, 1957). But it was not until after the promulgation of the 1977 Surface Mining Control and Reclamation Act, that surface subsidence and its remediation became a part of public law in the United States.

Accordingly, major surface subsidence research in the United States began in the late 1970s, and during the initial period, all subsidence theories were borrowed from those developed in Europe, especially the UK National Coal Board. As subsidence data accumulated, it was found that subsidence parameters for US coalfields differ from those in Europe. Subsequently, subsidence prediction models pertaining to US coalfields and structural damage mitigation techniques were developed in the late 1980s and early 1990s.

14.2 Characteristics of surface movement

14.2.1 Surface movement basin

During underground longwall mining, when the gob exceeds a certain size, strata movement reaches the surface and forms a low area of limited size above the gob. This is called the subsidence basin, or more exactly, the movement basin. When the seam is horizontal and the gob is rectangular in shape, the movement basin is approximately elliptical immediately above the gob (Fig. 14.2.1).

Defining the edge of the movement basin depends on the precision of the measurements. Currently, the contour line of 0.4 in. (10 mm) of subsidence is used to define the edge of the movement basin (Luo and Peng, 1997). The maximum subsidence (S_o), which is generally located at the center of the basin, increases with panel (opening) width (in two-dimension

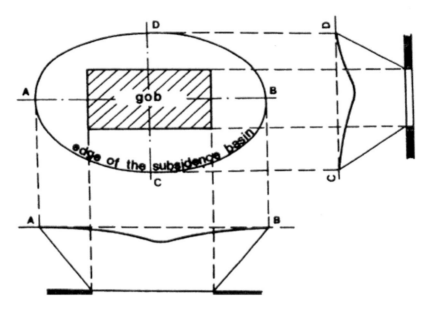

Figure 14.2.1 Surface subsidence basin

across the faceline) or gob dimensions (in three-dimension). When the panel width exceeds a critical value, the maximum subsidence reaches its maximum possible value (S_{max}) and begins to spread, rather than increases further with increasing panel width. The panel width at this time, when S_{max} starts to occur, is called the critical panel width. When the panel width and length of a rectangular mined out gob are less than the critical dimension, the subsidence basin formed is a subcritical panel width. Conversely, when the panel width and length of the mined-out gob are larger than the critical width, it forms a supercritical subsidence basin (Fig. 14.2.1), and the corresponding panel width is supercritical. According to Luo and Peng (1997) the critical panel width, W_c, is linearly related to cover depth, h, by

$$W_c = 100 + 1.048h \tag{14.2.1}$$

Surface movements and deformations consist of the following five components (Figs 14.2.2 and 14.2.3):

1   Subsidence, S – the vertical component of the surface movement vector is surface subsidence.
2   Displacement, U – the horizontal component of the surface movement vector is surface displacement.
3   Slope, i – the difference in surface subsidence between the two end points of a line section in the subsidence basin divided by the horizontal distance between the two end points is called the surface slope of the section.
4   Curvature, k – the difference in surface slope between two adjacent sections divided by the average length of the two sections is called surface curvature. All those that are convex are positive curvature while those that are concave are negative curvature.

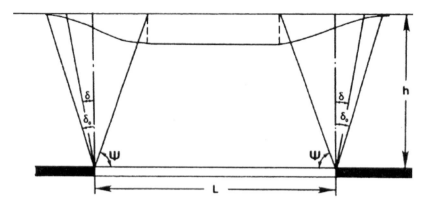

Figure 14.2.2 Terminology used for subsidence profile

5   Horizontal strain, ε – within the movement basin, the difference in horizontal displacement between any two points divided by the distance between these two points is called horizontal strain. If the distance between the two points is lengthening, it is tensile strain, which is positive. Conversely, if the distance is shortening, it is compressive strain, which is negative.

6   Twisting, T – this is the difference in slope between two parallel line sections separated by a horizontal distance of a unit length.

7   Shearing strain, γ – this is the difference in horizontal displacement along two parallel line sections separated by a horizontal distance of a unit length.

Within the movement basin, the areas where buildings or surface structures are subjected to damage are the disturbed zones. The remaining areas, in which there is deformation but no harm is done to buildings or surface structures, are the nondisturbed zones. The critical deformations for damage to various buildings and surface structures vary with the quality of the structural elements and types of structures.

The ratio of maximum possible subsidence, S_{max}, to mining height, H, is called the subsidence factor, a. The subsidence factor is a key parameter always used in subsidence prediction equations. Note that it is applicable to only cases involving critical or supercritical panel widths. For subcritical width, it is called apparent subsidence factor, or S_o/H. Subsidence factor is also related to cover depth, h, by:

$$a = 0.6815519 \times 0.9997398^{h} \tag{14.2.2}$$

On a major cross-section of the subsidence basin, which is the cross-section at the center of the subsidence basin either along the mining or face line direction, the point dividing the convex and concave portions of the subsidence profile is called the inflection point. At the inflection point, the surface slope is at its maximum and the curvature is zero. Generally, the inflection point lies above the gob edge and leans toward the gob area. The distance between the inflection point and the gob edge is the offset distance of the inflection point, d. It is also a function of cover depth, h, and can be determined by

$$d = 0.38123he^{-0.000699h} \tag{14.2.3}$$

When the gob has reached the critical size or nearly so, on the major cross-section of a movement basin, the angle between the vertical line at the panel edge and the line connecting the panel edge and the edge of the movement basin is the angle of draw, δ_o (Fig. 14.2.3). The maximum angle of draw is 24° with 95% of it less than 20°. The angle of draw is fairly constant when the cover is less than 800 ft (244 m) and then increases with cover depth, h, such that (Peng et al., 1995).

$$\delta_o = 6.87 - 0.0072h + 8.872 \times 10^{-6} h^2 \qquad (14.2.4)$$

The angle between the vertical line at the panel edge and the line connecting the panel edge and the point of critical deformation on the surface is the angle of critical deformation, δ (Fig. 14.2.2). Critical deformation is the amount of deformation above which it will cause structural damages. Critical deformation depends on the types of deformation and the types of surface structures. According to Peng et al. (1995), δ decreases with increasing cover depth, h, or

$$\delta = 27.96 - 0.02426h + 6.9 \times 10^{-6} h^2 \qquad (14.2.5)$$

When the gob has reached the critical size on the major cross-section of the subsidence basin, the acute angle between the coal seam roofline and the line connecting the panel edge and the edge of the flat bottom of the subsidence basin is the angle of full subsidence, ϕ (Fig. 14.2.2). It can be determined by (Peng et al., 1995):

$$\phi = 42.55 + 0.0417h - 2.16 \times 10^{-5} h^2 \qquad (14.2.6)$$

The angle of major influence, β, is defined as the angle between the horizontal line at the mining level and the line connecting the edge of the subsidence basin and the vertically projected point of the inflection point on the coal seam. It can be determined by:

$$\beta = 58.89 + 0.03089h - 1.84 \times 10^{-5} h^2 \qquad (14.2.7)$$

In critical and supercritical conditions, β can also be defined by the inflection point and the point of full subsidence.

When the gob has reached the critical size or nearly so, the elapsed time at the point of maximum possible subsidence from the initiation of surface movement to the final stabilized condition is the movement or subsidence period. The surface movement is considered to have initiated when subsidence reaches 0.4 in. (10 mm), and it is considered to have stabilized if the subsidence accumulated in six months does not exceed 1.2 in. (30 mm). The movement period is related to seam depth, rate of face advance, and overburden strata properties.

14.2.2 Surface movements and deformation during longwall mining

Under normal conditions, when the face has advanced for a distance approximately one-sixth to one-third of the seam depth from the setup room, depending on the seam depth and the physical properties of the overburden strata, the movement of the overburden strata reaches the surface, whereupon the surface also begins to deform. Thereafter, the surface point where

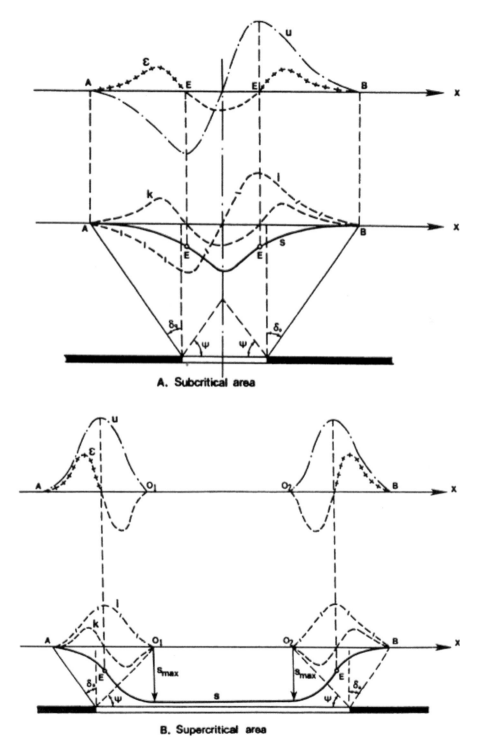

Figure 14.2.3 Surface movement and deformation distribution along major cross-section

the movement initiates is always located at a fixed distance ahead of the face. The angle between the vertical line at the faceline and the line connecting the movement initiation point on the surface and the faceline is the angle of advance influence, w. The angle of advance influence can be used to determine the distance, ℓ_1, which is the influence zone ahead of the face. ℓ_1 for the northern Appalachian coalfield can be determined by the following empirical equation:

$$\ell_1 = \frac{0.113h}{1+0.1825\sqrt{v}} \qquad (14.2.8)$$

Where v is the face advance rate in ft d^{-1}.

Surface movement as a result of longwall mining is categorized in two ways: along the faceline direction and along the mining direction.

14.2.2.1 Along the mining direction

The size of the surface subsidence basin increases with the increase in gob size. In sub-critical conditions, the maximum subsidence in the subsidence basin increases with face advance. But when it reaches the critical condition, the maximum subsidence reaches the maximum possible value and will not increase further even though the face continues to advance. Figure 14.2.4 shows the development of subsidence profiles along the face advancing direction (Wade and Conroy, 1980). The face reaches the critical condition between F and G, because before that, the maximum subsidence increases with face advance and after that the maximum subsidence does not increase any longer. It must be emphasized that all of these subsidence profiles represent the instantaneous profiles for the corresponding face positions. If the face stops, the subsidence will increase further except for those points that have reached maximum subsidence. The final subsidence profile is in most cases at least 10% larger than the instantaneous one.

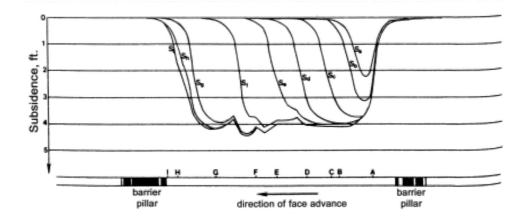

Figure 14.2.4 Development of surface subsidence profile as the face advances through the panel

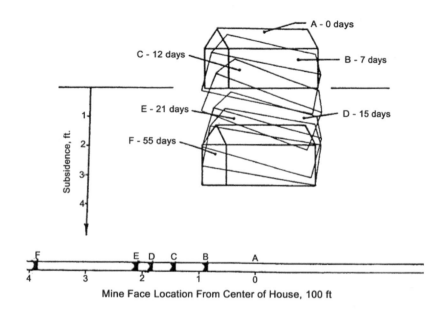

Figure 14.2.5 Sequence of movement of a house as the face approaches and passes under it

As subsidence develops following the face advance, structures on the surface tilt, rotates, and returns to the original pre-mining conditions except the elevation drops down by the amount of subsidence. Figure 14.2.5 shows the sequence of a house's behavior as the face advances.

Whether the subsidence for a surface point has fully developed or not depends largely on face location. For instance, as shown in Figure 14.2.6 (Peng and Cheng, 1981), when the face moves to a horizontal distance of approximately 0.8 times the seam depth, the surface at point p starts to subside. As the face moves on, the subsidence gradually increases. When the face is directly under p, the subsidence reaches only 6–9% of the total amount of subsidence. The subsidence at p accelerates after the face has passed it. The maximum amount of subsidence is reached, and the subsidence completed, when the face has passed the point p for a distance between 1.2 and 1.7 times the seam depth.

Figure 14.2.6 also shows that subsidence development curves vary with seam depth. There is no significant difference before the face reaches a point directly under the point of interest, p, on the surface. But when the face has passed it, surface subsidence develops much faster and, thus, finishes sooner for a seam depth at 800 ft (244 m) than that at 400 ft (122 m). In other words, a shallower seam requires a relatively shorter distance of face advance to reach the maximum possible subsidence than a deeper one.

There are cases in which the surface ahead of the faceline heaves. It may or may not disappear when the faceline has passed under it. Some of them permanently remain in the final subsidence profiles.

14.2.2.2 Along the faceline direction

The progression of surface movements and deformations on the major cross-section perpendicular to the face advance direction (Fig. 14.2.7) differs from those parallel to the face

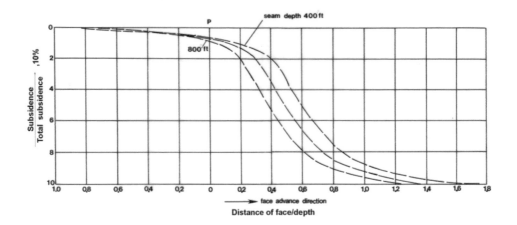

Figure 14.2.6 Subsidence development curve

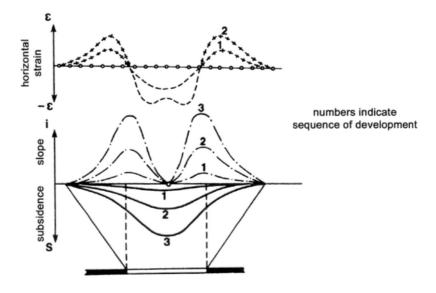

Figure 14.2.7 Variation of surface movements and deformations along the faceline direction

advance direction. On the cross-section parallel to the faceline, surface movement and deformation increase gradually from small to the maximum value.

Finally, it must be noted that after the surface movement has reached the stage of full development, although there are no deformations at the flat bottom of the subsidence basin, each point there has been subjected to the sequence of progression as described previously.

14.2.3 Surface subsidence velocity

The subsidence velocity differs for each point on the major cross-section of the subsidence basin. When the surface movement has not been fully developed, subsidence velocity at each point and the maximum subsidence velocity increase with an increase in the gob size. When the face advances to A or B (Fig. 14.2.8), the corresponding subsidence velocity curves are V_a or V_b, respectively. When the gob has developed to critical size and the surface movement is fully developed (face at C), the subsidence velocity for each surface point reaches a fixed maximum value under the prevailing conditions. Thereafter, subsidence velocity curves are basically the same regardless of face location. The relative position of the face and the shape of the subsidence velocity curves does not change at this time. Similarly, the point where the maximum subsidence velocity occurs falls behind the face at a fixed distance and remains so. The acute angle between the seam roofline and the line connecting the point of maximum subsidence velocity and the face line is the angle of delay, φ. The angle of delay can be used to determine the zone of most active movement during face advance.

When the face stops, the subsidence velocity for each surface point does not increase. Instead, it decreases gradually until stabilized.

The subsidence velocity reflects the movement intensity of the overburden strata above the gob area. For flat seams, when the subsidence velocity at a surface point reaches the maximum value, it indicates that the caved and the fractured zones underneath that point have reached their maximum heights.

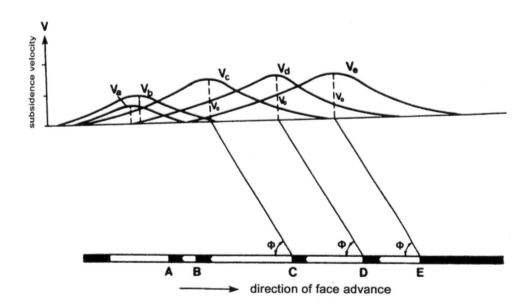

Figure 14.2.8 Variation of surface subsidence velocity curves as the face advances

14.3 Effects of geological and mining factors

Surface and overburden strata movements are the combined results of mining activities and geological conditions. These factors are illustrated here:

1   Properties of the overburden strata – the physical properties of the overburden strata are some of the most important factors that control the magnitude of maximum subsidence. When the strata are strong and hard, the maximum subsidence is small. Conversely when the strata are weak and soft, the maximum subsidence is large.

The shape of the subsidence profile is related to rock properties and determined by the location of the inflection point, which, in turn, varies with rock properties. When the strata are stronger, they overhang farther into the gob. As a result, the inflection point moves farther into the gob.

2   Seam inclination – seam inclination is a very important factor that controls the characteristics of strata movement and failure. In flat and slightly inclined seams, the direction of movement is mainly perpendicular to the beddings. But, in inclined and steeply inclined seams, the direction of movement is the resultant of two components of movements, one is parallel and the other is perpendicular to the beddings. Also, caved rock fragments tend to slide down along the floor.

3   Mining depth and mining height – as mining depth increases, the area where the induced surface movements occur expands. Since the maximum subsidence in supercritical and critical panels is not related to mining depth, the surface movement basin becomes milder, and all types of deformations decrease as the mining depth increases. Thus, every other thing being equal, surface deformation is inversely proportional to mining depth.

When mining height increases, surface deformation also increases. This is why the mining-depth-to-mining-height ratio is so commonly used for determining the effects of mining conditions on surface movement. Obviously, when the depth-to-height ratio gets larger, surface deformations become smaller and milder. When the depth-to-height ratio is very small, large surface cracks, steps, and even sink holes could occur on the surface.

Mining depth considerably affects the velocity and period of surface movement. Surface movements last longer when mining depth is larger. Maximum surface subsidence velocity is inversely proportional to mining depth. When the mine is very deep, surface subsidence velocity is small and surface movement is more uniform, but it is slower and lasts longer.

4   Gob size – before surface movement reaches the stage of full development, surface subsidence increases with an increase in gob size. The required gob size for full development of surface movement relies very much on the overburden rock properties. In order to evaluate the degree of development of surface movements, the following two development coefficients are used:

$$n_1 = C\frac{L_1}{h} \text{ and } n_2 = C\frac{L_2}{h} \qquad (14.3.1)$$

Where C is a rock property influence coefficient. It is larger for hard rocks. L_1 and L_2 are the length and width of the gob, respectively, and h is the seam or mining depth. When

$n_1 < 1$ or $n_2 < 1$, the gob is subcritical in size, and the surface movement has not reached the full development stage. When both n_1 and n_2 are larger than one, the gob is critical or supercritical in size, and the surface movement has reached the full development stage.

5 Multiple-panel mining – traditionally, mines are laid out so that there are generally three or more panels parallel to each other but separated by panel gateroads, which range from 120 to 300 ft (36.6–91.4 m) wide. The surface subsidence caused by one panel tends to overlap those induced by the adjacent panels if the mining height is large.

Theoretically, subsidence profiles caused by two or more panels of mining in the same neighborhood are similar to each other, and a final subsidence profile from multiple-panel mining can be obtained by simple superposition of the subsidence profile created by single-panel mining. In practice, however, the true final subsidence may deviate considerably from that predicted by simple superposition. Figure 14.3.1 shows the development of the final subsidence profile as a result of mining two adjacent panels (Peng and Cheng, 1981). The individual subsidence profiles for panel one and two are not similar, which is reflected in the final subsidence profile. The additional subsidence increases with the ratio of the depth to width of the chain pillar system.

6 Faults and other planes of weakness – since rock in and around the faults is generally much weaker than the surrounding area, shear displacements along the fault planes are likely to occur. Therefore, at the fault outcrops, strata deformation, such as cracks or even steps are concentrated.

The intensity with which a fault controls the strata or surface movement depends on fault inclination, fault size, fault strength, and its relative position with respect to the gob.

7 Topography – if there are steep slopes within the surface movement basin, its stability will be affected once the bedrock begins to move. As a result, landslides could occur on the steep slopes. Generally speaking, the upper portion of the steep slope tends to slide

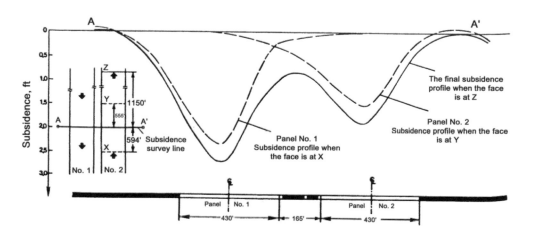

Figure 14.3.1 Development of surface subsidence profile in multiple-panel mining

along the planes of weakness. If the strata beddings coincide with the steep slopes, the slopes will be more likely to slide along strata beddings.

Gentry (1979) found that subsidence was greater at topographic highs and less in areas of topographic lows due to a piling up of overburden in the low areas. Conroy and Gyarmaty (1982) found, however, that horizontal displacement at the panel center moved toward the panel edges and coincided with the down dip of the surface slopes which ranged from 30 to 180 percent. But mild slope ($\leq 20\%$) has little effect on surface subsidence (Peng and Cheng, 1981).

14.4 Subsidence prediction, assessment, and mitigation

14.4.1 Introduction

From mid 1980s to early 2000s, hundreds of case studies, mainly in the Pittsburgh seam of the northern Appalachian coalfield (Luo and Peng, 1997; Luo *et al.*, 2003) had led to the development of a systematic approach in dealing with surface and subsurface structural damages due to underground longwall mining. These case studies cover all categories of surface and subsurface structures, including small to very large houses of all types of construction (wood, concrete block, brick, stone, and log); pipelines of all types (steel, plastic, concrete, and cast iron) and sizes with and without internal pressure for transmission of gas, oil, and water; precision cell/telephone transmission towers; high voltage power transmission towers; municipal water towers; farm crop storage silos; railroads and bridges; highways (county, state and interstate) and bridges (wood and concrete) of all types; creeks/ponds/lakes/tailing impoundments; overland belt conveyors; refuse disposal facilities; underground gas storage chambers; cemeteries; and others. The approach is simple, fairly inexpensive, and has proven to be effective in all cases. It involves the following three basic steps:

1 Subsidence prediction – final and dynamic surface movements and deformations in the area of the structures are predicted.
2 Influence assessments – the predicted final and dynamic deformations are then compared to the critical deformations tolerable by the structures without significantly degrading functionality, stability, and structural integrity. The material design and construction are also considered in the assessment. The locations and severity of the anticipated problems are determined.
3 Design and implementation of mitigation measures – mitigation measures designed and implemented for surface structures are based on the following principles: (1) reduce surface deformation by better mine design and mining efforts, (2) reduce the deformations transmitted to the structures, and (3) reinforce the structures to make them more tolerable to the deformations.

14.4.2 Assessment of subsidence influences

The main tasks in assessing subsidence influences include the following: (1) to quantify the disturbance potential to a surface structure associated with a mining subsidence event, (2) to estimate the locations and intensity of the structural disturbance, and (3) to identify the deformation-generating ground. The assessment involves the three steps, which are discussed in the following subsections.

14.4.2.1 Pre-mining site visit and information collection

The first step is to collect geological and mining information for subsidence prediction and to gather the details of the structures for influence assessment. The required geological and mining information normally includes the following:

- Contour maps of surface topography and the bottom elevation of the coal seam.
- Stratigraphy of the overburden strata with emphasis on hard rock layers.
- Layout of the longwall panel and its spatial relation to the structures, and current mining practices including the average face advance rate.
- During the site visit, more detailed information about the localized changes in surface topography, groundwater, drainage, depth of soil zone, and the underlying rock in the area of the structures are reviewed. For the structures, the following information is collected: material, construction, and layout of the foundation and super-structures are examined. The depth and construction of the foundation/basement should always be investigated. Structural dimensions are measured. Number and locations of chimneys are noted. Strengths and weaknesses (e.g., doors and windows, joint lines, pre-existing cracks) of the structures are recorded.
- Space consideration for candidate mitigation measures, such as trees, bushes, flower gardens, driveways, walkways, retention walls, etc., located around the structures.

14.4.2.2 Subsidence prediction

Dynamic and final surface movements (i.e., subsidence and horizontal displacement) and deformations (i.e., slope, strain, and curvature) in the location of the structures can be predicted using the Comprehensive and Integrated Subsidence Prediction Model, CISPM (Peng and Luo, 1992). The prediction model is based on the influence function method, as well as findings from extensive subsidence research and field monitoring. CISPM contains six modular programs for various tasks in dealing with mining subsidence. Among them, programs such as LWSUB and DYNSUB can be used to predict the final and dynamic subsidence associated with longwall operations. The theories leading to the development of these prediction programs have been detailed elsewhere (Luo, 1989; Luo and Peng, 1990, 1991, 1992; Peng and Luo, 1992). Among many of the complicated mathematical expressions, two basic but important ones are briefly described in this section. The first one is the expression for final subsidence at a surface point caused by mining a rectangular longwall panel, S_f.

$$S_f(x,y) = \frac{ma}{R_2} \int_{d_1-x}^{L-d_2-x} e^{-\pi\left(\frac{x'}{R}\right)^2} dx' \int_{d_3-y}^{W-d_4-y} e^{-\pi\left(\frac{y'}{R}\right)^2} dy' \qquad (14.4.1)$$

Where
   x, y – Coordinates of the prediction point with respect to the lower left corner of the longwall panel
    m – Mining height
    a – Subsidence factor
    R – Radius of major influence
    d_n – Offsets of inflection points along the four edges of the longwall panel, $n = 1, 2, 3,$ and 4
  L, W – Length and width of the longwall panel

A method has also been developed for predicting the final subsidence induced by the convergence of the chain pillar system between two adjacent longwall panels so that the effects of mining in multiple longwall panels can be considered (Luo and Peng, 1990). When a structure is located on or near a steep slope (>15°), additional surface movements could be induced by the interaction between surface topography and mining subsidence. An integrated approach has been proposed for predicting final subsidence and potential for landslide in hilly terrains (Luo and Peng, 1999).

The dynamic subsidence process associated with a longwall mining operation is much more complicated than the final subsidence. The complete dynamic subsidence development process mainly consists of the following three phases:

Subsidence initiation phase – this phase occurs at the beginning of mining a longwall panel. In this phase, there is no or very little surface subsidence until the longwall face has reached a critical distance away from the panel set-up room. Once the critical distance is reached, a rapid subsidence process follows. The entire phase takes place when the longwall face is still within 1.5 to 2.0 times the overburden depth away from the panel setup room.

Normal subsidence process – this phase occurs when the longwall face is moving in the remaining length of the longwall panel with a fairly constant advance rate. The characteristics of the normal subsidence phase are that the moving front of the subsidence basin travels at the same pace with the longwall face and its shape remains basically unchanged. In this phase, the development of subsidence velocity at a surface point can be represented by a bell-shaped mathematical distribution function. Based on this assumption, the derived mathematical expression for subsidence at a surface point at a given time on the dynamic subsidence front, S_d, is shown as:

$$S_d(x_d, y) = \frac{S_f(x, y)}{2} + \sqrt{\frac{2}{\pi}} \frac{S_f(x, y)}{l + l_1} \int_{x_d}^{-l} e^{-2\left(\frac{x'+l}{l+l_1}\right)^2} dx' \tag{14.4.2}$$

Where

x_d – distance between the prediction point and the longwall face (+ ahead and – behind the face)

l – offset of subsidence velocity peak occurring behind the longwall face

l_1 – offset of subsidence initiation point ahead of the longwall face where ground surface begins to subside

Residual subsidence phase – this phase, a decaying process, takes place after the longwall face has stopped advancing.

In Equations (14.4.1) and (14.4.2), a, R, and d_n are called final subsidence parameters while l and l_1 are the dynamic subsidence parameters. Upon the establishment of these mathematical models, the accuracy of the subsidence predictions depends on the final and dynamic subsidence parameters. These parameters are available through a large number of collected longwall subsidence cases (Peng et al., 1995).

In each subsidence study, final surface movements and deformations are predicted in the area of the structure. For those cases where the structure is located over the "central" portion of the longwall panels, dynamic surface movements and deformations have to be predicted. For the structures located near a steep slope, the effects of surface topography have to be assessed.

Table 14.4.1 Critical deformations for residential structures

Deformations	Critical Value	Effects
Strain	2×10^{-3} ft/ft	Cause integrity problems to structural parts in direct
	3×10^{-3} ft/ft	contact with ground (basement and foundation)
Curvature	6×10^{-5} 1/ft	Cause integrity problems to super-structures
Slope	1%	Make structures uncomfortable to live

14.4.2.3 Assessing subsidence influences

Surface deformations (i.e., slope, strain, and curvature), if sufficient, will affect the integrity and functionality of common residential structures. Among the deformations, final and dynamic surface strain, particularly the tensile strain, is frequently responsible for most problems observed on residential structures. For common residential structures, if one of the subsidence-induced surface deformations exceeds the respective critical deformation for the type of structure, there is a high probability that the subsidence will cause problems to the structure. Critical deformations have been deducted from the results of an extensive program monitoring the responses to the subsidence process of various surface structures. Table 14.4.1 lists a number of the critical deformations for common residential structures.

The materials (i.e., wood, brick, concrete blocks, or a combination of them), layout (rectangular vs. more complicated shapes), construction methods, as well as the orientation of the house, length, and direction to the mining/panel edge are used as adjustment factors in the influence assessment. Special attention is paid to the weak spots (e.g., large doors and windows, pre-existing cracks) of the structure in assessing the locations and severity of the anticipated structural problems.

Based on the predicted dynamic and final subsidence, the locations from which surface deformations, particularly the strain, originate are identified. If the natural slope in and near the area of the structure is steep and the ground poorly drained, the potential for these conditions to trigger additional ground movements and deformations should be assessed. Special attention is placed on structures located near the toe of a long steep hill.

For most residential and farm structures, longwall subsidence is unlikely to cause stability problems. However, if tall structures such as free-standing chimneys or farm crop storage silos are undermined, the structural stability under the influence of final and dynamic slope should be assessed. In these cases, analysis should be performed on both the short-term and the long-term stability with and without considering lateral wind force.

14.4.3 Mitigation measures

If the assessment indicates that subsidence influence may cause integrity problems to the structures, proper mitigation measures are recommended to reduce the severity of the disturbance. The following three basic approaches can be applied to design the mitigation measures:

1   Reduction of surface deformations by better mine design and mining operations. For example, placing important structures away from major influence zones of the final subsidence basin (beyond the panel edges or in the flat bottom of a super-critical subsidence basin) or maintaining a faster, non-stop longwall face movement within a critical

distance. Through such measures, the maximum final and dynamic surface deformations in the area of the structures can be reduced.

2   Reduction of the transmission of surface deformation from the ground to structures. This can be done by absorbing the surface deformations with artificial weak planes or by cutting off or weakening the means of deformation transmission. Through such reduction, the structures will experience smaller impact.

3   Structure reinforcement. Reinforced structures can tolerate larger deformations. The reinforcement can also compensate the final or dynamic tensile strain that is the most disturbing to common surface structures.

Based on these approaches, many mitigation measures have been developed and tested. Among these mitigation measures, the most commonly applied are shown in Table 14.4.2.

1   Compensation trench method

A properly designed and constructed compensation trench creates a weak plane between the structure and the ground so that a reduction of the transmission of surface strain from the ground to the structure can be realized. This method is used to reduce final and dynamic strains sustained by structural parts having direct contact with the ground, including the lower part of the structure. Numerous applications have shown that this method is more effective in reducing compressive rather than tensile strain.

2   Tension cable method

In this method, the structure is wrapped with pre-tensioned steel wire cables. When properly designed and implemented, the tension in the steel cables helps do the following: (1) reinforce the structure so that it can tolerate larger deformations, (2) create a compressive stress field in the structure for compensating the tensile stress in the structure induced by final and/or dynamic surface tensile strain, and (3) offset some of the moments created by surface curvature so that the super-structure can be effectively protected. The tension cable method is good for stone, brick, and concrete block structures. The tension rope method is similar to tension cable method for wood structures.

Table 14.4.2 Common mitigation measures for residential structures

Method	Purpose	Used for
Compensation trench	Reduce strain transmission	Structural parts in direct contact with the ground
Slotting	Absorb deformations	Hard floor pavements
Plane fitting	Reduce transmission of deformations upwards	Wood or similar super-structures
Tension cable	Reinforce structures, compensate tensile stress	Brick, stone, and block structures
Tension rope	Reinforce structures, compensate tensile stress	Wood structures
Internal and external bracing	Reinforces structures at weak spots	Irregularly shaped structures, including doors and windows

3   Plane-fitting method

This method is similar to the leveling method in preventing deformations in the foundation/basement walls from propagating upward to the super-structures. However, it differs by maintaining the super-structure on a time-dependent, inclined plane, using height-adjustable devices placed under the super-structure rather than a level plane. By doing so, the required amount of separation between the super-structure and the foundation/basement to keep the super-structure on a plane can be greatly reduced. This method is good for residential houses with wood frame super-structures. When the overburden is small (e.g., < 300 ft or 91.5 m) and the subsidence-induced curvature is large, the plane-fitting method is most effective for reducing problems on super-structures. This method requires an extensive subsidence monitoring program, often one survey a day, during the active subsidence period. The measured subsidence around the structure beside its exterior walls is used to determine the proper plane and the required adjustments on the height-adjustable devices.

4   Internal bracing method

This method is normally used with the tension cable method to reinforce the weak spots of the structure, such as large-sized doors and windows.

In order to ensure success, the implementation of the mitigation measures should be carried out by trained personnel. Careful coordination with the longwall operation is required to maintain fast and non-stop face advance when the longwall face is moving between 0.1 h inby and 0.9 h outby the structure to be mitigated (h is overburden depth). The performance of the mitigation measures should be monitored during the active part of the subsidence process. If the structure is located in the major influence zone of the final subsidence basin where permanent deformations still exist after the subsidence process, extra care should be exercised to remove the mitigation devices and to reclaim the site so that the residual deformations on the structure can be further reduced.

14.4.4 A mitigation example

Since 1994, two longwall mines in the Pittsburgh seam had mined under or near more than 160 residential structures. Subsidence influence assessment has been performed for about 120 residences, some with multiple structures. Mitigation measures had been implemented on about 80 residential houses. Many light, small, and simple structures, such as mobile homes and frame houses, were not mitigated due to insignificant influences.

In this section, an example of protecting an old residential house is given. Figure 14.4.1 shows the longwall panel located under the house and three farm structures on the property.

The panel was about 1075 ft (327.7 m) wide and mining height was 6.5 ft (1.98 m). All the structures were located along the ridge of a hill. The slope on the right side of the house was about 19°. The overburden depth at the house was about 818 ft (249.4 m). The house was located about 496 ft (151.2 m) away from the panel headgate.

The two-story stone house (Fig. 14.4.2) was built in 1843, one of the oldest in the area. The plan view of the house was U-shaped. The front part and left wing of the house (the original part) were constructed with cut stones (Fig. 14.4.3). A brick addition was attached to the house to form the right wing. The joint line between the original part and the addition was one of the weak spots of the structure. The house was about 49 ft (14.94 m) wide and

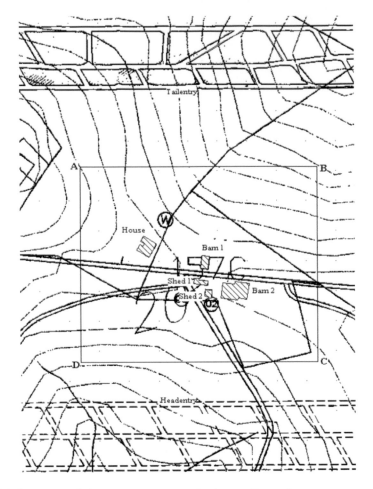

Figure 14.4.1 Location of the structures over the longwall panel

Figure 14.4.2 Front view of the house

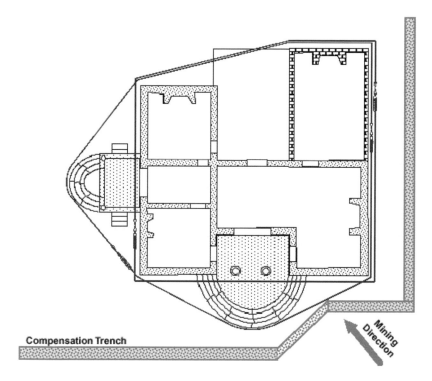

Compensation Trench

Mining Direction

Figure 14.4.3 Recommended mitigation measures for the house

48 ft (14.63 m) deep. Four chimneys were located at different ends of the house as shown in Fig. 14.4.3. Two semi-circular stone porches were attached to the front and left sides, respectively. A field stone paved back yard was sandwiched between the left and right wings of the house. A full basement was located under the front part of the house. The exterior basement walls were of block stone type, and their interior parts were built with field stones.

14.4.4.1 Predicted final subsidence and influences

CISPM was used to predict the final subsidence. The predicted final surface subsidence in a rectangular area around the structures is plotted in Figure 14.4.4. It shows that the house was located in the narrow flat bottom portion of the final subsidence basin to be formed over the longwall panel. The maximum subsidence at the house was about 3.75 ft (1.14 m) without any significant differential subsidence across the house after the subsidence process was over.

The predicted final surface strain (the component along the panel's transverse direction) is plotted in Figure 14.4.5. The negative values indicate that the ground surface would experience compressive strain in the area of the structures. The final surface strain at the house range from zero, under most parts of the house, to -4.0×10^{-4} ft/ft at the front left corner. The predicted final compressive strains at these structures are all insignificant. The predicted final surface horizontal displacement, slope and curvature at the location of the house were all found to be insignificant.

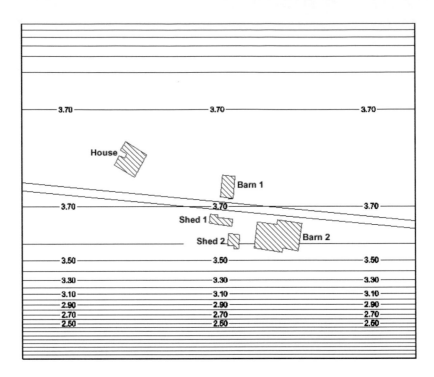

Figure 14.4.4 Predicted final subsidence in the area of the structures

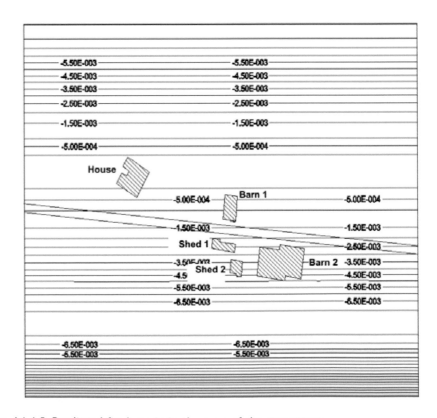

Figure 14.4.5 Predicted final strain in the area of the structures

14.4.4.2 Predicted dynamic subsidence and influences

Although the house would not experience any significant final surface deformations because of its location with respect to the panel edges, it still has to go through a complicated dynamic subsidence process. Using CISPM, the dynamic subsidence has been predicted at the center point of the house. Five face advance rates ranging from 60 to 100 ft d^{-1} (18.29–30.48 m d^{-1}) are achievable by the longwalls. The dynamic surface movements and deformations presented are their components along the panel longitudinal direction.

The predicted development curves for surface dynamic subsidence at the selected surface point (the center of the house) are shown in Figure 14.4.6. There would be very little subsidence before the longwall face reached directly under the surface point. The subsidence process would accelerate afterwards and reach about one half of its final subsidence when the longwall face was about 320 to 350 ft (97.53–106.7 m) outby the surface point. At this time, the subsidence process is most active and the subsidence velocity is the largest. A decelerating process would follow and the ground surface would regain its stable condition when the face is about 830 ft (253 m) outby the surface point of interest.

The predicted development curves of surface dynamic slope (the longitudinal component) at the house are plotted in Figure 14.4.7. The maximum dynamic slope ranging from 0.64 percent to 0.77% would occur between 310 and 340 ft (94.5 and 103.6 m) behind the longwall face. The ground surface at the location of the house would regain its level condition after the longwall face is 900 ft (274.3 m) passed the house.

14.4.4.3 Assessment of subsidence influences on the house

The final subsidence predicted by CISPM showed that the house was located on the flat bottom portion of the final subsidence basin. Although surface elevations at the house would be lowered by an amount of about 3.75 ft (1.14 m), the final surface deformations would be too

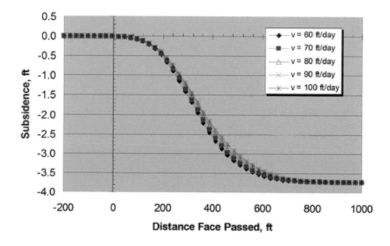

Figure 14.4.6 Dynamic subsidence development curves at the house

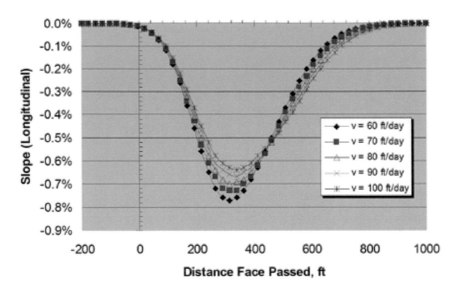

Figure 14.4.7 Dynamic slope development curves at the house

small to affect the stability and integrity of the house. Therefore, after the subsidence process was over, the house would not be subjected to any significant permanent deformations, and its structural integrity and functionality should not be affected.

The dynamic subsidence prediction indicated that only the dynamic tensile strain in the first half of the dynamic subsidence process would possibly cause structural integrity problems to the house. However, such influence would occur during a time interval when the face was between 90 ft (27.44 m) past the house's front right corner and 270 ft (82.3 m) past its rear left corner as shown in Figure 14.4.8. The dynamic tensile strain could induce cracks in the foundations and lower part of the house walls along the left and right sides if it was not properly protected. Probable locations for these cracks would be near the windows, doors, and pre-existing lines of structural weakness. If not properly protected, the total width of the cracks along either the left or the right house wall could be up to two inches (50.8 mm) wide.

The maximum dynamic slope is still within the critical limit listed in Table 14.4.1. Therefore, the living condition (functionality) of the house would not be affected during the active subsidence period.

14.4.4.4 Mitigation measures

As indicated in the influence assessment, only the dynamic tensile strain would cause significant structural integrity problems to the house. The house would experience no permanent surface deformations after the subsidence process was over. Therefore a number of mitigation measures were recommended to reduce the severity of the anticipated problems caused by the dynamic strain.

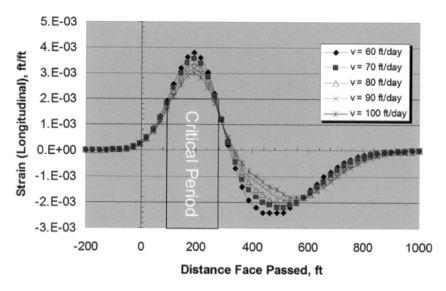

Figure 14.4.8 Dynamic strain development curves at the house

A. COMPENSATION TRENCH METHOD

An L-shaped compensation trench was dug on the front and right sides of the house (Fig. 14.4.3). The trench was intended to reduce the severity of the anticipated problems in the foundation, the basement walls and the lower portion of the house – the structural parts having direct contact with the ground surface. Through such a reduction of strain transmission, the strain experienced and the problems expected on the lower part of the house would be reduced.

The trench was kept well drained by burying a perforated pipe at the bottom with gravel and filling its remaining space with loose hay. The trench was securely covered with wood boards and fenced during its service.

B. TENSION CABLE METHOD

In this method, the house was wrapped with pre-tensioned steel wire cables. The tension cable method is suitable for protecting those structural parts that have higher compressive strength than tensile strength, such as stone, block, and brick structures. Steel wire cables in good working condition were used as the tension cable. The required tensions in the wire cables varied from 2.5 tons to 3.5 tons (2.3 to 3.2 mt) from the top cable to the bottom. Force distributors built with wooden boards were inserted between the tension cables and the house corners so that the force applied by the tension cables would be evenly distributed over a large area. Two tension/spring devices were inserted on each of the three cables at the locations shown in Figure 14.4.3. The cable tensions were periodically monitored during active subsidence and adjusted if necessary. The cables and the tension/spring devices were covered or fenced in traffic areas.

Wood bracings were placed between the indented portion of the front stone porch and in the back courtyard between the left and right wings of the house as shown in Figures 14.4.3 and 14.4.9.

The longwall face was under the house and continuously mined for 10 days without stop as the house experienced the active part of the subsidence process. The subsidence process ended one and a half months later. The tension cables were then removed, and the trench was refilled about two months after undermining. The mitigation efforts were successful, without any noticeable damage observed on the house.

14.5 Surface and groundwater effects

As stated in Section 3.2, longwall mining disturbs the overburden from the coal seam level to the surface and changes surface topography. Therefore, depending on the stratigraphy and the location with respect to longwall panels, surface and underground water bodies may or may not be affected.

In the northern Appalachian coalfield where modern longwall mining began and has been practiced extensively during the past 60 years, numerous surface and underground water bodies, as well as surface structures, have been undermined. Extensive data are available for analysis, although most are proprietary. Data presented in this section pertain to the northern Appalachian coalfield, but the results are applicable to other regions with similar geological and topographical settings.

Figure 14.4.9 Bracing placed between left and right wings of the house

14.5.1 General concepts of longwall mining and its effect on groundwater

As stated in Section 3.2 (p. 45), after longwall mining, the overburden is disturbed to different degrees. Based on the location from the coal seam, there are four zones in ascending order: caved zone with its thickness approximately two to eight times mining height; fractured zone with a thickness approximately 30 to 60 times mining height; continuous deformation zone covering the remainder of the overburden except the soil zone; and the soil zone, which is approximately 30 to 50 ft (9.1–15.2 m) deep from the surface, though in some areas it can be up to 150 ft (45.7 m) deep, has been identified.

If surface water wells reach or aquifers are located in the caved and fractured zones, well water or groundwater in the aquifers will be lost permanently. On the other hand, if they are located in the continuous deformation zone, longwall mining should not affect well water or groundwater. And, if it does, it will be temporary. In the soil zone, as mining progresses, cracks open and close, and water levels will decrease and return. However, cracks of various depths parallel to the panel edges are expected to remain permanently, and if surface wells are located in this region, then water is likely to disappear permanently.

Since mining height in the Pittsburgh seam ranges from 6.5 to 8 ft (2.0–2.4 m), the fractured zone height ranges from 325 to 400 ft (99.1–122 m). So, conservatively speaking, if the seam depth is less than 400 ft (122 m), the potential for loss of water for wells and surface streams exists. However, surface topography above the Pittsburgh seam is such that the terrain is full of valleys and hills, and the relief between valleys and hilltops are 200 to 500 ft (61–152.4 m). If both the valleys and hilltops are above the fractured zone, the behavior of well water and surface streams may or may not be related to the effects of longwall mining. Rather, local topography and its natural fracture systems, and water recharge and discharge are the more important factors. Since most rural water wells and springs are shallow, dewatering issues related to longwall mining will have to be investigated from this perspective.

14.5.2 Longwall mining and its effects on surface and groundwater

14.5.2.1 Groundwater flow and fracture systems in hilly terrains

Groundwater movement in an aquifer is controlled by aquifer lithology, structure, permeability, water level, and topographic setting. Hilltop aquifers are typically perched or semi-perched. The hill is made up of a series of interbedded permeable aquifers (i.e., sandstone) and non- or less-permeable aquitards (i.e., shale or clay). If there is a water table in each aquifer, the aquifer is perched. Otherwise, it is a semi-perched aquifer (Fig. 14.5.1) (Coe and Stowe, 1984).

According to Wyrick and Borchers (1981), stress-relief fracturing follows a distinctive pattern (Fig. 14.5.2) and controls groundwater flow in an Appalachian valley. The most permeable fracture systems are located beneath valley streams. Fracture systems developed along the valley margins and in the hillslopes are fairly permeable. But on hilltops and their cores, fracture systems are less developed due to compression.

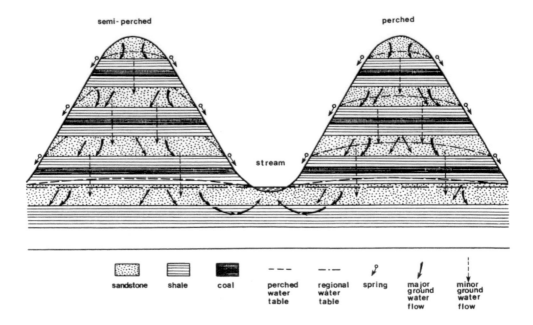

Figure 14.5.1 Idealized groundwater flow in hilly terrain

Source: Modified from Coe and Stowe (1984)

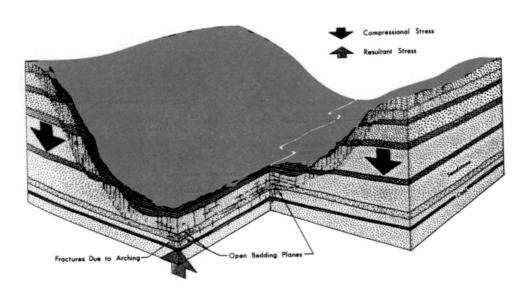

Figure 14.5.2 Generalized features of stress-relief fracturing

Source: Modified from Ferguson (1974)

14.5.2.2 Effects of longwall mining on shallow water wells

Leavitt (1992) analyzed 174 domestic water supplies, including 14 dug wells, 137 drilled wells, and 23 springs in northern West Virginia and southwestern Pennsylvania subjected to longwall mining (Table 14.5.1). He found that 43% of them were not affected, 17% were temporarily affected in quantity, and 4% were temporarily affected in quality. The remaining 36%, or 62 wells/springs, were affected by longwall mining, and among them, 57 wells/springs required intervention to maintain their quantity and 5 required intervention to maintain quality. Of the 174 sources, 112 of them were ultimately returned to service for the landowners, while the remaining 62 were replaced either by new or deepened wells or by connection to municipal water supplies. With the exception of those 24 connections to city water, more than 99% of the wells/springs continued to rely on the same groundwater sources after mining.

The effect of surface topography is shown in Figure 14.5.3. Of the 65 wells/springs located in the valleys, 53, or 82%, were ultimately returned to service; of the 74 located on the

Table 14.5.1 Depth and quantity of the 174 water sources investigated by Leavitt (1992)

Wells (numbers)	Depth, ft		
	Shallowest	Deepest	Average
Dug (14)	12	34.3	22.6
Drilled (137)	20	245	82.5
	Flow, gpm		
Springs (23)	<0.1	1.79	–

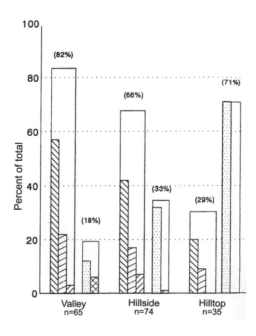

Figure 14.5.3 Effect of topography on water wells/springs due to longwall mining

hillslope, 49, or 66%, were returned to service; and only 10, or 29%, of the 35 located on the hilltops were returned for the landowners' use. Wells/springs located in valleys were less affected than those on the hillslopes, which, in turn, were less affected than those on hilltops.

Figure 14.5.4 shows that drilled wells were least affected, probably due to the greater depth that provides more areas of water inflow. Conversely, dug wells depend more on recent precipitation. For springs, more than 50% remained in service, and in a few cases, after a spring ceased to flow, a new spring emerged down slope from the old site. This indicated that the source of the spring remained, only the flow path had changed due to changes of slope resulting from subsidence.

Within the surface subsidence basin, surface movement is concentrated within the quarter-panel, whereas in a supercritical panel, the mid-panel is free of permanent surface displacement. Surface movement is minimal over the gateroad areas. Figures 14.5.5, 14.5.6, and 14.5.7 show the effects on wells/springs located in quarter-panels, mid-panels, and over gateroads, respectively. The term "quarter-panel" means one-quarter of the panel width from both headgate and tailgate, while mid-panel is the remainder half at the center of the panel.

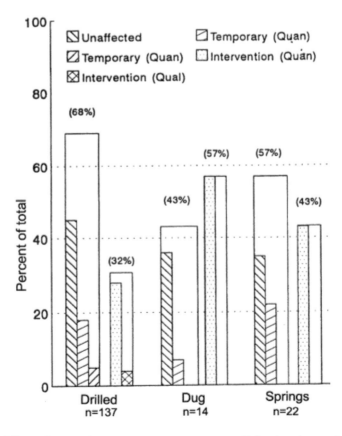

Figure 14.5.4 Effect of type of water sources on water wells/spring due to longwall mining

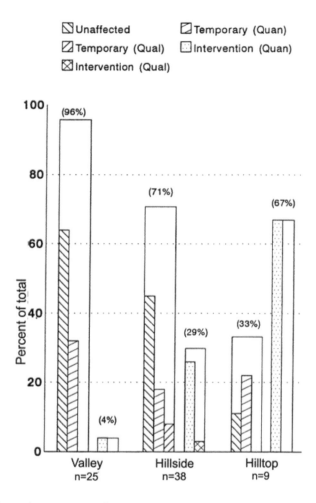

Figure 14.5.5 Effect of quarter-panel sources on water wells/springs due to longwall mining

Within the quarter-panel, the wells/springs in the valleys showed the least affect, followed by the hillslope, and the worst affected are those on the hilltops. Similar patterns exist for those in the mid-panel (Fig. 14.5.6) and over the gateroads (Fig. 14.5.7). Over the gateroads, those located on hilltops nearly all required intervention, mainly because they are located on the edges of a subsidence basin, where the subsidence induced slope, plus the natural slope of the hillside accelerates the discharge of groundwater at the hilltop.

Walker (1988) investigated 10 shallow water wells, all 150 ft (45.7 m) deep over four longwall panels with a coal seam depth of 400–600 ft (122–183 m). He found that well water levels fluctuated as mining progressed. Fluctuation was a function of well panel location and the proximity of the longwall face. He also found that water levels in shallow wells would fully return or be near pre-mining levels after mining and that wells located in stream

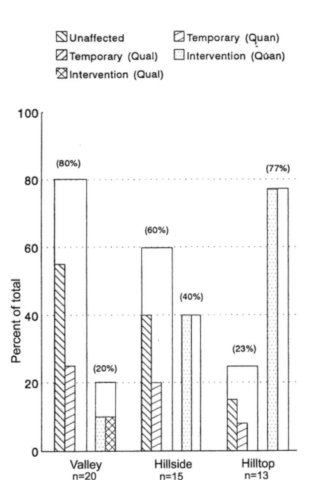

Figure 14.5.6 Effect of mid-panel sources on water wells/springs due to longwall mining

valleys exhibited a lesser response to mining. Fluctuation of well water level began when the longwall face was approximately one seam depth inby. Fluctuation continued as the face progressed. When the face was 200 ft (61 m) outby, the rate of water level fall increased sharply. Fluctuation ceased when the face was approximately one seam depth outby.

Water level fluctuation is related to the subsidence-induced surface strain. As the face progresses, the surface strain changes from tensile when the face is in and around the wells, to compressive when the wells are far behind the face. When the wells are located in the tensile zone, the water level drops. Its rate of drop decreases when it is in the compression zone.

Matetic and Trevits (1990, 1991, 1992) and Trevits and Matetic (1991) investigated 20 water observation wells in Greene and Cambria Counties, Pennsylvania, and Vinton County, Ohio. Pre- and post-mining tests were performed at each well to observe the hydrologic parameters changes such as specific capacity and transmissivity.

Specific capacity of a well is the productivity of the well, usually expressed as gallons per minute per foot of drawdown. The larger the specific capacity, the more productive the

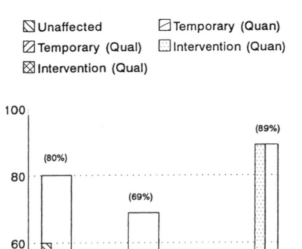

Figure 14.5.7 Effect of gateroad sources on water wells/springs due to longwall mining

well. The transmissivity is the product of hydraulic conductivity and saturated thickness of an aquifer or water bearing formation. Hydraulic conductivity is sometimes referred to as the coefficient of permeability.

They found that a dramatic increase in specific capacity was measured at the well which was located at the bottom of the valley before and after mining. This might be a highly fractured area with more avenues for ground water to flow and thus the rock units might have increased permeability. Also, it was believed to be related to the change in precipitation over the study area. The specific capacity for other wells appeared to be unaffected by mining. It should be noted that the effect of regional drought or precipitation was not considered.

Pre- and post-mining transmissivity values did not change and appeared have no correlation with mining.

The water level began to decline with the ground surface going into tension. The maximum decline in water was observed at the point of maximum tension. The water level began to recover at the point of maximum compressive strain, and the final value of water level was less than the initial value.

Yields of hilltop wells before and after longwall mining, studied by Johnson (1992), remained the same. He attributed it to two factors: sustained compression and reduced post-mining water levels. Since the core of the hilltop is under compression, mining-induced fractures were closed and would not affect the strata permeability. Water level was generally lower after mining, which perhaps offset any increase in permeability resulting in unchanged well yields.

Therefore, longwall mining does not drain the near-surface groundwater, which is the sources of rural domestic water supplies, and the flow pattern of which, in a stress-relief fracturing valley, consists of recharge at the topographic high areas moving toward the valley bottoms. Longwall mining merely changes the individual flow paths and flow rates.

Topographic setting is the dominant factor controlling the response to longwall mining. Valley wells/springs are the least affected due to abundant fractures providing access to large areas of groundwater water sources, including those from the hillslope and hilltop. Those located in the hilltop are the most affected among the three topographic settings: valley bottom, hillslope, and hilltop. They depend solely on recharge from rain falls, which is limited to the area bounded by the elevation contour corresponding to the well bottom. Furthermore, since hilltop aquifers are either perched or semi-perched, if the aquitards are disturbed by longwall mining, groundwater could migrate downward or emerge as springs. Consequently, recovery of hilltop wells is much slower than those in the valley bottom (Johnson, 1992).

The location of a well/spring within the panel is another factor. Those in the quarter-panel are more affected than those at the mid-panel. Those over the gateroads are least affected and those off the panel are not affected (Leavitt, 1992; Johnson, 1992).

Well water quality is seldom affected by longwall mining regardless of location and topographic setting (Leavitt, 1992; Johnson, 1992; Walker, 1988).

14.5.2.3 Effects of longwall mining on surface streams

Johnson (1997) measured streamflow in four streams both before and after longwall mining in the Pittsburgh seam of 500 to 600 ft (152–183 m) deep. All four streams were perennial with pre-mining streamflow from 500 to 4600 gpm (1900–17,480 liters per minute). He found that pre- and post-mining streamflows were virtually the same and that streamflow at the headgate and tailgate remained approximately the same before and after mining, i.e., longwall mining did not affect the streamflow as it passed through the undermined panel, even though the flow quantity was less after mining.

During mining, flow reduction may occur as the traveling tensile strain accompanying the moving longwall face arrives. Streamflow will recover gradually as the compressive strain approaches and stays. The length of time required for recovery depends on the streambed. For instance, sandstone will need a longer time to recover. Flow reduction may occur over the headgate and tailgate development sections where permanent tensile strain exists. Conversely, flow increase may occur inside the panel resulting in no net gain or loss over the whole panel.

Chapter 15

Longwall top coal caving mining

15.1 Introduction

Longwall top coal caving mining (LTCC) is one of the mining methods used for thick coal seams, 20 ft (6 m) thick or more. It has been employed widely in China since the mid-1980s. Australia began to employ LTCC in 2005 (Ringleff and Rutherford, 2007). It reached a maximum of three mines in the mid-2010s and has been reduced to a single mine today.

This chapter summarizes the major components of LTCC emphasizing its major differences from the regular longwall mining which has been illustrated in exhaustive detail in this book.

15.1.1 Definition

In LTCC mining, it divides the seam into two layers, lower and upper. The lower layer employs the regular longwall mining techniques in the front side of the face as described elsewhere in this book. In the upper layer, the top coal, having been crushed to broken coal by the subsiding action of the roof strata due to the regular longwall mining (Fig. 15.1.1), is drawn through gravity periodically from below the tail shield of shield support and dumped on the AFC in the rear side of the face.

Therefore, in LTCC, there are two AFCs, one in the front and the other in the rear of the shield support. The front AFC moves the coal cut by the shearer just like the regular longwall mining, while the rear AFC moves the coal drawn from the top coal. The coals from both AFCs are fed into a common stage loader and crusher at the headgate T-junction and then onto the panel belt conveyor.

15.1.2 Benefits

There are other methods such as large mining height (up to 27.6 ft or 8.4 m) and multi-slice methods available for thick seam mining. Comparing to those two methods, LTCC has the following advantages:

1   It resembles one-slice mining, requiring only one time panel and equipment set up and thereby reducing investment and saving cost. It also avoids the stress concentration created in multi-slice mining.
2   It adapts easily to wide ranging seam dip angles (0–60°) and variation of seam thickness 20–82 ft or 6–25 m). It enables high recovery and safety.

(A) 3D Artist's overall view

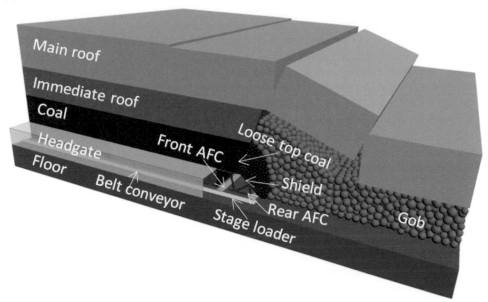

(B) 2D cross-section of the face (Courtesy Caterpillar)

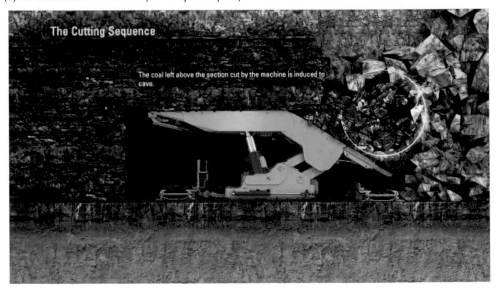

Figure 15.1.1 Principle of LTCC mining

3   In large mining height mining, the roof supports are extremely heavy, requiring very high capital investment and ensuing difficulty in underground transportation. It offers less flexibility in varying geological conditions that are likely to be encountered during the roof support's service life. The extreme face height presents a high risk for miners due to possible face sloughage, roof fall, and face collapse.

15.1.3 Geological conditions suitable for LTCC

The key to the success of LTCC lies in whether the top coal can be easily released and recovered at high rate. Factors influencing top coal recovery rate include seam thickness, seam inclination, coal seam strength, location and thickness of partings, and methane content.

15.1.3.1 Seam inclination

LTCC adapts easily to coal seam inclination, being suitable for flat, gently inclined (8–25°), large inclination (25–45°) and even steeply inclined (larger than 45°) coal seams. For LTCC advancing along the strike direction, it has been successfully applied in seams up to 60°. Very stable roof supports have been developed for large and steeply inclined seams by using larger and more cylinders on side shields, or installing adjustable bottom plate that increases its self-adjusting ability in the seam dip direction. In multi-slice LTCC, seams up to 90° have been successfully mined.

15.1.3.2 Coal seam strength

The medium-hard coal seams with uniaxial compressive strength (UCS) 1450–4350 psi (10–30 MPa) are suitable for LTCC, because it's good for top coal release and roof stability. Top coal recovery rate is closely related to the development of fractures in the coal seam. For hard coal seams with poorly developed fracture systems, top coal release is poor and artificial fracturing may be required to weaken it.

For seams with UCS larger than 4350 psi (30 MPa), top coal is either difficult to cave or, if it does, the broken fragments may be too large to be withdrawn, resulting in low top coal recovery. Therefore, this type of hard coals generally requires artificial means to increase fracture density before it can be released. For seams with UCS less than 1450 psi (10 MPa), the loose and broken coal easily fall off in front of or between the supports, causing face spall and leading to poor support-roof interaction.

15.1.3.3 Location and thickness of partings

Partings influences coal recovery and reject rates. The strength, thickness, and number and location of partings will influence top coal's fracture and release. When the partings are located at the middle and upper portions of the coal seam and clay partings are less than 19.7 in. (50 cm) thick, its influence on top coal fracture and release will be minor: when the partings is 3.28 ft (1 m) thick, its influence must be considered with other factors relating to top coal's fracture and release: when the partings are 4.9–6.3 ft (1.5–2.0 m) thick, it's not suitable for LTCC. If the partings are the high-strength igneous rocks and more than 11.8 in. (30 cm) thick, it will affect top coal's fracture. When it is thicker than 0.2 in. (0.5 cm), it will affect top coal recovery rate. The greater the number of layers, and thicker, of the partings, the more rejects there will be in the recovered coal.

15.1.3.4 Methane content

Methane content affects the application of LTCC. In gassy mines, special measures must be implemented before LTCC can be employed. When the mining height in LTCC increases, so is the methane emission from the top coal, even if it is a less gassy coal seam. Therefore, in LTCC, proper methane drainage measures must be implemented.

15.1.3.5 Other conditions

In LTCC, remanent coal pillars or debris are often left in the gob. If the coal is liable to spontaneous combustion and the incubation time is short, self-burning could easily initiate if the face is advancing slowly. The normal practice is to keep the face moving steadily. If this cannot be done or if done, it is not effective, then injection of inert materials such as nitrogen is recommended.

15.2 Panel and equipment layouts

15.2.1 Typical layout

Panel layout in LTCC is similar to that of regular longwall mining, but equipment arrangement is somewhat different due to the addition of rear AFC (Fig. 15.2.1).

Panel width ranges from 492 ft (150 m) to 984 ft (300 m) with most being less than 820 ft (250 m). Panels are either developed by single entry (in China) or two-entry gateroad system (in both China and Australia).

15.2.2 Caving to cutting ratio

Caving to cutting ratio refers to the ratio of top coal drawing thickness on the top to mining height in the bottom.

Coal production in LTCC derives from two parts, shearer's cutting in the bottom and drawing from top coal. Therefore, properly increasing shearer's cutting height not only can increase cutting recovery rate but also is good for top coal drawing. Selection of a proper cutting height not only needs to consider factors such as reserves, drawing to cutting ratio, ventilation, walkway, and recovery ratio, it also needs to consider the required equipment and stability of face and top coal. The higher the cutting height is, the easier face spalling and the higher capital equipment investment cost will be. However, increasing cutting height and decreasing the drawing-to-cutting ratio will increase total coal recovery. Increasing the top coal drawing space will increase the release of large pieces of top coal. Currently the cutting height for coal seams less than 49.2 ft (15 m) is 8.2–12.5 ft (2.5–3.8 m), i.e., the maximum working height of shields is generally less than 12.5 ft (3.8 m), while the typical top coal caving height is 8.2–26.2 ft (2.5–8 m). In recent years, the large mining height has been developed for coal seams more than 49.2 ft (15 m) thick, in which the cutting height is 11.5–16.4 ft (3.5–5.0 m). LTCC with large mining height can mine seams up to 65.6 ft (20 m) thick.

15.3 Face equipment

As stated in previous sections, mining of LTCC front lower layer employs, and its equipment and layout are similar to that of, a regular longwall mining method. The major differences are in the use of a rear AFC at the rear end of shield for top coal transportation. In addition, in order to accommodate the rear AFC, LTCC shields differ from those of regular longwall mining. Furthermore, the gateend shields are much longer because they are designed to protect the two AFCs in tandem in a single unit, unlike that in the regular longwall mining in which a single AFC pan is located under the shield canopy in front of shield legs.

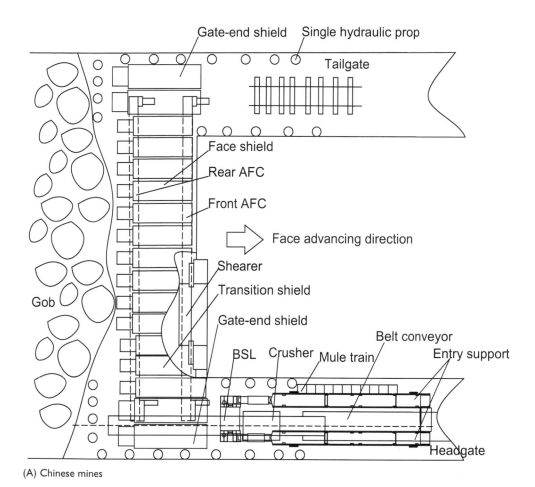

(A) Chinese mines

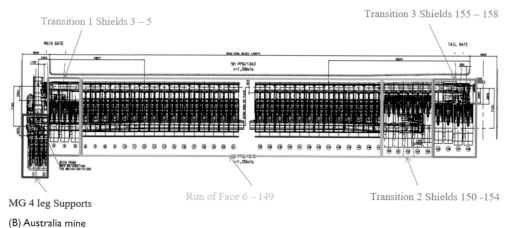

(B) Australia mine

Figure 15.2.1 Typical LTCC face equipment layouts

Source: B – courtesy of BHP Billiton Mitsubishi Alliance

15.3.1 Face shields

The caving shield in LTCC shield has an extended rear canopy with a retractable short canopy or beam at the bottom end (Fig. 15.3.1). When fully extended, the extended rear canopy and retractable canopy covers the rear AFC under it, completely separating the rear AFC from the top coal above it. During top coal drawing (Fig. 15.3.2), the retractable canopy is fully retracted into the extended rear canopy first (Fig. 15.3.2 B), followed by closure of the extended rear canopy (Fig. 15.3.2 C), exposing the rear AFC fully to receive the top coal to be withdrawn.

The rear extended canopy is controlled by a double-acting cylinder, the extension and retraction of its piston causes the rear extended canopy to swing between two extreme positions of opening (Fig. 15.3.2 B) and closing (Fig. 15.3.2 C) the rear AFC. During top coal drawing, rapid and repeated swing action creating vibration helps with unlocking the top coal onto the rear AFC.

15.3.2 Transition shields

Transition shields are located near the headgate and tailgate T-junctions between gateend shields and face shields. In these areas, the head and tail drives as well as CST soft start take up a lot of space such that the LTCC face shields are not fully operable. In order to accommodate this special equipment layout and protect the top coal and roof strata in these areas, the transition shields were developed. The transition shields are larger than the face shields in overall dimension, but the space for legs and stabilizing devices is limited to allow more space for equipment coverage, ventilation, and walkway (Fig. 15.3.3).

Therefore, in a LTCC operation, top coal caving and drawing do not apply to the full panel width. Actual top coal caving width occurs between the innermost transition shields on both face ends leaving a 32.8–49.2 ft (10–15 m) coal pillar on each side to protect the headgate and tailgate, respectively. Similarly, coal pillars are also left on the setup room and recovery room side.

Figure 15.3.1 LTCC face shield

Source: Courtesy of Caterpillar

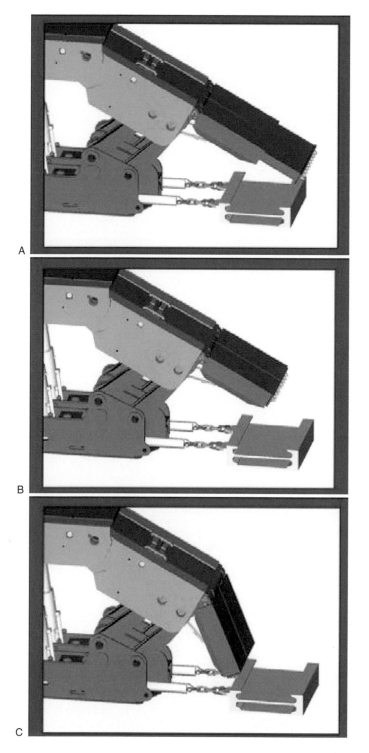

Figure 15.3.2 Three photos showing how the top coal is released and loaded onto the rear AFC

Source: Courtesy of Caterpillar

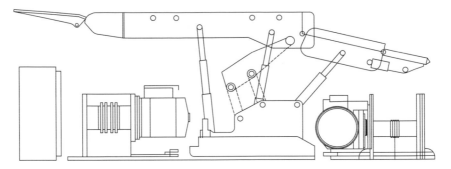

Figure 15.3.3 Transition shields

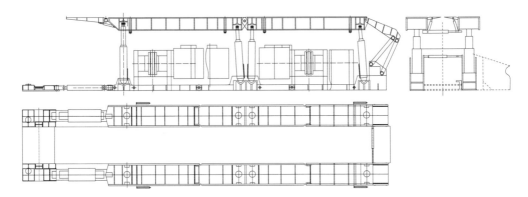

Figure 15.3.4 An example of headgate gateend shield

15.3.3 Gateend shields

Gateend shields are located at the headgate and tailgate T-junctions for the protection of AFC head drives and stage loader, and AFC tail drive, as well as the stability of headgate and tail-gate T-junctions. Since there are two AFCs with two drives at both ends, the gateend shields are designed to cover and protect them. Consequently, the gateend shields are structurally very large (Figs 15.3.4 and 15.3.5)

15.3.4 Selection of face shield

In LTCC, top coal drawing or top coal release has always been to utilize shield's various designs to implement the operation. Top coal drawing or top coal release went through a high-level, a medium-level, to the current low-level system. In the high-level system, a chute was constructed toward the rear on the shield canopy and top coal was drawn from top of canopy and followed the chute to dump on the front AFC. In this system there was no rear AFC. Next came the medium-level system in which a door was made in the lower portion of the caving shield. As the door is opened, the top coal flows onto the rear AFC laid down on the base plate in the rear of the shield. With this system, the recovery rate was low; the

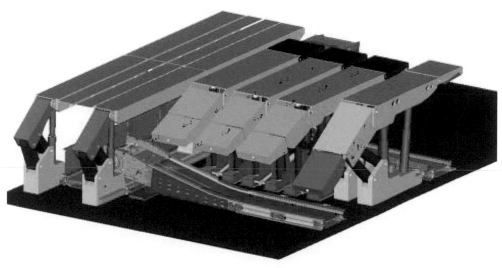

(A) Chinese mine (courtesy of Caterpillar)

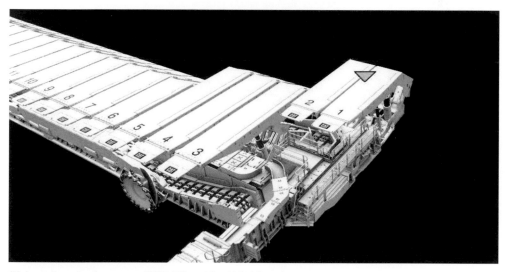

(B) Australian mine (courtesy of BHP Billiton Mitsubishi Alliance)

Figure 15.3.5 Examples of headgate layout

space in the rear of the shield was small and not convenient for maintenance and clean float coal. In addition, the base toe pressure was high, causing floor digging and making shield advance difficult.

The latest model is the low-level top coal drawing mechanism. Figure 15.3.6 shows four types of shield models being used in China. In Figure 15.3.6 A, the rear AFC sits on the shield's base plate below the tail canopy. For coal seams more than 29.5 ft (9 m) thick, the large mining height shield (Fig. 15.3.6 B) was designed to mine it in one slice. In this type of shield, the mining height and supporting density have been increased significantly. Since coal seams are thick, the power capacity and physical volume of the rear AFC also increases. In order to increase the coverage space under the tail canopy and top coal drawing opening, the tail canopy in a large mining height shield has a retractable beam. During top coal drawing, the retractable beam is retracted first and then swing down the tail canopy to open the drawing opening. The addition of retractable beam greatly increases the top coal drawing opening allowing for rapid top coal release.

For the four-leg chock shields, the four legs are generally subject to non-uniform loading: the front two legs are always larger than the rear two legs. Sometime the rear two legs are even subjected to tension, greatly decreasing the supporting density of the support. In this respect, the two-leg shields with a solid canopy have apparent advantages.

The two-leg shields were developed for convenience of employing electrohydraulic control system (Fig. 15.3.6 C). Figure 15.3.6 D shows the shield designed for steeply inclined coal seams with its side shields capable of ensuring smooth contact between shields.

15.3.5 Selection and design of AFC

The selection of front AFC follows the principle of regular longwall mining, mainly to match the cutting capacity of the shearer.

The carrying capacity of the rear AFC can be determined by:

$$Q_x = \rho h l s \tag{15.3.1}$$

Where Q_x = amount of top coal release in a drawing cycle, ρ = density of top coal, h = top coal thickness, l = panel width, and s = top coal drawing interval.

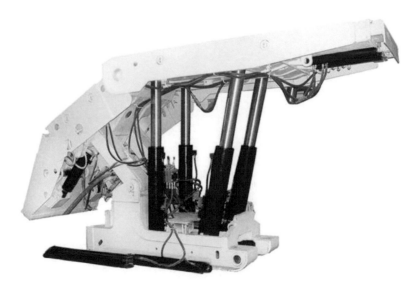

(A) Four-leg chock shield

Figure 15.3.6 Four types of shields of low level top coal drawing models in China

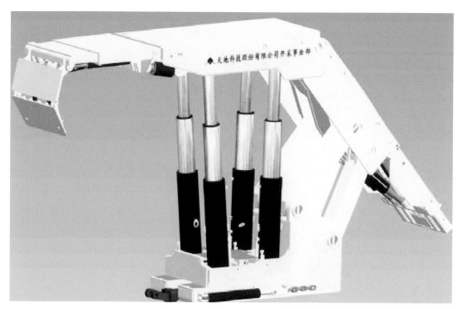

(B) Chock shield for large mining height LTCC

(C) Two-leg shield

Figure 15.3.6 (Continued)

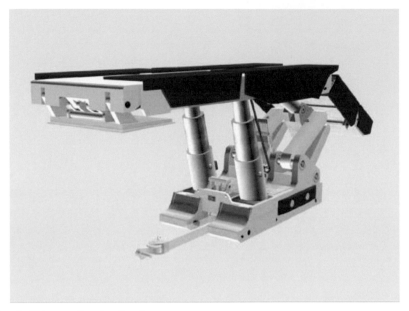

(D) Two-leg shield for steeply inclined coal seams

Figure 15.3.6 (Continued)

The carrying capacity of rear AFC, Q_s, is then obtained by:

$$Q_s = k_f Q_x / T_x \qquad\qquad (15.3.2)$$

Where K_f = uneven coefficient of top coal drawing and T_x = duration of top coal drawing.

15.4 LTCC Mining technique

15.4.1 Basic flow scheme

The major task sequence in LTCC is this:

Coal cutting > shield advance > push the front AFC > top coal drawing > pull the rear AFC.

The first three tasks are performed in the regular longwall mining that have been described elsewhere in this book. The last two steps, top coal drawing and pull the rear AFC, pertain to LTCC.

15.4.1.1 Top coal drawing

The interval for top coal drawing depends on support type, location and dimension of drawing, and thickness and fracture conditions of top coal. The "one cut one drawing" or "two cuts one drawing" is most frequently employed, starting from one gateend, shield by shield, in sequential order. If top coal is thick, multi-round drawing may be used. Single-round drawing can be more cleanly controlled because drawing stops as soon as waste rocks appear.

If large pieces of coal that are difficult to be drawn appear, the retractable beam may be retracted and extended for a few times, plus the tail canopy may be simultaneously vibrated slightly, assisting to break up the large coal pieces for ease of drawing.

15.4.1.2 Pull the rear AFC

After completion of top coal drawing, the rear AFC is pulled sequentially forward. Do not pull the rear AFC from both ends. Note that each pan is connected to the shield in front of it by two DA rams, one on each side, with a chain for pulling the rear AFC (Fig. 15.3.2).

15.4.2 Methods of top coal drawing

Among the five sequential tasks in LTCC, top coal drawing is the most important factor controlling top coal recovery rate. Consequently, the method of top coal drawing must be selected based on seam conditions.

15.4.2.1 Interval of top coal drawing

The interval of top coal drawing affects top coal recovery rate, reject rate, and coal seam recovery. In general, most employ one cut and one drawing. If the interval is too large, residual coal will be left in the gob, decreasing coal recovery and increasing the liability for spontaneous combustion.

15.4.2.2 Method of top coal drawing

A. SINGLE ROUND SEQUENTIAL DRAWING

For flat or near flat seams, this is the most common method. It starts from one end of the face and in each shield, top coal is drawn until waste rocks appear. After closing the drawing opening, top coal drawing for the next shield begins. This process repeated shield by shield until it reaches the other end of the face.

B. MULTIPLE ROUND ALTERNATE DRAWING

For thick seams, due to thick top coal, multiple alternate drawing is used. Take two round alternate drawings for example, in the first round drawing, the top coal is drawn for the odd-numbered shields for ½ or 1/3 of the total top coal drawn, followed by drawing for the even-numbered shields for similar ½ or 1/3 of the total top coal drawn. The first round drawing will loosen the top coal. Once the first round has completed for more than 10 shields, the second round begins to draw the remaining ½ or 2/3 of top coal. The time interval between the first and second rounds should be separated by 10–15 shields.

15.5 Automation of LTCC

Automation of LTCC similarly consists of two parts, the regular longwall mining in the front and the top coal drawing in the rear part of the shield. Automation of regular longwall mining has been addressed in Chapter 9 "Automation of Longwall Components and Systems." This section will cover only the automation of top coal drawing and the system operation related to it.

Since top coal drawing involves operation and movement of tail canopy, retractable beam, pull of rear AFC, the cylinders that control these three devices are each individually equipped with a reed rod sensor to control its movement accurately (Fig. 15.5.1) (Jones, 2017). Furthermore since the location of tail canopy and retractable beam are related to shield heights and angle of shield canopy inclination, sensors are also installed to monitor these two components.

Several automated LTCC cutting and drawing operations are available (Caterpillar, 2018). Two examples are shown in Figures 15.5.2 and 15.5.3.

In general, the automated bi-di LTCC operation consists of the following major steps (Fig. 15.5.2): the shearer passes the shield area of interest, followed by first the advance of shield of interest and then operation of the LTCC top coal drawing openings several shields behind or a safety zone, which is normally about six shields. One or two top coal drawing openings must open simultaneously within a group of four shields but with no overlapping.

The recommended uni-di LTCC cutting and drawing sequence is the staggered procedure as shown in Figure 15.5.3. The top coal drawing openings are in a group of seven within which the automated system controls multiple drawing openings simultaneously in alternate shields but not adjacent shields. The procedure starts by the shearer's cutting from headgate to tailgate in uni-di mode, followed by a shearer's clean up and then top coal drawing operation from headgate to tailgate. In this procedure, top coal drawing is independent of actual shearer position enabling highest production.

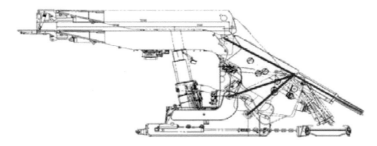

(A) Top coal drawing door closed

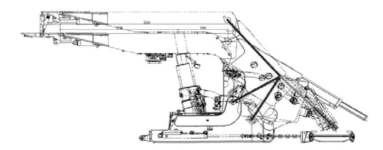

(B) Top coal drawing door open

Figure 15.5.1 Sensors installed for devices for operating top coal drawing operations

Source: Courtesy of BHP Billiton Mitsubishi Alliance

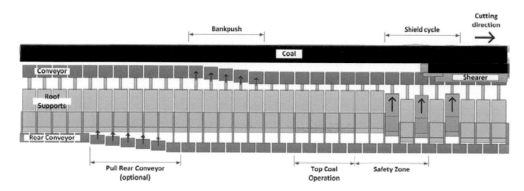

Figure 15.5.2 Automated LTCC bi-di cutting and drawing sequence

Source: Courtesy of Caterpillar

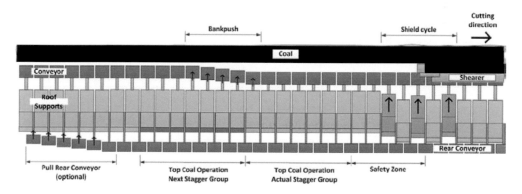

Figure 15.5.3 Automated LTCC uni-di cutting and drawing sequence

Source: Courtesy of Caterpillar

The current most productive automated LTCC operation is in Australia. It has a panel width of 1049.6 ft (320 m) by 6560–9840 ft (2–3 km) long with 23 ft (7 m) seam thickness. The top coal caving width is 1000 ft (305 m). The shearer's cutting height is 13.1 ft (4 m) and the top coal caving height is 9.8 ft (3 m) (Fig. 15.5.4). The face equipment are as follows:

Shearer: EL3000, 8.2 ft (2.5 m) drums, 31.5 in. (800 mm) web, 2125 hp (1540 kW) installed power

AFC: front – 1080 mm wide pan, 2 × 1656 hp (2 × 1200 kW) drive power, 3600 tph
Rear – 1080 mm wide pan, 1656 and 1449 hp (1200 and 1050 kW) drive power, 2800tph

BSL: 1600 mm wide pan, 966 hp (700 kW) drive power, 5300 tph

Shield (Fig. 15.5.5): face – two-leg, 2.05 m wide, 1450 tons yield

Gateend: three versions, three-leg, 1630 ton yield. Buttress shields, four-leg, 1820 tons yield

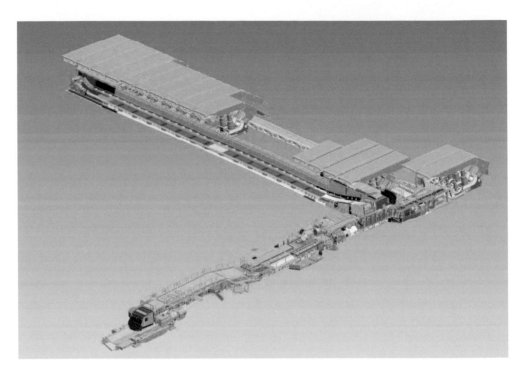

Figure 15.5.4 3D view of LTCC face equipment layout

Source: Courtesy of BHP BILLITON Mitsubishi Alliance

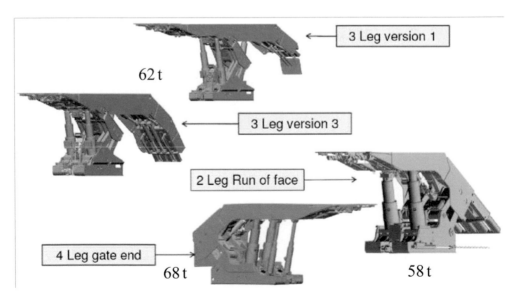

Figure 15.5.5 Types of shields used in Australian LTCC

Source: Courtesy of BHP Billiton Mitsubishi Alliance

Both cutting and top coal drawing operations are fully automated, independent of each other, and run simultaneously. There are no miners at the face although the shift crew consists of six miners working outby or checking for problems or perform routine inspection at the face when needed.

The duration for top coal drawing based on initial experiments is set for 10 seconds for each shield and runs in alternate shields such that it allows for sharper drawing profile behind each shield. The mine's annual production is around 8 million mtons ROM with 84–87% clean coal recovery.

15.6 Health and safety issues

Due to caving and drawing of top coal, respirable dust is obvious a very critical issue with LTCC. Fortunately water spraying technique for dust suppression and isolation in shields is well developed and has been applied successfully [see Sections 9.6.4 and 11.4.2.3(C)(2)] for regular longwall mining. The only modifications if applicable is the gob sprays, possibly addition and/or modification of sprays in the lemniscate links and/or tail canopy.

In LTCC, portions of top coal will always be left behind in the gob. If the coal tends to be self-heating, it is liable to spontaneous combustion somewhere in the gob when the required conditions are met. For safety reason, its incubation period must be determined in advance. The control methods include the following: (1) maintain a proper face advance rate to prevent the coal oxidation behind the shields to catch up the critical temperature; (2) develop a ventilation system that will eliminate creation of critical velocity zone for spontaneous combustion to occur; and (3) develop a monitoring program for detection of spontaneous heating for early mitigation measures.

15.7 Theory of withdrawn body in LTCC

After the top coal breaks and falls off onto the tail canopy and retractable beam, the movement characteristics of the withdrawn body is the key to top coal drawing. Unlike the regular longwall mining, miners in LTCC cannot observe the actual process of top coal release and, therefore, top coal drawing has been practiced based on miner's subjective observation. Therefore, the investigation into how the withdrawn top coal body (or withdrawn body) is initiated, developed, and withdrawn and in the process how roof rock intrusion occurs affecting the total coal recovery are the most important research area in LTCC.

According to Wang and Zhang (2015) the key top coal drawing or release is the relationship between the withdrawn body and the contact plane between rock and top coal. The top coal withdrawn body is the space occupied by it before it is withdrawn (Fig. 15.7.1).

Figure 15.7.1 D is the shape of the withdrawn body at the end of coal drawing (i.e., when rocks begins to enter into the drawing opening). The origin and development of this withdrawn body at any moment can be traced back by using numerical modeling (Fig. 15.7.2). The final withdrawn body is a "cutting influenced ellipse" cut off by the caving shield. Its front part develops faster due to the effects of shield's caving shield (the friction coefficient between caving shield and top coal is smaller than the coefficient of internal friction of coal).

Figure 15.7.2 shows clearly how during top coal drawing the rocks enter into the drawing opening. The rocks enter into the withdrawn body initially at the middle of the withdrawn

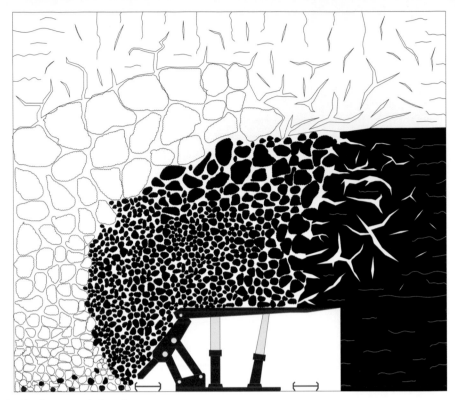

(A) Coal/rock interface before drawing

Figure 15.7.1 Schematic drawings showing the coal/rock interface and the withdrawn body
Source: Wang and Zhang (2015)

(B) Coal/rock interface after drawing

(C) The withdrawn coal

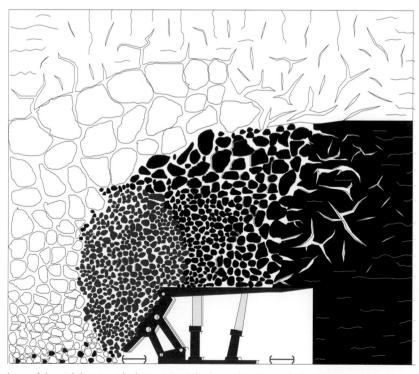

(D) The shape of the withdrawn coal when putting it back to where it was before drawing

Figure 15.7.1 (Continued)

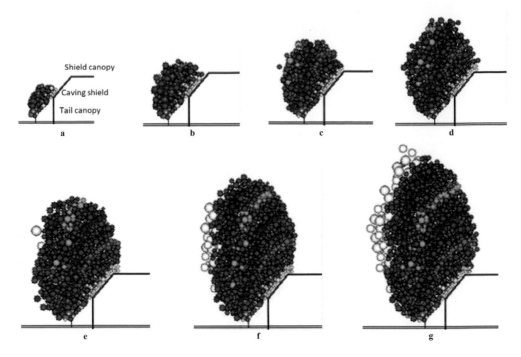

Figure 15.7.2 The sequence of development of the withdrawn body

Source: Wang and Zhang (2015)

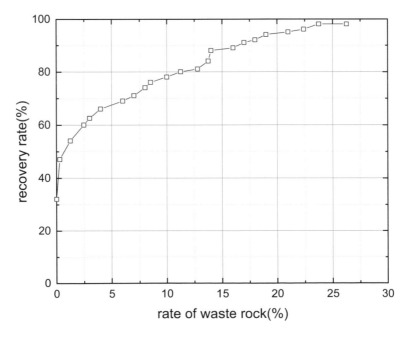

Figure 15.7.3 Relationship between top coal recovery rate and rate of waste rock

Source: Wang and Zhang (2015)

body on the gob side. It then develops up and down but mainly expanding upward. In fact, rock intrusions may start in the rear of the adjacent shield where coal drawing has just completed.

The starting point of rock intrusion begins with the first point of intersection between the withdrawn body and the rock/coal interface. Because since the curvature of the withdrawn body is larger than that of coal/rock interface, the point of largest lateral dimension is always where it first makes contact with the coal/rock, forming the point of rock intrusion, i.e., the rocks located in gob side more easily enter into the withdrawn body. For multiple cut drawing such as "three cut one drawing," the coal/rock interface extends wider on the gob side covering more top coal such that the rocks enter the withdrawn body and become the "waste rocks." In order to increase top coal recovery, lowering waste rock content, it must make the boundary of withdrawn body contact the coal/rock interface at multiple areas, shearing each other at the gob side. This task can be achieved by top coal drawing techniques and controlling factors.

For example, when the coal is very thick, multiple drawings may be used to reduce the number of coal/rock contact areas. However, the rear AFC must be wider capable of carrying a much larger capacity.

Increasing top coal recovery rate and decreasing waste rock rejects are two conflicting events. During the initial period of coal drawing, pure top coal drawing can be achieved and the withdrawn body is all coal (Fig. 15.7.2 a–d). But as drawing proceeds, the withdrawn body grows bigger, allowing the broken immediate roof rocks to enter it, evolving into a withdrawn body mixed with waste rocks. At this time, parts of top coal remain in the roof without being withdrawn (Fig. 15.7.2 e–g). Therefore, the more coal is withdrawn the more is the rock enter into it resulting in higher rejects. Figure 15.7.3 shows the relationship between top coal recovery rate and rate of waste rock.

It can be seen from Figure. 15.7.3 that before coal recovery rate reaches 50%, rejects are less than 2%; when coal recovery rate reaches 70%, the rejects percentage increases quickly. When rejects reach 11%, coal recovery rate approaches 80%. When rejects reach 15%, coal recovery rate reaches 88% (Fig. 15.7.2 e–g). Therefore, in practice, it is recommended that proper amounts of rejects should be allowed in coal drawing in order to achieve higher coal recovery.

Furthermore, coal recovery is different at different elevation. In general, the recovery of the middle part of top coal is higher, while the recovery for both the upper and lower portions is lower. From the shape of the withdrawn body, when its middle portion makes first contact with the coal/rock interface, rock begins to enter it. If top coal drawing stops at this moment, the top coal above and below this intersection point will stop moving and remain in the gob as parts of the top coal are lost, as demonstrated by both laboratory experiments and underground monitoring (Wang et al., 2014).

References

Acharya, B.L. (1982) *In-Mine Assessment of Pillar Stress and Entry Load in Main and Tail Entries in Retreating Longwall Mining at Quarto No. 4 Mine*. M.S. Thesis, Department of Mining Engineering, College of Engineering and Mining Resources, West Virginia University, Morgantown, WV. p. 101.

Adam, R.F.J., Pimentel, R.A. & Shoff, W.E. (1984) A handbook for face-to-face moves of longwall equipment. Final report submitted to US Department of Energy. p. 114.

Afrouz, A. (1975) Yield and bearing capacity of coal mine floor. *International Journal of Rock Mechanics and Mining Sciences & Geomechanics Abstracts*, 12(8), 241–253.

Allekotte, G. (Electrical Engineer). (Personal communication, 2006).

Aul, G. & Ray, R, Jr. (1991) Optimizing methane drainage systems to reduce mine ventilation requirements. *Proceedings of the 5th U.S. Mine Ventilation Symposium*, Y.J. Wang, ed., Society for Mining, Metallurgy, and Exploration, Inc., Littleton, CO, pp. 638–646.

Barczak, T.M. (1987) Rigid-body and elastic solutions to shield mechanics. Final report submitted to US Bureau of Mines, RI 9144. p. 20.

Barczak, T.M. (1989) Advanced guidelines for performance testing of two-legged longwall shields. Final report submitted to US Bureau of Mines, 1C 9231. p. 149.

Barczak, T.M. (1990) Shield mechanics and critical load studies for unsymmetric contact configurations. Final report submitted to US Bureau of Mines, RI 9290. p. 81.

Barczak, T.M. (1999) Design methodology for standing secondary roof support in longwall tailgates. In: Peng, S.S. (ed) *Proceedings of 18th International Conference on Ground Control in Mining, 1999, Morgantown, WV*. pp. 136–148.

Barczak, T.M. (2001a) Design considerations for the next generation of longwall shields. *Proceedings of Longwall USA Conference and Exhibit, June 2001, Pittsburgh, PA*. pp. 17–38.

Barczak, T.M. (2001b) Updating the NIOSH support technology optimization program (STOP) with new support technologies and additional design features. In: Peng, S.S. (ed) *Proceeding of the 20th International Conference on Ground Control in Mining, 2001, Morgantown, WV*. pp. 337–346.

Barczak, T.M. & Garson, R.C. (1986) Shield mechanics and resultant load vector studies. Final report submitted to US Bureau of Mines, RI 9027. p. 48.

Barczak, T.M. & Gearhart, D. (1998) Performance and safety considerations of hydraulic support systems. In: Peng, S.S. (ed) *Proceedings of 17th International Conference on Ground Control in Mining, August 1998, Morgantown, WV*. pp. 176–186.

Barczak, T.M. & Oyler, D.O. (1991) A model of shield-strata interaction and its implications for active shield setting requirements. Final report submitted to US Bureau of Mines, RI 9394. p. 13.

Barczak, T.M. & Schwemmer, D.E. (1988) Horizontal and vertical load transferring mechanisms in longwall roof supports. Final report submitted to US Bureau of Mines, RI 9188. p. 24.

Barczak, T.M. & Tadolini, S. (2008) Pumpable roof supports: An evolution in longwall roof support technology. *Society for Mining, Metallurgy & Exploration Annual Meeting, Feb. 2008*. p. 13.

Barczak, T., Tadolini, S. & Zhang, P. (2007) Evaluation of support and ground response as longwall face advanced into and widens pre-drive recovery room. In: *Proceedings of the 26th International Conference on Ground Control in Mining, Morgantown, WV*. pp. 160–172.

Barry, A.J. & Nair, O.B. (1970) In situ tests of bearing capacity of roof and floor in selected bituminous coal mines. Final report submitted to US Bureau of Mines, RI 7406. p. 20.

Barry, A.J., Nair, O.B. & Miller, J.S. (1969) Specifications for selected hydraulic-powered roof supports. Final report submitted to US Bureau of Mines, IC 8424. p. 15.

Bassier, R. & Migenda, P. (2003) Shield your performance. *World Coal*, 12(12), 25–28.

Bates, J.J. (1978) An analysis of powered support behavior. *Mining Engineer*, 137(203), 681–692.

Bauer, E.R., Reeves, E.R. & Durr, T.M. (2005) Testing and evaluation of an engineering noise control on a longwall stage loader. *Society for Mining, Metallurgy & Exploration Annual Meeting, 2005, Salt Lake City, UT*. p. 12.

Beck, K., Mishra, M., Chen, J.S., Morsy, K., Gadde, M. & Peng, S.S. (2004) Effect of the approaching longwall faces on barrier and entry stability. In: Peng, S.S. (ed) *Proceedings of 23rd International Conference on Ground Control in Mining, August 2004, Morgantown, WV*. pp. 20–26.

Bessinger, S.L. (1991) CONSOL refines automated longwall. *COAL*, 96(12), 41–43.

Bessinger, S.L. (1996) *A Review of CONSOL Innovations*. Paper presented at the Longwall USA Conference and Exhibit, June, Pittsburgh, PA. p. 14.

Bessinger, S.L. (2001) Let's talk longwall mining perspectives for both worlds. *Joy Machinery*, 7(2).

Bessinger, S.L. (Research Engineer). (Personal communication, 2006).

Bomcke, A. (1998) Highly optimized chain performance. *Proceeding of the Longwall USA Conference and Exhibit, June, Pittsburgh, PA*. Mining Media, Jacksonville, FL, pp. 157–161.

Bowles, J.E. (1965) *Foundation Analysis and Design*. McGraw-Hill, New York. pp. 42–103.

Broadfoot, A.R. & Betz, R.E. (1997) Power requirements prediction for armored face conveyors. *IEEE Transactions on Industry Application*, 33(1), A93–A101.

Brooker, C.M. (1979) Theoretical and practical aspects of cutting and loading by shearer drums. *Colliery Guardian Coal International*, 227(1), 9–16 and 227(4), 0–27.

Bryja, J. & Beck, K. (2005) Considerations in evaluating wider longwall faces. *Proceedings of Longwall USA Conference and Exhibit, 2005, Pittsburgh, PA*. p. 11.

Burkhart, J.E., McCawley, M.A. & Wheeler, R.W. (1987) Particle size distribution in underground coal mines. *American Industrial Hygiene Association Journal*, 48(2), 122–126.

Camargo, H.E., Azman, A.S. & Alcorn, L. (2016) Development of noise controls for longwall shearer cutting drums. *Noise Control Engineering Journal*, 64(5), 573–585.

Camargo, H.E., Yantek, D.S., Randolph, R., Schwartz, K., Ravetta, P. & Burdisso, R. (2013) Noise source identification on a stageloader system. *NOISE-CON, 26–28 August 2013, Denver, Colorado*. p. 9.

Campoli, A.A., Barton, T.M., Van Dyke, F.C. & Gauna, M. (1990) *Mitigating Destructive Longwall Bumps Through Conventional Gate Entry Design*. U.S. Bureau of Mines, RI 9325. p. 45.

Campoli, A.A., Barton, T.M., Van Dyke, F.C. & Gauna, M. (1993) *Gob and Gate Road Reaction to Longwall Mining in Bump-Prone Strata*. U.S. Bureau of Mines, RI 9445. p. 95.

Castronova, V. (2000) From coal mine dust to quartz: Mechanisms of pulmonary pathogenicity. *Inhalation Toxicology*, 12(3), 7–14.

Caterpillar Inc. (2018) Efficient mining of high seams with automated LTCC operations. Caterpillar Inc. p. 15.

Caudill, D. (2005) Quality and pride key to century. *American Longwall Magazine*, 18–21.

Cecala, A.B. & Jankowski, R.A. (1986) How to lower methane concentration around longwall shearer. *Proceedings of Longwall USA Conference and Exhibit, June 1986, Pittsburgh, PA*. pp. 358–362.

CFR (Code of Federal Regulations) (2017) 797. Title 30 Mineral Resources, Part 1–199.

Chang, Y.Y. (1971) Investigations into the relation between percentage of coal lumps produced, loading capacity, and the types of cutting drums. *Translated from Coal Mining Mechanization and Automation*, Poland, No. 1, 1971. pp. 1–25 (in Chinese).

Chen, J.S., Mishra, M., Zabrosky, C. & DeMichiei, J. (1999) Experience of longwall face support selection at RAG American coal company. *Proceedings of Longwall USA Conference and Exhibit, June 1999, Pittsburgh, PA*. pp. 53–77.

Chen, J.S. & Peng, S.S. (1999) Design of longwall face support by use of neural network models. *Translated from the Institute of Mining and Metallurgy, Section A, Mining Industry*, A143–A152.

Chen, J.S. & Peng, S.S. (2003) Analysis of longwall face roof activity using shield leg pressure. *Proceedings of Longwall USA Conference and Exhibit, June 2003, Pittsburgh, PA*. pp. 163–174.

Cheng, J.Y. (2015) Theory and technique of roof weighting prediction and roof fall warning at fully-mechanized longwall face[D]. China University of Mining and Technology.

Cheng, J.Y. & Peng, S.S. (2017) *Study on the Factors Influencing the Load Capacity of Shield, In Advances in Coal Mine Ground Control*, Woodhead Publishing, Cambridge, UK, 2017. Chapter 2. pp. 35–65. doi:10.1016/B978-0-08-101225-3.00002-5

Cheng, J.Y., Wan, Z.J. & Ji, Y.L. (2018) Shield-roof interaction in longwall panels: Insights from field data and their application to ground control. *Advances in Civil Engineering*, 12, 1–18. doi:10.1155/2018/3031714.

Cheng, J.Y., Wan, Z.J., Peng, S.S., Liu, S.F. & Ji, Y.L. (2015) What can the changes in shield resistance tell us during the period of Shearer's cutting and shields' advance? *International Journal of Mining Science and Technology*, 25(3), 361–367.

Chiang, H.S. (1980) *Fully-Mechanized Longwall Coal Mining*. Shandong Scientific and Technical Publishing Co., China. pp. 172–235. (in Chinese).

Chiang, H.S., Peng, S.S. & Yang, S.L. (1996) Measurements and 3-D/2-phase modeling of respirable dust distribution in longwall faces. *MINING Technology*, 79(906), 52–56.

Chugh, Y.P., Arti, A. & Dougherty, J. (1989) Laboratory and field characterization of immediate Floor Strata in Illinois Basin coal mines. *Proceedings of 30th U.S. Symposium on Rock Mechanics, 1989, West Virginia University, WV*. pp. 47–54.

Coal. (1988–1996) US Longwall Census, February each year, Mining Media International.

Coal Age. (1976–1987, 1997–2019) US Longwall Census, February each year. Mining Media International.

Coal Mining. (1984–1987) US Longwall Census, February each year. Mining Media International.

Coal Mining and Processing. (1982–1983) US Longwall Census, February each year. Mining Media International.

Coates, D.F.(1978) *Rock Mechanics Principles. Canadian Department of Energy, Mineral and Resources*. Chapter 7. p. 78.

Coates, D.F. & Gyenge, M. (1965) Plate-load testing on rock for deformation and strength properties. *Testing Techniques for Rock Mechanics, ASTM, STP 402*, 19–40.

Coe, C.J. & Stowe, S.M. (1984) Evaluating the impact of longwall coal mining on the hydrologic balance. *Proceedings of 1984 Symposium on Surface Mining, Hydrology, Sedimentation, and Reclamation on Surface Mining, Lexington, KY*. pp. 395–403.

Colwell, M. (2006) *A Study of the Mechanics of Coal Mine Rib Deformation and Rib Support as a Basis for Engineering Design*. Ph.D. dissertation, University of Queensland, Queensland, Australia. p. 167.

COMINEC (1976) Conceptual design of an automated longwall mining system. Final report submitted to US Bureau of Mines. pp. 50–75.

Conroy, P. (1977) Rock mechanics studies: Longwall demonstration, old ben mine no. 24, Benton, Illinois, phase III. Preliminary Report Panel 1. Preliminary report submitted to US Bureau of Mines, p. 39.

Conroy, P. (1979) Rock mechanics studies: Longwall demonstration, old ben mine no. 24, Benton, Illinois, Phase III, Preliminary Report Panel 2. Preliminary report submitted to US Bureau of Mines, p. 18.

Conroy, P. & Curth, E.A. (1981) Longwall mining in Illinois. In: Ramani, R.J. *et al.* (eds) *Longwall-Shortwall Mining State-of-the-Art*, Society for Mining, Metallurgy & Exploration Annual Meeting, 1981. pp. 191–200.

Conroy, P. & Gyarmaty, J. (1982) Planning subsidence monitoring program over longwall panels. In *State-of-the-Art of Ground Control in Longwall Mining and Mining Subsidence*, Society for Mining, Metallurgy & Exploration Annual Meeting, September 1982. pp. 225–234.

Cooper, J.R. (1999) Effective AFC control systems and the use of high tensile steels in face conveyor. *Proceedings of Longwall USA Conference and Exhibits*, June 1999, Pittsburgh, PA.

Courtney, W.R. & Cheng, L. (1977) Control of respirable dust by improved water sprays. Final report submitted to US Bureau of Mines, IC 8753. pp. 92–106.

CSIRO (2005) Landmark longwall automation project C10100: ACRP. Final report. p. 176.

Curl, S.J. (1978) *Methane Prediction in Coal Mines*. IEA Coal Research, London. pp. 23–24.

Dahl, H.D. & Parsons, R.C. (1972) Ground control Studies in Humphrey No.7 Mine. Trans. SME-AIME, 252, June 1972. pp. 2–11.

Dahl, H. D. & Vonschonfeldt, H.A. (1976) Rock mechanics elements for coal mine design. *Proceedings of 17th U.S. Symp. on Rock Mech, Univ. of Utah, 1976, Paper No. 3A1*. p. 9.

Davis, J.G. & Krickovic, S. (1973) Gob degasification research-a case history in methane control in Eastern U.S. coal mines. Final report submitted to US Bureau of Mines, 1C 8621. pp. 62–72.

Demichiei, J. & Beck, K. (2001) Continuous miner vs longwall productivity measures. *Proceedings of Longwall USA Conference and Exhibits, June 2001, Pittsburgh, PA*. pp. 39–49.

Dey, A. (1985) *Orthogonal Fractional Factorial Design*. Wiley, New York. pp. 1–44.

Diamond, W.P. (1993) Methane control for underground coal mines. Final report submitted to US Bureau of Mines, IC 9395. p. 55.

Eichbaum, F. & Bendmayr, H. (1974) Vergleichende Untersuchungen and Schramwalzen. *Gluckauf*, 110(23), 45–51.

Esterhuizen, G.S., Gearhart, D.F., Klemetti, T., Dougherty, H. & Van Dyke, M. (2018) Analysis of gateroad stability at two longwall mines based on field monitoring research and numerical model analysis. In: *Proceedings of the 37th International Conference on Ground Control in Mining, Morgantown, WV*. SME. pp. 141–150.

Fama, J.P. (2005) Longwall Mining. [Lecture] a lecture note for MSHA Academy, Beckley, WV, 2005.

Fayol, M. (1913) Effects of coal mining on the surface. *Colliery Engineering*, 33, 548–552, 617–622.

Ferguson, H. F. (1974) Geologic observations and geotechnical effects of valley stress relief in the Allegheny Plateau. *The ASCE, Water Resources Engineering Meeting, 1974, Los Angeles, USA*. p. 31.

Fiscor S. (2017) Improved safety and productivity through advanced shearer automation. *Coal Age*, 22(7), 16–23.

Forrelli, J. (1999) *Making used longwall equipment work*. Paper presented at the Longwall USA Conference and Exhibit, June 1999, Pittsburgh, PA. p. 20.

Foster-Miller Associates, Inc. (1982) Evaluate longwall dust sources and control technology. Interim technical progress report submitted to US Bureau of Mines, HO202016. p. 85.

Fourie, W. (2019) *The Digital Mine Eco-System*. Komatsu Mining Corporation, p. 8.

Franks, J.R. (1996) Analysis of audigrams for a large cohort of noise-exposed miners. Interim Report submitted to NIOSH, Cincinnati, OH.

Fusser, B. (2005) Requirements for an optimum hydraulic system resulting from the productivity increase in longwall faces. *Proceedings of Longwall USA Conference and Exhibit, June 2005, Pittsburgh, PA*.

Garson, R.C., Yavorsky, P.M., Barczak, T.M. & Maayeh, F.S. (1982) State of the art testing of powered roof supports. *Proceedings of 2nd Conference on Ground Control in Mining, July 1982, West Virginia University, Morgantown, WV*. pp. 64–77.

Gearhart, D.F., Esterhuizen, G.S. & Tulu, I.B. (2017) Change in stress and displacement caused by longwall panel retreats. In: *Proceedings of the 36th International Conference on Ground Control in Mining. Morgantown, WV*, SME. pp. 313–320.

Gearhart, D.F., Esterhuizen, G.S. & Zhang, P. (2018) The effect of passing longwall face on the roadway, pillars, and standing supports in Northern West Virginia. In: *Proceedings of the 37th International Conference on Ground Control in Mining. Morgantown, WV*, SME. pp. 359–365.

Gentry, D.W. (1979) Rock mechanics instrumentation program for Kaiser Steel corporation's demonstration of shield-type longwall supports at York Canyon Mine, Raton, New Mexico. Final Tech. Report submitted to U.S. Department of Energy. May, 1979, p. 318.

Graham, J.J. (1978) *Control Valve Systems*. Colliery Guardian Coal International, June. pp. 41–46.

Gregor, M. (1969) Effects of cutting speeds, bit configurations, rake angle, clearance angle and bit conditions on cutting force and transverse reaction force. *Gluckauf-Furs-Chungsheft*, 30(1), 13–19 (in German).

Guillon, P. & Pechalat, F. (1976) Study of shearer cutting. *Mining Industry*, 65–91. (in French).

Hafera, P., Peng, S.S., Chiang, H.S. & Zabrosky, C.E. (1989) Underground evaluation of four types of shield supports. *Proceedings of the Longwall USA Conference and Exhibit, June 1989, Pittsburgh, PA*. pp. 45–85.

Haramy, K. & Fejes, A. (1992) Characterization of overburden response to longwall mining in the western united states. In: *Proceedings of 11th International Conference on Ground Control in Mining, the University of Wollongong, NSW*. pp. 334–344.

Hart, W.M. (1988) Integration of demand/design concepts for improving longwall productivity. American Mining Congress, Session Papers, Vol.1 Technical. pp. 153–200.

Heasley, K., Worley, P. & Zhang, Y. (2003) Stress analysis and support design for longwall mine-through entries (a case study). *Proceeding of 22nd International Ground Control Conference in Mining, August 2003, Morgantown, WV*. pp. 11–18.

Hendon, G. (1998) Gateroad pillar extraction experience at Jim Walter resources. In: Peng, S.S. (ed) *Proceedings of 17th International Conference on Ground Control in Mining, August 1998, Morgantown, WV*. pp. 1–10.

Hsiung, S.M. & Peng, S.S. (1985) First caving and its effects – a case study. In: Peng, S.S. (ed) *Proceedings of 4th International Conference on Ground Control in Mining, July 1985, Morgantown, WV*. pp. 83–93.

Hsiung, S.M. & Peng, S.S. (1987) Design guidelines for multiple seam mining. *Coal Mining*, Part 1, 24(9), 42–46 and Part 2, 24(10), 48–50.

Huczko, R. (sales manager). (Personal communication, 2006).

Hutchinson, T.L. (product manager, Swanson Industries). (Personal communication, 2006).

Hutchinson, T.L. (2013) Fluid power generation and transportation systems for longwalls. *Proceedings of Longwall USA Conference and Exhibit, 2013, Pittsburgh, PA*. p. 6.

Hutchinson, T.L. (consultant). (Personal communication, 2019).

Ingram, G.R. (1994) Longwall automation at consolidation coal company's Blacksville No. 2 mine. *Proceedings of Longwall USA Conference and Exhibit, June 1994, Pittsburgh, PA*. pp. 27–41.

Jackson, D.J.H. (1979) The testing of shield supports. *The Mining Engineer*, 138(211), 763–771.

Jacobi, O. (1976) *Praxis der Gebirgsbeherrschung*. Verlag-Gluckauf GMBH, Essen, W. Germany. p. 494.

Jayaraman, N.I. & Grigal, D. (1977) Dust control on a longwall face with shearer-mounted dust collector. Final report submitted to US Bureau of Mines, RI 8248. p. 10.

Jayaraman, N.I., Jankowski, R.A. & Kissell, F.N. (1985) Improved shearer-clearer system for double-drum shearers on longwall faces. Final report submitted to U.S. Department of the Interior, Bureau of Mines, RI 8963. p. 56.

Jenkins, J.D. (1955) Mechanics of floor penetration in mines. *Iron and Coal Trade Review*, 171, 541–547.

Jenkins, J.D. (1957, 1958) Some investigation into the bearing capacity of floors in the North Cumberland and Durham Coalfields. *Trans. Institute of Mining Engineers (London)*, 117, 726–736.

Jiang, J.M., Peng, S.S. & Chen, J.S. (1989) DEPOWS – a powered support selection model. *Proceedings of 30th U.S. Sym. on Rock Mechanics, June 1989 Morgantown, WV*. pp. 141–148.

Johann-Essex, V., Keles, C., Rezaee, M., Scaggs-Witte, M. & Sarver, E. (2017) Respirable coal mine dust characteristics in samples collected in Central and Northern Appalachia. *International Journal of Coal Geology*, 182, 85–93.

Johnson, K.L. (1992) Influence of topography on the effect of longwall mining on shallow aquifers in the Appalachian coalfield. In: Peng, S.S. (ed) *Proceedings of 2nd Workshop on Surface Subsidence Due to Underground Mining, 1992, Morgantown, WV*. pp. 197–203.

Johnson, K.L. (1997) Effects of longwall mining on streamflow in the Pittsburgh Seam Basin. In: Peng, S.S. (ed) *Proceedings of 16th International Conference on Ground Control in Mining, 1997, Morgantown, WV*. pp. 25–32.

Jones, M. (2017) Longwall top coal caving at broadmeadow mine. Presentation at the 2017 Longwall USA & Exhibit, Pittsburgh, PA. 25 ppts.

Junker, M. & Lemke, M. (2018) *Technical Developments in Coal Winning*. Vulkan Verlag, Essen, Germany. p. 522.

Kelly, M. (2005) Outcome of the landmark longwall automation project with reference to ground control issues. In: Peng, S.S. *et al.* (eds) *Proceedings of 24th International Conference on Ground Control in Mining, August 2005, Morgantown, WV*. pp. 66–73.

Kidybinski, A. (1977) Methods of investigations, estimation and classification of roof in mines in the U.S.A. for the selection of suitable mechanized supports for longwalls. Final report submitted to US Department of Energy, DOE/TIC-11483. p. 116.

Kissell, F.N., Colinet, J.F. & Organiscak, J.A. (2003) Longwall dust control. In: Kissel, F.N. (ed) *The Handbook for Dust Control in Mining*. NIOSH IC 9465. pp. 39–57.

Kissell, F.N., Jayaraman, N.I., Taylor, C. & Jankowski, R. (1981) Reducing dust at longwall shearer by confining the dust cloud to the face. Final report submitted to US Bureau of Mines Tech Progress Report, 111. p. 21.

Komatsu Mining. (2018) *Longwall Controls and Automation-Longwall System* (unpublished ppts). 12 ppts.

Kopex Group. (2013) *Experience from Operating an Automated Longwall System Pniowek Coal Mine*, PPT presentation at the USA Longwall Conference and Exhibit, Pittsburgh, PA, 2013.

Kratzsch, H. (1983) *Mining Subsidence Engineering*. Springer, Berlin, Germany. p. 543.

Kravits, S. & Dubois, G. (2014) Coal bed methane-from prospect to pipeline. In: Thakur, P. *et al.* (eds) *Horizontal Coalbed Methane Wells Drilled from Surface*. Elsevier, Amsterdam, The Netherlands. pp. 137–153.

Kravits, S. & Yearsley, M. (2017) Today's directional drilling. *Proceedings of Longwall USA Conference and Exhibit, 2017, Pittsburgh, PA*, 31 ppts.

Kuzyniazouv, G.N. (1954) The mechanics of interactions between surrounding strata and supports on longwall faces in slightly-inclined coal seam. *In Colloq. on Investigation into the Suitability of Mine Pressures on Mechanized Supports, Moscow*. p. 40.

Langefeld, O. & Paschedag, U. (2018) Longwall mining – developments and transfer. *The SOMP Regional Conference on German Hard Coal Mining -Technical Footprint of German Hard Coal Mining, October 18, Technical University of Applied Sciences Georg Agricola, Bochum*.

Larson, M.K. & Whyatt, J.K. (2012) Load transfer distance calibration of a coal panel scale model: A case study. In: Barczak *et al.* (eds) *Proceedings of the 31st International Conference on Ground Control in Mining, July 31–August 2, West Virginia University, Morgantown, WV*. pp. 195–205.

Leavitt, B.R. (1992) Effects of longwall coal mining on rural water supplies and stress-relief fracture flow systems. In: Peng, S.S. (ed) *Proceedings of 2nd Workshop on Surface Subsidence due to Underground Mining, Morgantown, WV*. pp. 228–236.

Lee, R.D. (1961) Testing mine floors. *Colliery Engineering*, 38, 255–261.

Listak, J.M. & Zelanko, J.C. (1987) An assessment of the effects of longwall chain pillar configuration on gate road stability. 28th US symposium on Rock Mechanics, Tucson.

Lu, J. & Hasenfus, G. (2018) Challenges of moving the first right-handed longwall panel in a new reserve block in Pittsburgh seam. In: *Proceedings of the 37th International Conference on Ground Control in Mining. Morgantown, WV*, SME. pp. 53–62.

Lu, P.H. (1982) Rock mechanics instrumentation and monitoring for ground control around longwall panels. *Proceedings of Symp. on State-of-the-Art of Ground control in Longwall Mining and Mining Subsidence, Society for Mining, Metallurgy & Exploration Annual Meeting, 1982*. pp. 159–166.

Luo, Y. (1989) *Integrated Computer Model for Predicting Surface Subsidence Due to Under-Ground Coal Mining – CISPM*. Ph.D. Dissertation, West Virginia University, Morgantown, WV, UMI order No. 9020385. p. 168.

Luo, Y. & Peng, S.S. (1990) A mathematical model for predicting subsidence over chain pillars between mined-out longwall panels. In: Elifrits, C.D. (ed) *Proceedings of AEG National Symposium on Mine Subsidence – Prediction and Control. Association of Engineering Geologists*. pp. 247–257.

Luo, Y. & Peng, S.S. (1991) Some new findings from surface subsidence monitoring over longwall panels. *Mining Engineering, Society for Mining, Metallurgy & Exploration Annual Meeting, 1991, Littleton, CO*. pp. 1261–1264.

Luo, Y. & Peng, S.S. (1992) A comprehensive dynamic subsidence prediction model for longwall operations. In: Aziz, N.I. & Peng, S.S. (eds) *Proceedings of 11th International Conference on Ground Control in Mining, 1992, University of Wollongong, Wollongong, Australia*. pp. 511–516.

Luo, Y. & Peng, S.S. (1997) Subsidence prediction influence assessment and damage control. In: Peng, S.S. (ed) *Proceedings of 16th International Conference on Ground Control in Mining, August, Morgantown, WV*. pp. 50–57.

Luo, Y. & Peng, S.S. (1999) Integrated approach for predicting mining subsidence in Hilly Terrain. *Mining Engineering, Society for Mining, Metallurgy & Exploration Annual Meeting, 1999, Littleton, CO*. pp. 100–104.

Luo, Y., Peng, S.S., Preusse, A., Herzog, C., Wings, R. & Sroka, A. (2001) Effects of face advance rate on the characteristics of subsidence process associated with US and German longwall mining operations. In: Peng, S.S. (ed) *Proceedings of 20th International Conference on Ground Control in Mining, August, Morgantown, WV*. pp. 140–148.

Luo, Y., Peng, S.S., Zabrosky, C. & Cole, J. (2003) Mitigating subsidence influences on residential structures caused by longwall mining operations. In: Peng, S.S. *et al.* (eds) *Proceedings of 22nd International Conference on Ground Control in Mining, Morgantown, WV*. pp. 352–359.

Lyman, J.A. (2017) Protecting your leg pockets with Loctite MR 5898. *Proceedings of Longwall USA Conference and Exhibit, 2017, Pittsburgh, PA*. 8 ppts.

Maleki, H. & Agapito, J. (1988) In-situ pillar strength determination for two-entry longwall gates. *Proceedings of 7th Conference on Ground Control in Mining, Morgantown, WV*. pp. 10–19.

Maleki, H., Fleck, K., Semborski, C. & Lafrentz, L. (2009) Regional two-seam stress analyses at the deer creek mine. *Proceedings of 28th Conference on Ground Control in Mining, Morgantown, WV*. pp. 299–304.

Mark, C. (1987) *Analysis of Longwall Pillar Stability*. Ph.D. Thesis, The Pennsylvania State University, Department of Mineral Engineering. p. 255.

Mark, C. (1990) Pillar design methods for longwall mining. Final report submitted to US Bureau of Mines, IC 9247. p. 53.

Mark, C. & Agioutantis, Z. (2018) Analysis of coal pillar stability (ACPS): A new generation of pillar design software. *Proceedings of the 37th International Conference on Ground Control in Mining, Morgantown, WV*. pp. 1–6.

Mark, C., Listak, J. & Bieniawski, Z.T. (1988) Yielding coal pillars- Field measurements and analysis of design methods. In: Cundall *et al.* (eds) *Key Questions in Rock Mechanics*, Bakema, Rotterdam, ISBN 9061918359. pp. 88–98.

Mark, C. & Mucho, T.P. (1994) Longwall mine design for control of horizontal stress. In New Technology for Longwall Ground Control. *Proceedings of Final Report Submitted to US Bureau of Mines Technology Transfer Seminar, 1994, SP 01–94*. pp. 53–76.

Martin, E., Carr, F. & Hendon, G. (1988) Strata control advance at Jim Walter resources, mining division. In: Peng, S.S. (ed) *Proceedings of 7th International Conference on Ground Control in Mining, 1988, Morgantown, WV*. pp. 66–71.

Matetic, R.J. & Trevits, M.A. (1990) Case study of longwall mining effects on water wells. *Society for Mining, Metallurgy & Exploration Annual Meeting, 1990, Salt Lake City, Utah*. pp. 1–7.

Matetic, R.J. & Trevits, M.A. (1991) *A Case Study of Longwall Mining and Near-Surface*. Session Papers from the American Mining Congress Coal Convention, Pittsburgh. pp. 446–472.

Matetic, R.J. & Trevits, M.A. (1992) Longwall mining and its impact on ground water quantity and quality at a mine site in the Northern Appalachia Coalfield. From The Stephen B. Thacker CDC Library. pp. 573–587.

McShannon, G. (2006) Cowless, radial shearer drums are the future. *Coal International*, 20–24.

Mills, K., Jeffrey, R. & Jones, D. (2000) Successful application of hydraulic fracturing to control wind-blast hazard at Moonee Colliery, NSW, Australia. In: Peng, S.S. (ed) *Proceedings of 19th International Conference on Ground Control in Mining, August 2000, Morgantown, WV*. pp. 45–50.

Morley, L.A. (1990) Mine power systems. Final report submitted to US Bureau of Mines, IC 9258. p. 437. Available from: http://hyperphysics.phy-astr.gsu.edu/hbase/Tables/wirega.html#c1.

Morsy, K. & Peng, S.S. (2003) New approach to evaluate the stability of yield pillars. In: Peng, S.S. (ed) *Proceedings of 22nd International Conference on Ground Control in Mining, August 2003, Morgantown, WV*. pp. 371–381.

Morsy, K. & Peng, S.S. (2006) Detailed stress analysis of longwall panels. In: Peng, S.S. *et al.* (eds) *Proceedings of 25th International Conference on Ground Control in Mining, August 2006, Morgantown, WV*. pp. 18–33.

Mowrey, G.L. (1991, 1992) Horizon control holds key to automation. *Coal*, Part I, 44–48 and Part II, 47–51.

MSHA (Mine Safety and Health Administration). (1999) Mine electricity. CI 7, 76.

MSHA (Mine Health and Safety Administration). (2000) *Longwall Dust Control Plan Parameters and Dust Sample Results*. Airlington, VA. [Online] Available from: www.msha.gov/sandHINFO/LONGWALL/LWMENU.HTM.

MSHA (Mine Health and Safety Administration). (2002) Bleeder and gob ventilation systems. ventilation specialists training course, course text, revised 2002. p. 163.

MSHA (Mine Health and Safety Administration). (2019). Practical ways to reduce exposure to coal dust in longwall mining – a toolbox. Available from: https://arlweb.msha.gov/sandhinfo/longwall/lwtoolbox.pdf.

Mucho, T.P., Barczak, T.M. & Dolinar, D.R. (1999) Design methodology for standing secondary roof support in longwall tailgates. In: Peng, S.S. (ed) *Proceedings of 18th International Conference on Ground Control in Mining, August 1999, Morgantown, WV*. pp. 136–148.

National Academies of Sciences, Engineering, Medicine. (2018) Monitoring and sampling approaches to assess underground coal mine dust exposures. A consensus study report of the National Academies of Sciences, Engineering, Medicine, Washington DC. p. 150.

Nelson, M.G. (1989) *Simulation of Boundary Coal Thickness Sensor*. Ph.D. Dissertation, Department of Mining Engineering, College of Engineering and Mining Resources, West Virginia University, Morgantown, WV. p. 153.

Niederriter, E. (2005) Cost-effective way to lower longwall cutting height. *Proceedings of Longwall USA Conference and Exhibit, June 7–9, 2005, Pittsburgh, PA*.

Niewiadomski, G.E., Jankowski, R.A. & Kissel, F.N. (1982) Ten ways to reduce longwwall dust. *Mining Congress Journal*, 46–49.

NIOSH (National Institute for Occupational Safety and Health). (1995) Criteria for a recommended standard. Occupational Exposure to Respirable Coal Mine Dust, Cincinnati, OH. p. 336.

NIOSH (2016) Ground stress in mining (part 1): Measurements and observations at two Western U.S. longwall mines. By Larson, M.K., Lawson, H.E., Zahl, E.G. & Jones, T.H. U.S. Department of Health and Human Services, Spokane, WA, CDC.

Ondrey, R.S., Haney, R.A. & Tomb, T.F. (1995) Dust control parameters necessary to control dust on longwall and continuous mining operations. *Society for Mining, Metallurgy & Exploration Annual Meeting, 1995, Denver, CO, Preprint No. 95–145*. p. 6.

Organiscak, J.A. & Colinet, J.F. (1999) Influence of coal properties and dust control parameters on longwall respirable dust level. *Mining Engineering*, 51(9), 41–48.

Ostermann, W. (1966) Die Schramleistung von Walzenschramladern in Abhangigkeit von der Schnittbreite und der Drehzahl der Walzen. *Gluckauf*, 102(4).

Pack, D. (1991) Longwall emulsion characteristics. *Mining Technology*, 115–118.

PA DEP, State of Pennsylvania. (1957) Gas well pillar study. Department of Mines and Mineral Industries, Oil and Gas Division. p. 56.

Park, D.W., Jiang, J.M., Carr, F. & Hendon, G.W. (1992) Analysis of longwall shields and their interaction with surrounding strata. *Proceedings of 11th Conference on Ground Control in Mining, 1992, Morgantown, WV*. pp. 109–116.

Patrick, C. (1992) *Investigation, Analysis, and Modeling of Longwall Face-to-Face Transfer*. PhD dissertation, Virginia Tech. p. 269.

Peng, S.S. (1976) Roof control studies at Olga No. 1 Mines, Coalwood, WV. Final report submitted to the U.S. Bureau of Mines, No. J0155125. p. 22.

Peng, S.S. (1980) *Roof Falls in Underground Coal Mines*. Department of Mining Engineering, West Virginia University, Morgantown, WV, TR 80–4. p. 36.

Peng, S.S. (1986) *Coal Mine Ground Control*. 2nd edition. Wiley, New York. p. 476.

Peng, S.S. (1987) Longwall automation grows. *Coal Mining*, 24(5), 48–61.

Peng, S.S. (1990) Design of active horizontal force for shield supports for controlling roof falls. *Mining Engineer*, 149(345), 457–461.

Peng, S.S. (1995) Rapid development. *COAL*, 100(1), 45–47.

Peng, S.S. (1998) What can a shield leg pressure tell us? *Coal Age*, 103(3), 41–43.

Peng, S.S. (2000) Cutting through open entries require proper support. *Coal Age*, 105(6), 37–40.

Peng, S.S. (2006) *Longwall Mining*. 2nd edition. S.S. Peng Publisher, Morgantown, WV. p. 550.

Peng, S.S. (2008) *Coal Mine Ground Control*. 3rd edition. S.S. Peng Publisher, Morgantown, WV, p. 750.

Peng, S.S. (2013) Squire Jim longwall mine design. Final Report submitted to Squire Jim Mining Co. p. 354.

Peng, S.S. & Chen, J.S. (1991) Control method of the unsupported area between canopy tip and faceline in longwall faces under weak roof conditions. *Mining Engineer*, 179–183.

Peng, S.S. & Chen, J.S. (1992) Torsional strength of the caving shield subject to a bias loading on the canopy. *Journal of Mining Research (India)*, 1(3), 39–48.

Peng, S.S. & Chen, J.S. (1993) Determination of floor pressure under the base plate of powered support. *Mining Engineer*, 246–252.

Peng, S.S., Cheng, J.S., Du, F. & Xu, Y. (2019) Underground ground control monitoring and interpretation, and numerical modeling, and shield capacity design. *International Journal of Mining Science and technology*, 29(1), 79–85.

Peng, S.S. & Cheng, S.L. (1981) Predicting surface subsidence for damage prevention. *Coal Mining and Processing*, 18(5), 84–95.

Peng, S.S. & Chiang, H.S. (1982) Roof stability in longwall coal face. *Proceedings of 1st Int. Conf. on Underground Stability, August 1982, Vancouver, Canada*. pp. 295–335.

Peng, S.S. & Chiang, H.S. (1984) *Longwall Mining*. 1st edition. Wiley, New York. p. 879.

Peng, S.S. & Chiang, H.S. (2000) *U.S. Longwall Mining Technique*. Lecture note. West Virginia University, December. p. 90.

Peng, S.S., Chiang, H.S. & Lu, D.F. (1982) Roof behaviors and support requirements for the shield-supported longwall faces. *Proceedings of Symposium on State-of-the-Art of Ground Control in Longwall Mining and Mining Subsidence, Society for Mining, Metallurgy & Exploration Annual Meeting, September 1982*. pp. 107–130.

Peng, S.S., Chiang, H.S., Yang, S.L. & Zhao, G.J. (1996) Field measurements and 2-D/2-phase modeling of respirable dust in longwall faces using a bi-directional cutting pattern. *Applied Occupational and Environmental Hygiene*, 11(7), July, 669–676.

Peng, S.S., Chiang, H.S., Zhu, D.R. & Jiang, Y.M. (1989) Model for the active horizontal forces of shield support. *Trans. Society for Mining, Metallurgy & Exploration Annual Meeting, 286, 1989.* pp. 1868–1873.

Peng, S.S. & Finfinger, G. (2001, December) Geology, roof control and mine design. *Coal Age*, 106(12), 29–31.

Peng, S.S., Hsiung, S.M. & Jiang, J.M. (1987) Method of determining the rational load capacity of shield support at longwall faces. *Mining Engineer*, 147(313), 161–167.

Peng, S.S. & Luo, Y. (1992) Comprehensive and integrated subsidence prediction model – CISPM (V2.0). In: Peng, S.S. (ed) *Proceedings of 3rd Workshop on Surface Subsidence Due to Underground Mining, West Virginia University, Morgantown, WV*. pp. 22–31.

Peng, S.S. & Luo, Y. (1995) Computer model analyzes longwall drum bit wear. *Coal*, 100(10), 58–60.

Peng, S.S., Luo, Y. and Zhang, Z.M. (1995) Subsidence parameters – their definitions and determination. *Society for Mining, Metallurgy & Exploration Annual Meeting Transactions, 1995, Vol. 300, Littleton, CO*. pp. 60–65.

Peng, S.S., Morsy, K., Zhang, Y.Q., Luo, Y. & Heasley, K. (2002) Technique for assessing the effects of longwall mining on gas wells – two case studies. *Society for Mining, Metallurgy & Exploration Annual Meeting, 2002, Preprint No.02–195*. p. 16.

Peng, S.S. & Park, D.W. (1977) Rock mechanics study for the shortwall mining at the valley Camp No. 3 mine, Triadelphia, WV. Final report submitted to the U.S. Bureau of Mines, No. J0155125. p. 30.

Peng, S.S. & Su, D.W.H. (1983) The causes of cyclic excessive convergence in the longwall tailentry. *International Journal of Mining Engineering*, 1(1), 27–41.

Peng, S.S. & Tsang, P. (1994) Panel width effects on powered supports and gate entries by using 3-D computer modeling techniques. *Proceedings of Symposium of the Application of Numerical Modeling in Geotechnical Engineering, Sept. 1994, SANGORM, Pretoria, South Africa*. pp. 81–88.

Platt, J. (1956) Floor distortion. *Colliery Guardian*, 193, 303–305.

Pothini, R., Chiang, H.S. & Peng, S.S. (1992) Pressure distribution of 2-leg shield supports. In: Aziz, A.I. & Peng, S.S. (eds) *Proceedings of 11th International Conference on Ground Control, July, Australia*. pp. 160–169.

Price, R.J. & Pickering, M.H.B. (1981) Application of higher setting loads to powered supports in South Nottinghamshires area. *Mining Engineer*, 140, 841–848.

Pula, O., Chugh, Y.P. & Pytel, W.M. (1990) Estimation of weak floor strata properties and related safety factors for design of coal mine layout. *Proceedings of 31st Symposium on Rock Mechanics, 1990, Colorado School of Mines*. pp. 111–124.

Ralston, J.C., Reid, D.C., Dunn, M.T. & Hainsworth, D.W. (2015) Delivering enabling technology to achieve safer and more productive underground mining. *International Journal of Mining Science and Technology*, 25(6), 865–876.

Rashed, G. (2019) (Personal communication), unpublished data from NIOSH.

REI Drilling, Inc. (2006) [Online] Available from: www.reidrilling.com.

Reid, B. (1991) (managing editor, Coal News), (Personal Communication 1991).

Reid, W.J. (1991) American longwall mining. *Mining Engineer*,150(356), 1–9.

Richards, D. (1998) Longwall moves using MRS's. Unpublished report, Peabody Coal Company, Well Complex. p. 50.

Rider, J.P. & Colinet, J.F. (2007) Current dust control practices in U.S. longwalls. *Proceedings of Longwall USA Conference and Exhibit, 5–7 June 2007, Pittsburgh, PA*.

Rider, J.P. & Colinet, J.F. (2010) Controlling respirable dust on longwall mining operations. In: Colinet *et al.* (eds) *Chapter 3 in Best Practices for Dust Control in Coal Mining*. NIOSH IC 9517. pp. 17–39.

Rider, J.P., Colinet, J.F. & Prokop, A.E. (2001) Impact of control parameters on shearer-generated dust levels. *Society for Mining, Metallurgy & Exploration Annual Meeting, 2001, Denver, CO, Preprint No. 01–184*. p. 9.

Ringleff, H. & Rutherford, A. (2007) Australia's first LTCC face operation. Presentation at Mine Automation Conference, Aachen, Germany, May, 2007. 32 ppts.

Robinson, R., Hasenfus, G., Su, D. & Perr, R. (2007) Rehabilitation of underground surge bunker following massive rib wall failure. *Proceeding of 26th International Conference on Ground Control in Mining, July, 2007, Morgantown, WV*. pp. 104–112.

Roscoe, M.S. & Hartshorn, B.A. (1980) *Longwall Instrumentation – Capco Longwall, Inter Office Correspondence*. The North American Coal Corp., Powhatan Point, OH. p. 14.

Rowland, S. (2002) Moranbah North coal -3 years and 15 MT later. *AJM*. Available from: www.ibcoz.com.au.

Ruggieri, S.K., Muldoon, T.L., Schroeder, W., Babbitt, C. & Rajan, S. (1983) Optimizing water sprays for dust control on longwall shearer faces. Foster-Miller, Inc. U.S. Bureau of Mines, J0308019. NTIS No. PB 86–205408. p. 123.

Russell, R. (Consultant). (Personal communication, 2006).

Rutherford, A. (2001) Half-web benefits remain untapped. *Australia's Longwalls*, 42–44.

Rutherford, A. (2005) Cutting edge. *Australia's Longwalls*, 13–16.

Sanda, A.P. (1996) Demystifying emulsions. *Coal*, 45–48.

Schuerger, M.G. (1985) An investigation of longwall pillar stress history. In *Proceedings of 4th Conference on Ground Control in Mining, Morgantown, WV*. pp. 41–49.

Scovazzo, V.A. (2018) Mining effects on gas and oil wells pad NV-35 field experiment field monitoring. In *Proceedings of the 37nd International Conference on Ground Control in Mining, West Virginia University, Morgantown, WV*. pp. 30–43.

Seiler, J.P., Valoski, M.P. & Crivaro, M.A. (1994) Noise exposure in U.S. coal mines. MSHA, IR 1214. p. 15.

Smart, B.D.G. & Aziz, N. (1986) The influence of caving in the hirst and buli seams on powered support rating. *Ground Movement and Control Related to Coal Mining Symposium, 1986, Wollongong, Australia*. pp. 182–193.

Song, Z.C. & Deng, T.L. (1982) Manifestations of mine pressures and its relations to overlying strata movements. In: Peng, S.S. (ed) *Proceedings of 2nd Conference on Ground Control in Mining, July 1982, Morgantown, WV*. pp. 22–35.

Steffenino, J. (Product Manager) (Personal communication, February 12, 2019).

Su, D.W.H. (2016) Effects of longwall-induced stress and deformation on the stability and mechanical integrity of shale gas wells drilled through a longwall abutment pillar. *Proceedings of the 35th International Conference on Ground Control in Mining*. Morgantown, WV. pp. 119–125.

Su, D.W.H., McCaffrey, J., Barletta, L., Thomas, E. & Toothman, R.C. (2001) Hydraulic fracturing of sandstone and longwall roof control – implementation and evaluation. In: Peng, S.S. (ed) *Proceedings of 20th International Conference on Ground Control in Mining, August 2001, Morgantown, WV*. pp. 1–10.

Tadolini, S., Zhang, Y. & Peng, S.S. (2002) Pre-driven experimental longwall recovery room under weak roof conditions – design, implementation, and evaluation. In: Peng, S.S. (ed) *Proceedings of 21st International Conference on Ground Control in Mining, 2002, Morgantown, WV*. pp. 1–10.

Target Drilling Inc. (2006) [Online]. Available from: www.targetdrilling.com.

Thakur, P.C. (2006) Coal seam degasification. In: Kissel, F. (ed) *The Handbook of Methane Gas*. NIOSH. p. 18.

Thakur, P.C. (2014) Coal bed methane-from prospect to pipeline. In: Thakur, P. *et al.* (eds) *Coal Seam Degasification*. Elsevier, Amsterdam, The Netherlands. pp. 155–175.

Thakur, P.C. (2017) Degasification and ventilation of super-wide longwall panels in the Pittsburgh coal seam. *Proceedings of Longwall USA Conference and Exhibit, 2017, Pittsburgh, PA*. 20 ppts.

Trackemas, J.D. & Peng, S.S. (2013) Factors considered for increasing longwall panel width. *Coal Age*, 32–43.

Trevits, M.A. & Matetic, R.J. (1991) A study on the relation between saturated zone response and longwall mining-induced ground strain. *From The Stephen B. Thacker CDC Library, 1991*. pp. 1100–1109.

Trueman, R., Lyman, G. & Callan, M. (2005) Fitness for purpose of longwall powered support. Australian Coal Association Research Program (ACARP). Project No. C12007. p. 87.

Tsang, P., Peng, S.S. & Biswas, K. (1996) Current practice of pillar design in U.S. coal mines. *Mining Engineering*, 48(12), 55–60.

US Congress. (1969) *PL 91–173, Mine Safety and Health Act of 1969 as amended by in 1977 as PL 95–164*, p. 88.

U.S. Department of Energy. (1976–1980) *US Longwall Production Statistics*. Mining Research Group, Germantown, MD.

Wade, L.V. (1977) Longwall support load predictions from geological information. *Trans. Society for Mining, Metallurgy & Exploration Annual Meeting, 262, September 1977*. pp. 209–213.

Wade, L.V. & Conroy, P.J. (1980) Rock mechanics study of a longwall panel. *Mining Engineering*, 32(12), 1728–1735.

Walker, J.S. (1988) Case study of the effects of longwall mining induced subsidence on shallow groundwater sources in the Northern Appalachian Coalfield. Final report submitted to US Bureau of Mines, RI 9198. p. 17.

Wang, J.C., Yang, S.L., Li, Y., Wei, L.K. & Liu, H.H. (2014) Caving mechanisms of loose top-coal in longwall top-coal caving mining method. *International Journal of Rock Mechanics & Mining Sciences*, 71, 160–170.

Wang, J.C. & Zhang, J.W. (2015) BBR study of top-coal drawing law in longwall top-coal caving mining. *Journal of China Coal Society*, 40, 487–493.

Watson, G. & Hussey, D. (2001) Cost preventive system to control unstable roof in main line haulage entries. In: Peng, S.S. (ed) *Proceedings of 20th International Conference on Ground Control in Mining, August 2001, Morgantown, WV*. pp. 267–273.

Westfalia Lunen. (1990) Factors to be Considered in the Selection of PanzerÒ Conveyor. p. 23.

Westman, E., Lu, J., Haycock, C. & Karmis, K. (1997) Ground control criteria for coal reserve optimization in multiple-seam mines. In: Peng, S.S. (ed) *Proceedings of 16th International Conference on Ground Control in Mining, August, Morgantown, WV*. pp. 311–315.

Whipkey, K. (2005) Productivity improvement for longwall development. *Coal Age*, 110(8), 28–31.

WHO (World Health Organization). (1999) Hazard prevention and control in work environment: Airborne Dust, WHO/SDE/OEH/99.14. Geneva, Switzerland.

Wilson, G. (1991) Mobile roof support for retreat mining. In: Peng, S.S. (ed) *Proceedings of 10th International Conference on Ground Control in Mining, August 1991, Morgantown, WV*. pp. 103–114.

Wilson, A.H. (1975) Support load requirements on longwall faces. *Mining Engineer*, 134, 479–491.

Wright, F.R., Howell, R.C. & Dearinger, J.A. (1979) Rock mechanics study of shortwall mining. Final Technical Report submitted to the U.S. Department of Energy, No. ET-73-C-01-9010. p. 122.

Wyrick, G.G. & Borchers, J.W. (1981) Hydrologic effects of stress-relief fracturing in an Appalachian Valley. *USGS, Geological Survey Water-Supply Paper*. p. 51.

Yu, Z., Chugh, Y.P., Miller, P.E. & Yang, G. (1993) A study of ground behavior in longwall mining through field instrumentation. *International Journal of Rock Mechanics and Mining Sciences & Geomechanics Abstracts*, 30(7), 1441–1444.

Zamora, A. & Trackemas, J. (2013) Longwall automation: Making mining safer through technology. In: *Proceedings of 2013 Longwall USA Conference and Exhibit. Pittsburgh, PA*. pp. 8–13.

Zupanick, J.A. (2005) Coal mine methane drainage utilizing multilateral horizontal wells. *Society for Mining, Metallurgy & Exploration Annual Meeting, Salt Lake City, UT*. p. 10.

Zhang, P., Dougherty, H., Su, D. & Trackemas, J. (2019) Influence of longwall mining on the stability of gas wells in chain pillars. *Proceedings 38th International Conference on Ground Control in Mining, Morgantown, WV, July, 2019*. pp. 217–228.

Zhang, P., Milam, M., Mishra, M., Hudak, W. & Kimutis, R. (2012) Requirements and performance of pumpable cribs in longwall tailgate entries and bleeders. *Proceedings of 31st International Conference on Ground Control in Mining, Morgantown, WV, July, 2012*.

Index